Proceedings of Symposia in PURE MATHEMATICS

Volume 96

String-Math 2015

String-Math 2015
December 31, 2015 – January 4, 2016
Tsinghua Sanya International
Mathematics Forum, Sanya, China

Si Li
Bong H. Lian
Wei Song
Shing-Tung Yau
Editors

American Mathematical Society
and
International Press of Boston

2010 *Mathematics Subject Classification.* Primary 14N35, 51P05, 53D37, 53D45, 53Z05, 81T20, 81T30, 81T60.

Library of Congress Cataloging-in-Publication Data

Names: String-Math (Conference) (2015: Sanya, China) | Li, Si, (Geometry) editor.
Title: String-Math 2015: December 31, 2015 – January 4, 2016, Tsinghua Sanya International Mathematics Forum, Sanya, China / Si Li, Bong H. Lian, Wei Song, Shing-Tung Yau, editors.
Description: Providence, Rhode Island: American Mathematical Society, [2017] | Series: Proceedings of symposia in pure mathematics; volume 96 | Includes bibliographical references.
Identifiers: LCCN 2017021722 | ISBN 9781470429515 (alk. paper)
Subjects: LCSH: Geometry, Algebraic–Congresses. — Quantum theory–Mathematics–Congresses. | AMS: Algebraic geometry – Projective and enumerative geometry – Gromov-Witten invariants, quantum cohomology, Gopakumar-Vafa invariants, Donaldson-Thomas invariants. msc | Geometry – Geometry and physics (should also be assigned at least one other classification number from Sections 70–86) – Geometry and physics (should also be assigned at least one other classification number from Sections 70–86). msc | Differential geometry – Symplectic geometry, contact geometry – Mirror symmetry, symplectic aspects; homological mirror symmetry; Fukaya category. msc | Differential geometry – Symplectic geometry, contact geometry – Gromov-Witten invariants, quantum cohomology, Frobenius manifolds. msc | Differential geometry – Applications to physics – Applications to physics. msc | Quantum theory – Quantum field theory; related classical field theories – Quantum field theory on curved space backgrounds. msc | Quantum theory – Quantum field theory; related classical field theories – String and superstring theories; other extended objects (e.g., branes). msc | Quantum theory – Quantum field theory; related classical field theories – Supersymmetric field theories. msc
Classification: LCC QA564 .S77 2015 — DDC 516.3/5–dc23 LC record available at https://lccn.loc.gov/2017021722

DOI: http://dx.doi.org/10.1090/pspum/096

∞ The paper used in this book is acid-free and falls within the guidelines
established to ensure permanence and durability.
Visit the AMS home page at http://www.ams.org/

10 9 8 7 6 5 4 3 2 1   22 21 20 18 19 17

Contents

Preface

String theory is a broad subject. It plays a central role in theoretical physics as a candidate for quantum theory, unifying gravity with other interactions. Since its birth, string theory has had profound connections with broad branches of modern mathematics. In the last decades, the prosperous interaction—built upon joint efforts from both mathematicians and physicists—has given rise to marvelous deep results in supersymmetric gauge theory, topological string, M-theory and duality on the physics side, as well as in algebraic geometry, differential geometry, algebraic topology, representation theory and number theory on the mathematics side.

The interplay is twofold. The mathematics has provided powerful tools to fulfill the physical interconnection of ideas and to clarify physical structures to understand the nature of string theory. On the other hand, ideas from string theory and quantum field theory have been a source of significant inspirations to reveal surprising mathematical structures and to create new directions in mathematics. The String-Math conference is an annual international event to bring together researchers working at the rapidly developing interface of these two academic fields to exchange current significant ideas and to explore future directions.

String-Math 2015 took place from December 31, 2015, to January 4, 2016, at Tsinghua Sanya International Mathematics Forum (TSIMF) in Sanya, China. The conference brought together around 150 talented participants, covering a variety of fast-developing topics including homological mirror symmetry, supersymmetric gauge theory, singularity theory, wall-crossing phenomenon, large N-duality, modular forms, string compactification and non-commutative geometry.

The current volume of proceedings of String-Math 2015 focuses on topics that are developing quickly, which were fruitfully discussed during this meeting. One of the main themes of the current proceedings collects frontier research on Calabi-Yau manifolds and mirror symmetry. Another main theme is concerned with the development of non-perturbative methods in supersymmetric gauge theories. The collected articles in this volume present state-of-the-art developments in these topics. We hope that their timely publication will help promote communications and broader interests in the physics and mathematics communities.

Si Li
Bong H. Lian
Wei Song
Shing-Tung Yau

Superstring compactifications to all orders in α'

Katrin Becker and Melanie Becker

ABSTRACT. This is the written version of the talk given at the String-Math 2015 conference at Sanya.

1. Introduction

The equations of motion of eleven-dimensional supergravity are solved by a space-time of the form

$$\mathbb{R}^{1,3} \times M, \tag{1.1}$$

i.e. a product of four dimensional Minkowski space and M, where M is a compact seven-dimensional Ricci flat manifold. Moreover, the background four-form flux vanishes. Supersymmetry is unbroken if there is a parallel spinor along M. In this case the metric on M has G_2 holonomy. We are interested in full G_2 holonomy and not a proper subgroup.

Supergravity is the low-energy limit of string theory (or M-theory). We expect an infinite series of corrections to the two-derivative action which are suppressed at large volume. Large volume is the regime in which the Einstein equations can be used. Then the following question arises: Can the classical metric of G_2 holonomy be modified to compensate the α' corrections to the equations of motion and supersymmetry transformations?

This is the outline of my talk:

1) Introduction
2) Type II string theory on G_2 manifolds
   2.1) General remarks.
   2.2) Leading order correction.
   2.3) All orders in α'.

   In section 2) I will describe the existence proof of a solution of type II string theory to all orders in α'. We will focus on the G_2 case. To be concrete I will focus on the case that the holonomy is G_2 and we will discuss how to prove the existence of a vacuum solution of string theory to all orders in α'. We will be solving a partial differential equation and the solvability is not for free. A solution will exist if certain cohomology conditions hold.

2010 *Mathematics Subject Classification.* Primary 83E15, 83E30.

This work was supported in part by NSF Grant #PHY-1214344.

3) Interpretation in 3d and 4d field theory.
   In part section 3) we will interpret these cohomology conditions in terms of 3d and 4d field theory.
4) Massive Kaluza-Klein modes in superspace.
   In section 4) we will dig deeper into properties of this field theory.
   Sections 2) and 3) are already published while the material presented in section 4) is more recent (in part work in progress).

2. Type II string theory on G_2 manifolds

2.1. General remarks. The basic tool to understand spaces on which unit spinors exist are tensors constructed as spinor bilinears. On a seven-dimensional spin manifold M there is a unit spinor η which defines a globally defined three-form according to

$$\varphi_{abc} = \eta^T \Gamma_{abc} \eta, \tag{2.1}$$

and a four-form

$$\psi_{abcd} = \eta^T \Gamma_{abcd} \eta, \tag{2.2}$$

related to the three-form by Hodge duality.

Such a three-form is more fundamental than the metric since it determines the metric according to

$$g_{ab} = g_{ab}[\varphi] = (\det s)^{-1/9} s_{ab}, \tag{2.3}$$

where

$$s_{ab} = -\frac{1}{144} \varphi_{amn} \varphi_{bpq} \varphi_{rst} \epsilon^{mnpqrst}, \tag{2.4}$$

where $\epsilon = \pm 1$. So we will be treating φ as fundamental object we wish to solve for and not the metric.

2.2. Leading order correction. If the spinor is covariantly constant

$$\nabla_\alpha \eta = 0, \tag{2.5}$$

the classical solution emerges

$$d\phi = d\psi = 0. \tag{2.6}$$

Once α' corrections are included the gravitino supersymmetry transformation becomes schematically

$$\delta \psi_\alpha = \nabla_\alpha \eta + A_\alpha \eta + i B^\beta_\alpha \Gamma_\beta \eta, \tag{2.7}$$

where the first term on the right hand side describes the classical contribution and the second and third terms encode the α' corrections. In seven dimensions irreducible spinors have eight real components and a basis of the space of spinors is given by $(\eta, i\Gamma_\alpha \eta)$ with $a = 1, \ldots, 7$. The last two terms in eqn. 2.7 describe an arbitrary spinor expanded into this basis. The coefficients of the expansion are tensors depending implicitly on the three-form

$$A_\alpha = A_\alpha[\phi], \qquad B^b_\alpha = B^b_\alpha[\phi]. \tag{2.8}$$

For example, to order α'^3

$$A_\alpha = 0, \qquad B^b_\alpha = \alpha'^3 \phi_{acd} \nabla^c Z^{db}, \tag{2.9}$$

where Z is a tensor constructed from three Riemann tensors. But as previously mentioned, we do not need to know the concrete expressions for A and B.

Our goal is to solve the spinor equations order by order in α'. To order α'^3 we find

$$\delta\psi_\alpha = \nabla'_\alpha \eta' + A_\alpha \eta + iB_\alpha^\beta \Gamma_b \eta = 0, \tag{2.10}$$

where we used primes to denote the α' corrected quantities. Since A and B have an explicit overall α'^3 they are constructed from the uncorrected three-form ϕ.

Instead of solving eqn. 2.10 directly it is more convenient (but equivalent) to solve the equations satisfied by the tensors

$$\phi'_{abc} = \eta'^T \Gamma'_{abc} \eta', \qquad \psi' = \star' \phi'. \tag{2.11}$$

Which equations determine ϕ' and ψ'? It turns out that we can get a complete set of equations by differentiating and anti-symmetrizing the previous expression. This leads to

$$\begin{aligned} d\phi' &= \alpha, \\ d\psi' &= \beta, \\ \psi' &= \star' \phi'. \end{aligned} \tag{2.12}$$

The forms α and β can be written in terms of A and B using eqn. 2.10, see ref. [1] for details:

$$\begin{aligned} \alpha_{abcd} &= 8A_{[a}\varphi_{bcd]} - 8B_{[a}{}^e \psi_{bcd]e}, \\ \beta_{abcde} &= 10A_{[a}\psi_{bcde]} - 40B_{[ab}\varphi_{cde]}. \end{aligned} \tag{2.13}$$

We can then ask: given α and β, does a solution of eqn. 2.12 with a globally defined ϕ' and ψ' exist? These equations are only solvable if α and β are exact. So we need

$$\begin{aligned} d\phi' &= \alpha = d\chi, \\ d\psi' &= \beta = d\xi. \end{aligned} \tag{2.14}$$

But why would α and β be exact? To order α'^3 we have explicit expressions for α and β and we can verify that they are indeed exact. At order α'^n, $n > 3$ we shall use space-time physics to argue that α and β are exact, as described in detail in the next section. For now we proceed assuming α and β to be exact.

The solution of the first equation in 2.12 is then

$$\phi' = \phi + \chi + db, \tag{2.15}$$

where b is a two-form. The equation determining ψ' then becomes

$$d\psi' = d \star' \phi' = d\xi, \tag{2.16}$$

and by linearizing one obtains

$$\star' \phi' = \star\phi + \star \left(\frac{4}{3}\pi_1 + \pi_7 - \pi_{27} \right) \delta\phi. \tag{2.17}$$

Here $\pi_{\mathbf{r}}$ is a projection of differential forms onto the subspace which transforms as a representation $\mathbf{r}$ of G_2. For instance, for the three-forms we have a decomposition

$$\Lambda^3 := \Lambda^3 T^* M = \Lambda^3_1 \oplus \Lambda^3_7 \oplus \Lambda^3_{27}. \tag{2.18}$$

The condition (2.17) is a second order linear partial differential equation for b. Explicitly

$$d\star\left(\frac{4}{3}\pi_1+\pi_7-\pi_{27}\right)(\chi+db)=d\xi, \tag{2.19}$$

where

$$\begin{aligned}\chi_{abc}&=-c\varphi_{abc}Z+3c\varphi_{[ab}{}^{d}Z_{c]d},\\ \xi_{abcd}&=-4c\psi_{[abc}{}^{e}Z_{d]e}\\ Z^{ab}&=\frac{1}{32g}\epsilon^{ac_1\dots c_6}\epsilon^{bd_1\dots d_6}R_{c_1c_2d_1d_2}R_{c_3c_4d_3d_4}R_{c_5c_6d_5d_6}.\end{aligned} \tag{2.20}$$

It turns out that this is a Laplace equation! To see this decompose $b\in\Lambda^2=\Lambda^2_7\oplus\Lambda^2_{14}$ and note that the Λ^2_7 contribution corresponds to a diffeomorphism and can be discarded. We can then take[1] $b\in\Lambda^2_{14}$ and $d^\dagger b=0$ to obtain the partial differential equation

$$\delta b=d^\dagger\rho,\qquad b\in\Lambda^2_{14},\qquad d^\dagger b=0, \tag{2.21}$$

with

$$\rho=-\star\xi-(\pi_{27}-\pi_7-\frac{4}{3}\pi_1)\chi. \tag{2.22}$$

As discussed in ref. **[1]** the $d^\dagger\rho$ source satisfies $d^\dagger\rho\in\Lambda^2_{14}$ and is orthogonal to the zero modes of the Laplacian (since it is co-exact). Consequently the Laplace equation is solvable.

This completes the existence proof and shows that to order α'^3 there exists a background solving the α' corrected equations. What about the order α'^n, $n>3$? We address this question next.

2.3. All orders in α'. In the same way as we cannot turn a Möbius strip into a cylinder by a small deformation, if the manifold is spin classically it will be so after a small perturbation. Consequently after a small perturbation there still exists a unit spinor η'. The supersymmetry transformation of the gravitino takes then the schematic form

$$\delta\psi'_\alpha=\nabla'_\alpha\eta'+A'_\alpha[\varphi']\eta'+iB'^{b}_{\alpha}[\varphi']\Gamma'_b\eta', \tag{2.23}$$

However, now we need to put primes on everything since all quantities are corrected. The tensors constructed as spinor bilinears are then again

$$\begin{aligned}\varphi'_{abc}&=\eta'^T\Gamma'_{abc}\eta',\\ \psi'_{abcd}&=\eta'^T\Gamma'_{abcd}\eta',\\ \psi'&=\star'\varphi'.\end{aligned} \tag{2.24}$$

Again, the vanishing of the gravitino variation then implies a set of conditions

$$\begin{aligned}d\varphi'&=\alpha[\varphi'],\\ d\psi'&=\beta[\varphi'],\\ \psi'&=\star'\varphi',\end{aligned} \tag{2.25}$$

[1]See ref. **[1]** for more details.

with

$$\begin{aligned} \alpha_{abcd} &= 8A'_{[a}\varphi'_{bcd]} - 8B'{}_{[a}{}^{e}\psi'_{bcd]e}, \\ \beta_{abcd} &= 10A'_{[a}\psi'_{bcde]} - 40B'_{[ab}\varphi'_{cde]}. \end{aligned} \tag{2.26}$$

Expanding to order n in α leads to

$$\begin{aligned} d\varphi' \mid_n &= \alpha[\varphi'] \mid_n, \\ d\psi' \mid_n &= \beta[\varphi'] \mid_n, \\ \psi' &= \star\varphi'. \end{aligned} \tag{2.27}$$

where $|_n$ indicates order n. A necessary condition for solvability is that order by order in α', α and β are exact, *i.e.* that there is globally defined χ_n and ξ_n such that

$$\begin{aligned} \alpha[\varphi'] \mid_n &= d\chi_n, \\ \beta[\varphi'] \mid_n &= d\xi_n. \end{aligned} \tag{2.28}$$

The exactness of α and β can be checked explicitly to order α'^3. To any order beyond α'^3 we need to assume exactness of α and β for now. This is an assumption about the form of the supersymmetry transformations. Below we will rephrase this assumption into properties of the space-time action.

It turns out that the exactness of α and β is not only a necessary condition for solvability of the supersymmetry condition but it is also sufficient as we prove in ref. [**1**].

3. Interpretation in 3*d* and 4*d* field theory

In this section our goal is to show that the exactness of α and β is equivalent to the conditions for unbroken space-time supersymmetry. It is more convenient (but equivalent) to turn to M-theory at this point because the space-time physics arising from M-theory on G_2 manifolds is simpler than its string theory counterpart. Space-time fields arise from fluctuations around the background $\mathbb{R}^{1,3} \times M$, where M is a compact G_2 manifold.

There are infinitely many space-time fields with only a finite set corresponding to the massless fields and infinitely many massive modes. We use two guiding principles to construct the four-dimensional quantum field theory. First, supersymmetry in space-time. We assemble all fields into four-dimensional superfields, keeping supersymmetry in four dimensions manifest. Second, we keep explicit locality along $\mathbb{R}^{1,3}$ and M. Fields in four dimensions depend on the space-time coordinates. This is locality in four dimensions. Besides this we also keep locality along M explicit. So fields are functions, forms or tensors on M, but we want to avoid doing any explicit spectral decomposition. We treat the coordinates on M as continuous labels.

The eleven-dimensional fields (labeled by capital letters) decompose into several four-dimensional fields (labeled by greek letters)

$$\begin{aligned} C_{MNP} &\to C_{abc}, C_{ab\mu}, C_{a\mu\nu}, C_{\mu\nu\rho}, \\ G_{MN} &\to g_{ab}, g_{a\mu}, g_{\mu\nu}. \end{aligned} \tag{3.1}$$

Some fields are space-time scalars: C_{abc} and g_{ab}, others are vectors, $C_{ab\mu}$, $g_{a\mu}$ and others anti-symmetric tensors of different types, $C_{a\mu\nu}$ and $C_{\mu\nu\rho}$. There is also an infinite collection of spin 2 fields in space-time $g_{\mu\nu} = g_{\mu\nu}(x, y)$.

So far only pieces of the four-dimensional action with the above properties are known. For example, the action for all bosonic fields can be found in ref. [**2**] but the full action with manifest supersymmetry is work in progress [**3**], [**4**], [**5**]. However, what we know suggests that the cohomology conditions for α and β follow from this action as conditions for unbroken supersymmetry in four dimensions.

To explain the above statements we construct the most general action with the above properties including chiral and vector superfields. Moreover, we consider only global four dimensional supersymmetry.

We denote the coordinates on flat superspace in four dimensions by x^μ and θ, the latter being a four component Majorana spinor. Superfields are then functions of these coordinates. So, for example, chiral or antichiral superfields satisfy

$$\overline{D}_{\dot\alpha}\Phi(x,\theta)=0, \qquad D_\alpha\Phi(x,\theta)=0, \tag{3.2}$$

respectively. Here $\overline{D}_{\dot\alpha}$ and D_α are the spinorial derivatives in superspace. Here and in the following we are using the conventions and notation of ref. [**6**]. There is an infinite family of chiral superfields

$$\Phi_{abc}(x,\theta;y)=\mathcal{C}_{abc}(x;y)+\ldots, \tag{3.3}$$

which have discrete labels $a,b,c=1,\ldots,7$ and a continuous label y, which is the coordinate on M. We denote the lowest component of the chiral superfield by $\mathcal{C}$ and the dots represent θ dependent terms. $\mathcal{C}$ is a complex field and there is a natural choice

$$\mathcal{C}_{abc}=\hat\phi_{abc}+iC_{abc}, \tag{3.4}$$

where C_{abc} is one of the components of the eleven-dimensional three-form and $\hat\phi_{abc}$ is related but not identical to the G_2-structure three form on M.

A supersymmetric action is constructed in the usual way

$$I=\frac{1}{2}\int d^4x d^4\theta K(\Phi,\Phi^\dagger)+\left(\int d^4x d^2\theta f(\Phi)+c.c.\right) \tag{3.5}$$

To illustrate the present framework we write the Lagrangian density for bosonic fields explicitly

$$\begin{aligned} L=&-\int d^7y d^7y' \frac{\delta^2 K}{\delta\mathcal{C}(y)\delta\bar{\mathcal{C}}(y')}\left[\partial_\mu\mathcal{C}(y)\partial^\mu\bar{\mathcal{C}}(y')-\mathcal{F}(y)\mathcal{F}^\star(y')\right]\\ &+2\mathrm{Re}\int d^7y\frac{\delta f(\mathcal{C})}{\delta\mathcal{C}(y)}\mathcal{F}(y), \end{aligned} \tag{3.6}$$

where we are suppressing the indices a,b,c. Here δ is a functional derivative. The above expression will be local (it is equivalent to a single integral over M of an expression built of local operators on M) if $K(\Phi,\Phi^\dagger)$ and $f(\Phi)$ are themselves local. This follows since, for example, the second functional derivative of K will produce a Dirac delta function $\delta^7(y-y')$.

There is a natural choice for the superpotential that is local in this sense, namely

$$f(\Phi)\sim\int_M \Phi\wedge d\Phi, \tag{3.7}$$

where d is the exterior derivative along M. In a supersymmetric ground state

$$\frac{\delta f}{\delta\Phi}=0, \tag{3.8}$$

which implies $d\Phi = 0$ or

$$d\hat{\phi} = 0, \quad G = 0. \tag{3.9}$$

This result is α' independent. To make contact to the previous discussion it is natural to identify

$$\hat{\phi} = \phi' - \chi. \tag{3.10}$$

The exactness of α is then equivalent to the existence of the field redefinition between $\hat{\phi}$ and ϕ'. To obtain more information we need to consider the Kähler potential described in terms of $\mathcal{C}$; these are coordinates of an infinite-dimensional complex manifold, which is the space of complex valued three-forms. K defines a Kähler metric and the Kähler form on this space is

$$J = \int_{M\times M} d^7y d^7y' \frac{\delta^2 K}{\delta\mathcal{C}(y)\delta\bar{\mathcal{C}}(y')}\delta\mathcal{C}(y) \wedge \delta\bar{\mathcal{C}}(y') = J_{A\bar{B}'} dy^A \wedge d\bar{y}^{\bar{B}}. \tag{3.11}$$

We have introduced de-Witt notation in the second equality [**1**]. More explicitly, the index A includes the index a and the label y and a sum over A also includes an integral over y. The Kähler metric has isometries. In particular, the shifts

$$\mathcal{C} \to \mathcal{C}' = \mathcal{C} + id\Lambda, \tag{3.12}$$

leave the Kähler metric invariant. Since J is a differential form (on our infinite-dimensional complex manifold)

$$J \to J' = J + d(\iota_{d\Lambda}J) + \iota_{d\Lambda}(dJ), \tag{3.13}$$

where ι is the contraction with the vector field $d\Lambda$. According to the Poincaré lemma there is a Killing potential P_Λ such that

$$\iota_{d\Lambda}J = dP_\Lambda, \tag{3.14}$$

where P_Λ is a map from the space of two-forms modulo closed two-forms into $\mathbb{C}$, and d on the right-hand-side is the exterior derivative on our infinite-dimensional Kähler manifold (as opposed to d in the subscript on the left-hand-side, which is the exterior derivative on M). We can then write

$$P_\Lambda = \langle \Lambda, \mu \rangle, \tag{3.15}$$

where $\langle \cdot, \cdot \rangle$ is the inner product on forms and μ is the moment map. This is a beautiful and extremely useful tool we borrow from symplectic geometry. For us, the moment map is important since the space-time action includes a term of the form

$$I \sim \int D_{ab}\mu^{ab}, \tag{3.16}$$

where D_{ab} is the auxiliary field of the vector multiplet containing $C_{ab\mu}$. Integrating out D_{ab} and requiring unbroken supersymmetry requires the moment map to vanish, which is equivalent to

$$\nabla_\alpha \left(\frac{\delta K}{\delta \mathcal{C}_{abc}} \right) = 0. \tag{3.17}$$

This is again valid to all orders in α' and it is equivalent to the existence of a co-exact two-form, or by Hodge duality, to an exact five-form. This is what we wanted to show. A similar analysis for compactifications on $Spin(7)$ manifolds can be found in ref. [**1**].

4. Massive Kaluza-Klein modes in superspace

The discussion in the previous sections leaves us with an interesting question or an interesting prediction. Does this space-time action really display the needed properties? Or maybe, can we treat the cohomology conditions as predictions about the space-time action and can we use these predictions to understand more about it. Let me then come to the final topic which is "what do we know about massive Kaluza-Klein modes and in particular what do we know about massive Kaluza-Klein modes in superspace?"

4.1. Bosonic fields. Lets start with the bosonic theory. The 11d fields are a 3-form C_{MNP} and a metric G_{MN}. The 11d fields decompose the following way:

$$
\begin{aligned}
&C_{MNP} \to C_{abc}, C_{ab\mu}, C_{a\mu\nu}, C_{\mu\nu\rho}, \\
&G_{MN} \to \begin{pmatrix} h_{\mu\nu} + g_{cd}A_\mu^c A_\nu^d & g_{bc}A_\mu^c \\ g_{ac}A_\nu^c & g_{ab} \end{pmatrix}
\end{aligned}
\tag{4.1}
$$

The three-form gives rise to space-time scalars, vectors and diverse anti-symmetric tensors. The metric gives rise to space-time scalars, vector and symmetric rank two tensors. In the same way the eleven-dimensional symmetries, which are gauge invariance under shifts of C by exact forms and diffeomorphism invariance, give rise to space-time symmetries.

From the space-time point of view this is a very complicated system given the number of different fields and the way they interact. However, given the simplicity of the eleven-dimensional theory we know the space-time system in a closed form. We know all symmetries and here is, for example, the lagrangian (see ref, [**2**] for details and notation)

(4.2)

$$
\begin{aligned}
S = &-\frac{1}{8\kappa^2}\int dv h^{\alpha\beta}\left(\frac{1}{2}g^{ab}g^{cd} + g^{ac}g^{bd}\right)\mathcal{D}_\alpha g_{ab}\mathcal{D}_\beta g_{cd} \\
&+\frac{1}{2\kappa^2}\int dv\left(h^{\beta\mu}h^{\gamma[\rho}h^{\alpha\nu} - \frac{1}{2}h^{\alpha\mu}h^{\beta[\nu}h^{\gamma]\rho}\right)\mathcal{D}_\alpha h_{\beta\gamma}\mathcal{D}_\mu h_{\nu\rho} \\
&+\frac{1}{4\kappa^2}\int dv f\Big[g^{ab}h^{\alpha[\beta}h^{\mu]\nu}\hat{\nabla}_a h_{\alpha\beta}\hat{\nabla}_b h_{\mu\nu} - h^{\alpha\beta}\left(\frac{1}{2}g^{ab}g^{cd} + g^{ac}g^{bd}\right)\hat{\nabla}_a h_{\alpha\beta}\hat{\nabla}_b g_{cd} \\
&+\left(g^{pt}g^{qu}g^{rs} - \frac{1}{2}g^{ps}g^{qt}g^{ru} + g^{pr}g^{qu}g^{st}\right)\hat{\nabla}_r g_{pq}\hat{\nabla}_u g_{st}\Big] - \frac{1}{8\kappa^2}\int dv f^{-1}\left(\mathcal{F}_{\mu\nu}^a\right)^2 \\
&-\frac{1}{24\kappa^2}\int dv\left[\left(\mathcal{D}_\mu \mathsf{C}_{abc} - 3\partial_{[a}\mathsf{C}_{bc]\mu}\right)^2 + 4f\left(\hat{\nabla}_{[a}\mathsf{C}_{bcd]}\right)^2\right] \\
&-\frac{1}{16\kappa^2}\int dv f^{-1}\left(\mathcal{F}_{\mu\nu ab} + \mathcal{F}_{\mu\nu}^c\mathsf{C}_{abc}\right)^2 - \frac{1}{24\kappa^2}\int dv\left[f^{-2}\left(\mathsf{F}_{\mu\nu\rho a}\right)^2 + \frac{f^{-3}}{4}\left(\mathsf{F}_{\mu\nu\rho\sigma}\right)^2\right] \\
&-\frac{55}{2^8 3^3\kappa^2}\int dx^{\mu\nu\rho\sigma}dy^{abcdefg}\mathsf{F}_{[\mu\nu\rho\sigma}\mathsf{F}_{abcd}\mathsf{C}_{efg]}.
\end{aligned}
$$

4.2. Tensor hierarchy in 4*d* superspace. Embedding the system described in the previous section into 4d superspace is work in progress. Here we describe some parts of the story. To start with we work our way around $g_{\mu\nu}(x, y)$ and consider a $4d$ theory in flat superspace with global supersymmetry. Let's also first consider a system that is more general than the system obtained from dimensional reduction from eleven to four dimensions (even though all is inspired in this case).

Let's consider a toy model. Given is a collection of bosonic fields in d space-time dimensions

$$\phi^{I_0}_{[0]}, \phi^{I_1}_{[1]}, \ldots, \phi^{I_p}_{[p]}, \ldots. \tag{4.3}$$

These fields are degree p differential forms in space-time. The degree is explicitly quoted in the square bracket. So $\phi^{I_0}_{[0]}$ is a set of scalars in space-time labelled by the index I_0, $\phi^{I_1}_{[1]}$ is a set of space-time vectors and so on. Here are a few comments:

- The set of fields obtained from the reduction of the eleven-dimensional three-form C_{MNP} to $4d$ is of this type.
- $\phi^{I_p}_{[p]}$ is a space-time p-form taking values in vector space V_p. In the example of the eleven to four reduction V_p is the vector space of differential forms on M (a $7d$ manifold) of some degree.
- We take the space-time to be flat $h_{\mu\nu} = \eta_{\mu\nu}$. Including a non-trivial space-time metric is work in progress [**5**].

4.3. Gauge transformations. We subject the above fields to a set of gauge transformations

$$\delta\phi^{I_p}_{[p]} = d\Lambda^{I_p}_{[p-1]} + \left(q^{(p)}\right)^{I_p}_{J_{p+1}} \Lambda^{J_{p+1}}_{[p]} \tag{4.4}$$

Here $\Lambda^{I_p}_{[p-1]}$ is a $p-1$ form taking values in a vector-space V_p, $\Lambda^{I_{p+1}}_{[p]}$ is a p form taking values in V_{p+1} and

$$q^{(p)} : V_{p+1} \to V_p, \tag{4.5}$$

are linear maps. These gauge transformations are inspired in the example of the dimensional reduction from eleven to four dimensions where, for example

$$\delta C_{\mu ab} = \partial_\mu \Lambda_{ab} - 2\partial_{[a}\Lambda_{b]\mu}. \tag{4.6}$$

However, the gauge transformations in eqn. 4.4 are more general.

4.4. Field strength. Associated to the potentials $\phi^{I_p}_{[p]}$ there are the field strengths

$$F^{I_p}_{[p+1]} = d\phi^{I_p}_{[p]} - \left(q^{(p)}\right)^{I_p}_{J_{p+1}} \phi^{J_{p+1}}_{[p+1]}, \tag{4.7}$$

which are only gauge invariant if

$$q^{(p)} \circ q^{(p+1)} = 0. \tag{4.8}$$

It is then natural to interpret q as a boundary operator of a chain complex.

$$V_\bullet = \left\{ \cdots \xrightarrow{q^{(p+1)}} V_{p+1} \xrightarrow{q^{(p)}} V_p \xrightarrow{q^{(p-1)}} \cdots \xrightarrow{q^{(0)}} V_0 \right\}, \tag{4.9}$$

Here we formulated the system abstractly but the system obtained by dimensional reduction from eleven to four dimensions is an example with a linear operator q corresponding to the differential operator d on M.

4.5. Higgs mechanism. We can also use this language to classify the different fields. Some are pure gauge, some are massless and some might be massive. For example, in $4d$

$$V_4 \xrightarrow{q^{(3)}} V_3 \xrightarrow{q^{(2)}} V_2 \xrightarrow{q^{(1)}} V_1 \xrightarrow{q^{(0)}} V_0. \tag{4.10}$$

At V_2, for example we have

$$q^{(1)} \circ q^{(2)} = 0, \tag{4.11}$$

and consequently

$$\mathrm{Im}q^{(2)} \subset \mathrm{Ker}q^{(1)} \subset V_2. \tag{4.12}$$

Each V_p can then be written as a direct sum of vector subspaces. For example, we can take a basis of $\mathrm{Im}q^{(2)}$ and extend it to a basis of $\mathrm{Ker}q^{(1)}$ and then take a basis of $\mathrm{Ker}q^{(1)}$ and extend it to a basis of V_2. Diagramatically

$$V_2 \simeq \mathrm{Im}q^{(2)} \oplus \mathrm{Ker}q^{(1)}/\mathrm{Im}q^{(2)} \oplus V_2/\mathrm{Ker}q^{(1)}. \tag{4.13}$$

Elements in $\mathrm{Im}q^{(2)}$ are pure gauge, those in $\mathrm{Ker}q^{(1)}/\mathrm{Im}q^{(2)}$ are massless physical states and those in $V_2/\mathrm{Ker}q^{(1)}$, are physical states which are in general massive.

4.6. $4d$ superfields. The component fields can be embedded into $4d$ superfields. For example, the bosonic potentials are embedded into superfields which we call prepotentials according to

$$\begin{aligned}
a^A &= \frac{1}{2}\left(\Phi^A + \bar{\Phi}^A\right)\Big|, \\
A^I_\alpha &= -\frac{1}{4}(\sigma_\alpha)_{\alpha\dot{\alpha}}\left[D^\alpha, \bar{D}^{\dot{\alpha}}\right]V^I\Big|, \\
B^M_{ab} &= -\frac{i}{2}\left[(\sigma_{ab})_\alpha{}^\beta D^\alpha \Sigma^M_\beta - (\bar{\sigma}_{ab})^{\dot{\alpha}}{}_{\dot{\beta}}\bar{D}_{\dot{\alpha}}\bar{\Sigma}^{M\dot{\beta}}\right]\Big|., \\
C^S_{abc} &= \frac{1}{8}\epsilon_{abcd}\sigma^d_{\alpha\dot{\alpha}}\left[D^\alpha, \bar{D}^{\dot{\alpha}}\right]X^S\Big|, \\
D^X_{abcd} &= \frac{i}{8}\epsilon_{abcd}\left(D^2\Gamma^X - \bar{D}^2\bar{\Gamma}^X\right)\Big|.
\end{aligned} \tag{4.14}$$

Here Φ^A and Γ^X are chiral superfields, V^I and X^S are real superfields and Σ^M_α are spinor chiral superfields. Here we quote only the result. See ref. [**3**] for details and notation. In this section space-time indices are labelled by small Latin indices from the beginning of the alphabet since Greek indices are reserved for spinor indices.

4.7. Gauge transformations in superspace.

$$\begin{array}{rclcl}
\delta\Phi^A &=& & + & (q\cdot\Lambda)^A \\
\delta V^I &=& \frac{1}{2i}\left(\Lambda^I - \bar{\Lambda}^I\right) & + & (q\cdot U)^I \\
\delta\Sigma^M_\alpha &=& -\frac{1}{4}\bar{D}^2 D_\alpha U^M & + & (q\cdot\Upsilon_\alpha)^M \\
\delta X^S &=& \frac{1}{2i}\left(D^\alpha\Upsilon^S_\alpha - \bar{D}_{\dot{\alpha}}\bar{\Upsilon}^{S\dot{\alpha}}\right) & + & (q\cdot\Xi)^S, \\
\delta\Gamma^X &=& -\frac{1}{4}\bar{D}^2\Xi^X & &
\end{array} \tag{4.15}$$

Here Λ^I is a chiral superfield, U^M and Ξ^X are real scalar superfields and Υ^S_α is a chiral spinor superfield.

4.8. Field strengths. The gauge invariant field strengths are defined by

$$\begin{array}{rclcl} E^Z & = & & - & (q\cdot\Phi)^Z \\ F^A & = & \frac{1}{2i}\left(\Phi^A - \bar{\Phi}^A\right) & - & (q\cdot V)^A \\ W^I_\alpha & = & -\frac{1}{4}\bar{D}^2 D_\alpha V^I & - & (q\cdot\Sigma_\alpha)^I \\ H^M & = & \frac{1}{2i}\left(D^\alpha\Sigma^M_\alpha - \bar{D}_{\dot\alpha}\bar{\Sigma}^{M\dot\alpha}\right) & - & (q\cdot X)^M \\ G^S & = & -\frac{1}{4}\bar{D}^2 X^S & - & (q\cdot\Gamma)^S . \end{array} \tag{4.16}$$

Again, see [3] for details.

4.9. Action. We can now construct actions. For example,

$$\begin{aligned} S = \int d^4x \Big[\mathrm{Re}\left(\int d^2\theta \tau_{IJ} W^{I\alpha} W^J_\alpha\right) \\ + \int d^4\theta \left(g_{AB} F^A F^B - \frac{1}{4} g_{MN} H^M H^N + g_{ST} \bar{G}^S G^T \right)\Big] \end{aligned} \tag{4.17}$$

This action is manifestly gauge invariant since the field strength superfields are gauge invariant and supersymmetric if we drop total derivatives. Consequently actions can be constructed using the gauge invariant field strengths either as real scalar combinations integrated over superspace or chiral combinations integrated over half of superspace

However, to make contact with the $4d$ space-time field theory obtained from dimensional reduction from eleven dimensions we need contributions to the Lagrangian density linear in prepotentials. The above action is not of this type.

There exists a Chern-Simons superfield action linear in the prepotentials which is again supersymmetric after dropping total derivatives but is only gauge invariant when many terms are combined and when integrated over superspace. See refs. [3] and [4] for details.

References

[1] Katrin Becker, Guangyu Guo, and Daniel Robbins, *Disc amplitudes, picture changing and space-time actions*, J. High Energy Phys. **1** (2012), 127, 45. MR2949272

[2] Katrin Becker, Melanie Becker, and Daniel Robbins, *Kaluza-Klein theories without truncation*, J. High Energy Phys. **2** (2015), 140, front matter+17. MR3321313

[3] Katrin Becker, Melanie Becker, William D. Linch III, and Daniel Robbins, *Abelian tensor hierarchy in 4D, N = 1 superspace*, J. High Energy Phys. **3** (2016), 052, front matter+29. MR3501730

[4] Katrin Becker, Melanie Becker, William D. Linch III, and Daniel Robbins, *Chern-Simons actions and their gaugings in 4D, N = 1 superspace*, J. High Energy Phys. **6** (2016), 097, front matter+34. MR3538258

[5] K. Becker, M. Becker, W. D. Linch and D. Robbins, work in progress.

[6] Julius Wess and Jonathan Bagger, *Supersymmetry and supergravity*, 2nd ed., Princeton Series in Physics, Princeton University Press, Princeton, NJ, 1992. MR1152804

George P. and Cynthia W. Mitchell Institute for Fundamental Physics and Astronomy, Texas A& M University, College Station, Texas 77843-4242

E-mail address: `kbecker@physics.tamu.edu`

George P. and Cynthia W. Mitchell Institute for Fundamental Physics and Astronomy, Texas A& M University, College Station, Texas 77843-4242

E-mail address: `mbecker@physics.tamu.edu`

Supersymmetric partition functions on Riemann surfaces

Francesco Benini and Alberto Zaffaroni

ABSTRACT. We present a compact formula for the supersymmetric partition function of 2d $\mathcal{N}=(2,2)$, 3d $\mathcal{N}=2$ and 4d $\mathcal{N}=1$ gauge theories on $\Sigma_g \times T^n$ with partial topological twist on Σ_g, where Σ_g is a Riemann surface of arbitrary genus and T^n is a torus with $n=0,1,2$, respectively. In 2d we also include certain local operator insertions, and in 3d we include Wilson line operator insertions along S^1. For genus $g=1$, the formula computes the Witten index. We present a few simple Abelian and non-Abelian examples, including new tests of non-perturbative dualities. We also show that the large N partition function of ABJM theory on $\Sigma_g \times S^1$ reproduces the Bekenstein-Hawking entropy of BPS black holes in AdS$_4$ whose horizon has Σ_g topology.

1. Introduction

Supersymmetric partition functions (*i.e.* path-integrals) of quantum field theories on Euclidean compact manifolds are an extremely powerful tool to study non-perturbative properties of those theories. On the one hand, as one varies the compact manifold where the theories are placed and the supersymmetric background in which they are immersed (*i.e.* the supersymmetric sources turned on), one gets access to a big deal of physical information such as correlation functions of operators and spectra of operators and states. On the other hand, keeping some supersymmetry in the process allows one to apply techniques, such as supersymmetric localization [**36, 41**], to explicitly and exactly compute those partition functions. This makes the program quantitative, not just qualitative.

In this note we study the partition function of two, three and four-dimensional supersymmetric gauge theories (with $\mathcal{N}=(2,2)$, $\mathcal{N}=2$ and $\mathcal{N}=1$ supersymmetry, respectively) on a Riemann surface Σ_g of genus g times a torus T^n (with $n=0,1,2$ respectively). In order to preserve (half of the) supersymmetry, we simply perform topological twist [1] on Σ_g, *i.e.* we turn on a background vector field coupled to the R-symmetry, equal and opposite to the spin connection. A technical assumption is then that the theories have a non-anomalous $U(1)_R$ R-symmetry. In fact this note is the natural generalization of [**11**]—where the case of the sphere, $g=0$, was studied—to higher genus. We will often refer to [**11**] for further details.

The partition functions become particularly interesting when the theories have some flavor symmetry. In that case we can turn on a background for the bosonic fields in external vector multiplets coupled to the flavor symmetry. It turns out

2010 *Mathematics Subject Classification.* Primary 81T60; Secondary 81T20, 81T45, 58D30.

[1] In 2d, or upon reduction to 2d, we perform A-type topological twist on Σ_g.

that we can turn on a (quantized) magnetic flux on Σ_g and a complex "fugacity" (whose definition depends on the dimension) along the Cartan subalgebra of the flavor group. [2] We parametrize them by

$$\mathfrak{n} = \frac{1}{2\pi}\int_{\Sigma_g} F^{\text{flav}} \qquad \text{and} \qquad v\,. \tag{1.1}$$

In 2d $v = \sigma^{\text{flav}}$, the complex scalar in the vector multiplet that gives rise to twisted masses; in 3d $v \simeq A_t^{\text{flav}} + i\sigma^{\text{flav}}$ where A_t^{flav} is a flat connection along S^1 and σ^{flav} is a real scalar (that gives real masses); in 4d $v \sim A_{\bar{z}}^{\text{flav}}$ a flat connection on T^2. The partition function is a function of these parameters, meromorphic in v.

We can also decorate the partition functions by the inclusion of certain operators. In the 2d case we can make arbitrary insertions of local twisted chiral operators $\Sigma(x)$. In 3d we can make arbitrary insertions of Wilson line operators $\mathcal{W}_{\mathcal{R}}[\gamma]$ that wrap S^1 at a fixed position on Σ_g. Probably in 4d one could easily add surface operators wrapping T^2, but we do not consider them here.

Notice that in 2d we compute the standard A-twisted partition function [**42**] of gauge theories. In 3d we more precisely compute a Witten index

$$Z_{\text{3d}} = \operatorname{Tr}_{\mathcal{H}} (-1)^F e^{-\beta H} e^{iA^{\text{flav}} J^{\text{flav}}}\,. \tag{1.2}$$

Here H is the Hamiltonian of the theory on Σ_g, with the prescribed R- and flavor symmetry fluxes, and the real masses generated by σ^{flav}. Then $\mathcal{H}$ is the Hilbert space of states of H, and J^{flav} are the flavor charge operators. From the supersymmetry algebra and the standard Witten index argument, one obtains that the states contributing to the index are those with

$$0 = \mathcal{Q}^2 = H - \sigma^{\text{flav}} J^{\text{flav}}\,. \tag{1.3}$$

We sometimes call Z_{3d} as the "higher-genus topologically twisted index". The 4d partition function could be put in the form of an index as well.

We compute the partition functions with localization techniques. The computation parallels the one in [**11**] (see also [**14**]) for the case $g = 0$, which in turn adapts the computation of the elliptic genus in [**6, 7**] to the present situation and reduces the path-integral to the evaluation of Jeffrey-Kirwan residues [**30**]. The novelty for $g > 0$ is that there are more bosonic and fermionic zero-modes. After taking them into account one can put the result in the following schematic form:

$$Z_{\Sigma_g \times T^n} = \sum_{u=u_{(\alpha)}} Z_{\text{cl},11}\Big|_{\mathfrak{m}=0} \left(\det_{ab} \frac{\partial B_a}{\partial u_b}\right)^{g-1}, \tag{1.4}$$

where $Z_{\text{cl},11}$ (classical and one-loop contribution) is a function of the flavor parameters $(v, \mathfrak{n})$ and similar gauge parameters $(u, \mathfrak{m})$, obtained from the quadratic expansion of the action around BPS configurations, then

$$iB_a = \frac{\partial \log Z_{\text{cl},11}}{\partial \mathfrak{m}_a}\,, \tag{1.5}$$

and $u_{(\alpha)}$ are a set of solutions to the equations, named Bethe Ansatz Equations (BAEs),

$$e^{iB_a} = 1 \tag{1.6}$$

[2] One could also turn on flat connections on Σ_g, but they do not affect the answer.

such that a Vandermonde determinant is not zero. This expression has appeared before, for instance in [**21**,**31**,**33**–**35**] (where the name BAEs was given). We give a derivation of the formula within the Jeffrey-Kirwan residue framework, from which one can extract a precise prescription for what solutions to the BAEs should be kept, even in theories with a complicated matter content. [3] To the best of our knowledge, the 4d case has not appeared before.

We present a few simple examples and applications of the formula in various dimensions, comparing with known results when they are available. In three dimensions we perform new tests of non-perturbative dualities, in particular of Aharony duality [**1**] and the so-called "duality appetizer" [**29**]. The higher-genus index proves to be useful to identify topological sectors involved in the dualities. In four dimensions we perform a simple check of Seiberg duality [**37**]. We leave a more detailed analysis for the future.

Specialized to $g = 1$ and turning off the flavor fluxes, $\mathfrak{n} = 0$, the partition function becomes independent of the flavor fugacities v and it computes the Witten index [**40**]. This result should be taken with care, though. First, our computation is only valid when there is a continuous non-anomalous R-symmetry, even at $g = 1$. Thus, in particular, we cannot use our formula to reproduce the Witten index of 4d super-Yang-Mills (SYM). Second, our formula computes the Witten index of the theory with generic fugacities v, which might be different from the index of the theory with $v = 0$. [4] With this proviso, we reproduce the Witten indices computed in [**27**] in the cases we consider.

One of the other purposes of this work is to generalize the counting of microstates of a class of black holes in AdS_4 performed in [**8**]. We then compute the higher-genus index of the three-dimensional ABJM theory [**2**] in the large N limit at leading order. We will use the large N solution of the BAEs found in [**8**] to evaluate the index and its dependence on g. It was shown in [**8**] that the $g = 0$ twisted index reproduces the Bekenstein-Hawking entropy of certain BPS black holes in AdS_4; we show here that the $g > 0$ index correctly reproduces the entropy of BPS black holes whose horizon has the topology of Σ_g.

The note is organized as follows. In section 2 we derive the formula of the higher-genus topologically twisted index of 3d theories. In sections 3 and 4 we give the generalization to 2d and 4d. In section 5 we present various examples. In section 6 we compare the large N index with the black hole entropy.

2. 3d theories on $\Sigma_g \times S^1$

We consider three-dimensional $\mathcal{N} = 2$ gauge theories with a $U(1)_R$ R-symmetry on $\Sigma_g \times S^1$, where Σ_g is an arbitrary Riemann surface of genus g. This setup is the higher-genus generalization of the "topologically twisted index" studied in [**11**]. On Σ_g we take vielbein $e^{1,2}$ and on S^1 we take $e^3 = \beta\, dt$ with $t \cong t + 1$. In order to preserve some supersymmetry, we perform a partial topological twist on Σ_g, *i.e.* we turn on a background connection V_μ on Σ_g coupled to the R-symmetry current,

[3]The formulation as Jeffrey-Kirwan residues is more general, while the expression given above is only valid if the roots of the BAEs are simple.

[4]This is because we treat the superpotential as a $\mathcal{Q}$-exact deformation, but one should be careful in doing that.

such that it cancels the spin connection for half of the supercharges:

$$(2.1)\qquad V = -\frac{1}{2}\omega^{12} , \qquad W = dV = -\frac{R_s}{4}\, e^1 \wedge e^2 , \qquad \frac{1}{2\pi}\int_{\Sigma_g} W = g - 1 .$$

Here ω_μ^{ab} is the spin connection, while R_s is the scalar curvature on Σ_g. Since the BRST-invariant sector of the twisted theory becomes independent of the metric on Σ_g [**42**], we will not need to specify it.

The supersymmetry parameters ϵ, $\tilde\epsilon$ are commuting spinors, have R-charge -1, and solve the twisted Killing spinor equation

$$(2.2)\qquad D_\mu\epsilon = D_\mu\tilde\epsilon = 0 .$$

Since $D_\mu\epsilon = (\partial_\mu - i\omega_\mu^{12}P_-)\epsilon$ (and similarly for $\tilde\epsilon$) where $P_\pm = (1\pm\gamma_3)/2$ are the chiral projectors on Σ_g, it follows that ϵ_+, $\tilde\epsilon_+$ are constant while $\epsilon_- = \tilde\epsilon_- = 0$. The only exception is the flat torus, namely $g = 1$, in which case also constant ϵ_-, $\tilde\epsilon_-$ are solutions: obviously T^2 does not break any supersymmetry. However as soon as we turn on flavor fugacities or fluxes, or we take a curved metric on T^2, those supercharges get broken.

We are interested in gauge theories of vector and chiral multiplets. They are specified by a gauge group G, a matter representation $\mathfrak{R}$ of G for chiral multiplets, and a superpotential $W(\phi)$ which is a holomorphic function, homogeneous of degree 2 with respect to a choice of R-charges q_ϕ. The choice of R-charges controls the coupling to the R-symmetry background, moreover the presence of a net R-symmetry flux requires the choice to satisfy the quantization condition

$$(2.3)\qquad q_{\mathcal{O}}\,(g-1) \in \mathbb{Z}$$

for all gauge-invariant operators $\mathcal{O}$. In three dimensions, besides the super-Yang-Mills Lagrangian we can also add a supersymmetric Chern-Simons (CS) term, which is a quantized Killing form k for G: the parameters are one integer for each simple factor, and an integer matrix for the Abelian factors. Thus, the Lagrangian of the theory is

$$(2.4)\qquad \mathcal{L} = \mathcal{L}_{\text{YM}} + \mathcal{L}_{\text{CS}} + \mathcal{L}_{\text{mat}} + \mathcal{L}_W .$$

The various terms (spelled out in appendix A) are essentially equal to those on flat space, the only difference being that they are coupled to the background.

The supersymmetry variations in our notation are given in appendix A. They realize the superalgebra $\mathfrak{su}(1|1)$, whose bosonic subalgebra $\mathfrak{u}(1)$ generates rotations of S^1 mixed with gauge and flavor rotations:

$$(2.5)\qquad \{Q,\widetilde{Q}\} = -i\mathcal{L}_v^A - \delta_{\text{gauge}}(\tilde\epsilon^\dagger\epsilon\,\sigma) , \qquad Q^2 = \widetilde{Q}^2 = 0 , \qquad v^\mu = \tilde\epsilon^\dagger\gamma^\mu\epsilon .$$

Here $\mathcal{L}_v^A$ is the gauge-covariant Lie derivative (including the R-symmetry connection) along the covariantly constant (and Killing) vector field v^μ. [5] In fact $v = \beta^{-1}\tilde\epsilon^\dagger\epsilon\,\partial_t$.

The supercharges are compatible with a non-trivial background for the flavor symmetries, *i.e.* background values for external vector multiplets coupled to the flavor currents (as in [**11**]). In three dimensions, flavor symmetries include the topological symmetries generated by $J_\mu^{\text{T}} = (*F)_\mu$ for each Abelian gauge field. It

[5] The explicit expression of the Lie derivative of fields of various spins can be found in appendix B.1 of [**10**].

turns out that, up to flavor rotations, one can turn on a constant flat connection A_3^{flav} along S^1, a constant σ^{flav}, and a flux F_{12}^{flav} on Σ_g accompanied by

$$D^{\text{flav}} = iF_{12}^{\text{flav}} , \tag{2.6}$$

all taking values in the Cartan subalgebra of the flavor symmetry group G_F. It is easy to check from (A.1) that the supersymmetry conditions $Q\lambda^{\text{flav}} = \widetilde{Q}\lambda^{\text{flav}} = 0$, and similarly for $\lambda^{\dagger\,\text{flav}}$, are met. We parametrize the background by

$$v = \beta\big(A_3^{\text{flav}} + i\sigma^{\text{flav}}\big) , \qquad \mathfrak{n} = \frac{1}{2\pi}\int_{\Sigma_g} F^{\text{flav}} . \tag{2.7}$$

The "complexified flat connection" v actually takes values on the complexified maximal torus of G_F, and the flux $\mathfrak{n}$ is GNO quantized (meaning that $\gamma(\mathfrak{n}) \in \mathbb{Z}$ for the weights γ of all representations of G_F, *i.e.* $\mathfrak{n}$ is in the coroot lattice). These parameters make the partition function on $\Sigma_g \times S^1$ an interesting function.

We can also include Wilson line operators $\mathcal{W}_\mathcal{R}$, where $\mathcal{R}$ are representations of G, along S^1 and sitting at arbitrary points on Σ_g. They are constructed in the standard way as

$$\mathcal{W}_\mathcal{R} = \mathrm{Tr}_\mathcal{R}\, \mathrm{Pexp} \oint_{S^1} \big(iA - \sigma\, e^3\big) , \tag{2.8}$$

where Pexp is the path-ordered exponential.

For genus $g > 0$, one could turn on flavor flat connections on Σ_g as well. Since such connections do not show up in the supersymmetry algebra (2.5)—*i.e.* they are not central charges—they do not affect the partition function and we will not introduce them.

The quantity of interest, that we can call a "higher-genus topologically twisted index", is the partition function of the theory, possibly with Wilson line insertions, on $\Sigma_g \times S^1$:

$$Z_{\Sigma_g\times S^1}(t_j, \mathcal{R}_\alpha) = \int \mathcal{D}\varphi \left(\prod\nolimits_\alpha \mathcal{W}_{\mathcal{R}_\alpha}\right) e^{-S[\varphi, t_j]} , \tag{2.9}$$

where t_j are collectively the parameters of the background.

2.1. The derivation of the formula. We wish to compute the partition function (2.9) on $\Sigma_g \times S^1$ with localization techniques [**42, 44**]. In doing the computation, one encounters a system of bosonic and fermionic zero-modes that can be dealt with using the method in [**6, 7**] (see also [**11, 14, 15, 22**]), originally envisaged to compute the elliptic genus of two-dimensional gauge theories. In fact, one can follow almost verbatim the computation in [**11**], with a few modification that we will discuss here.

We perform localization with respect to the supercharge $\mathcal{Q} = Q + \widetilde{Q}$ where one takes $\epsilon = \tilde{\epsilon}$. The BPS equations give a moduli space $\widetilde{\mathcal{M}}_{\text{BPS}}$ of complexified BPS configurations modulo gauge transformations

$$\widetilde{\mathcal{M}}_{\text{BPS}} = \Big\{ D = iF_{12} \,,\; F_{13} = F_{23} = 0 \,,\; D_\mu \sigma = 0 \Big\} / \mathcal{G} . \tag{2.10}$$

As argued in [**11**], only configurations with constant D contribute to the real path-integral, and we call $\mathcal{M}_{\text{BPS}}$ such a moduli space. For $g = 0$, $\mathcal{M}_{\text{BPS}}$ is parametrized by "complexified flat gauge connections" u and gauge fluxes $\mathfrak{m}$,

$$u = \beta(A_3 + i\sigma) , \qquad \mathfrak{m} = \frac{1}{2\pi}\int_{\Sigma_g} F , \tag{2.11}$$

where $\mathbb{R}\mathrm{e}\, u$ is along the maximal torus H of G, $\mathbb{I}\mathrm{m}\, u$ and $\mathfrak{m}$ are along the Cartan subalgebra $\mathfrak{h}$, $\mathfrak{m}$ is GNO quantized to the coroot lattice $\Gamma_{\mathfrak{h}}$, and we mod by Weyl transformations. In other words, $\mathcal{M}_{\text{BPS}}^{g=0} = (H \times \mathfrak{h} \times \Gamma_{\mathfrak{h}})/W$. The novelty for $g > 0$ is that we also have flat connections on Σ_g. For generic u, the flat connections on Σ_g should be in the maximal torus H, giving an extra factor H^{2g}. Along special complex-codimension-1 hyperplanes in the u plane where the commutant of u in G is non-Abelian, one can have non-Abelian flat connections on Σ_g and their space is larger. As we will see, those hyperplanes do not contribute to the path-integral and can be removed. This is similar to the behavior of chiral multiplets already observed for $g = 0$. For generic values of u, the BPS equations set the chiral multiplets to zero; along special hyperplanes there can be non-trivial solutions, however those hyperplanes do not contribute.

Thus, for generic values of u, the bosonic zero-modes take values on

$$\mathfrak{M} \times H^{2g} \qquad \text{where} \qquad \mathfrak{M} = H \times \mathfrak{h}\,. \tag{2.12}$$

There are also fermionic zero-modes, from the Cartan gaugini. It turns out that they form 0d off-shell supermultiplets, which can be thought of as the dimensional reduction of 2d $\mathcal{N}{=}(0,2)$ supermultiplets.

The fermionic zero-modes are constant on S^1 and satisfy $\not{D}_{(\Sigma)}\lambda = 0$ on the Riemann surface. In components this is $\partial_{\bar{z}}\lambda_+ = 0$ and $(\partial_z - i\omega_z^{12})\lambda_- = 0$, where $e^z = e^1 + ie^2$ is the holomorphic vielbein, in particular λ_+ is a constant (the unique holomorphic function) and $\lambda_-^\dagger$ is a holomorphic differential. Therefore for any value of g there is a 0d vector multiplet—comprised of $(u, \bar{u}, \lambda_0, \lambda_0^\dagger, D_0)$ where u, $\bar{u}$, D_0 are bosonic and λ_0, $\lambda_0^\dagger$ are fermionic—that parametrizes the constant zero-modes. Besides u, $\bar{u}$, the other ones are

$$\lambda_0 = \beta\, \tilde{\epsilon}^\dagger \lambda\,, \qquad \lambda_0^\dagger = \beta\, \lambda^\dagger \epsilon\,, \qquad D_0 = \beta\, \tilde{\epsilon}^\dagger \epsilon\, (D - iF_{12}) \tag{2.13}$$

where we implicitly took the constant mode in each case. We get the supersymmetry algebra
(2.14)

$$\begin{array}{lllll} Qu = 0 & Q\bar{u} = i\lambda_0^\dagger & Q\lambda_0 = -D_0 & Q\lambda_0^\dagger = 0 & QD_0 = 0 \\ \widetilde{Q}u = 0 & \widetilde{Q}\bar{u} = i\lambda_0 & \widetilde{Q}\lambda_0 = 0 & \widetilde{Q}\lambda_0^\dagger = D_0 & \widetilde{Q}D_0 = 0\,. \end{array}$$

The mode D_0 is an auxiliary zero-mode. Notice that $Q^2 = \widetilde{Q}^2 = \{Q, \widetilde{Q}\} = 0$ on the zero-mode subspace, since the zero-modes are translationally invariant and commute with σ.

For $g > 0$ there are g 0d chiral multiplets—comprised of $\big(a^{(\alpha)}, \bar{a}^{(\alpha)}, \eta_0^{(\alpha)}, \eta_0^{(\alpha)\dagger}\big)$ where the first two are bosonic and the last two fermionic—that parametrize the modes proportional to the g holomorphic differentials ω_α on Σ_g. It is convenient to define the following 1-forms on Σ_g:

$$\eta_j = \frac{i}{2} \tilde{\epsilon}^\dagger \gamma_j \lambda \;\; \text{with} \;\; \eta_z = 0\,, \qquad \eta_j^\dagger = \frac{i}{2} \lambda^\dagger \gamma_j \epsilon \;\; \text{with} \;\; \eta_{\bar{z}}^\dagger = 0 \tag{2.15}$$

where $j = 1, 2$ only. Since $\tilde{\epsilon}^\dagger \gamma_{\bar{z}} \lambda = \tilde{\epsilon}_+^\dagger \lambda_-$ and $\lambda^\dagger \gamma_z \epsilon = \lambda_-^\dagger \epsilon_+$, those forms contain the negative-chirality zero-modes in λ. Using that all modes are independent from

t and along the Cartan subalgebra, from (A.1) we obtain the transformations

$$QA_j = \eta_j^\dagger\,, \qquad Q\eta_j^\dagger = 0\,, \qquad Q\eta_{\bar z} = i\tfrac{\bar\epsilon^\dagger\epsilon}{\beta} D_{\bar z} u$$
$$\widetilde{Q}A_j = \eta_j\,, \qquad \widetilde{Q}\eta_j = 0\,, \qquad \widetilde{Q}\eta_z^\dagger = i\tfrac{\bar\epsilon^\dagger\epsilon}{\beta} D_z u\,. \tag{2.16}$$

We can parametrize the flat connections on Σ as

$$A_{(\Sigma)} = \sum_{\alpha=1}^{g} a^{(\alpha)}\,\omega_\alpha + A_z^{\text{ref}}dz + \sum_{\alpha=1}^{g} \bar a^{(\alpha)}\,\bar\omega_\alpha + A_{\bar z}^{\text{ref}}d\bar z\,, \tag{2.17}$$

where A^{ref}, chosen such that $dA^{\text{ref}} = 2\pi\mathfrak{m}\, d\text{vol}_{\Sigma_g}/\text{Vol}(\Sigma_g)$, is a reference background connection. Similarly we parametrize

$$\eta^\dagger = \sum\nolimits_\alpha \eta_0^{(\alpha)\dagger}\omega_\alpha\,, \qquad \eta = \sum\nolimits_\alpha \eta_0^{(\alpha)}\bar\omega_\alpha\,. \tag{2.18}$$

We thus obtain the superalgebra

$$Qa^{(\alpha)} = \eta_0^{(\alpha)\dagger}\,, \qquad Q\bar a^{(\alpha)} = 0\,, \qquad Q\eta_0^{(\alpha)} = Q\eta_0^{(\alpha)\dagger} = 0$$
$$\widetilde{Q}a^{(\alpha)} = 0\,, \qquad \widetilde{Q}\bar a^{(\alpha)} = \eta_0^{(\alpha)}\,, \qquad \widetilde{Q}\eta_0^{(\alpha)} = \widetilde{Q}\eta_0^{(\alpha)\dagger} = 0\,. \tag{2.19}$$

The on-shell classical action has been computed in [11]. Let us distinguish the background fields $(v,\mathfrak{n})$ for "standard" flavor symmetries from those $(w,\mathfrak{t})$ for the topological symmetries, as they appear differently in the Lagrangian. We use the notation $y = e^{iv}$ to mean that $y^\gamma = e^{i\gamma(v)}$ for the flavor weights γ, and similarly $\xi = e^{iw}$. The CS action contributes $Z_{\text{CS}} = e^{ik(u,\mathfrak{m})}$, where k is the CS Killing form. In particular each simple factor G_I in G gives

$$Z_{\text{CS}}^{\text{n-Ab}} = x^{k\mathfrak{m}} = \prod_{i=1}^{\text{rank}\,G_I} x_i^{k_I\mathfrak{m}_i}\,, \tag{2.20}$$

while the Abelian factors in G give

$$Z_{\text{CS}}^{\text{Ab}} = \prod\nolimits_{i,j} x_i^{k_{ij}\mathfrak{m}_j}\,. \tag{2.21}$$

These expressions are valid, with the substitution $(u,\mathfrak{m}) \to (v,\mathfrak{n})$ or $(u,\mathfrak{m}) \to (w,\mathfrak{t})$, for gauge-flavor and flavor-flavor CS terms as well. Gauge-R or flavor-R CS terms give

$$Z_{\text{CS}}^{\text{R}} = \prod\nolimits_i x_i^{(g-1)k_{iR}}\,, \tag{2.22}$$

since the R-symmetry flux is fixed by supersymmetry. The coupling of topological currents J_μ^{T} to background vector multiplets leads to a factor

$$Z_{\text{CS}}^{\text{T}} = x^{\mathfrak{t}}\xi^{\mathfrak{m}} \tag{2.23}$$

for each topological symmetry. Finally, each Wilson line operator insertion gives a factor

$$\mathcal{W}_{\mathcal{R}} = \text{Tr}_{\mathcal{R}}\, x = \sum\nolimits_{\rho\in\mathcal{R}} x^\rho \tag{2.24}$$

which is the character of $\mathcal{R}$.

The one-loop determinants of small quadratic fluctuations around the BPS configurations have been computed in [**21, 32, 34**] (see also [**11**]). From chiral multiplets we get

$$Z^{\text{chiral}}_{\text{1-loop}} = \prod_{\rho\in\mathfrak{R}} \Big(\frac{x^{\rho/2} y^{\gamma/2}}{1 - x^\rho y^\gamma} \Big)^{\rho(\mathfrak{m}) + \gamma(\mathfrak{n}) + (g-1)(q_\rho - 1)} , \tag{2.25}$$

where the product is over the weights of the gauge representation $\mathfrak{R}$. From vector multiplets we get

$$Z^{\text{gauge}}_{\text{1-loop}} = \prod_{\alpha\in G} (1 - x^\alpha)^{1-g} (i\, du)^{\operatorname{rank} G} , \tag{2.26}$$

where the product is over the roots of G. $Z^{\text{gauge}}_{\text{1-loop}}$ is a holomorphic top-form on $\mathfrak{M}$. We will compactly call

$$Z_{\text{cl,1l}}(x, \mathfrak{m}, y, \mathfrak{n}, \xi, \mathfrak{t})$$

the product of all classical (including the Wilson line operators) and one-loop contributions: this is a holomorphic top-form on $\mathfrak{M}$. Note that there is no dependence on the flat connections on Σ_g.

At this point we should integrate over all zero-modes. The integral over the modes $a^{(\alpha)}$, $\bar a^{(\alpha)}$ is trivial because $Z_{\text{cl,1l}}$ does not depend on them. They parametrize the space H^{2g} of Abelian flat connections on Σ_g, therefore the integral gives a volume factor. [6]

Let us denote by $\mathcal{Z}(u, \bar u, \lambda_0, \lambda_0^\dagger, D_0, a, \bar a, \eta_0, \eta_0^\dagger; \mathfrak{m})$ the effective partition function of the zero-modes, obtained by integrating out all non-zero-modes in a given flux sector $\mathfrak{m}$. Setting to zero all fermionic and the auxiliary mode D_0, it coincides with the classical and one-loop contribution we have computed above:

$$\mathcal{Z}\Big|_{\lambda_0 = \lambda_0^\dagger = \eta_0 = \eta_0^\dagger = D_0 = 0} = Z_{\text{cl,1l}}(u, \bar u; \mathfrak{m}) . \tag{2.27}$$

The integral over fermionic zero-modes is a derivative, that we need to compute:

$$\int d\lambda_0 d\lambda_0^\dagger \Big(\prod_{\alpha=1}^g d\eta_0^{(\alpha)} d\eta_0^{(\alpha)\dagger} \Big) \mathcal{Z} = \Big(\prod_{\alpha=1}^g \frac{\partial^2}{\partial\eta_0^{(\alpha)} \partial\eta_0^{(\alpha)\dagger}} \Big) \frac{\partial^2}{\partial\lambda_0 \partial\lambda_0^\dagger} \mathcal{Z} . \tag{2.28}$$

Such a derivative is fixed by supersymmetry, using some tricks in [**11, 14, 18, 34**]. For the sake of clarity, let us consider the case that $\operatorname{rank} G = 1$. Since the effective action is topological on Σ_g, it takes the generic form

$$\mathcal{Z} = \mathcal{A} \exp \int_{\Sigma_g} \Big[\mathcal{B} F + c\, \mathcal{G}\, \eta^\dagger \wedge \eta \Big] \tag{2.29}$$

where $\mathcal{A}$, $\mathcal{B}$ and $\mathcal{G}$ are functions of $u, \bar u, \lambda_0, \lambda_0^\dagger, D_0$ and the constant $c = \big(\frac{\bar\epsilon^\dagger \epsilon}{\beta} \big)^{-1}$ has be inserted for later convenience. In particular the logarithm of $\mathcal{Z}$ is linear in $\mathfrak{m}$ (which is indeed the case for $Z_{\text{cl,1l}}$) and there is no dependence on the flat connections on Σ_g. From the fact that $\mathcal{Z}$ is supersymmetric and using (2.14) and (2.16):

$$0 = Q\mathcal{Z} = i\lambda_0^\dagger \frac{\partial\mathcal{Z}}{\partial\bar u} - D_0 \frac{\partial\mathcal{Z}}{\partial\lambda_0} + \mathcal{Z} \cdot \int_{\Sigma_g} \Big[\mathcal{B}\, d\eta^\dagger - i\, \mathcal{G}\, \eta^\dagger \wedge du \Big] . \tag{2.30}$$

[6] Since the final result is proportional to a normalization constant that we need to fix in one known example, say $U(1)$ Chern-Simons theory, we need not compute the volume factor here.

The only terms containing $\eta^\dagger$ but not η are in the integral. Integrating by parts we conclude

$$\mathcal{G} = \frac{\partial \mathcal{B}}{\partial iu} \,. \tag{2.31}$$

We also have $\mathcal{B} = \partial \log \mathcal{Z}/\partial \mathfrak{m}$, therefore

$$\mathcal{G} = \frac{\partial^2 \log \mathcal{Z}}{\partial iu\, \partial \mathfrak{m}} \,. \tag{2.32}$$

Expanding η, $\eta^\dagger$ into the zero-modes and assuming a normalized basis of holomorphic differentials on Σ_g, we find

$$\prod_{\alpha=1}^{g} \frac{\partial^2 \mathcal{Z}}{\partial \eta_0^{(\alpha)} \partial \eta_0^{(\alpha)\dagger}} = c^g\, \mathcal{G}^g\, \mathcal{Z}\Big|_{\eta=\eta^\dagger=0} = c^g \left(\frac{\partial^2 \log \mathcal{Z}}{\partial iu\, \partial \mathfrak{m}} \right)^g \mathcal{Z}\Big|_{\eta=\eta^\dagger=0} \,. \tag{2.33}$$

The only terms containing $\lambda_0^\dagger$ but not λ_0 are the first two in (2.30), which should cancel each other. Taking a derivative with respect to $\lambda_0^\dagger$ we get

$$D_0 \frac{\partial^2 \mathcal{Z}}{\partial \lambda_0 \partial \lambda_0^\dagger} = -i \frac{\partial \mathcal{Z}}{\partial \bar{u}}\Big|_{\lambda_0=\lambda_0^\dagger=0} \,. \tag{2.34}$$

We thus find, up to an unimportant multiplicative constant:

$$\left(\prod_{\alpha=1}^{g} \frac{\partial^2}{\partial \eta_0^{(\alpha)} \partial \eta_0^{(\alpha)\dagger}} \right) \frac{\partial^2}{\partial \lambda_0 \partial \lambda_0^\dagger} \mathcal{Z} = -\frac{i}{D_0} \frac{\partial}{\partial \bar{u}} \left(\frac{\partial^2 \log \mathcal{Z}}{\partial iu\, \partial \mathfrak{m}} \right)^g \mathcal{Z}\Big|_{\lambda_0=\lambda_0^\dagger=\eta_0=\eta_0^\dagger=0} \,. \tag{2.35}$$

Then we should integrate over the bosonic zero-modes u, $\bar{u}$, D_0. As in [**11**], on the domain $\mathfrak{M}$ of $u, \bar{u}$ there are singular hyperplanes (in the rank $G = 1$ case, those are just points) H_i where some chiral multiplet becomes massless:

$$H_i = \left\{ u \in \mathfrak{M} \,\middle|\, e^{i\rho_i(u) + i\gamma_i(v)} = 1 \right\} . \tag{2.36}$$

There are also hyperplanes where some W-boson becomes massless (whose wavefunction on Σ_g is a holomorphic differential):

$$H_\alpha = \left\{ u \in \mathfrak{M} \,\middle|\, e^{i\alpha(u)} = 1 \right\} . \tag{2.37}$$

These are precisely the zeros of the Vandermonde determinant $\prod_{\alpha \in G}(1 - x^\alpha)$. As explained in [**11**] (following [**6**]), using (2.35) and Stoke's theorem, the integral over $\mathfrak{M}$ reduces to a Cauchy contour integral around the hyperplanes H_i, H_α and around the regions at infinity. Then, as a result of the integration over D_0, the contour integral is taken at $D_0 = 0$, and only certain residues are picked up. Which ones depends on the choice of an auxiliary parameter $\eta \in \mathfrak{h}^*$, and it is controlled by the Jeffrey-Kirwan residue [**12, 30, 38**]. The computation that leads to (2.35) is easily generalized to the higher-rank case, as well the reduction to a contour integral (this is more intricate, and we refer to [**7**] and [**11**]). The result is

$$\begin{aligned} Z_{\Sigma_g \times S^1} = \frac{1}{|W|} \sum_{\mathfrak{m} \in \Gamma_\mathfrak{h}} \sum_{u_* \in \mathfrak{M}^*_{\text{sing}}} \underset{u=u_*}{\text{JK-Res}} \left(\mathsf{Q}_{u_*}, \eta \right) \left(\det_{ab} \frac{\partial^2 \log Z_{\text{cl},11}}{\partial iu_a\, \partial \mathfrak{m}_b} \right)^g Z_{\text{cl},11} \\ + \text{bound. contrib.} \end{aligned} \tag{2.38}$$

Here $|W|$ is the order of the Weyl group; $\mathfrak{M}^*_{\text{sing}}$ is the collection of singular points in $\mathfrak{M}$ where at least rank G linearly independent singular hyperplanes meet; Q_{u_*} is the set of charge covectors in $\mathfrak{h}^*$ of the singular hyperplanes passing through u_*.

The Jeffrey-Kirwan residue operation performs integration along selected middle-dimensional contours around each of the points u_*, based on the choice of η, that satisfy the following conditions:

$$\underset{u=0}{\text{JK-Res}}(\mathrm{Q}_*,\eta)\,\frac{dQ_{j_1}(u)}{Q_{j_1}(u)}\wedge\cdots\wedge\frac{dQ_{j_r}(u)}{Q_{j_r}(u)} = \begin{cases}\operatorname{sign}\det(Q_{j_1}\dots Q_{j_r}) & \text{if } \eta\in\operatorname{Cone}(Q_{j_1}\dots Q_{j_r})\\ 0 & \text{otherwise}\end{cases} \tag{2.39}$$

where Cone denotes the cone spanned by the vectors in the argument. The boundary contributions are integrals of the same form along middle-dimensional contours around infinity, constructed in [**11**]. Fortunately, one can usually avoid to evaluate the boundary contours by commuting sum and integration in (2.38): this brings all singular points in the interior of $\mathfrak{M}$. More details can be found in [**7, 11**].

In the rank $G = 1$ case, the formula simplifies to

$$Z_{\Sigma_g\times S^1} = \frac{1}{|W|}\sum_{\mathfrak{m}\in\Gamma_{\mathfrak{h}}}\ \sum_{x_*\in\mathfrak{M}_{\text{sing}},0,\infty} \underset{x=x_*}{\text{JK-Res}}\left(\mathrm{Q}(x_*),\eta\right)\left(\frac{\partial^2\log Z_{\text{cl},11}}{\partial\log x\,\partial\mathfrak{m}}\right)^g Z_{\text{cl},11}\,, \tag{2.40}$$

where we mapped the cylinder to the punctured Riemann sphere by $x = e^{iu}$. The boundary contributions are captured by the very same JK residue, assigning to the points $x = 0, \infty$ charges proportional to the effective CS coupling at infinity on the Coulomb branch:

$$k_\pm \equiv k_{\text{eff}}(u=\pm\infty) = k + \frac{1}{2}\sum_i Q_i^2\operatorname{sign}(\pm Q_i)\,,\qquad Q_0 = -k_+\,,\qquad Q_\infty = k_-\,. \tag{2.41}$$

Here Q_i are the charges of chiral multiplets if $G = U(1)$, and the weights if $G = SU(2)$. In particular, choosing $\eta > 0$ one picks the residues from poles with associated positive charge; choosing $\eta < 0$ one picks minus the residues from poles with associated negative charge. The two choices give the same answer.

For Abelian gauge groups these formulæ give the final answer. In the non-Abelian case, instead, the formulæ are not yet complete. One of the assumptions in the derivation of the JK residue in [**6, 7, 11**] is that at each point $u_* \in \mathfrak{M}^*_{\text{sing}}$, the singular hyperplane arrangement is "projective". In the rank-one case this means that all fields that become massless at u_* have charges with the same sign; in the higher-rank case this means that the weights ρ of the massless fields at u_* lie in a half-space. This condition is not met by the hyperplanes H_α from W-bosons, since for each H_α there is a coincident $H_{-\alpha}$. In order to correctly evaluate the contribution from the W-bosons, we need to resolve their singularities.

Among the various possibilities, we choose the following regularization of the vector multiplet one-loop determinant. Such determinant is equal to the inverse of the determinant for chiral multiplets of R-charge 0 transforming as the roots α of G. [7] This follows from the Higgs mechanism. Suppose we add to the theory an adjoint chiral multiplet of R-charge 0. Its diagonal components can get a VEV without breaking the R-symmetry. A W-boson along a root α combines with the

[7] In [**11**] we have argued that the one-loop determinants suffer from a sign ambiguity, which shows up when there are Abelian factors in G and it can be reabsorbed by a sign redefinition of ξ. Hence, to make our formulæ lighter, we have included a factor $(-1)^{2\delta(\mathfrak{m})}$ into the vector multiplet determinant, where $\delta = \frac{1}{2}\sum_{\alpha>0}\alpha$ is the Weyl vector.

component α of the chiral multiplet and they become massive. The product of the two one-loop determinants should then be 1:

$$Z^{\mathrm{W}}_{\text{1-loop}}(\alpha) = \frac{1}{Z^{\Phi,\,q=0}_{\text{1-loop}}(\alpha)} \,. \tag{2.42}$$

Indeed, $(-1)^{2\delta(\mathfrak{m})}\prod_\alpha 1/Z^{\text{chiral},\,q=0}_{\text{1-loop}} = \prod_\alpha (1-x^\alpha)^{1-g}$. Our regularization consists in introducing a complexified flat connection v for a would-be symmetry that acts on the extra g modes from the vector multiplet as it would on a chiral multiplet. This would be a symmetry of the effective quantum mechanics on S^1 obtained by KK reduction on Σ_g in each flux sector, if we switched some interactions off. Thus we take as regularized gauge one-loop determinant:

$$Z^{\text{gauge, reg}}_{\text{1-loop}} = (-1)^{2\delta(\mathfrak{m})} \prod_{\alpha\in G} (1-x^\alpha)\Big(\frac{x^{\alpha/2}y^{1/2}}{1-x^\alpha y}\Big)^{-\alpha(\mathfrak{m})+g} \,. \tag{2.43}$$

Here $\delta = \frac{1}{2}\sum_{\alpha>0}\alpha$ is the Weyl vector, and $y = e^{iv}$. For $y\to 1$ we recover $Z^{\text{gauge}}_{\text{1-loop}}$, however for $y \neq 1$ the singularities are "projective" and we can safely compute the JK residue. At the end of the computation we should take $y \to 1$.

To see what happens, let us consider the simple case of $SU(2)$ Chern-Simons theory at level k. We do not need to introduce matter, since we already know how to treat the singularities from chiral multiplets. Applying the rules described above, the classical and regularized one-loop contribution is

$$Z_{\text{cl},11} = -\frac{(1-x^2)^2}{x^2}\Big(\frac{1-x^2y}{x^2-y}\Big)^{2\mathfrak{m}}\Big(\frac{x^2y}{(1-x^2y)(x^2-y)}\Big)^g x^{2k\mathfrak{m}} \,. \tag{2.44}$$

The factor from fermionic zero-modes is

$$\frac{\partial^2 \log Z_{\text{cl},11}}{\partial \log x\, \partial\mathfrak{m}} = 2k - \frac{4x^2}{x^2-y} - \frac{4x^2y}{1-x^2y} \,. \tag{2.45}$$

After a redefinition $x^2 = z$, the expression for the partition function is

$$Z_{\Sigma\times S^1} = -\frac{1}{2}\sum_{\mathfrak{m}\in\mathbb{Z}} \int_{\text{JK}} \frac{dz}{2\pi i z}\frac{(1-z)^2}{z}\Big(\frac{1-zy}{z-y}\Big)^{2\mathfrak{m}}\Big(\frac{zy}{(1-zy)(z-y)}\Big)^g \times z^{k\mathfrak{m}}\Big(2k - \frac{4z}{z-y} - \frac{4zy}{1-zy}\Big)^g \tag{2.46}$$

and the contour is the one prescribed by the JK residue. We choose $\eta < 0$, *i.e.* we should take minus the residues at $z = 0$ and $z = y$ (and not the residues at $z = y^{-1}$ and $z = \infty$). The poles at $z = 0$ are confined in the domain $\mathfrak{m} \le M - 1$ (we can take M large and positive), and the poles at $z = y$ in the domain $\mathfrak{m} \ge -M$. First we compute the sum of the residues at $z = 0$, for $\mathfrak{m} \le M-1$. Let us define

$$e^{iB} = z^k\Big(\frac{1-zy}{z-y}\Big)^2 \,. \tag{2.47}$$

The root of the geometric series is $(e^{iB})^{-1}$. Thus, we should find a contour $\mathcal{C}_a$ that includes only $z = 0$ (and not the other poles at $z = y, y^{-1}, \infty$), while being inside the region $|e^{iB}| > 1$. On such a contour we have uniform convergence, so we can commute sum and integration. We find

$$Z^{(a)}_{\Sigma\times S^1} = \sum_{z_i}^{(a)} \operatorname*{Res}_{z=z_i} \frac{(1-z)^2}{2z^2}\Big(\frac{zy}{(1-zy)(z-y)}\Big)^g\Big(2k - \frac{4z}{z-y} - \frac{4zy}{1-zy}\Big)^g \frac{1}{e^{iB}-1} \,, \tag{2.48}$$

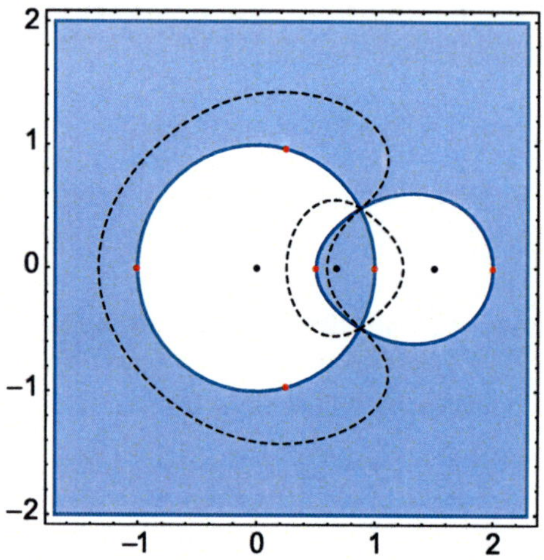

FIGURE 1. Analytic structure of e^{iB} in the complex z plane, for $k = 4$ and $y = 0.7$. The black dots (darker) are the points $\{0, y, y^{-1}\}$ from left to right. The red dots (lighter) are the solutions to $e^{iB} = 1$. The blue delimiter is the line $|e^{iB}| = 1$, while the shaded and white regions have $|e^{iB}| > 1$ and < 1, respectively. The dashed contour inside the shaded region is $\mathcal{C}_a$, the one inside the white region is $\mathcal{C}_b$.

where z_i are the solutions to the BAE

$$e^{iB} = 1 \tag{2.49}$$

inside the contour $\mathcal{C}_a$. Then we compute the sum of the residues at $z = y$, for $\mathfrak{m} \geq -M$. The root of the geometric series is e^{iB}, the inverse than before. This time we should find a contour $\mathcal{C}_b$ that includes only $z = y$ and lies inside the region $|e^{iB}| < 1$. Resumming, we find the same expression as before, but with opposite sign:

$$Z^{(b)}_{\Sigma\times S^1} = -\sum_{z_i}^{(b)} \operatorname*{Res}_{z=z_i} \frac{(1-z)^2}{2z^2} \Big(\frac{zy}{(1-zy)(z-y)} \Big)^g \Big(2k - \frac{4z}{z-y} - \frac{4zy}{1-zy} \Big)^g \frac{1}{e^{iB}-1} \,. \tag{2.50}$$

This time z_i are the solutions to the BAE inside the contour $\mathcal{C}_b$. In practice, solutions that are inside or outside both contours do not contribute—the contour is effectively $\mathcal{C}_a - \mathcal{C}_b$.

We should understand the analytic structure of e^{iB}. It turns out that we can find the desired contours if we take $y = 1 - \epsilon$ with $\epsilon > 0$. See Figure 1. In this case $\mathcal{C}_a$ contains the $k - 1$ approximate roots of $z^k = 1$ with $z \neq 1$, as well as a solution along the real interval $(0, y)$; instead $\mathcal{C}_b$ contains the latter solution and $z = 1$. The residue at $z = 1$ is zero because of the Vandermonde determinant. Therefore $Z_{\Sigma\times S^1}$ equals the sum of the residues at the $k - 1$ approximate roots of $z^k = 1$ with $z \neq 1$.

At this point the limit $y \to 1$ is taken easily, because all possible subtleties happen at $z = 1$ that is excluded, hence the limit is smooth. We find

$$Z_{\Sigma\times S^1} = -\sum_{z_i}{}' \operatorname*{Res}_{z=z_i} 2k \frac{\left[(1-z)(1-z^{-1})\right]^{1-g}}{2z} \frac{1}{e^{iB}-1}, \tag{2.51}$$

where

$$e^{iB} = z^k \tag{2.52}$$

and the sum is over the solutions to the BAE $e^{iB} = 1$ such that the Vandermonde determinant $\prod_\alpha (1-z^{\alpha/2})$ does not vanish. The expression in (2.51) is precisely what one would have obtained from the general formula in (2.40) using the unregularized one-loop determinants, by taking the JK residues from the matter singularities H_i (and not H_α), resumming over $\mathfrak{m}$, and then discarding the solutions to the BAEs for which the Vandermonde determinant vanish. Although we have done the computation for $SU(2)$ only, the result is general. It is also clear that the presence of chiral multiplets does not change the argument, as long as their poles are kept away from the zeros of the Vandermonde determinant by a generic choice of flavor background parameters.

2.2. The final formula. Summarizing, we find the following prescription. One uses (2.38)—where only the poles from chiral multiplets and those at infinity are taken into account, as in [**11**]—to produce a geometric series in $\mathfrak{m}$ and a contour of uniform convergence. One should resum the geometric series, obtaining a sum of residues at the roots of the BAEs $e^{iB_a} = 1$, but only the roots for which the Vandermonde determinant does not vanish should be kept.

In case all acceptable roots of the BAEs are simple, after some manipulations one obtains, schematically:

$$Z_{\Sigma_g\times S^1} = \frac{(-1)^{\operatorname{rank} G}}{|W|} \sum_{x=x_{(i)}} Z_{\text{cl},11}\Big|_{\mathfrak{m}=0} \left(\det_{ab} \frac{\partial B_a}{\partial u_b}\right)^{g-1}, \tag{2.53}$$

where

$$iB_a = \frac{\partial \log Z_{\text{cl},11}}{\partial \mathfrak{m}_a} \tag{2.54}$$

and $x_{(i)}$ are a set of solutions to the BAEs,

$$x_{(i)} \qquad \text{such that} \qquad e^{iB_a} = 1 \qquad \text{and} \qquad \prod\nolimits_{\alpha\in G} \left(1 - x_{(i)}^\alpha\right) \neq 0\,. \tag{2.55}$$

Notice that in (2.53) we could equivalently write the matrix $\partial e^{iB_a}/\partial \log x_b$, since the expression is evaluated at $e^{iB_a} = 1$.

This expression has appeared before, for instance in [**21,33–35**]. We have given a precise contour-integral prescription for (2.53), which in all cases determines what specific subset of solutions to the BAEs should be kept. Moreover, if the roots of the BAEs are not simple, the more general JK prescription (2.38) should be used.

2.3. The Witten index. For $g = 1$ and setting the flavor fluxes $\mathfrak{n} = 0$, the twisted index reduces to the standard Witten index [**40**], *i.e.* the supersymmetric partition function of the $\mathcal{N} = 2$ theory on T^3. All four supercharges are unbroken, even in the presence of generic twisted masses and flavor flat connections. Hence, it follows from the supersymmetry algebra (2.5) that the only contributing states are those with $H \pm \sigma^{\text{flav}} J^{\text{flav}} = 0$, *i.e.* the zero-energy ground states.

As always, we should be careful to interpret the path-integral on T^3 as a Witten index (see also the discussion in [**27**]): this is true only if the theory is well-behaved and it has a finite number of zero-energy states (in particular, no moduli space). We perform our computation with no superpotential, and then treat the superpotential as a $\mathcal{Q}$-exact deformation that does not affect the path-integral: but this is true only if the theory without superpotential is well-behaved. Thus, we can compute the Witten index of a theory with generic real masses σ^{flav}, provided that those real masses give a finite number of vacua even turning off the superpotential.

This remark is important. Consider the example of a single chiral multiplet Φ with superpotential $W = \Phi^{N+1}$. This theory has N vacua and the Witten index is $I_{\text{W}} = N$. The theory has a discrete $\mathbb{Z}_{N+1}$ flavor symmetry and an R-symmetry with $q_\Phi = 2/(N+1)$. The twisted index from (2.25) is simply

$$Z = \left(\frac{y}{1-y}\right)^{\frac{1-N}{1+N}(g-1)} \qquad \text{with} \qquad y^{N+1} = 1 \,. \tag{2.56}$$

For $g = 1$ and $y = 1$ the formula is not defined, indeed without superpotential the theory has a flat direction and the path integral is not well-defined. Instead for $y \neq 1$ we find, up to a sign, $Z = -1$. This is the correct "flavored Witten index" $\operatorname{Tr}(-1)^F y^J$, since the N vacua have charges $J = 1, \ldots, N$. It is also confirmed by the computation of the elliptic genus in the analogous 2d $\mathcal{N} = (2,2)$ case [**46**]. On the other hand, if we switch off the superpotential the theory has a $U(1)_F$ flavor symmetry, we can turn on the associated real mass σ^{flav} and then the theory has a single vacuum, $I_{\text{W}} = 1$, in agreement with our formula for $g = 1$ and generic y.

By the standard argument, the Witten index is independent from continuous deformations of the real masses, and by holomorphy also from the flavor fugacities. Indeed the formula (2.53)-(2.55) specialized to $g = 1$ gives

$$I_{\text{W}} = \frac{1}{|W|} \; \# \text{ solutions to BAEs} \,, \tag{2.57}$$

the number of acceptable solutions to the BAEs (modded by Weyl transformations).

Another remark is that our computation is reliable only if the theory has a non-anomalous $U(1)_R$ R-symmetry. This is obvious for $g \neq 1$, since we need a background for the R-symmetry to preserve supersymmetry; it is not obvious for $g = 1$, *i.e.* on flat $T^2 \times S^1$, because there is no background for the R-symmetry and the $U(1)_R$ is not needed to preserve supersymmetry. The point is that in our computation we have assumed that the number of zero-modes equals the one predicted by the index theorem. This is true if we put a generic curved metric on T^2, accompanied by a background for the R-symmetry. On the other hand, on flat $T^2 \times S^1$ there are more zero-modes that should be taken into account. This comment is particularly important in the 4d case, where pure SYM breaks the continuous R-symmetry and our formula cannot be applied.

3. 2d theories on Σ_g

In the two- and four-dimensional cases the analysis is essentially the same, and we will not repeat it here. We will quote the final results.

We consider two-dimensional $\mathcal{N} = (2,2)$ gauge theories of vector and chiral multiplets with a $U(1)_R$ vector R-symmetry, placed on Σ_g with A-twist (same background as in section 2). The Lagrangian is [**45**]

$$\mathcal{L} = \mathcal{L}_{\text{YM}} + \mathcal{L}_{\text{mat}} + \mathcal{L}_{\widetilde{W}} + \mathcal{L}_W \,, \tag{3.1}$$

where the terms are the SYM Lagrangian, the matter kinetic Lagrangian, the twisted superpotential and superpotential interactions, respectively. The parameters of the background are the flux $\mathfrak{n} = \frac{1}{2\pi}\int_\Sigma F^{\text{flav}}$ and a constant VEV v for the complex scalar σ^{flav} in the external vector multiplet. Similarly there are parameters $(\mathfrak{m}, u)$ for the dynamical gauge fields.

We are interested in the partition function of the theory, as well as in correlators of local twisted chiral operators $\mathcal{O}(x)$ [**42–44**]. These are gauge-invariant polynomial functions P of the complex scalar σ in the vector multiplet, which is also the bottom component of a twisted chiral multiplet Σ. Thus, we are interested is the path-integrals

$$\big\langle \mathcal{O}_1(x_1)\dots\mathcal{O}_s(x_s)\big\rangle_g = \int \mathcal{D}\varphi\, P_1\big(\sigma(x_1)\big)\dots P_n\big(\sigma(x_n)\big)\, e^{-S} \tag{3.2}$$

on Σ_g, where $\mathcal{O}_i = P_i(\Sigma)$. These are usually called "amplitudes", they are topological and do not depend on the positions x_i.

The classical and one-loop contribution $Z_{\text{cl},1\text{l}}(u,\mathfrak{m},v,\mathfrak{n})$ is constructed as follows. A gauge-invariant twisted superpotential $\widetilde{W}(\sigma)$ gives

$$Z_{\widetilde{W}} = e^{4\pi \widetilde{W}'(u)\cdot\mathfrak{m}} \,, \tag{3.3}$$

and in the special case that the twisted superpotential is linear and it represents a complexified FI term it becomes

$$Z_{\text{FI}} = q^{\operatorname{Tr}\mathfrak{m}} \tag{3.4}$$

where $q = e^{-\zeta - i\theta}$. The one-loop determinant for chiral multiplets is

$$Z^{\text{chiral}}_{\text{1-loop}} = \prod_{\rho\in\mathfrak{R}} \left[\frac{1}{\rho(u)\,\gamma(v)}\right]^{\rho(\mathfrak{m})+\gamma(\mathfrak{n})+(g-1)(q_\rho-1)} \,, \tag{3.5}$$

where γ are the flavor weights. For gauge multiplets we have

$$Z^{\text{gauge}}_{\text{1-loop}} = (-1)^{\sum_{\alpha>0}\alpha(\mathfrak{m})} \prod_{\alpha\in G} \alpha(u)\;\cdot\; du^{\operatorname{rank} G} \,. \tag{3.6}$$

For each operator insertion $\mathcal{O}_i = P_i(\Sigma)$ we have

$$P_i(u) \,.$$

We denote the product of these factors by $Z_{\text{cl},1\text{l}}$, which is a meromorphic top form on $\mathfrak{M} = \mathfrak{h}_{\mathbb{C}}$, the complexified Cartan subalgebra. The contribution from the extra fermionic zero-modes on Σ_g is

$$\left(-\det_{ab}\frac{\partial^2 \log Z_{\text{cl},1\text{l}}}{\partial u_a\,\partial\mathfrak{m}_b}\right)^g ,$$

where the sign has been inserted for convenience.

In the Abelian case, the final formula is

$$\begin{aligned} Z_{\Sigma_g} = \frac{1}{|W|}\sum_{\mathfrak{m}\in\Gamma_{\mathfrak{h}}}\sum_{u_*\in\mathfrak{M}^*_{\text{sing}}} \operatorname*{JK-Res}_{u=u_*}\left(\mathsf{Q}_{u_*},\eta\right)\left(-\det_{ab}\frac{\partial^2\log Z_{\text{cl},1\text{l}}}{\partial u_a\,\partial\mathfrak{m}_b}\right)^g Z_{\text{cl},1\text{l}} \\ + \text{bound. contrib.} \end{aligned} \tag{3.7}$$

The contour at the boundary is controlled by the beta functions for the FI terms. In particular for $G = U(1)$ the boundary contribution is the JK residue at $u = \infty$, assigning charge

$$Q_\infty = -\sum_i Q_i \tag{3.8}$$

to that point.

In the non-Abelian case the contributions from W-bosons are correctly taken into account by first resumming over $\mathfrak{m}$, and then excluding the roots of the associated BAEs for which the Vandermonde determinant is zero. One obtains a formula as in (2.53)-(2.55).

4. 4d theories on $\Sigma_g \times T^2$

We consider four-dimensional $\mathcal{N} = 1$ gauge theories of vector and chiral multiplets with a non-anomalous $U(1)_R$ R-symmetry, placed on $\Sigma_g \times T^2$ with twist on Σ_g. The torus T^2 has modulus τ. The Lagrangian is

$$\mathcal{L} = \mathcal{L}_{\text{YM}} + \mathcal{L}_{\text{mat}} + \mathcal{L}_W \,, \tag{4.1}$$

where the terms are the standard SYM Lagrangian, the matter kinetic Lagrangian, and the superpotential interactions, respectively. For Abelian factors in the gauge group, we can also consider a Fayet-Iliopoulos term

$$\mathcal{L}_{\text{FI}} = -i\frac{\zeta}{2\pi} D \,. \tag{4.2}$$

The parameters of the background are the flux $\mathfrak{n} = \frac{1}{2\pi}\int_\Sigma F^{\text{flav}}$ and the flavor flat connection

$$v = 2\pi \oint_{\text{A-cycle}} A - 2\pi\tau \oint_{\text{B-cycle}} A \,. \tag{4.3}$$

Each component of such a variable has identifications $v_a \cong v_a + 2\pi \cong v_a + 2\pi\tau$, therefore it lives on a copy of the spacetime T^2. Similarly there are parameters $(\mathfrak{m}, u)$ for the dynamical gauge fields, and $\mathfrak{M} = H^2 \cong T^{2r}$ where $r = \operatorname{rank} G$. It is convenient to define

$$q = e^{2\pi i\tau} \,, \qquad x = e^{iu} \,, \qquad y = e^{iv} \tag{4.4}$$

with the same notation as before (2.20). In particular $x \cong qx$.

The classical and one-loop contribution $Z_{\text{cl},1l}$ consists of the following pieces. The only classical-action contribution is from the FI term:

$$Z_{\text{FI}} = e^{-\text{Vol}(T^2)\,\zeta\,\mathfrak{m}} \,. \tag{4.5}$$

The one-loop determinant for chiral multiplets is

$$Z^{\text{chiral}}_{\text{1-loop}} = \prod_{\rho\in\mathfrak{R}} \left(\frac{i\eta(q)}{\theta_1(x^\rho y^\gamma; q)} \right)^{\rho(\mathfrak{m}) + \gamma(\mathfrak{n}) + (g-1)(R_\rho - 1)} \,, \tag{4.6}$$

where the elliptic functions are $\eta(q) = q^{1/24}\prod_{n=1}^\infty (1-q^n)$ and

$$\theta_1(x;q) = -iq^{\frac{1}{8}}x^{\frac{1}{2}}\prod_{k=1}^\infty (1-q^k)(1-xq^k)(1-x^{-1}q^{k-1}) \tag{4.7}$$

We used R for the R-charges, not to make confusion with the modular parameter q. The one-loop determinant for off-diagonal vector multiplets is

$$Z_{\text{1-loop}}^{\text{gauge, off}} = (-1)^{\sum_{\alpha>0}\alpha(\mathfrak{m})} \left(\prod_{\alpha\in G} \frac{\theta_1(x^\alpha;q)}{i\eta(q)} \right)^{1-g} , \tag{4.8}$$

while the contribution from vector multiplets along the Cartan generators is

$$Z_{\text{1-loop}}^{\text{gauge, Cartan}} = \eta(q)^{2r(1-g)} \, (i\, du)^r \; . \tag{4.9}$$

The contribution from the fermionic zero modes on Σ_g is

$$\left(\det_{ab} \frac{\partial^2 \log Z_{\text{1loop}}}{\partial i u_a \, \partial \mathfrak{m}_b} \right)^g .$$

In the Abelian case the final formula is

$$Z_{\Sigma_g\times T^2} = \frac{1}{|W|} \sum_{\mathfrak{m}\in\Gamma_{\mathfrak{h}}} \sum_{u_*\in\mathfrak{M}^*_{\text{sing}}} \underset{u=u_*}{\text{JK-Res}} \left(\mathsf{Q}_{u_*},\eta\right) \left(\det_{ab} \frac{\partial^2 \log Z_{\text{cl},11}}{\partial i u_a \, \partial \mathfrak{m}_b} \right)^g Z_{\text{cl},11} \; . \tag{4.10}$$

In particular there are no boundary contributions since the integration domain $\mathfrak{M}$ is compact.

In the non-Abelian case the contributions from W-bosons are correctly taken into account by first resumming over $\mathfrak{m}$, and then excluding the roots of the associated BAEs for which the Vandermonde determinant is zero. [8] One obtains a formula as in (2.53)-(2.55).

5. Examples

In this section we present various examples, to illustrate the use of the formula. We perform new checks of non-perturbative dualities, and compare with known results when available. We will compactly call Z the partition function $Z_{\Sigma_g\times S^1}$.

5.1. $U(1)_k$ Chern-Simons theory. Consider $U(1)$ supersymmetric Yang-Mills-Chern-Simons (YMCS) theory at level k. At low energies this is equivalent to bosonic Chern-Simons at level k. The classical and one-loop contribution is $Z_{\text{cl},11} = x^{\mathfrak{t}}\xi^{\mathfrak{m}} x^{k\mathfrak{m}}$, turning on a background for the topological symmetry. Thus

$$Z = \sum_{\mathfrak{m}\in\mathbb{Z}} \int_{\text{JK}} \frac{dx}{2\pi i x} \, k^g x^{k\mathfrak{m}+\mathfrak{t}} \xi^{\mathfrak{m}} \; . \tag{5.1}$$

The charges of the points at infinity are $Q_0 = -k$ and $Q_\infty = k$. Assuming $k > 0$, we can choose $\eta < 0$ and pick minus the residue at $x = 0$. This gives

$$Z = \begin{cases} -k^g \, \xi^{-\mathfrak{t}/k} & \text{if } \mathfrak{t} = 0 \pmod{k} \; , \\ 0 & \text{otherwise} \; . \end{cases} \tag{5.2}$$

For $\mathfrak{t} = 0$ and up to an ambiguous sign, this gives the known result k^g, which is the number of ground states of $U(1)_k$ Chern-Simons on Σ_g.

[8] In the four-dimensional case there is a small subtlety: the $U(1)$ symmetry associated to an adjoint chiral multiplet, used in the regularization argument we gave in section 2.1, is anomalous. We can neglect this problem because we only need to turn on an infinitesimal deformation, and then turn it off at the end of the computation.

5.2. $U(1)_{1/2}$ with one chiral multiplet. We consider the following duality [4,17]. The "electric theory" is a $U(1)$ YMCS theory at level $k=\frac{1}{2}$ with one chiral multiplet of gauge charge 1 and R-charge 1. The "magnetic theory" is a free chiral multiplet with flavor charge 1 under $U(1)_T$ and R-charge 0, and global CS terms $k_{TT}=-\frac{1}{2}$, $k_{RT}=-\frac{1}{2}$.

In the electric theory, the classical and one-loop contribution is

$$Z_{\text{cl},11} = x^{\mathfrak{t}}(-\xi)^{\mathfrak{m}} x^{\mathfrak{m}/2}\Big(\frac{x^{1/2}}{1-x}\Big)^{\mathfrak{m}} , \tag{5.3}$$

where we have redefined the sign of ξ for later convenience. Thus, the index is given by

$$Z = \sum_{\mathfrak{m}\in\mathbb{Z}} \int_{\text{JK}} \frac{dx}{2\pi i x} \frac{(-\xi)^{\mathfrak{m}} x^{\mathfrak{t}+\mathfrak{m}}}{(1-x)^{\mathfrak{m}+g}} . \tag{5.4}$$

There are poles at $x=0,1,\infty$, with associated charges $-1,1,0$, respectively. We can choose $\eta=-1$, therefore we pick minus the residues at $x=0$. We resum $\sum_{\mathfrak{m}\leq M-1}$ for some large M. If we take a contour around $x=0$ and $|\xi|\gg 1$, we have uniform convergence on the contour and we can exchange summation and integration. The resummed function has a unique pole inside the contour, at $x=(1-\xi)^{-1}$. Hence:

$$Z = -\oint_{x=\frac{1}{1-\xi}} \frac{dx}{2\pi i} \frac{x^{\mathfrak{t}-1}}{(1-x)^{g-1}} \frac{\big(\frac{\xi x}{x-1}\big)^M}{(1-\xi)x-1} = (-1)^g \frac{\xi^{1-g}}{(1-\xi)^{\mathfrak{t}+1-g}} . \tag{5.5}$$

Up to the ambiguous sign, this is precisely the index of the magnetic theory:

$$Z = (-1)^g \Big(\frac{\xi^{1/2}}{1-\xi}\Big)^{\mathfrak{t}+1-g} \xi^{-\mathfrak{t}/2}\xi^{-(g-1)/2} . \tag{5.6}$$

5.3. Witten index of $U(1)_k$ with matter. Consider $U(1)_k$ with N chiral multiplets of charges Q_i and R-charge 1 (in order to avoid parity anomalies in the R-symmetry). The classical and one-loop contribution is

$$Z_{\text{cl},11} = \xi^{\mathfrak{m}} x^{k\mathfrak{m}+\mathfrak{t}} \prod_{i=1}^N \Big(\frac{x^{Q_i/2}y_i^{1/2}}{1-x^{Q_i}y_i}\Big)^{Q_i\mathfrak{m}+\mathfrak{n}_i} , \tag{5.7}$$

where $(y_i,\mathfrak{n}_i)$ control the background for flavor symmetries up to one combination that could be reabsorbed into $(x,\mathfrak{m})$, and $(\xi,\mathfrak{t})$ controls the background for the topological symmetry. In order to avoid a gauge-gauge parity anomaly (*i.e.* in order for $Z_{\text{cl},11}$ to be a single-valued function of x) the condition

$$k+\frac{1}{2}\sum\nolimits_i Q_i^2 \in\mathbb{Z} \qquad \text{equivalent to} \qquad k+\frac{1}{2}\sum\nolimits_i Q_i\in\mathbb{Z} \tag{5.8}$$

should be met. In general there are gauge-flavor parity anomalies: one could cure them with the addition of gauge-flavor CS terms, however to the purpose of computing the Witten index we will simply set $\mathfrak{n}_i=\mathfrak{t}=0$. In general there are also flavor-flavor parity anomalies, however they are not inconsistencies of the theory: they simply imply that Z is not a single-valued function of y_i.

The poles are at $x=0,y_i^{-1/Q_i},\infty$. Depending on the choice of $\eta=\pm1$ one will have to collect different classes of residues. In any case, the sum over $\mathfrak{m}$ generates the expression in (2.53):

$$Z = - \sum_{x=x_{(\alpha)}} Z_{\mathrm{cl},11}\Big|_{\mathfrak{m}=0} \left(\frac{\partial e^{iB}}{\partial \log x} \right)^{g-1} , \tag{5.9}$$

where $x_{(\alpha)}$ are the roots of the BAE

$$e^{iB} = \xi\, x^k \prod_{i=1}^{N} \left(\frac{x^{Q_i/2} y_i^{1/2}}{1 - x^{Q_i} y_i} \right)^{Q_i} = 1 \,. \tag{5.10}$$

If we specialize to $g = 1$ and $\mathfrak{n}_i = \mathfrak{t} = 0$, the index Z_{T^3} is the Witten index I_{W} of the theory. On the other hand (5.9) becomes, up to the sign, the number of solutions to the BAE.

To compute the number of solutions, we divide the chiral multiplets into two groups: I_+ are those with $Q_i > 0$, I_- are those with $Q_i < 0$. Then the equation can be written as

$$\Big(\xi \prod\nolimits_i y_i^{Q_i/2} \Big)\, x^{k + \frac{1}{2}\sum_{i\in I_+} Q_i^2 - \frac{1}{2}\sum_{i \in I_-} Q_i^2} \prod_{i \in I_-} \left(x^{|Q_i|} - y_i \right)^{|Q_i|} = \prod_{i\in I_+} \left(1 - x^{Q_i} y_i\right)^{Q_i} . \tag{5.11}$$

We define the non-negative numbers

$$n_+ = \frac{1}{2} \sum_{i\in I_+} Q_i^2 \,, \qquad n_- = \frac{1}{2} \sum_{i \in I_-} Q_i^2 \,. \tag{5.12}$$

Then the number of solutions to (5.11) is

$$I_{\mathrm{W}} = \begin{cases} \max\left(k + n_+ + n_- \,,\, 2n_+\right) & \text{if } k + n_+ - n_- \geq 0 \,, \\ \max\left(-k + n_+ + n_- \,,\, 2n_-\right) & \text{if } k + n_+ - n_- \leq 0 \,. \end{cases} \tag{5.13}$$

After some algebra, this expression can be rewritten as

$$I_{\mathrm{W}} = \begin{cases} |k| + n_+ + n_- & \text{if } |k| \geq |n_+ - n_-| \,, \\ \max(2n_+ \,,\, 2n_-) & \text{if } |k| \leq |n_+ - n_-| \,. \end{cases} \tag{5.14}$$

This reproduces the Witten index computed in [**27**].

5.4. $U(N_c)$ SQCD with N_f flavors and Aharony duality. Consider a $U(N_c)$ theory, with N_f chiral multiplets Q_a in the fundamental and $\tilde{Q}_b$ in the antifundamental representations, and no CS interactions. The topologically twisted index of this theory for $g = 0$ has been computed in [**11**], and matched with the index of its Aharony dual theory [**1**]. Here we generalize that computation to arbitrary g.

For simplicity, we only introduce backgrounds for the R-symmetry, the topological symmetry and the axial $U(1)_A$ subgroup of the flavor symmetry acting with the same charge on all chiral fields. We use $(\xi, \mathfrak{t})$ for the fugacity and background flux for the topological symmetry, and $(y, \mathfrak{n})$ for the axial flavor symmetry. We assign R-charge 1 to the chiral fields. Hence:

	$U(N_c)$	$U(1)_T$	$U(1)_A$	$U(1)_R$
Q_a	$\mathbf{N_c}$	0	1	1
$\tilde{Q}_b$	$\overline{\mathbf{N}}_\mathbf{c}$	0	1	1
$M_{ab} = Q_a\tilde{Q}_b$	$\mathbf{1}$	0	2	2
T	$\mathbf{1}$	1	$-N_f$	$-N_c+1$
$\tilde{T}$	$\mathbf{1}$	-1	$-N_f$	$-N_c+1$

Here $T, \tilde{T}$ are the monopole operators $V_\mathfrak{m}$ corresponding to the magnetic fluxes $\mathfrak{m} = (1, 0, \ldots, 0)$ and $\mathfrak{m} = (0, \ldots 0, -1)$, respectively.

The classical and one-loop contribution is

$$Z_{\text{cl},1\text{l}} = (-1)^{N_c} \prod_{a=1}^{N_c} x_a^{\mathfrak{t}} \big((-1)^{N_f}\xi\big)^{\mathfrak{m}_a} \prod_{a\neq b}^{N_c} \Big(1 - \frac{x_a}{x_b}\Big)^{1-g} \times \left(\frac{x_a^{1/2} y^{1/2}}{1 - x_a y}\right)^{N_f(\mathfrak{m}_a+\mathfrak{n})} \left(\frac{x_a^{-1/2} y^{1/2}}{1 - y/x_a}\right)^{N_f(\mathfrak{n}-\mathfrak{m}_a)} . \tag{5.15}$$

We have redefined $\xi \to (-1)^{N_f}\xi$ and included a sign $(-1)^{N_c}$ for later convenience. We directly apply the formula in (2.53)-(2.55). The quantities B_a are given by

$$iB_a = \frac{\partial \log Z_{\text{cl},1\text{l}}}{\partial \mathfrak{m}_a} = \log(-1)^{N_f}\xi + N_f \log\left(\frac{x_a^{1/2} y^{1/2}}{1 - x_a y}\right) - N_f \log\left(\frac{x_a^{-1/2} y^{1/2}}{1 - y/x_a}\right) . \tag{5.16}$$

The BAEs are

$$1 = e^{iB_a} = \xi \left(\frac{y - x_a}{1 - x_a y}\right)^{N_f} . \tag{5.17}$$

Since the BAEs are all equal and decoupled, the solutions are simply collections of roots of the polynomial

$$\mathcal{P}(x) = \xi(y - x)^{N_f} - (1 - xy)^{N_f} \equiv (-1)^{N_f}(\xi - y^{N_f}) \prod_{\alpha=1}^{N_f} (x - x_\alpha) , \tag{5.18}$$

where in the last expression we have written $\mathcal{P}$ in terms of its N_f roots x_α. According to the prescription, one should only retain solutions in which the N_c selected roots $\{x_\alpha\}$ are all different (otherwise the Vandermonde determinant vanishes). This leads to a sum over ordered collections of N_c distinct roots out of N_f. The contribution from the fermionic zero-modes on Σ_g follows from the matrix

$$\frac{\partial iB_a}{\partial \log x_b} = \delta_{ab} N_f \frac{1 - y^2}{(1 - x_a y)(1 - y/x_a)} . \tag{5.19}$$

Putting all pieces together, we find the expression

$$Z = \frac{y^{N_c N_f \mathfrak{n}} N_f^{N_c(g-1)}}{(1 - y^2)^{N_c(1-g)}} \sum_I \prod_{\alpha \in I} \frac{x_\alpha^{\mathfrak{t}+(N_c-1)(g-1)}}{\big[(1 - x_\alpha y)(1 - y/x_\alpha)\big]^{N_f\mathfrak{n}+g-1}} \prod_{\substack{\beta\in I \\ (\beta\neq\alpha)}} (x_\alpha - x_\beta)^{1-g} . \tag{5.20}$$

The sum over I runs over the set $C_{N_c}^{N_f}$ of unordered combinations of N_c different integers in $\{1, \ldots, N_f\}$, while I^c denotes the complementary set $\{1, \cdots, N_f\} \setminus I$

belonging to $C^{N_f}_{N_f-N_c}$. We used that the summands in (5.20) are invariant under permutations of the roots, to reduce the sum to unordered combinations and cancel the factor $N_c!$.

The number of combinations is

$$I_{\mathrm{W}} = \# C^{N_f}_{N_c} = \binom{N_f}{N_c} \qquad \text{if} \quad N_f \geq N_c \tag{5.21}$$

and zero otherwise, which gives the Witten index of the theory (when well-defined), setting $g = 1$ and $\mathfrak{n} = \mathfrak{t} = 0$. In fact the theory breaks supersymmetry whenever $N_f \leq N_c - 2$, is IR free (then the Witten index is not well-defined) for $N_f = N_c - 1$, and has isolated vacua for $N_f \geq N_c$.

To make contact with [11], we notice the following identities:

$$\prod_{\alpha=1}^{N_f} x_\alpha = \frac{\xi y^{N_f} - 1}{\xi - y^{N_f}}\,, \qquad \prod_{\alpha=1}^{N_f} (1 - x_\alpha y) = \frac{\xi(1-y^2)^{N_f}}{\xi - y^{N_f}}\,, \tag{5.22}$$

and

$$\prod_{\beta(\neq\alpha)}^{N_f} (x_\alpha - x_\beta) = \frac{N_f \xi (1-y^2)}{\xi - y^{N_f}} \; \frac{x_\alpha^{N_f-1}(1 - y/x_\alpha)^{N_f-1}}{1 - x_\alpha y}\,. \tag{5.23}$$

They can be used to recast the index into the form

$$Z = \frac{(-1)^{N_c N_f(\mathfrak{n}+g-1)} \xi^{N_c \mathfrak{n}} y^{N_c N_f \mathfrak{n}}}{(\xi - y^{N_f})^{N_c(1-g)}} \sum_I \prod_{\alpha \in I} \frac{x_\alpha^{\mathfrak{t} + N_c(g-1) + N_f \mathfrak{n}} \prod_{\beta \in I^c} (x_\alpha - x_\beta)^{g-1}}{(1 - x_\alpha y)^{N_f(2\mathfrak{n}+g-1)}}\,. \tag{5.24}$$

For $N_f < N_c$, the expression above obviously vanishes. If $N_f = N_c$, there is only one I while $I^c = \emptyset$. We immediately get

$$Z = \frac{(-1)^{\mathfrak{t}}\, y^{N_c^2(3\mathfrak{n}+2g-2)}\, \xi^{\mathfrak{t}}}{(1-y^2)^{N_c^2(2\mathfrak{n}+g-1)}\,(1 - \xi y^{-N_c})^{N_c(1-g-\mathfrak{n})+\mathfrak{t}}\,(1 - \xi^{-1} y^{-N_c})^{N_c(1-g-\mathfrak{n})-\mathfrak{t}}}\,. \tag{5.25}$$

The dual theory for $N_f = N_c$ is given by the fields M_{ab}, T and $\tilde{T}$, coupled through the superpotential $W = T\tilde{T} \det M$ [3]. The partition function of the dual theory is then

$$Z_{\text{dual}} = \left(\frac{y}{1-y^2}\right)^{(2\mathfrak{n}+g-1)N_c^2} \left(\frac{\xi^{\frac{1}{2}} y^{-\frac{N_c}{2}}}{1 - \xi y^{-N_c}}\right)^{N_c(1-g-\mathfrak{n})+\mathfrak{t}} \left(\frac{\xi^{-\frac{1}{2}} y^{-\frac{N_c}{2}}}{1 - \xi^{-1} y^{-N_c}}\right)^{N_c(1-g-\mathfrak{n})-\mathfrak{t}}\,. \tag{5.26}$$

This agrees with (5.25), up to an ambiguous sign $(-1)^{\mathfrak{t}}$.

The expression (5.24) for $N_f > N_c$ is more complicated but we can use it to check Aharony dualities [1]. The dual theory is a $U(N_f - N_c)$ gauge theory with N_f fundamentals q_a, N_f anti-fundamentals $\tilde{q}_b$ and $N_f^2 + 2$ singlets M_{ab}, T and $\tilde{T}$, corresponding to the mesons and monopoles of the original theory, with a superpotential $W = M_{ab} q_a \tilde{q}_b + v_- T + v_+ \tilde{T}$, where $v_\pm$ are monopoles of the dual theory. We assign the charges consistently with the original theory:

	$U(N_f-N_c)_g$	$U(1)_T$	$U(1)_A$	$U(1)_R$
q_a	$\mathbf{N_f-N_c}$	0	-1	0
$\tilde{q}_b$	$\overline{\mathbf{N_f-N_c}}$	0	-1	0
M_{ab}	0	0	2	2
T	0	1	$-N_f$	$-N_c+1$
$\tilde{T}$	0	-1	$-N_f$	$-N_c+1$
v_+	0	1	N_f	N_c+1
v_-	0	-1	N_f	N_c+1

Notice that the dual quarks have R-charge zero and axial flavor charge -1.

The partition function of the dual theory is obtained by multiplying the contribution of the gauge sector for the quarks $q_a, \tilde{q}_b$ with the contribution of the singlets M_{ab}, T and $\tilde{T}$. The first contribution is the partition function for a $U(N_f-N_c)$ theory with quarks $q_a, \tilde{q}_b$ which we can read from (5.24). According to our assignment of charges, we need to replace the background charge and fugacity for the flavor symmetry by $y \leftrightarrow y^{-1}$ and $\mathfrak{n} \leftrightarrow 1-g-\mathfrak{n}$, as well as $N_c \leftrightarrow N_f - N_c$. We find

$$(5.27)\quad Z_{q\tilde{q}} = \frac{y^{(N_f-N_c)N_f(\mathfrak{n}+g-1)}\xi^{(N_f-N_c)(1-g-\mathfrak{n})}}{(-1)^{(N_f-N_c)N_f\mathfrak{n}}(\xi-y^{-N_f})^{(N_f-N_c)(1-g)}}\times \sum_J \prod_{\beta\in J} \frac{\tilde{x}_\beta^{-N_f\mathfrak{n}+N_c(1-g)+\mathfrak{t}}}{(1-\tilde{x}_\beta y^{-1})^{N_f(1-g-2\mathfrak{n})}\prod_{\alpha\in J^c}(\tilde{x}_\beta-\tilde{x}_\alpha)^{1-g}}$$

and the sum is over $J \in C^{N_f}_{N_f-N_c}$. Here the $\tilde{x}_\beta$ are the roots of the polynomial $\tilde{\mathcal{P}}(\tilde{x}) = \xi(y^{-1}-\tilde{x})^{N_f} - (1-\tilde{x}y^{-1})^{N_f} = 0$, and in fact $\tilde{x}_\beta = 1/x_\beta$. We can thus rewrite $Z_{q\tilde{q}}$ in terms of x_β, and convert the products over J into products over J^c using the full products in (5.22). We get:

$$(5.28)\quad Z_{q\tilde{q}} = \frac{(-1)^{N_f(N_f-N_c)(1-g-\mathfrak{n})}y^{N_f(N_f-N_c)(1-g-\mathfrak{n})}\xi^{N_f\mathfrak{n}+N_c(\mathfrak{n}+g-1)}}{(1-y^2)^{N_f^2(1-g-2\mathfrak{n})}(\xi-y^{N_f})^{N_f\mathfrak{n}-\mathfrak{t}}(\xi y^{N_f}-1)^{N_f\mathfrak{n}-N_c(1-g)+\mathfrak{t}}}\times \sum_J \prod_{\alpha\in J^c} \frac{x_\alpha^{N_f\mathfrak{n}-N_c(1-g)+\mathfrak{t}}}{(1-x_\alpha y)^{N_f(2\mathfrak{n}+g-1)}\prod_{\beta\in J}(x_\alpha-x_\beta)^{1-g}}\,.$$

The contribution of the gauge singlets is

$$(5.29)\quad Z_{MT\tilde{T}} = \frac{(-1)^{\mathfrak{t}-N_f\mathfrak{n}+N_c(1-g)}y^{N_f(N_f(\mathfrak{n}+g-1)+N_c(1-g))}\xi^{-N_f\mathfrak{n}+N_c(1-g)}}{(1-y^2)^{N_f^2(2\mathfrak{n}+g-1)}(\xi-y^{N_f})^{\mathfrak{t}-N_f\mathfrak{n}+N_c(1-g)}(\xi y^{N_f}-1)^{-\mathfrak{t}-N_f\mathfrak{n}+N_c(1-g)}}\,.$$

Then the partition function of the dual theory, $Z_{\text{dual}} = Z_{q\tilde{q}}Z_{MT\tilde{T}}$, equals the one of the electric theory up to $(-1)^{(N_f-N_c)(g-1)+\mathfrak{t}}$.

Since Giveon-Kutasov duality [**19**] can be derived from Aharony duality with an RG flow [**4, 19**] (and viceversa [**27**]), we have also implicitly verified that Giveon-Kutasov dual theories have the same higher-genus index.

5.5. $SU(2)_k$ Chern-Simons theory. $SU(2)$ supersymmetric YMCS theory at level $k \geq 3$ is equivalent, at low energies, to bosonic $SU(2)$ CS at level $\bar{k} = k-2$ (while for $k = 0, 1$ it breaks supersymmetry, and for $k = 2$ it confines; we assume $k \geq 0$).

The classical and one-loop contribution is

$$Z_{\text{cl},11} = x^{2k\mathfrak{m}}\left[(1-x^2)(1-x^{-2})\right]^{1-g} . \tag{5.30}$$

The naive expression for the partition function is

$$Z = \frac{(-1)^{g-1}}{2}\sum_{\mathfrak{m}\in\mathbb{Z}}\int_{\text{JK}}\frac{dx}{2\pi i x}\,(2k)^g x^{2k\mathfrak{m}}\left[\frac{(1-x^2)^2}{x^2}\right]^{1-g} , \tag{5.31}$$

however this expression does not correctly capture the contribution from W-bosons. The effective charges at the boundary are $Q_0 = -k$ and $Q_\infty = k$. We choose $\eta < 0$, therefore we should pick minus the residues at $x = 0$. A contour of uniform convergence for the geometric series in $\mathfrak{m}$ is in $|x| > 1$. According to the prescription in section 2.2, we should first resum over $\mathfrak{m} \leq M - 1$ for some large M, and then pick the residues inside the contour for which the Vandermonde determinant does not vanish. The geometric series has poles at the roots of $x^{2k} = 1$. Thus, after some manipulations, we find

$$Z = (-1)^g\frac{(2k)^{g-1}}{2}\sum_{x=x_{(i)}}\left[\frac{(1-x^2)^2}{x^2}\right]^{1-g} , \tag{5.32}$$

where $x_{(i)}$ are the roots of $x^{2k} = 1$ with $x_{(i)}^2 \neq 1$. This is precisely the expression in (2.53)-(2.55), using $B = 2ku$.

The index can be recast, up to the ambiguous sign, in the form

$$Z = \left(\frac{\bar{k}+2}{2}\right)^{g-1}\sum_{j=1}^{\bar{k}+1}\left(\sin\frac{\pi i}{\bar{k}+2}j\right)^{2-2g} , \tag{5.33}$$

which is the standard Verlinde formula for $SU(2)_{\bar{k}}$.

5.6. $SU(2)_k$ with matter and the "duality appetizer". Let us consider $SU(2)_k$ with matter. First, we take one chiral multiplet Φ in the adjoint representation (this setup has been studied at length in [**21**]). There is a flavor $U(1)_F$ symmetry that rotates the adjoint, and we indicate by $(y, \mathfrak{n})$ the corresponding background. To cancel a parity anomaly we introduce a flavor-flavor CS term $k_{FF} = \frac{1}{2}$. We assign R-charge 1 to Φ. The classical and one-loop contribution is

$$Z_{\text{cl},11} = -\frac{(1-x^2)^{2-2g}}{x^{2-2g}}\,x^{2k\mathfrak{m}}\,y^{\mathfrak{n}/2}\left(\frac{xy^{1/2}}{1-x^2y}\right)^{2\mathfrak{m}+\mathfrak{n}}\left(\frac{y^{1/2}}{1-y}\right)^{\mathfrak{n}}\left(\frac{x^{-1}y^{1/2}}{1-x^{-2}y}\right)^{-2\mathfrak{m}+\mathfrak{n}} , \tag{5.34}$$

where we included a sign $(-1)^g$ for later convenience. The associated BAE is

$$e^{iB} = x^{2k}\left(\frac{x^2-y}{1-x^2y}\right)^2 = 1 . \tag{5.35}$$

This equation has $2|k| + 4$ solutions, however two of them are ± 1 and should be discarded. The index takes the general form (2.53)-(2.55).

For $k = 1$ we can evaluate the index explicitly. The four acceptable solutions to the BAE are $x^2 = \frac{1}{2}\left(y^2 + 2y - 1 \pm (y+1)\sqrt{y^2+2y-3}\right)$, and substituting we obtain

$$Z = 2^g\, y^{2\mathfrak{n}}\,(y^2-1)^{1-g-2\mathfrak{n}} . \tag{5.36}$$

In fact, it has been argued in [**27**,**29**] that the theory is dual to a free chiral multiplet $Y = \operatorname{Tr}\Phi^2$ with flavor charge 2 and R-charge 2, plus an R-flavor CS term $k_{RF} = -1$ and a topological sector $U(1)_2$. The index of the dual theory is

$$Z_{\text{dual}} = 2^g\, y^{1-g} \Big(\frac{y}{1-y^2} \Big)^{2\mathfrak{n}+g-1} , \tag{5.37}$$

which is the same as before up to an ambiguous sign. Notice how the higher-genus index captures, through the factor 2^g, the topological sector.

We can consider other types of matter content. For instance, let us take $2N_f$ chiral multiplets in the fundamental representation (to cancel Witten's anomaly [**39**] the number should be even). Introducing for simplicity a background for the axial $U(1)_A$ flavor symmetry only, and assigning to the flavors R-charge 1, the classical and one-loop contribution is

$$Z_{\text{cl},11} = x^{2k\mathfrak{m}} \big[(1-x^2)(1-x^{-2}) \big]^{1-g} \Big(\frac{x^{\frac12} y^{\frac12}}{1-xy} \Big)^{2N_f(\mathfrak{m}+\mathfrak{n})} \Big(\frac{x^{-\frac12} y^{\frac12}}{1-x^{-1}y} \Big)^{2N_f(-\mathfrak{m}+\mathfrak{n})} . \tag{5.38}$$

The associated BAE is

$$e^{iB} = x^{2k} \Big(\frac{x-y}{1-xy} \Big)^{2N_f} = 1 . \tag{5.39}$$

This equation has $2|k| + 2N_f$ solutions, however two of them are ± 1 which should be discarded because are zeros of the Vandermonde determinant. We thus have $2|k| + 2N_f - 2$ acceptable solutions $x_{(i)}$. Expanding out (2.53)-(2.55) we get

$$Z = \frac{(-1)^g}{2} \sum_{x = x_{(i)}} \frac{(1-x^2)^{2-2g}}{x^{2-2g}} \frac{y^{2N_f\mathfrak{n}} x^{2(N_f+k)\mathfrak{n}}}{(1-xy)^{4N_f\mathfrak{n}}} \Big(2k + \frac{2N_f(1-y^2)}{(1-xy)(1-x^{-1}y)} \Big)^{g-1} . \tag{5.40}$$

Specializing to $g=1$ and $\mathfrak{n}=0$, we obtain the Witten index of the theory equal to half the number of acceptable solutions to the BAE:

$$I_{\text{W}} = |k| + N_f - 1 \tag{5.41}$$

or $I_{\text{W}} = 0$ if the number on the right-hand-side is negative, in agreement with [**27**].

In case of arbitrary matter content $\mathfrak{R} = \bigoplus \mathfrak{R}_i$, where $\mathfrak{R}_i$ is the spin I_i representation, it is easy to repeat the computation of the Witten index. Each representation $\mathfrak{R}_i$ brings a factor

$$\prod_{j=0}^{\lfloor I_i - \frac12 \rfloor} \Big(\frac{x^{2(I_i - j)} - y}{1 - x^{2(I_i-j)} y} \Big)^{2(I_i-j)}$$

to the BAE. Thus half the number of acceptable solutions is

$$I_{\text{W}} = |k| + \frac12 \sum_i T_2(\mathfrak{R}_i) - 1 , \tag{5.42}$$

where T_2 is the quadratic Casimir $T_2(\mathfrak{R}_i) = \frac23 (2I_i+1) I_i (I_i+1)$. This agrees with [**27**].

5.7. Two-dimensional A-twisted $\mathbb{CP}^{N-1}$ model. Consider a 2d $\mathcal{N}=(2,2)$ $U(1)$ theory with N chiral multiplets with charge 1 (and R-charge 0). In the IR this theory realizes a NLSM on $\mathbb{CP}^{N-1}$. We want to compute correlators of the field-strength twisted chiral multiplet Σ, which represents the Kähler class (also known as hyperplane class) of $\mathbb{CP}^{N-1}$. With s insertions of Σ, the classical and one-loop contribution is

$$Z_{\text{cl},1l} = u^s\, q^{\mathfrak{m}}\, \frac{1}{u^{(\mathfrak{m}+1-g)N}}\ . \tag{5.43}$$

Therefore we find the expression

$$\Big\langle \Sigma(x_1)\ldots\Sigma(x_s)\Big\rangle_g = \sum_{\mathfrak{m}\in\mathbb{Z}} \int_{\text{JK}} \frac{du}{2\pi i}\, \frac{N^g\, q^{\mathfrak{m}}}{u^{(\mathfrak{m}+1-g)N+g-s}}\ . \tag{5.44}$$

We can choose $\eta > 0$, then we should take the residues of the poles at $u = 0$ from chiral multiplets. The result is

$$\big\langle \Sigma_1\ldots\Sigma_s\big\rangle_g = \begin{cases} N^g\, q^{\frac{s+(N-1)(g-1)}{N}} & \text{if } s = g-1 \pmod N\ , \\ 0 & \text{otherwise}\ , \end{cases} \tag{5.45}$$

where we indicated $\Sigma(x_p)$ simply as Σ_p since the amplitudes are independent from the positions.

For $g = 0$ the shortest non-vanishing correlator is $\langle\Sigma_1\ldots\Sigma_{N-1}\rangle_{g=0} = 1$, corresponding to the fact that the intersection of $N-1$ hyperplanes in $\mathbb{CP}^{N-1}$ is a single point, while higher-point correlators are determined by the quantum cohomology (or chiral ring) relation $\Sigma^N = q$. For $g = 1$ the shortest non-vanishing correlator is

$$\langle 1\rangle_{g=1} = N\ , \tag{5.46}$$

which reproduces the Witten index of $\mathbb{CP}^{N-1}$.

All other correlators follow, since we have a topological field theory [**43**]. The N states of the theory on S^1 are realized by insertions of Σ^k with $k = 0,\ldots,N-1$, and their duals by Σ^{N-k-1}. Indeed the correlator $\langle\Sigma^j\,\Sigma^{N-k-1}\rangle_{g=0} = \delta_{jk}$ (with $j,k = 0,\ldots,N-1$) is interpreted as the propagator. Higher genus correlators are obtained by adding two insertions of Σ^k, Σ^{N-k-1} and gluing with a propagator. The unpunctured torus is $\langle 1\rangle_{g=1} = \sum_k\langle\Sigma^k\,\Sigma^{N-k-1}\rangle_{g=0} = N$. The case of $g = 2$ with one puncture is $\langle\Sigma\rangle_{g=2} = \sum_k\langle\Sigma\,\Sigma^k\,\Sigma^{N-k-1}\rangle_{g=1} = N^2 q$, where we used the chiral ring relation. All other correlators can be reproduced this way.

This example could be generalized in many ways. For instance one could study a $U(N_c)$ gauge theory with N_f chiral multiplets in the fundamental representation—which flows to a NLSM on the complex Grassmannian $Gr(N_c,N_f)$—and possibly N_a in the antifundamental—that represent N_a copies of the tautological bundle. It would be interesting to test the non-perturbative dualities of [**5, 9, 20, 23**].

5.8. Four-dimensional SQCD and Seiberg duality. We consider a simple example of Seiberg duality for SQCD in four dimensions. The simplest model to study is $USp(2)$ SQCD with $2N_f = 6$ flavors, whose global symmetry is $SU(6)\times U(1)_R$. The magnetic dual is a Wess-Zumino model of fifteen chiral multiplets, transforming as the antisymmetric tensor M_{ij} $(i,j = 1,\ldots,6)$ of $SU(6)$, interacting through the cubic superpotential $W = \Lambda^{-3}\,\text{Pf}\,M$ [**28**]. The M_{ij} correspond to the mesons of the electric theory. We now check, at lowest order in q, that the partition functions of the two theories on $\Sigma_g\times T^2$ coincide.

Consider first the $USp(2)$ model. We need to satisfy the quantization condition (2.3) for the R-charges. The quarks have R-charge $r = \frac{1}{3}$ and the gauge invariants are the mesons M_{ij}, therefore for $g-1 \in 3\mathbb{Z}$ the R-charges are correctly quantized. On the other hand, for generic g we need to mix the exact R-symmetry of the IR fixed point with some flavor symmetry. As in [**11**], we choose the non-anomalous and integer R-symmetry $U(1)'_R = \text{diag}(1,1,0,0,0,0)$, which is a combination of the exact R-symmetry and an Abelian subgroup of the $SU(6)$ flavor symmetry. The latter is then broken to $SU(4)\times SU(2)\times U(1)$. For simplicity, we also choose a magnetic flavor flux along $U(1)_A = \text{diag}(-2,-2,1,1,1,1)$, thus preserving the residual flavor symmetry. We use the compact notation

$$f_\chi(b,a,r) = \left(\frac{i\eta(q)}{\theta_1(x^b y^a;q)}\right)^{b\mathfrak{m}+a\mathfrak{n}+(g-1)(r-1)} . \tag{5.47}$$

Then the classical and one-loop contribution to the partition function $Z_{\Sigma_g\times T^2}$ is

$$Z_{\text{cl},11} = (i\,du)\,\eta(q)^{2(1-g)}\left(\frac{\theta_1(x^2;q)}{i\eta(q)}\frac{\theta_1(x^{-2};q)}{i\eta(q)}\right)^{1-g}\prod_{b=\pm1} f_\chi(b,-2,1)^2\, f_\chi(b,1,0)^4 . \tag{5.48}$$

Formally the partition function is given by

$$Z_{\Sigma_g\times T^2} = \frac{1}{2}\sum_{\mathfrak{m}\in\mathbb{Z}}\frac{1}{2\pi i}\int_{\text{JK}} Z_{\text{cl},11}\left(\frac{\partial^2\log Z_{\text{cl},11}}{i\partial u\,\partial\mathfrak{m}}\right)^g , \tag{5.49}$$

however in order to correctly take into account the contribution from W-bosons we should first sum over $\mathfrak{m}$ and then take the residues, avoiding the poles at the roots of the BAE. Choosing $\eta>0$ we need to collect the poles at $x=y^2$ and $x=y^{-1}$, and this dictates that we should resum over $\mathfrak{m}\geq -M$ for some large positive integer M. The resulting BAE is

$$e^{iB} = \frac{\theta_1(x^{-1}y^{-2};q)^2\,\theta_1(x^{-1}y;q)^4}{\theta_1(xy^{-2};q)^2\,\theta_1(xy;q)^4} = 1 , \tag{5.50}$$

to be solved on the torus $x\cong qx$.

We compute the partition function at the lowest order in q, in the limit $q\to 0$. Then the relevant solutions to the BAE are at the two roots of the polynomial equation

$$2y^2(1+x^2) - x(1+2y-2y^2+2y^3+y^4) = 0 .$$

We do not take the poles at $x=+1$ where the 1-loop gauge determinant vanishes, nor at $x=0$ which is outside the domain. We expand $Z_{\text{cl},11}$ at the lowest order in q and apply (2.53)-(2.55). The result is

$$Z_{\Sigma_g\times T^2} = -q^{5(g-1)/12}\Big(y^{4(1-g)}(1+y)^{-8\mathfrak{n}}(1-y^2)^{-5(1-g)}(1+y^2)^{4\mathfrak{n}+1-g} + \mathcal{O}(q)\Big) . \tag{5.51}$$

It is easy to check that this is precisely the expansion of the partition function of the dual theory

$$Z_{\text{dual}} = f_\chi(0,-4,2)\, f_\chi(0,-1,1)^8\, f_\chi(0,2,0)^6 , \tag{5.52}$$

up to a factor $(-1)^g$. With some effort, the analysis can be similarly extended to higher order in q, again with perfect agreement. It would clearly be desirable a proof of the equality for generic q, based on identities of theta functions. We leave such an analysis for future work.

6. Large N limit and black hole entropy

The topologically twisted index of the three-dimensional ABJM theory [**2**] for $g = 0$ has been computed in [**11**], and its large N limit matched with the Bekenstein-Hawking entropy of a class of AdS_4 static supersymmetric black holes with horizon $\mathrm{AdS}_2 \times S^2$ and an embedding into M-theory [**13, 16, 26**]. Similar black holes exist with horizon $\mathrm{AdS}_2 \times \Sigma_g$. Here we evaluate the index for generic g and show that the matching with the entropy extends to arbitrary genus.

In $\mathcal{N} = 2$ notation, the ABJM theory is a $U(N)_k \times U(N)_{-k}$ supersymmetric three-dimensional Chern-Simons theory (the subscripts are the CS levels) with bi-fundamental chiral multiplets A_i and B_j, $i, j = 1, 2$, transforming in the $(N, \overline{N})$ and $(\overline{N}, N)$ representations of the gauge group, respectively, and subject to the superpotential

$$W = \mathrm{Tr}\left(A_1 B_1 A_2 B_2 - A_1 B_2 A_2 B_1\right) . \tag{6.1}$$

We focus on $k = 1$ where the theory has $\mathcal{N} = 8$ superconformal symmetry and $SO(8)$ R-symmetry and it is dual to $\mathrm{AdS}_4 \times S^7$. From the point of view of an $\mathcal{N} = 2$ subalgebra, the flavor symmetry appears to be $SU(2) \times SU(2) \times U(1)$ and its three Cartan generators give charges $(1,0,0,-1)$, $(0,1,0,-1)$, $(0,0,1,-1)$ to (A_1, A_2, B_1, B_2) respectively, while we take R-charges $(0,0,0,2)$. We introduce fugacities $y_{1,2,3}$ and fluxes $-\mathfrak{n}_{1,2,3}$ for the three Cartan flavor symmetries of ABJM. The index can be written as

$$Z_g = \frac{1}{(N!)^2} \sum_{\mathfrak{m}, \widetilde{\mathfrak{m}} \in \mathbb{Z}^N} \int_{\mathrm{JK}} Z_{\mathrm{cl},11} \left(\det_{AB} \frac{\partial^2 Z_{\mathrm{cl},11}}{i\, \partial u_A\, \partial \mathfrak{m}_B} \right)^g \tag{6.2}$$

where [9]

$$Z_{\mathrm{cl},11} = \prod_{i=1}^N \frac{dx_i}{2\pi i x_i} \frac{d\tilde{x}_i}{2\pi i \tilde{x}_i} x_i^{\mathfrak{m}_i} \tilde{x}_i^{-\widetilde{\mathfrak{m}}_i} \times \prod_{i \neq j} \left(1 - \frac{x_i}{x_j}\right)^{1-g} \left(1 - \frac{\tilde{x}_i}{\tilde{x}_j}\right)^{1-g} \times$$
$$\prod_{i,j=1}^N \prod_{a=1,2} \left(\frac{\sqrt{\frac{x_i}{\tilde{x}_j} y_a}}{1 - \frac{x_i}{\tilde{x}_j} y_a} \right)^{\mathfrak{m}_i - \widetilde{\mathfrak{m}}_j - \mathfrak{n}_a + 1 - g} \prod_{b=3,4} \left(\frac{\sqrt{\frac{\tilde{x}_j}{x_i} y_b}}{1 - \frac{\tilde{x}_j}{x_i} y_b} \right)^{\widetilde{\mathfrak{m}}_j - \mathfrak{m}_i - \mathfrak{n}_b + 1 - g} , \tag{6.3}$$

and $u_A = (u_i, \tilde{u}_j), \mathfrak{m}_A = (\mathfrak{m}_i, \widetilde{\mathfrak{m}}_j)$ and $A, B = (i, j)$. In this formula $a = 1, 2$ refers to the fields A_1 and A_2 and $b = 3, 4$ to B_1 and B_2. The $\mathfrak{n}_a$ should be integers by the quantization condition:

$$\mathfrak{n}_a \in \mathbb{Z} . \tag{6.4}$$

In the expression above we have introduced the quantities y_4 and $\mathfrak{n}_4$ fixed by

$$\prod_{a=1}^4 y_a = 1 , \qquad \sum_{a=1}^4 \mathfrak{n}_a = 2(1-g) . \tag{6.5}$$

They are useful since the results will be manifestly invariant under the S_4 that permutes the $U(1)$ factors in the Cartan of $SO(8)$.

---

[9] As in [**11**], we chose a convenient parameterization for the set of independent fugacities and fluxes. Topological symmetries have been identified with a combination of flavor and gauge symmetries.

Choosing covectors $-\eta = \widetilde{\eta} = (1, \ldots, 1)$, we can resum the integrand of (6.3) with a cut-off $\mathfrak{m}_i \leq M-1$ and $\widetilde{\mathfrak{m}}_j \geq -M$ for some large integer M. We obtain the very same BAE as in [**11**], since the BAE does not depend on g:

(6.6)
$$e^{iB_i} = x_i^k \prod_{j=1}^N \frac{\big(1 - y_3 \frac{\tilde{x}_j}{x_i}\big)\big(1 - y_4 \frac{\tilde{x}_j}{x_i}\big)}{\big(1 - y_1^{-1} \frac{\tilde{x}_j}{x_i}\big)\big(1 - y_2^{-1}\frac{\tilde{x}_j}{x_i}\big)} \,, \qquad e^{i\widetilde{B}_j} = \tilde{x}_j^k \prod_{i=1}^N \frac{\big(1 - y_3 \frac{\tilde{x}_j}{x_i}\big)\big(1 - y_4 \frac{\tilde{x}_j}{x_i}\big)}{\big(1 - y_1^{-1} \frac{\tilde{x}_j}{x_i}\big)\big(1 - y_2^{-1}\frac{\tilde{x}_j}{x_i}\big)} \,.$$

The index is then obtained from (2.53). In the large N limit we expect that the sum in (2.53) will be dominated by the contribution of a single distribution $x_{(i)}$, $\tilde{x}_{(i)}$.

The large N solution to (6.6) for real chemical potentials $y_a = e^{i\Delta_a}$ has been analyzed in detail in [**11**]. The large N saddle-point eigenvalue distribution was found to be of the form

$$u_i = iN^{\frac{1}{2}} t_i + v_i \,, \qquad \tilde{u}_i = iN^{\frac{1}{2}} t_i + \tilde{v}_i \,. \tag{6.7}$$

In the large N limit one defines the continuous functions $t(i/N) = t_i$, $v(i/N) = v_i$, $\tilde{v}(i/N) = \tilde{v}_i$ and introduces the density of eigenvalues $\rho(t) = N^{-1} di/dt$, normalized as $\int dt\, \rho(t) = 1$. The solution for $\rho(t)$ and $\delta v(t) = v(t) - \tilde{v}(t)$ is of a characteristic piecewise form, divided into regions bounded by the transition points

$$t_{\ll} = -\frac{\mu}{\Delta_3} \,, \qquad t_< = -\frac{\mu}{\Delta_4} \,, \qquad t_> = \frac{\mu}{\Delta_2} \,, \qquad t_{\gg} = \frac{\mu}{\Delta_1} \,, \tag{6.8}$$

where we have chosen $\Delta_1 \leq \Delta_2$ and $\Delta_3 \leq \Delta_4$ without loss of generality. We are also taking $0 < \Delta_a < 2\pi$ and $\sum_a \Delta_a = 2\pi$. In the *left tail* we have

$$\begin{aligned} \rho &= \frac{\mu + t\Delta_3}{(\Delta_1 + \Delta_3)(\Delta_2 + \Delta_3)(\Delta_4 - \Delta_3)} \\ \delta v &= -\Delta_3 + e^{-\sqrt{N} Y_3} \,, \qquad Y_3 = \frac{-t\Delta_4 - \mu}{\Delta_4 - \Delta_3} \end{aligned} \qquad t_{\ll} < t < t_< \,. \tag{6.9}$$

In the *inner interval* we have

$$\begin{aligned} \rho &= \frac{2\pi\mu + t(\Delta_3\Delta_4 - \Delta_1\Delta_2)}{(\Delta_1 + \Delta_3)(\Delta_2 + \Delta_3)(\Delta_1 + \Delta_4)(\Delta_2 + \Delta_4)} \\ \delta v &= \frac{\mu(\Delta_1\Delta_2 - \Delta_3\Delta_4) + t\sum_{a<b<c} \Delta_a\Delta_b\Delta_c}{2\pi\mu + t(\Delta_3\Delta_4 - \Delta_1\Delta_2)} \end{aligned} \qquad t_< < t < t_> \tag{6.10}$$

and $\delta v' > 0$. In the *right tail* we have

$$\begin{aligned} \rho &= \frac{\mu - t\Delta_1}{(\Delta_1 + \Delta_3)(\Delta_1 + \Delta_4)(\Delta_2 - \Delta_1)} \\ \delta v &= \Delta_1 - e^{-\sqrt{N} Y_1} \,, \qquad Y_1 = \frac{t\Delta_2 - \mu}{\Delta_2 - \Delta_1} \end{aligned} \qquad t_> < t < t_{\gg} \,. \tag{6.11}$$

The normalization factor is $\mu = \sqrt{2\Delta_1\Delta_2\Delta_3\Delta_4}$. Notice that in the tails δv is constant up to exponentially vanishing contributions in N, which are nevertheless important for the evaluation of the index.

The index is then computed by evaluating the contribution of the saddle-point solution to the sum (2.53). The computation is essentially the same as in section

2.4 of [**11**] to which we refer for details. The final result is

$$\begin{aligned}\mathbb{R}\mathrm{e}\log Z_g(\mathfrak{n}) = -N^{\frac{3}{2}} \int dt\, \rho(t)^2 \Big[(1-g)\frac{2\pi^2}{3} + \sum_{\substack{a=3,4\,:\,+\\ a=1,2\,:\,-}} (\mathfrak{n}_a - 1 + g)\, g'_\pm\big(\delta v(t) \pm \Delta_a\big)\Big] \\ - N^{\frac{3}{2}} \sum_{a=1}^{4} \mathfrak{n}_a \int_{\delta v \approx \varepsilon_a \Delta_a} dt\, \rho(t)\, Y_a(t) \quad ,\end{aligned} \tag{6.12}$$

up to corrections of order $N \log N$. Here $g'_\pm(u) = \frac{u^2}{2} \mp \pi u + \frac{\pi^2}{3}$. The first contribution in the first line of (6.12) comes from the Vandermonde determinant and the second one from the matter contribution. The second line in (6.12) comes from the matter contribution and the fermionic zero-modes on Σ_g. Since the logarithm of the one loop determinant of the chiral fields is singular on the tail regions, we need to take into account the exponentially small corrections Y_a to the tails. The exponent $-\mathfrak{n}_a + 1 - g$ of the one-loop determinant is corrected to $-\mathfrak{n}_a$ by an analogous contribution from $\det(\partial B/\partial u)$ in (2.53).

By plugging ρ and δv into (6.12) we find

$$\mathbb{R}\mathrm{e}\log Z_g(\mathfrak{n}) = -\frac{N^{\frac{3}{2}}}{3}\sqrt{2\Delta_1\Delta_2\Delta_3\Delta_4} \sum_{a=1}^{4} \frac{\mathfrak{n}_a}{\Delta_a} \,. \tag{6.13}$$

Indeed, it is obvious from (6.12) that $\mathbb{R}\mathrm{e}\log Z_g(\mathfrak{n}) = (1-g)\,\mathbb{R}\mathrm{e}\log Z_{g=0}\big(\mathfrak{n}/(1-g)\big)$, at least for $g \neq 1$, and so the result for generic g follows from the one for $g = 0$ found in [**11**]. The expression in (6.13) looks independent from g, however one should recall that $\mathfrak{n}_a$ obey (6.5). The field theory entropy is obtained by extremizing (6.13) with respect to $\Delta_{a=1,2,3}$ with $\sum_a \Delta_a = 2\pi$. Other $\mathcal{N} = 2$ theories could be similarly analized at higher genus, generalizing the results in [**24, 25**].

We now compare the result with the entropy of magnetically-charged BPS black holes (from M-theory on S^7) with horizon $\mathrm{AdS}_2 \times \Sigma_g$ and asymptotic to AdS_4. We refer to appendices A and C of [**11**] for a detailed discussion of these black holes. The metric of the black holes is of the form

$$ds^2 = -e^{2U(r)}dt^2 + e^{-2U(r)}\Big(dr^2 + e^{2V(r)} ds^2_{\Sigma_g}\Big) \,, \tag{6.14}$$

with scalar fields depending on the radial coordinate. We choose a constant curvature metric $ds^2_\Sigma = e^{2h}(dx^2 + dy^2)$ on the Riemann surface with

$$e^{2h} = \begin{cases} \frac{4}{(1+x^2+y^2)^2} & \text{for } S^2 \\ 2\pi & \text{for } T^2 \\ \frac{1}{y^2} & \text{for } H^2 \end{cases} \,. \tag{6.15}$$

We have $R_s = 2\kappa$ where $\kappa = 1$ for S^2, $\kappa = 0$ for T^2, and $\kappa = -1$ for H^2. The range of coordinates are $(x, y) \in \mathbb{R}^2$ for S^2, $(x, y) \in [0, 1)^2$ for T^2, and $(x, y) \in \mathbb{R} \times \mathbb{R}_{>0}$ for H^2. In the H^2 case the upper half-plane has to be quotiented by a suitable Fuchsian group to get a compact Riemann surface $\Sigma_{g>1}$. The ranges are chosen in such a way that

$$\mathrm{Vol}(\Sigma_g) = \int e^{2h} dx\, dy = 2\pi\eta \,, \qquad \eta \equiv \begin{cases} 2|g-1| & \text{for } g \neq 1 \\ 1 & \text{for } g = 1 \end{cases} \tag{6.16}$$

where we defined the positive number η.

The explicit expression for the entropy of the black holes is given in appendix A of [11]. The expression for the horizon area (proportional to the Bekenstein-Hawking entropy) is

$$A = e^{2f_2(\hat{\mathfrak{n}})}\, 2\pi\eta\,, \tag{6.17}$$

in terms of magnetic charges $\hat{\mathfrak{n}}_a$. [10] The function e^{2f_2}, equal to e^{2V-2U} at the horizon, is homogeneous under a common positive rescaling of the $\hat{\mathfrak{n}}_a$, with weight 1. The charges $\hat{\mathfrak{n}}_a$ obey the BPS constraint $\sum_a \hat{\mathfrak{n}}_a = 2\kappa$ and the quantization condition $\hat{\mathfrak{n}}_a \in \frac{2}{\eta}\mathbb{Z}$. It follows that we can identify

$$\hat{\mathfrak{n}}_a = \frac{2}{\eta}\mathfrak{n}_a \tag{6.18}$$

with the magnetic charges $\mathfrak{n}_a$ used here. Then the horizon area, in terms of the $\mathfrak{n}_a$, is

$$A = 4\pi\, e^{2f_2(\mathfrak{n})}\,, \tag{6.19}$$

with no explicit dependence on g. Therefore, the matching of the field theory index with the black hole entropy for $g = 0$ exhibited in [11] implies matching for all values of g. The extremization of the quantity (6.13) is the field theory counterpart of the attractor mechanism for AdS_4 black holes [16].

Acknowledgements

We thank Giulio Bonelli and Wenbin Yan for useful discussions. FB is supported in part by the MIUR-SIR grant RBSI1471GJ "Quantum Field Theories at Strong Coupling: Exact Computations and Applications", by the INFN, and by the Royal Society as a Royal Society University Research Fellowship holder. AZ is supported by the INFN and the MIUR-FIRB grant RBFR10QS5J "String Theory and Fundamental Interactions".

Appendix A. Notation, Lagrangians and supersymmetry variations

The 3d $\mathcal{N} = 2$ vector multiplet $\mathcal{V}$ contains the fields $(A_\mu, \sigma, \lambda, \lambda^\dagger, D)$: they are the gauge field, a real scalar, a Dirac spinor and its conjugate, and the auxiliary real scalar, respectively, all in the adjoint representation of the gauge group G. In Euclidean signature they get complexified. They have R-charges $(0, 0, -1, 1, 0)$ respectively. The supersymmetry variations, specialized to constant positive chirality spinors ϵ, $\tilde{\epsilon}$, are: [11]

(A.1)

$$\begin{aligned}
&QA_\mu = \frac{i}{2}\lambda^\dagger\gamma_\mu\epsilon && Q\lambda = +\frac{1}{2}\gamma^{\mu\nu}\epsilon F_{\mu\nu} - D\epsilon + i\gamma^\mu\epsilon\, D_\mu\sigma \\
&\widetilde{Q}A_\mu = \frac{i}{2}\tilde{\epsilon}^\dagger\gamma_\mu\lambda && \widetilde{Q}\lambda^\dagger = -\frac{1}{2}\tilde{\epsilon}^\dagger\gamma^{\mu\nu}F_{\mu\nu} + \tilde{\epsilon}^\dagger D + i\tilde{\epsilon}^\dagger\gamma^\mu D_\mu\sigma \\
&QD = -\frac{i}{2}D_\mu\lambda^\dagger\gamma^\mu\epsilon + \frac{i}{2}[\lambda^\dagger\epsilon, \sigma] && Q\lambda^\dagger = 0 \qquad\qquad Q\sigma = -\frac{1}{2}\lambda^\dagger\epsilon \\
&\widetilde{Q}D = \frac{i}{2}\tilde{\epsilon}^\dagger\gamma^\mu D_\mu\lambda + \frac{i}{2}[\sigma, \tilde{\epsilon}^\dagger\lambda] && \widetilde{Q}\lambda = 0 \qquad\qquad \widetilde{Q}\sigma = -\frac{1}{2}\tilde{\epsilon}^\dagger\lambda
\end{aligned}$$

[10] The magnetic charges $\mathfrak{n}_a$ used in [11] are now called $\hat{\mathfrak{n}}_a$ to avoid confusion.

[11] The transformation are valid for generic covariantly constant spinors ϵ, $\tilde{\epsilon}$.

and then

$$QF_{\mu\nu} = iD_{[\mu}\lambda^\dagger\gamma_{\nu]}\epsilon\,, \qquad \widetilde{Q}F_{\mu\nu} = -i\tilde{\epsilon}^\dagger\gamma_{[\mu}D_{\nu]}\lambda\,. \tag{A.2}$$

The chiral multiplet Φ contains the fields (ϕ, ψ, F): a complex scalar, a Dirac spinor and the auxiliary complex scalar, in a representation $\mathfrak{R}$ of G. They have R-charges $(q, q-1, q-2)$. The anti-chiral multiplet $\overline{\Phi}$ contains $(\phi^\dagger, \psi^\dagger, F^\dagger)$ in representation $\overline{\mathfrak{R}}$ and with opposite R-charges. The supersymmetry variations are (A.3)

$$\begin{aligned}
&Q\phi = 0 && Q\psi = \big(i\gamma^\mu D_\mu\phi + i\sigma\phi\big)\epsilon && \widetilde{Q}\psi = \tilde{\epsilon}^c F \\
&\widetilde{Q}\phi = -\tilde{\epsilon}^\dagger\psi && \widetilde{Q}\psi^\dagger = \tilde{\epsilon}^\dagger\big(-i\gamma^\mu D_\mu\phi^\dagger + i\phi^\dagger\sigma\big) && Q\psi^\dagger = -\epsilon^{c\dagger}F^\dagger \\
&Q\phi^\dagger = \psi^\dagger\epsilon && QF = \epsilon^{c\dagger}\big(i\gamma^\mu D_\mu\psi - i\sigma\psi - i\lambda\phi\big) && \widetilde{Q}F = 0 \\
&\widetilde{Q}\phi^\dagger = 0 && \widetilde{Q}F^\dagger = \big(-iD_\mu\psi^\dagger\gamma^\mu - i\psi^\dagger\sigma + i\phi^\dagger\lambda^\dagger\big)\tilde{\epsilon}^c && QF^\dagger = 0\,.
\end{aligned}$$

We define charge conjugate spinors $\epsilon^c = C\epsilon^*$ and $\epsilon^{c\dagger} = \epsilon^{\mathsf{T}}C$, where C is the charge conjugation matrix such that $C\gamma^\mu C^{-1} = -\gamma^{\mu\mathsf{T}}$. We choose $C = \gamma_2$ so that $C = C^{-1} = C^\dagger = -C^{\mathsf{T}} = -C^*$.

The Lagrangian terms are essentially equal to those on flat space, with the only difference that they are coupled to the R-symmetry and flavor background. The Yang-Mills action is

$$\mathcal{L}_{\text{YM}} = \operatorname{Tr}\left[\frac{1}{4}F_{\mu\nu}F^{\mu\nu} + \frac{1}{2}D_\mu\sigma D^\mu\sigma + \frac{1}{2}D^2 - \frac{i}{2}\lambda^\dagger\gamma^\mu D_\mu\lambda - \frac{i}{2}\lambda^\dagger[\sigma,\lambda]\right]. \tag{A.4}$$

The Chern-Simons Lagrangian for each simple factor G_I in G is

$$\mathcal{L}_{\text{CS}}^{\text{n-Ab}} = -\frac{ik_I}{4\pi}\operatorname{Tr}\left[\epsilon^{\mu\nu\rho}\Big(A_\mu\partial_\nu A_\rho - \frac{2i}{3}A_\mu A_\nu A_\rho\Big) + \lambda^\dagger\lambda + 2D\sigma\right]. \tag{A.5}$$

In particular there is a separate integer Chern-Simons coupling k_I for each simple factor. The Chern-Simons Lagrangian for the Abelian factors is

$$\mathcal{L}_{\text{CS}}^{\text{Ab}} = -\frac{ik_{ij}}{4\pi}\left[\epsilon^{\mu\nu\rho}A_\mu^{(i)}\partial_\nu A_\rho^{(j)} + \lambda^{(i)\dagger}\lambda^{(j)} + 2D^{(i)}\sigma^{(j)}\right], \tag{A.6}$$

where k_{ij} is a symmetric integer matrix. The kinetic Lagrangian for chiral multiplets is

$$\begin{aligned}\mathcal{L}_{\text{mat}} = D_\mu\phi^\dagger D^\mu\phi + \phi^\dagger\big(\sigma^2 + iD - q_\phi W_{12}\big)\phi + F^\dagger F \\ + i\psi^\dagger(\gamma^\mu D_\mu - \sigma)\psi - i\psi^\dagger\lambda\phi + i\phi^\dagger\lambda^\dagger\psi\,.\end{aligned} \tag{A.7}$$

Notice that W_{12} couples to each field ϕ proportionally to its R-charge q_ϕ. Superpotential interactions are described by the Lagrangians

$$\mathcal{L}_W = iF_W\,, \qquad \mathcal{L}_{\overline{W}} = iF_W^\dagger\,, \tag{A.8}$$

where

$$F_W = \frac{\partial W}{\partial\Phi_i}F_i - \frac{1}{2}\frac{\partial^2 W}{\partial\Phi_i\partial\Phi_j}\psi_j^{c\dagger}\psi_i\,, \qquad F_W^\dagger = \frac{\partial\overline{W}}{\partial\Phi_i^\dagger}F_i^\dagger - \frac{1}{2}\frac{\partial^2\overline{W}}{\partial\Phi_i^\dagger\partial\Phi_j^\dagger}\psi_j^\dagger\psi_i^c \tag{A.9}$$

are the F-terms of the chiral multiplet $W(\Phi)$ and its antichiral partner, separately supersymmetric.

References

[1] Ofer Aharony, *IR duality in $d = 3$ $N = 2$ supersymmetric* $\mathrm{USp}(2N_c)$ *and* $\mathrm{U}(N_c)$ *gauge theories*, Phys. Lett. B **404** (1997), no. 1-2, 71–76, DOI 10.1016/S0370-2693(97)00530-3. MR1456971

[2] Ofer Aharony, Oren Bergman, Daniel Louis Jafferis, and Juan Maldacena, *$\mathcal{N} = 6$ superconformal Chern-Simons-matter theories, M2-branes and their gravity duals*, J. High Energy Phys. **10** (2008), 091, 38, DOI 10.1088/1126-6708/2008/10/091. MR2452954

[3] O. Aharony, A. Hanany, K. Intriligator, N. Seiberg, and M. J. Strassler, *Aspects of $N = 2$ supersymmetric gauge theories in three dimensions*, Nuclear Phys. B **499** (1997), no. 1-2, 67–99, DOI 10.1016/S0550-3213(97)00323-4. MR1468698

[4] Francesco Benini, Cyril Closset, and Stefano Cremonesi, *Comments on 3d Seiberg-like dualities*, J. High Energy Phys. **10** (2011), 075, 40, DOI 10.1007/JHEP10(2011)075. MR2909709

[5] Francesco Benini and Stefano Cremonesi, *Partition functions of $\mathcal{N} = (2,2)$ gauge theories on S^2 and vortices*, Comm. Math. Phys. **334** (2015), no. 3, 1483–1527, DOI 10.1007/s00220-014-2112-z. MR3312441

[6] Francesco Benini, Richard Eager, Kentaro Hori, and Yuji Tachikawa, *Elliptic genera of two-dimensional $\mathcal{N} = 2$ gauge theories with rank-one gauge groups*, Lett. Math. Phys. **104** (2014), no. 4, 465–493, DOI 10.1007/s11005-013-0673-y. MR3177993

[7] Francesco Benini, Richard Eager, Kentaro Hori, and Yuji Tachikawa, *Elliptic genera of 2d $\mathcal{N} = 2$ gauge theories*, Comm. Math. Phys. **333** (2015), no. 3, 1241–1286, DOI 10.1007/s00220-014-2210-y. MR3302634

[8] Francesco Benini, Kiril Hristov, and Alberto Zaffaroni, *Black hole microstates in* AdS_4 *from supersymmetric localization*, J. High Energy Phys. **5** (2016), 054, front matter+57. MR3521773

[9] Francesco Benini, Daniel S. Park, and Peng Zhao, *Cluster algebras from dualities of 2d $\mathcal{N} = (2,2)$ quiver gauge theories*, Comm. Math. Phys. **340** (2015), no. 1, 47–104, DOI 10.1007/s00220-015-2452-3. MR3395147

[10] Francesco Benini and Wolfger Peelaers, *Higgs branch localization in three dimensions*, JHEP **1405** (2014), 030.

[11] Francesco Benini and Alberto Zaffaroni, *A topologically twisted index for three-dimensional supersymmetric theories*, J. High Energy Phys. **7** (2015), 127, front matter+75. MR3383085

[12] Michel Brion and Michèle Vergne, *Arrangement of hyperplanes. I. Rational functions and Jeffrey-Kirwan residue* (English, with English and French summaries), Ann. Sci. École Norm. Sup. (4) **32** (1999), no. 5, 715–741, DOI 10.1016/S0012-9593(01)80005-7. MR1710758

[13] Sergio L. Cacciatori and Dietmar Klemm, *Supersymmetric* AdS_4 *black holes and attractors*, J. High Energy Phys. **1** (2010), 085, 22, DOI 10.1007/JHEP01(2010)085. MR2660799

[14] Cyril Closset, Stefano Cremonesi, and Daniel S. Park, *The equivariant A-twist and gauged linear sigma models on the two-sphere*, J. High Energy Phys. **6** (2015), 076, front matter+110. MR3370259

[15] Cyril Closset, Wei Gu, Bei Jia, and Eric Sharpe, *Localization of twisted $\mathcal{N}$=(0,2) gauged linear sigma models in two dimensions*, JHEP **03** (2016), 070.

[16] Gianguido Dall'Agata and Alessandra Gnecchi, *Flow equations and attractors for black holes in $\mathcal{N} = 2$ U(1) gauged supergravity*, J. High Energy Phys. **3** (2011), 037, 23, DOI 10.1007/JHEP03(2011)037. MR2821149

[17] Tudor Dimofte, Davide Gaiotto, and Sergei Gukov, *Gauge theories labelled by three-manifolds*, Comm. Math. Phys. **325** (2014), no. 2, 367–419, DOI 10.1007/s00220-013-1863-2. MR3148093

[18] Anton A. Gerasimov and Samson L. Shatashvili, *Higgs bundles, gauge theories and quantum groups*, Comm. Math. Phys. **277** (2008), no. 2, 323–367, DOI 10.1007/s00220-007-0369-1. MR2358287

[19] Amit Giveon and David Kutasov, *Seiberg duality in Chern-Simons theory*, Nuclear Phys. B **812** (2009), no. 1-2, 1–11, DOI 10.1016/j.nuclphysb.2008.09.045. MR2493130

[20] Jaume Gomis and Bruno Le Floch, *M2-brane surface operators and gauge theory dualities in Toda*, JHEP **04** (2016), 183.

[21] Sergei Gukov and Du Pei, *Equivariant Verlinde formula from fivebranes and vortices*, (2015).

[22] Kentaro Hori, Heeyeon Kim, and Piljin Yi, *Witten index and wall crossing*, J. High Energy Phys. **1** (2015), 124, front matter+105. MR3313918

[23] Kentaro Hori and David Tong, *Aspects of non-abelian gauge dynamics in two-dimensional $\mathcal{N} = (2,2)$ theories*, J. High Energy Phys. **5** (2007), 079, 41, DOI 10.1088/1126-6708/2007/05/079. MR2318130

[24] Seyed Morteza Hosseini and Noppadol Mekareeya, *Large N topologically twisted index: necklace quivers, dualities, and Sasaki-Einstein spaces*, J. High Energy Phys. **8** (2016), 089, front matter+33. MR3555653

[25] Seyed Morteza Hosseini and Alberto Zaffaroni, *Large N matrix models for 3d $\mathcal{N} = 2$ theories: twisted index, free energy and black holes*, J. High Energy Phys. **8** (2016), 064, front matter+36. MR3555628

[26] Kiril Hristov and Stefan Vandoren, *Static supersymmetric black holes in* AdS_4 *with spherical symmetry*, J. High Energy Phys. **4** (2011), 047, 24. MR2833272

[27] Kenneth Intriligator and Nathan Seiberg, *Aspects of 3d $\mathcal{N} = 2$ Chern-Simons-matter theories*, J. High Energy Phys. **7** (2013), 079, front matter+62. MR3106220

[28] K. Intriligator and P. Pouliot, *Exact superpotentials, quantum vacua and duality in supersymmetric* $\mathrm{Sp}(N_c)$ *gauge theories*, Phys. Lett. B **353** (1995), no. 4, 471–476, DOI 10.1016/0370-2693(95)00618-U. MR1339460

[29] Daniel Jafferis and Xi Yin, *A Duality Appetizer*, (2011).

[30] Lisa C. Jeffrey and Frances C. Kirwan, *Localization for nonabelian group actions*, Topology **34** (1995), no. 2, 291–327, DOI 10.1016/0040-9383(94)00028-J. MR1318878

[31] Nikita A. Nekrasov and Samsom L. Shatashvilli, *Bethe/Gause correspondence on curved spaces*, J. High Energy Phys. **1501**, 100 (2015). DOI: 10.1007/JHEP01(2015)100. arXiv:1405.6046 [hep-th].

[32] Kazutoshi Ohta and Yutaka Yoshida, *Non-Abelian Localization for Supersymmetric Yang-Mills-Chern-Simons Theories on Seifert Manifold*, Phys. Rev. **D86** (2012), 105018.

[33] Satoshi Okuda and Yutaka Yoshida, *G/G gauged WZW model and Bethe ansatz for the phase model*, J. High Energy Phys. **11** (2012), 146, front matter + 13. MR3036453

[34] Satoshi Okuda and Yutaka Yoshida, *G/G gauged WZW-matter model, Bethe Ansatz for q-boson model and Commutative Frobenius algebra*, JHEP **03** (2014), 003.

[35] Satoshi Okuda and Yutaka Yoshida, *Gauge/Bethe correspondence on $S^1 \times \Sigma_h$ and index over moduli space*, (2015).

[36] Vasily Pestun, *Localization of gauge theory on a four-sphere and supersymmetric Wilson loops*, Comm. Math. Phys. **313** (2012), no. 1, 71–129, DOI 10.1007/s00220-012-1485-0. MR2928219

[37] N. Seiberg, *Electric-magnetic duality in supersymmetric non-abelian gauge theories*, Nuclear Phys. B **435** (1995), no. 1-2, 129–146, DOI 10.1016/0550-3213(94)00023-8. MR1314365

[38] András Szenes and Michèle Vergne, *Toric reduction and a conjecture of Batyrev and Materov*, Invent. Math. **158** (2004), no. 3, 453–495, DOI 10.1007/s00222-004-0375-2. MR2104791

[39] Edward Witten, *An* SU(2) *anomaly*, Phys. Lett. B **117** (1982), no. 5, 324–328, DOI 10.1016/0370-2693(82)90728-6. MR678541

[40] Edward Witten, *Constraints on supersymmetry breaking*, Nuclear Phys. B **202** (1982), no. 2, 253–316, DOI 10.1016/0550-3213(82)90071-2. MR668987

[41] Edward Witten, *Topological quantum field theory*, Comm. Math. Phys. **117** (1988), no. 3, 353–386. MR953828

[42] Edward Witten, *Topological sigma models*, Comm. Math. Phys. **118** (1988), no. 3, 411–449. MR958805

[43] Edward Witten, *On the structure of the topological phase of two-dimensional gravity*, Nuclear Phys. B **340** (1990), no. 2-3, 281–332, DOI 10.1016/0550-3213(90)90449-N. MR1068086

[44] Edward Witten, *Mirror manifolds and topological field theory*, (1991). arXiv:hep-th/9112056.

[45] Edward Witten, *Phases of $N = 2$ theories in two dimensions*, Nuclear Phys. B **403** (1993), no. 1-2, 159–222, DOI 10.1016/0550-3213(93)90033-L. MR1232617

[46] Edward Witten, *On the Landau-Ginzburg description of $N = 2$ minimal models*, Internat. J. Modern Phys. A **9** (1994), no. 27, 4783–4800, DOI 10.1142/S0217751X9400193X. MR1296156

INTERNATIONAL SCHOOL FOR ADVANCED STUDIES (SISSA), VIA BONOMEA 265, 34136 TRIESTE, ITALY — INFN, SEZIONE DI TRIESTE – AND – BLACKETT LABORATORY, IMPERIAL COLLEGE LONDON, SOUTH KENSINGTON CAMPUS, LONDON SW7 2AZ, UNITED KINGDOM
E-mail address: fbenini@sissa.it

DIPARTIMENTO DI FISICA, UNIVERSITÀ DI MILANO-BICOCCA, I-20126 MILANO, ITALY

INFN, SEZIONE DI MILANO-BICOCCA, I-20126 MILANO, ITALY
E-mail address: alberto.zaffaroni@mib.infn.it

On the mathematics and physics of Mixed Spin P-fields

Huai-Liang Chang, Jun Li, Wei-Ping Li, and Chiu-Chu Melissa Liu

ABSTRACT. We outline various developments of affine and general Landau-Ginzburg models in physics. We then describe the A-twisting and coupling to gravity in terms of Algebraic Geometry. We describe constructions of various path integral measures (virtual fundamental class) using the algebro-geometric technique of cosection localization, culminating in the theory of "Mixed Spin P (MSP) fields" developed by the authors.

CONTENTS

1. Introduction

In this survey we will describe several mathematical and physical theories.

(1) The physical theory of generalized LG model (Guffin-Sharpe), and the mathematical theory of stable maps with P-fields and the hyperplane property in all genera (H.-L. Chang and J. Li).
(2) The Fan-Jarvis-Ruan-Witten (FJRW) theory of affine LG model, and the algebro-geometric construction of Witten's top Chern class in the narrow case (H.-L. Chang, J. Li and W.P. Li).

2010 *Mathematics Subject Classification.* Primary 14N35; Secondary 14J33.

(3) Witten's Gauged Linear Sigma Model (GLSM) which specializes to (1) in the Calabi-Yau (CY) phase and (2) in the Landau-Ginzburg (LG) phase as the Kähler parameter $r \to +\infty$ and $r \to -\infty$ respectively. The two phases are then linked via promoting the Kähler parameter to a ν-field, which leads to moduli spaces of Mixed Spin P (MSP) fields.

All the above three models (stable maps with P-fields, Witten's top Chern class, MSP fields) require Kiem-Li's cosection localization. A ghost P-field in the CY phase can be transformed continuously to a field in the LG phase determining the spin structure on the underlying curve. This phenomenon is named Landau-Ginzburg transition, and is responsible for the interaction between fields in the CY and LG phases. In the end we discuss how effective the MSP moduli could be used to attack various problems, including the enumeration of positive-genus curves in the quintic Calabi-Yau threefold.

The article is written by mathematicians, aiming to include the physics involved and the mathematics therefore stimulated, especially the algebro-geometric constructions by authors. In most sections we survey results in mathematics and in physics separately. In physics side, C is a compact Riemann surface (worldsheet), and K_C denotes its canonical line bundle. In mathematical side, C is an orbifold curve with markings with at worst nodal singularities, and ω_C denotes its dualizing sheaf. In physics part, L is a C^∞ complex line bundle over the compact Riemann surface C, while in mathematical part L is an algebraic line bundle over the algebraic curve C. Sections of a bundle or maps between spaces, if not being mentioned "C^∞", are assumed to be holomorphic, i.e. algebraic sections/maps in algebraic geometry. Finally, we let $W_5 = x_1^5 + \cdots + x_5^5$ be the Fermat quintic polynomial.

2. Mirror Symmetry and Gromov-Witten Invariants of Quintics

2.1. Physics. A 2d supersymmetric sigma model governs maps from a fixed Riemann surface Σ to a target manifold X. When the target is a Calabi-Yau manifold, Witten [**Wi2**] introduced two different ways to twist the standard supersymmetric sigma model, known as the A twist and the B twist, and obtained two different topological field theories, the A-model and B-model on X, denoted $A(X)$ and $B(X)$.

- In the A-model, the path integral over the infinite dimensional space of maps to X can be reduced to an integral over the space of holomorphic maps to X. The A-model correlation functions depend on the Kähler structure but not the complex structure on X.
- In the B-model, the path integral over the infinite dimensional space of maps of X can be reduced to an integral over the space of constant maps to X, i.e., an integral over X. The B-model correlation functions depend on the complex structure but not the Kähler structure on X.

Given a Calabi-Yau manifold X, the mirror $\check{X}$ of X is another Calabi-Yau manifold of the same dimension such that

$$A(X) \cong B(\check{X}), \quad B(X) \cong A(\check{X}). \tag{2.1}$$

The expected/virtual (complex) dimension of space of holomorphic maps from a closed Riemann surface to X is $\dim_{\mathbb{C}} X \cdot (1-g)$, where g is the genus of the Riemann surface. Therefore, we expect there to be no holomorphic maps from a fixed generic Riemann surface of genus $g > 1$ to X. By allowing the complex structure on the domain Riemann surface to vary, we obtain the 2d sigma model coupled with gravity. The A-model (resp. B-model) topological string theory on X is obtained by applying A twist (resp. B twist) to the sigma model on X coupled with gravity. In the rest of this paper we will always consider theories coupled with gravity, still denoted by $A(X)$ and $B(X)$. The mirror symmetry (2.1) is still expected. The equivalence $A(X) \cong B(\check{X})$ implies an equality of genus g topological string amplitudes:

$$F_g^{A(X)}(q(t)) = F_g^{B(\check{X})}(t) \tag{2.2}$$

where $t \mapsto q(t)$ is the mirror map from the moduli of complex structures on $\check{X}$ to the moduli of (complexified) Kähler classes on X. The genus zero B-model is determined by the classical variation of Hodge structures. In 1993, Bershadsky-Cecotti-Ooguri-Vafa (BCOV) developed the Kodaira-Spencer theory of gravity, which is a string field theory of higher genus B-model [**BCOV**].

A typical example is the quintic Calabi-Yau threefold Q, which is a degree 5 hypersurface in $\mathbb{P}^4$. Its mirror $\check{Q}$ is a degree 5 hypersurface in $\mathbb{P}^4/(\mathbb{Z}_5)^3$. In 1991, Candelas-de la Ossa-Green-Parkes [**COGP**] derived a formula for the genus zero B-model topological string amplitude $F_0^{B(\check{Q})}$ of $\check{Q}$ and the mirror map in terms of explicit hypergeometric series, and obtained a mirror formula of the genus zero A-model topological string amplitude $F_0^{A(Q)}$ of Q, which is a generating function of (virtual) numbers of rational curves in Q. Mirror symmetry predictions on higher genus A-model topological string amplitudes $F_g^{A(Q)}$ (counting genus g curves in Q) have been obtained by Bershadsky-Cecotti-Ooguri-Vafa at genus $g = 1, 2$ ([**BCOV**], 1993), by Katz-Klemm-Vafa at genus $g = 3, 4$ ([**KKV**], 1999), and at genus $g \leq 51$ by Huang-Klemm-Quackenbuch ([**HKQ**], 2007).

Using results of BCOV [**BCOV**] and Yamaguchi-Yau [**YY**] and assuming mirror symmetry, Huang-Klemm-Quackenbush [**HKQ**] provide an algorithm to determine $F_g^{A(Q)}(q(t)) = F_g^{B(\check{Q})}(t)$ for genus $g \leq 51$. When $g \geq 2$, the holomorphic anomaly equation determines $F_g^{A(Q)}(q)$ up to $3g - 2$ unknowns. The degree zero Gromov-Witten invariant $N_{g,d=0}$ is known, so we are left with $3g-3$ unknowns; the boundary conditions at the orbifold point (which corresponds to Landau-Ginzburg theory of the Fermat quintic polynomial in five variables) impose $\lceil \frac{3}{5}(g-1) \rceil$ constraints on the $3g-3$ unknowns, whereas the "gap condition" at the conifold point imposes $2g-2$ constraints on the $3g-3$ unknowns. In summary, the holomorphic anomaly equation and the boundary conditions determine $F_g^{A(Q)}$ up to $\lfloor \frac{2}{5}(g-1) \rfloor$ unknowns. When genus $g \leq 51$, the Gopakumar-Vafa conjecture (which relates Gromov-Witten invariants and Gopakumar-Vafa invariants) and the Castelnuovo bound (which implies vanishing of low degree Gopakumar-Vafa invariants) are sufficient to fix the remaining $\lfloor \frac{2}{5}(g-1) \rfloor$ unknowns.

2.2. Mathematics. Gromov-Witten theory can be viewed as a mathematical theory of the A-model topological string theory. There are two approaches to Gromov-Witten theory. Here we describe the algebro-geometric definition. For non-negative integers d, g, $\overline{M}_g(Q,d)$ denotes the moduli space of stable maps from genus g nodal curves to Q of degree d. Li-Tian [**LT**] and Behrend-Fantachi [**BF**] construct a degree zero cycle $[\overline{M}_g(Q,d)]^{\mathrm{vir}} \in A_0(\overline{M}_g(Q,d);\mathbb{Q})$, which is called the virtual cycle. Note that $\overline{M}_g(Q,d)$ is empty when $(g,d) \in \{(0,0),(1,0)\}$. For

$(g,d) \neq (0,0), (1,0)$, define genus g, degree d Gromov-Witten invariant of Q by

$$N_{g,d} := \deg[\overline{M}_g(Q,d)]^{\mathrm{vir}} \in \mathbb{Q}.$$

The genus-g Gromov-Witten potential of Q is given by

$$F_g^A(q) := \begin{cases} \frac{5}{6}(\log q)^3 + \sum_{d=1}^{\infty} N_{0,d}q^d, & g=0; \\ -\frac{25}{12}\log q + \sum_{d=1}^{\infty} N_{1,d}q^d, & g=1; \\ \sum_{d=0}^{\infty} N_{g,d}q^d, & g \geq 2. \end{cases}$$

One of the main unsolved problems in Gromov-Witten theory is to determine $F_g^A(q)$, which is a generating function of genus g Gromov-Witten invariants of Q.

Using the hyperplane property in genus zero, Kontsevich [**Ko**] proposed to use torus localization to calculate the genus zero Gromov-Witten invariants $N_{0,d}$. Givental [**Gi**] and Lian-Liu-Yau [**LLY**] proved the mirror formula of $F_0^A(q)$ predicted in [**COGP**]. The BCOV mirror formula of $F_1^A(q)$ was solved in 2000's. J.Li and A. Zinger [**LZ**] obtained a formula

$$N_{1,d} = N_{1,d}^{\mathrm{red}} + \frac{1}{12}N_{0,d} \tag{2.3}$$

where $N_{1,d}^{\mathrm{red}}$ is the genus one, degree d reduced GW-invariant of Q. Using (2.3) and $\mathbb{C}^*$-localization, Zinger proved the BCOV mirror formula of $F_1^A(q)$ in [**Zi2**]. Gathmann [**Gath**] provided an algorithm for $N_{1,d}$ using the relative GW-invariant formula.

Using degeneration, Maulik and Pandharipande [**MP**] found an algorithm which determines $N_{g,d}$ for all genus g and degree d: one degenerates the quintic of $\mathbb{P}^4$ to a quartic and a $\mathbb{P}^3$, and than degenerates the quartic to a cubic and a $\mathbb{P}^3$, etc. In [**MP**, Section 0.6], Maulik-Pandharipande described a second algorithm based on Gathmann's proposal. The second algorithm only requires one degeneration: one degenerates $\mathbb{P}^4$ to $\mathbb{P}^4$ and a $\mathbb{P}^1$ bundle over Q. Therefore, the second algorithm should be significantly more efficient than the first algorithm. Gathmann did the genus 0 and 1 cases. J. Li's degeneration formula and [**MP**, Theorem 1] (the quantum Leray-Hirsch) allow Maulik and Pandharipande to pursue Gathmann's proposal in all genera. Maulik-Pandharipande proved that the second algorithm determines all the genus 2 invariants $N_{2,d}$ and conjectured that it determines $N_{g,d}$ for all g,d; recently, L. Wu proved this conjecture in the genus 3 case.

We remark that the theories used by mathematicians to approach $N_{g,d}$ as above are essentially (1) "hyperplane property" for $g=0,1$, (2) torus localization formula, and (3) degeneration formula. They have intrinsic origin from theory of virtual cycles in mathematics.

It remains a central problem in Gromov-Witten theory to find new effective algorithms to calculate all genus Gromov-Witten invariants of Q, with structural properties compatible with physics treatment by mirror symmetry, such as (quasi-) modularity of F_g^A and finitely many holomorphic ambiguities with linear growth in g.

3. Witten's Gauged Linear Sigma Model (GLSM)

The same quintic polynomial $W_5 = x_1^5 + \ldots + x_5^5$ defines a map $\mathbb{C}^5 \to \mathbb{C}$. The corresponding physical theory is the Landau-Ginzburg theory for the pair $(\mathbb{C}^5, W_5)$. Since W_5 is invariant under the diagonal multiplicative action of $\mathbb{Z}_5$ on $\mathbb{C}^5$, it descends to give an orbifold LG model $([\mathbb{C}^5/\mathbb{Z}_5], W_5)$. In **[GLSM]**, Witten embedded Q into a larger background with superpotenial as follows. Let $\mathbb{C}^*$ act on $\mathbb{C}^6 = \mathbb{C}^5 \times \mathbb{C} = \{(x_1, \ldots, x_5, p)\}$ with weights $(1, \ldots, 1, -5)$. The quotient $[\mathbb{C}^6/\mathbb{C}^*]$ has two GIT quotients:

$$\begin{array}{rcl} ((\mathbb{C}^5 - \{\vec{0}\}) \times \quad \mathbb{C} \quad)/\mathbb{C}^* & = & K_{\mathbb{P}^4} \ , \\ (\quad \mathbb{C}^5 \quad \times (\mathbb{C} - 0))/\mathbb{C}^* & = & \mathbb{C}^5/\mathbb{Z}_5. \end{array}$$

Here $K_{\mathbb{P}^4}$ is the total space of the canonical line bundle $\mathcal{O}(-5)$ on $\mathbb{P}^4$. The polynomial $p(x_1^5 + \cdots + x_5^5)$ on $\mathbb{C}^6$ is invariant under the above $\mathbb{C}^*$ action, so it descends to a function $\widetilde{W} : [\mathbb{C}^6/\mathbb{C}^*] \to \mathbb{C}$. Thus one has a picture relating generalized Landau-Ginzburg models

(3.1)
$$\begin{array}{ccccc} & & ([\mathbb{C}^6/\mathbb{C}^*], \widetilde{W}) & & \\ & {\scriptstyle (x_1,\cdots,x_5)\neq 0} \nearrow & & \nwarrow {\scriptstyle p\neq 0} & \\ Q \hookrightarrow (K_{\mathbb{P}^4}, \mathbf{w}) & & & & ([\mathbb{C}^5/\mathbb{Z}_5], W_5) \end{array}$$

where the restriction $\mathbf{w}$ of $\widetilde{W}$ to $K_{\mathbb{P}^4}$ is the function induced by tensoring $x_1^5 + \cdots + x_5^5 \in H^0(\mathbb{P}^4, \mathcal{O}_{\mathbb{P}^4}(5))$ under the pairing $\mathcal{O}_{\mathbb{P}^4}(-5) \otimes \mathcal{O}_{\mathbb{P}^4}(5) \to \mathcal{O}_{\mathbb{P}^4}$. The critical locus of the superpotential $\mathbf{w}$ on $K_{\mathbb{P}^4}$ is the quintic Q embedded in $K_{\mathbb{P}^4}$ as the subvariety defined by $p = 0$ and $x_1^5 + \cdots + x_5^5 = 0$. The two skew arrows in Diagram (3.1) are open smooth subsets defined by $(x_1, \cdots, x_5) \neq 0$ and $p \neq 0$ respectively.

In 1993 Witten **[GLSM]** provides a theory called Gauged Linear Sigma Model (GLSM) which can be considered as a sort of "quantization" of $([\mathbb{C}^6/\mathbb{C}^*], \widetilde{W})$. Here the word quantization means promoting variables $(x_1, \cdots, x_5)$ to fields $(\varphi_1, \cdots, \varphi_5)$ on the worldsheet (which is a connected closed Riemann surface), and promoting $\mathbb{C}^*$ to a principal $\mathbb{C}^*$-bundle over the worldsheet, with a gauge field. Witten's GLSM theory is parameterized by a real number r called Fayet-Iliopoulos parameter, which is essentially the Kähler parameter of the symplectic quotient of $\mathbb{C}^6$ by the Hamiltonian $U(1)$-action with weights $(1, \ldots, 1, -5)$. When $r \to -\infty$ the GLSM is contributed by the Landau-Ginzburg model $([\mathbb{C}^5/\mathbb{Z}_5], W_5)$; when $r \to +\infty$ the GLSM is contributed by massless instantons in Q, the critical locus of $(K_{\mathbb{P}^4}, \mathbf{w})$. Witten also showed that for $r \to +\infty$ the GLSM is contributed by instantons of a sigma model on the quintic threefold Q, i.e. holomorphic curves in Q.

Witten's model suggests a few things. Firstly, a physics theory for the Landau-Ginzburg model $(K_{\mathbb{P}^4}, \mathbf{w})$ should be found to undertake the specialization from GLSM to sigma model on the quintic threefold Q. Secondly, when $r > 0$ is finite, contribution to GLSM is made by massive instantons in Q, where mass corresponds to common zeros of $\varphi_1, \cdots, \varphi_5$. In the language of mathematics, it foresees intermediate theories (obtained by varying the stability condition) other than Gromov-Witten (GW) theory of Q (at $r \to +\infty$) or Landau-Ginzburg (LG) theory of $([\mathbb{C}^5/\mathbb{Z}_5], W_5)$ (at $r \to -\infty$).

When [**GLSM**] first appeared, it was far from a mathematical theory: as Witten's [**GLSM**] is a gauged field theory defined by using path integral, mathematicians need to find conditions to ensure the convergence of the path integral and then substitute the infinite dimensional path integral measure with certain finite dimensional construction to obtain rigorous mathematical definitions. Moreover, theories for every spaces in (3.1) were neither twisted nor coupled to gravity.

Witten's GLSM tells us, once mathematicians can possibly achieve finite dimensional constructions which lead to rigorous mathematical definitions, the (massive or massless) theories for Q, $(K_{\mathbb{P}^4}, \mathbf{w})$, $([\mathbb{C}^5/\mathbb{Z}_5], W_5)$, and possibly even the universal $([\mathbb{C}^6/\mathbb{C}^*], \widetilde{W})$, should determine each other; namely, they should be "equivalent" theories. However, what are the explicit relations among amplitudes from any two different theories? Could these conjectural equivalences help determine all of them, or just reduce three sorts of mysteries to one that is still mysterious? In later sections of this paper, we will discuss solutions to the above questions on constructions and relations.

In the following we shall call $([\mathbb{C}^5/\mathbb{Z}_5], W_5)$ an affine Landau-Ginzburg model as $\mathbb{C}^5$ is an affine space, to be distinguished from $(K_{\mathbb{P}^4}, \mathbf{w})$, which is a general Landau-Ginzburg model.

4. Hyperplane Property, Ghost, and P-field

4.1. Physics: Guffin and Sharpe. Guffin and Sharpe consider the A-twisting of the LG model $(K_{\mathbb{P}^4}, \mathbf{w})$ suggested by Witten, and show its amplitudes are equal to genus zero GW invariants of Q. This can be viewed as the hyperplane property in physics, at least in genus zero. To describe the matter field $\varphi : C$" $\longrightarrow$ "$K_{\mathbb{P}^4}$ (where φ and C are smooth), in additional to classical $\varphi_1, \cdots, \varphi_5$ as C^∞ sections of a line bundle L on C, the noncompact direction needs to be twisted by the canonical line bundle K_C of C and considered as

$$p \in C^\infty(C, K_C \otimes L^{\otimes -5}), \tag{4.1}$$

so a term in the Lagrangian becomes a top form on C and can be integrated to make sense of the action. Guffin and Sharpe showed that , in the genus zero case, their integral (as an invariant associated to enumerating curves mapped to $(K_{\mathbb{P}^4}, \mathbf{w})$) is equal to

$$e(E_d) \cap [X_d] \in H_0(X; \mathbb{Q}) = \mathbb{Q}$$

where $e(E_d) \in H^{10d+2}(X_d; \mathbb{Q})$ is the Euler class of the finite complex vector bundle

$$E_d = \bigcup_{\varphi \in X_d} H^0(C, \varphi^* \mathcal{O}_{\mathbb{P}^4}(5))$$

over the smooth compact complex orbifold $X_d = \{\varphi : C \to \mathbb{P}^4 \mid \deg \varphi = d, g(C) = 0\}$, and $[X_d] \in H_{10d+2}(X_d; \mathbb{Q})$ is the fundamental class. By Kontsevich's hyperplane property, $e(E_d) \cap [X_d]$ is the genus zero, degree d GW invariant $N_{g=0,d}$ of Q.

4.2. Mathematics: Hyperplane Property. We explain the hyperplane problem in mathematics. Fixing the degree d, for each genus g, over the finite dimensional compact space

$$Y_{g,d} := \{\varphi : C \to \mathbb{P}^4 \ \text{ stable map } \mid \deg \varphi = d, g(C) = g\}$$

there are two unions of vector spaces

$$V_{g,d} := \bigcup_{\varphi \in Y_g} H^0(C, \varphi^* \mathcal{O}_{\mathbb{P}^4}(5)) \quad \text{and} \quad V'_{g,d} := \bigcup_{\varphi \in Y_g} H^1(C, \varphi^* \mathcal{O}_{\mathbb{P}^4}(5)),$$

defining two sheaves over $Y_{g,d}$. By Riemann-Roch formula, the difference of the dimensions $\dim V_{g,d}|_\varphi - \dim V'_{g,d}|_\varphi$ is $5d + 1 - g$, which is independent of $\varphi \in Y_{g,d}$.

If $g = 0$, one can show $V'_{0,d} = 0$. Thus $V_{0,d}$ has constant dimensional fiber over $Y_{0,d}$, namely $V_{0,d}$ is a complex vector bundle of rank $5d + 1$ over $Y_{0,d}$. The hyperplane property of Kontseviech says

$$N_{0,d} = \deg\left(e(V_{0,d}) \cap [Y_{0,d}]\right) \in \mathbb{Q}. \tag{4.2}$$

This reconstructs the enumeration of rational curves in Q from the information of $\mathbb{P}^4$. The identity (4.2) is an easy consequence of virtual cycle theory [**KKP**]. By (4.2), one may compute $N_{0,d}$ by torus localization as $V_{0,d}$ and $Y_{0,d}$ admit a $(\mathbb{C}^*)^4$-action inherited from that on $\mathbb{P}^4$.

When $g > 0$, everything above fails unfortunately. For example when $g = 1$, $Y_{g=1,d}$ contains essentially two kinds of components. The two components collect maps of different forms (see Figure 1 below).

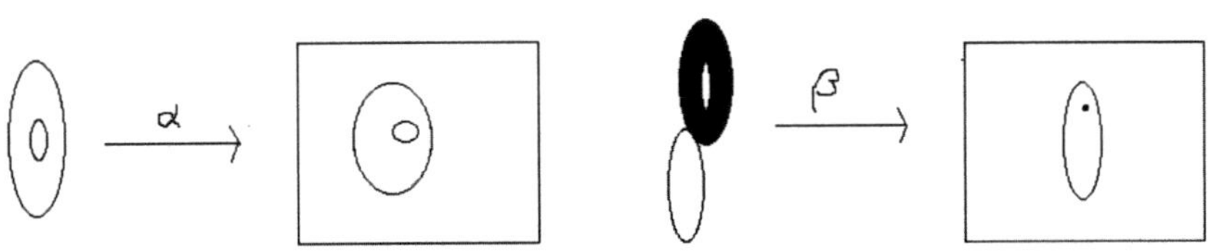

FIGURE 1. graphs for an honest map α and a ghost map β.

The main component of $Y_{1,d}$ consists of maps looking like α, which has positive degree on the genus one component of the curve. The other component of $Y_{1,d}$ consists of maps looking like β, which contracts the (black) genus one component to one point, and has positive degree on the (blank) genus zero $\mathbb{P}^1$. For the curve and maps being indicated as α, every element in $H^1(C, L^{\otimes 5}) = H^0(C, K_C \otimes L^{\otimes -5})^\vee$ vanishes since $L = \alpha^* \mathcal{O}(1)$ has positive degree and $K_C \otimes L^{\otimes -5}$ has negative degree. Thus the P-field must vanish for α (recall that $P \in H^0(C, K_C \otimes L^{\otimes -5})$), or equivalently, $V'_{1,d}|_\alpha = 0$. However, for β one can have nonzero one form on the elliptic component, which extends by zero to give a section of $K_C \otimes L^{\otimes -5}$. This corresponds to the fact that β contracts a genus one component of the curve C, for which reason we say β is a ghost map. The black genus one component is a "ghost" component, and *P-field can survive on a ghost.* One easily calculates $V'_{1,d}|_\beta \cong \mathbb{C}$.

We remark that in the approach determining $N_{1,d}$ in [**LZ, Zi**], a key issue is to locate the contribution of the ghost in the counting. In their formula $N_{1,d} = N^{\text{red}}_{1,d} + \frac{1}{12} N_{0,d}$, the term $N^{\text{red}}_{1,d}$ is the contribution from maps of type α, and $\frac{1}{12} N_{0,d}$ is the contribution from maps of type β, where the $\frac{1}{12}$ comes from integrating out all the P-fields living on the ghost (black) elliptic component of β. For our ultimate purpose to approach $N_{g,d}$ for larger g, locating the contribution of P-field (including ghosts) becomes very difficult and out of control.

Since $V'_{1,d}|_\alpha = 0$ and $V'_{1,d}|_\beta \cong \mathbb{C}$, $V'_{1,d}$ has fiber rank jumping over $X_{1,d}$, and by Riemann-Roch $V_{1,d}$ also does and is not a vector bundle over $Y_{1,d}$. As the

Euler class is only defined for vector bundles, $e(V_{1,d})$ no longer makes sense. It is natural to ask how the hyperplane property (4.2) should be modified, so that the information of $\mathbb{P}^4$ can be used to reconstruct enumeration of higher genus curves in Q in mathematics (namely only finite dimensional construction allowed).

After A-twisting, the topological string theory (with supersymmetry) admits a mathematical counterpart called "virtual cycle"[1]. As virtual cycle ([**LT**]) is governed by tangent-obstruction (deformation) theory (in physics words, after A-twisting, the zero mode of fermions over SUSY fixed loci, even if the loci is singular, recovers the path integral algebraically), we may view the above problem of higher genus hyperplane property in the following way. Let $f : C \to Q$ be a point in

$$f = [f_1, \cdots, f_5] \in \overline{M}_g(Q, d) := X_{g,d} \subset \overline{M}_g(\mathbb{P}^4, d) = Y_{g,d}.$$

The exact sequence $0 \to T_Q \to T_{\mathbb{P}^4}|_Q \to \mathcal{O}(5)|_Q \to 0$ induces the following long exact sequence

(4.3)
$$\begin{aligned} 0 \longrightarrow H^0(C, f^*T_Q) &\longrightarrow H^0(C, f^*T_{\mathbb{P}^4}) \longrightarrow H^0(C, f^*\mathcal{O}(5)) \\ \longrightarrow H^1(C, f^*T_Q) &\longrightarrow H^1(C, f^*T_{\mathbb{P}^4}) \longrightarrow H^1(C, f^*\mathcal{O}(5)) \longrightarrow 0. \end{aligned}$$

Every vector space above has geometric meanings, namely the sequence (4.3) is identical to

$$\begin{aligned} 0 \longrightarrow T_{f,X_{g,d}} &\longrightarrow T_{f,Y_{g,d}} \longrightarrow \mathcal{O}b_{f,X_{g,d}/Y_{g,d}} \\ \longrightarrow \mathcal{O}b_{f,X_{g,d}} &\longrightarrow \mathcal{O}b_{f,Y_{g,d}} \longrightarrow \mathcal{O}b^{\text{higher}}_{f,X_{g,d}/Y_{g,d}} \longrightarrow 0. \end{aligned} \tag{4.4}$$

where

- $T_{f,X_{g,d}}$ and $T_{f,Y_{g,d}}$ are the first order deformations of f in $X_{g,d}$ and $Y_{g,d}$ (relative to moduli space of genus g nodal curves) respectively;
- $\mathcal{O}b_{f,X_{g,d}}$ and $\mathcal{O}b_{f,Y_{g,d}}$ are obstructions to deforming f in $X_{g,d}$ and $Y_{g,d}$ respectively;
- $H^0(C, f^*\mathcal{O}(5))$, which contains the element $f_1^5 + \cdots + f_5^5$, is the obstruction[2] of an element in $Y_{g,d}$ to be in $X_{g,d}$, namely the relative obstruction $\mathcal{O}b_{f,X_{g,d}/Y_{g,d}}$;
- $H^1(C, f^*\mathcal{O}(5))$ is the *higher obstruction* of a point in $Y_{g,d}$ to lie in $X_{g,d}$.

Now recall that the tangent and obstruction theory would determine the virtual cycle (path integral measure), and the two terms in the left column in (4.4) are tangent and obstruction theories of $X_{g,d}$, therefore are responsible for the Gromov-Witten invariant $N_{g,d}$ of the quintic Calabi-Yau threefold Q. The two terms in the middle column are tangent and obstruction theories of $Y_{g,d}$ which parametrizes maps to $\mathbb{P}^4$. To solve the hyperplane property problem, one should combine the right column in (4.4) with the datum of $Y_{g,d}$. If this can be done then one may expect to recover $N_{g,d}$.

We observe that the last term (higher obstruction) $H^1(C, f^*\mathcal{O}(5))$ is dual to the space $H^0(C, K_C \otimes f^*\mathcal{O}(-5))$ of algebraic P-fields (c.f. (4.1)), which we may add it to the moduli space $Y_{g,d} = \overline{M}_g(\mathbb{P}^4, d)$ of stable maps to $\mathbb{P}^4$ to form the moduli

[1] also called virtual fundamental class

[2] because $f_1^5 + \cdots + f_5^5 = 0$ characterizes the $f \in Y_{g,d}$ lies in $X_{g,d}$

of stable maps to $\mathbb{P}^4$ with P-fields[3]

(4.5)

$$Y_{g,d}^p := \overline{M}_g(\mathbb{P}^4,d)^p = \{[\varphi : C \to \mathbb{P}^4] \in \overline{M}_g(\mathbb{P}^4,d)), \rho \in \Gamma(C, K_C \otimes \varphi^*\mathcal{O}(-5))\},$$

where ρ is called an algebraic "P-field" as its analogue in (4.1). As the obstructions to deforming φ and ρ lie in $H^1(C,\phi^*T_{\mathbb{P}^4})$ and $H^1(C, K_C \otimes \varphi^*(\mathcal{O}(-5))) = H^0(C,\varphi^*\mathcal{O}(5))^\vee$ respectively, the deformation theory of $Y_{g,d}^p$ is given by the middle and right columns in (4.3). If one is able to define a virtual cycle for $\overline{M}_g(\mathbb{P}^4,d)^p$, then it is expected to be "equivalent" to the virtual cycle of $X_{g,d}$, and the hyperplane problem is solved. Now the difficulty appears because $\overline{M}_g(\mathbb{P}^4,d)^p$ is *non-compact* due to the presence of *P*-fields: for example, over the ghost map β, the P-field can be any element in $\mathbb{C}$ that is unbounded. This difficulty is then overcome by the invention of "cosection localization" by Y.H. Kiem and Jun Li, along with H.L. Chang's observation that "*the supersymmetry variation of the superpotential on worldsheet defines a cosection, which solves Witten's equation in the general Landau-Ginzburg theory.*"

We now describe the algebro-geometric results of H.-L. Chang and J. Li [**CL1**] discovered based on the above reasoning. In the definition of (4.5), the data $([\varphi, C], \rho)$ is equivalent to the data $(C, L, \varphi_1, \cdots, \varphi_5, \rho)$, since the map φ is equivalent to the line bundle $L = \varphi^*\mathcal{O}_{\mathbb{P}^4}(1)$ with five sections $(\varphi_1, \cdots, \varphi_5)$ of L. We regard $\overline{M}_g(\mathbb{P}^4,d)^p$ as a space of "maps from curve to $K_{\mathbb{P}^4}$." The moduli stack $\overline{M}_g(\mathbb{P}^4,d)^p$ has a perfect obstruction theory relative to the smooth Artin stack $\mathcal{D} = \{(C, L)\}$. At $\xi = [(C, L, \varphi_i, \rho)] \in \overline{M}_g(\mathbb{P}^4,d)^p$, the (relative) obstruction space of deforming ξ is

$$\mathcal{O}b_{M/\mathcal{D}}|_\xi = H^1(L)^{\oplus 5} \oplus H^1(L^{\vee 5} \otimes \omega_C).$$

There exists a cosection

$$\sigma\colon \mathcal{O}b_{M/\mathcal{D}} \to \mathcal{O}_{\overline{M}_g(\mathbb{P}^4,d)^p}$$

constructed as follows. Let

$$(\dot\varphi_1, \ldots, \dot\varphi_5, \dot\rho) \in H^1(L)^{\oplus 5} \oplus H^1(L^{\vee 5} \otimes \omega_C) = \mathcal{O}b_{M/\mathcal{D}}|_\xi.$$

Define

$$\sigma|_\xi(\dot\varphi_1, \ldots, \dot\varphi_5, \dot\rho) \colon = \dot\rho \sum_{i=1}^5 \varphi_i^5 + \rho \sum_{i=1}^5 5\varphi_i^4 \dot\varphi_i.$$

The degeneracy locus $D(\sigma)$ of the cosection σ consists of ξ such that $\sigma|_\xi$ is zero, i.e., $\sigma|_\xi(\dot\varphi_1, \ldots, \dot\varphi_5, \dot\rho) = 0$ for all $\dot\varphi_i$ and $\dot\rho$. Thus

$$D(\sigma) = \{\xi \in \overline{M}_g(\mathbb{P}^4,d)^p \mid \rho = 0 \text{ and } \sum_{i=1}^5 \varphi_i^5 = 0\} = \overline{M}_g(Q,d) \subset \overline{M}_g(\mathbb{P}^4,d).$$

This corresponds to the fixed loci of supersymmetry (SUSY) in path integral. The expression of the cosection σ comes from supersymmetry variation δ (in physics) applied to $p \cdot W_5 = p(x_1^5 + \ldots + x_5^5)$ where p and x_i live on the worldsheet, via H.L. Chang's observation.

Since ρ is a section, the moduli space $\overline{M}_g(\mathbb{P}^4,d)^p$ is not proper (when $g \geq 1$) and hence cannot be used to define invariants. However, the degeneracy locus $D(\sigma)$ is the moduli space $\overline{M}_g(Q,d)$ of stable maps to the quintic threefold Q and thus proper. Using cosection localization developed by Y.H. Kiem and J. Li [**KL**],

[3]now allow $f : C \to Q$ to be more general $\varphi : C \to \mathbb{P}^4$

H.L. Chang and J. Li constructed [**CL1**] the cosection localized virtual cycle for Landau-Ginzburg theory

$$[\overline{M}_g(\mathbb{P}^4,d)^p]^{\mathrm{vir}}_{\mathrm{loc}} \in A_*D(\sigma) = A_*\overline{M}_g(Q,d).$$

As always one defines the P-fields GW invariants

$$N^p_{g,d} := \deg[\overline{M}_g(\mathbb{P}^4,d)^p]^{\mathrm{vir}}_{loc} \in \mathbb{Q}.$$

H.L Chang and J. Li proved the following.

THEOREM 4.1 (H.L. Chang - J. Li [**CL1**]). *The GW invariant of the quintic threefold Q equals the P-fields GW-invariant up to a sign:*

$$N_{g,d} = (-1)^{d+g+1}N^p_{g,d}.$$

The advantage of this result is that $F^A_g(q) = \sum_d N_{g,d}q^d$ now becomes the amplitude of a theory valued in $K_{\mathbb{P}^4} = \big((\mathbb{C}^5-\vec{0})\times\mathbb{C}\big)/\mathbb{C}^*$.

In conclusion, the invariant enumerating maps from curves to $(K_{\mathbb{P}^4},\mathbf{w})$ is equal to the invariant enumerating maps from curves to Q, up to a sign. This generalizes the genus zero case to the positive genus case, and solves the hyperplane property problem.

5. Fields Valued in Two GIT Quotients

5.1. Physics: GLSM. From §4, we see the curve-enumerating theories for Q and $(K_{\mathbb{P}^4},\mathbf{w})$ in Diagram (3.1) are both established in mathematics for all genera. Witten's [**GLSM**] suggests the theory for $([\mathbb{C}^5/\mathbb{Z}_5],W_5])$ at the lower right corner of (3.1) should also exist, and match the physical theory of A-twisted LG model coupled with gravity.

5.2. Mathematics. We now consider the space of maps from curves to each target in (3.1), viewed as a sort of "quantization" of (3.1).

The previous sections tells us the space $\overline{M}_g(\mathbb{P}^4,d)^p$ of all "maps to $(K_{\mathbb{P}^4},\mathbf{w})$" is the set of all (C,L,φ,ρ) where $\varphi=(\varphi_1,\cdots,\varphi_5)$ is section of $L^{\oplus 5}$, ρ is a section of $\omega_C\otimes L^{\otimes -5}$, and

$$(+)\qquad \varphi=(\varphi_1,\cdots,\varphi_5) \text{ has no zeros on } C$$

so that $\varphi_i's$ define an honest map to $\mathbb{P}^4$. Without the condition (+), one obtains a huge Artin stack $\mathcal{A}rt$ of all (C,L,φ,ρ) for arbitrary $\varphi\in\Gamma(C,L^{\oplus 5})$ and $\rho\in\Gamma(C,\omega_C\otimes L^{\otimes -5})$. The stack $\mathcal{A}rt=\{(C,L,\varphi,\rho)\}$ should be viewed as the moduli space of maps to $[\mathbb{C}^6/\mathbb{C}^*]_{(1,1,1,1,1,-5)}$. $\overline{M}_g(\mathbb{P}^4,d)^p$ is the open substack of objects in $\mathcal{A}rt$ subject to condition (+), which corresponds to the open substack $K_{\mathbb{P}^4}\subset[\mathbb{C}^6/\mathbb{C}^*]$ in (3.1) defined by $(x_1,\cdots,x_5)\neq(0,\cdots,0)$. After quantizing it translates to the requirement (+), as $\varphi_1,\cdots,\varphi_5$ are the five fields promoted from the five coordinates $x_1,\cdots,x_5$.

Parallelly, since the open substack $[\mathbb{C}^5/\mathbb{Z}_5]\subset[\mathbb{C}^6/\mathbb{C}^*]$ is defined by $p\neq 0$ in (3.1), to define a theory whose target is $([\mathbb{C}^5/\mathbb{Z}_5],W_5)$, one analogously expects to pick up the open substack of $\mathcal{A}rt$ subject to the condition

$$(-)\qquad \rho \text{ has no zeros on } C$$

as ρ is the field promoted from coordinate p in (3.1). Namely ρ trivializes $\omega_C\otimes L^{-5}$, or equivalently, gives an isomorphism $L^{\otimes 5} \xrightarrow{\cong} \omega_C$. One then expects the theory of $([\mathbb{C}^5/\mathbb{Z}_5], W_5)$ to start with the moduli space of all (C, L, φ) where

(1) L is a fifth root[4] of ω_C, and
(2) $\varphi = (\varphi_1, \cdots, \varphi_5)$ is an arbitrary section of $L^{\oplus 5}$.

We denote this moduli as $\overline{M}_g^{1/5,5p}$ where $1/5$ denotes the 5-spin structure and $5p$ indicates that an object consists of five sections $\varphi_1, \cdots, \varphi_5$ of L (by abuse of notation).

When one quantizes every space in (3.1), one then obtains two open substacks (subspaces) of the common huge Artin stack as follows:

(5.1)
$$\begin{array}{ccccc} & & \{(C, L, \varphi, \rho)\} & & \\ & \overset{\varphi \text{ nowhere } 0}{\nearrow} & & \overset{\rho \text{ nowhere } 0}{\nwarrow} & \\ \overline{M}_g(Q, d) \hookrightarrow \overline{M}_g(\mathbb{P}^4, d)^p & & & & \overline{M}_g^{1/5,5p} \end{array}$$

Naturally one wonders whether the substack $\overline{M}_g^{1/5,5p}$ at the bottom right corner has a virtual cycle, with which intersections represent invariants of the Landau-Ginzburg model $([\mathbb{C}^5/\mathbb{Z}_5], W_5)$ from physics, as Witten predicted. Coincidentally, around 2010 H. Fan, T. Jarvis, and Y. Ruan carried out a construction of an A-side theory of which the target may be any affine LG space $([\mathbb{C}^n/G], \mathbf{w})$, where G is a finite group and the "superpotential" $\mathbf{w}$ is a G-invariant polynomial on $\mathbb{C}^n$. Their approach to the affine Landau-Ginzburg model $([\mathbb{C}^n/G], \mathbf{w})$ originates from a different line in history, namely the gauged WZW model, Witten equation, and Hamiltonian Floer theory, which we brief in §6 below.

6. Affine LG Phase and Spin Structure

6.1. Physics: SUSY A-twisted LG Theory Coupled To Gravity. The classical Landau-Ginzburg theory on the A-side follows a different line of development in history. In [**Wi**] E. Witten conjectured that descendant integrals on moduli spaces of stable curves $\overline{M}_{g,n}$ satisfy the KdV equations, and the string equation (proved by Witten) and the KdV equations uniquely determine all descendant integrals from the initial value $\int_{\overline{M}_{0,3}} 1 = 1$. Witten's conjecture was first proved by Kontsevich [**Kon1**] by stratification of $\overline{M}_{g,n}$ and matrix model. For the purpose to generalize above to N matrix model, Witten in [**Wi1**] considered A-twisted gauged WZW model targeting $SU(2)/U(1)$ coupled with gravity, and obtained a topological theory which he conjectured [**Wi3**] (refining/twisting the minimal model of [**KLi**] et.al.) to solve the generalized KdV hierarchies (N-matrix model).

Witten's A-twisted theory is localized to the SUSY fixed locus consisting of objects almost definable in algebraic geometry. Let $\overline{M}_g^{1/r}$ denote the moduli space of Riemann surfaces C (with at worst nodal singularities) together with a line bundle L such that $L^{\otimes r} \cong K_C$. Mathematically $\xi = (C, L)$ is referred as a r-spin curve. One may also add orbifold marked points on C but we omit them here for

[4]sometimes called 5-spin structure on C;

simplicity. Witten roughly argued that $\overline{M}_g^{1/r}$ is smooth and compact. He set

$$M^\infty = \bigcup_{(C,L)\in\overline{M}_g^{1/r}} C^\infty(C,L)$$

where $C^\infty(C,L)$ is the space of C^∞ sections u of L.

The topological correlation function of Witten's theory amounts to counting the intersection number of the zero section of the (infinite rank) bundle

$$E^\infty = \bigcup_{(\xi,u)\in M^\infty} \Omega_C^{(0,1)}(L) \longrightarrow M^\infty \tag{6.1}$$

with the graph of the section

$$(\xi = (C,L),u) \mapsto s_W(\xi,u) := \overline{\partial} u + r(\bar{u})^{r-1} \tag{6.2}$$

and possibly with insertions (**[Wi3]**) such as gravitational descendants (if one adds markings on each C). Note that we may choose a Kähler metric on the Riemann surface C and a Hermitian metric on the line bundle L, so that $(\bar{u})^{r-1}$ becomes a section of $(\bar{L})^{\otimes(r-1)} \cong L^{\otimes(1-r)} \cong \overline{K}_C \otimes L$, where $\overline{\partial} u$ lives. In short the theory counts solutions of

$$\overline{\partial} u + r(\bar{u})^{r-1} = 0. \tag{6.3}$$

The Euler class of E^∞ localized by Witten's section s_W is then called "Witten's top Chern class", a core object in the definition of the theory.

For the purpose to interpret Witten's correlation function more directly, one may regard it as the A-twisted (and coupled to gravity) version of the "Landau-Ginzburg theories" defined in [**Vafa**], [**Ito**] (also c.f. [**Ce**]). Vafa, et.al.'s model build the Landau-Ginzburg structure directly in the Lagrangian. Namely, it is a path integral whose configuration space of fields is the set of maps from the worldsheet to $([\mathbb{C}^n/G],\mathbf{w})$, with fermions coupled with terms as

$$s_{\mathbf{w}} = (\overline{\partial} u_i + \overline{\partial_{u_i}\mathbf{w}(u_1,\cdots,u_n)})_{i=1}^n,$$

and the contribution to the theory comes from solutions of

$$\overline{\partial} u_i + \overline{\partial_{u_i}\mathbf{w}(u_1,\cdots,u_n)} = 0 \qquad \text{for all } i = 1,\cdots,n \tag{6.4}$$

generalizing (6.3) where $n = 1, \mathbf{w} = x^r$.

However, the theories in [**Ito**], [**Vafa**] are not coupled with gravity, and the group G is trivial $G = \{e\}$. It was then later understood (by Fan-Jarvis-Ruan etc.) that Witten's model is using $([\mathbb{C}/\mathbb{Z}_r], \mathbf{w} = x^r)$, whose state spaces are indexed by the monodromy weights of the r-spin bundle at markings.

6.2. Mathematics: FJRW Invariants. Based on Witten's infinite dimensional Euler class model (with section to be $s_{\mathbf{w}}$), Fan-Jarvis-Ruan [**FJR1, FJR2**] used analytic methods to construct the Witten's top Chern class, and defined correlators of a Cohomological Field Theory (CohFT) by capping the Witten's top Chern class with states of the Landau-Ginzburg model $([\mathbb{C}^n/G],\mathbf{w})$. Fan-Jarvis-Ruan's pioneer work is now known as FJRW invariants associated to the singularity $([\mathbb{C}^n/G],\mathbf{w})$. FJRW invariants of special ADE type singularities can be enumerated and are governed by the Kac-Wakimoto/Drinfeld-Sokolov hierarchies [**LRZ**], generalizing [**FSZ**]'s proof of Witten's r-spin conjecture.

In FJRW theory, Witten's top Chern class is constructed in differential geometry via perturbing (6.4). It can also be constructed in algebraic geometry without perturbing (6.4). The algebro-geometric constructions (in the narrow case) were carried out by Polishchuk-Vaintrob [**PV**], by Chiodo [**Chi**], and by H.L. Chang, J. Li and W.P. Li [**CLL**]. For our purpose to provide a field theory valued in $([\mathbb{C}^5/\mathbb{Z}_5], W_5)$, we brief the construction in [**CLL**] here, using the version with markings. Recall that the moduli $\overline{M}_g^{1/5,5p}$ for $([\mathbb{C}^5/\mathbb{Z}_5], W_5)$ requires a fifth root L of ω_C, which does not exist if $\deg \omega_C = 2g-2$ is not divisible by five. One thus extends the setup by allowing C to be a twisted curve with markings (which can be scheme points or stacky points). Thus our field valued in $\big(\mathbb{C}^5 \times (\mathbb{C}-0)\big)/\mathbb{C}^* = [\mathbb{C}^5/\mathbb{Z}_5]$ consists of

$$\xi = (\Sigma^C, C, L, \varphi_1, \ldots, \varphi_5, \rho)$$

where (Σ^C, C) is a pointed twisted curve with markings Σ^C possibly stacky, L is an invertible sheaf on C, $\varphi_i \in H^0(L)$, and $\rho \in H^0(L^{\vee 5} \otimes \omega_C^{\log})$ with $\omega_C^{\log} = \omega_C(\Sigma^C)$, and the corresponding property of (–)

(*) the section ρ is nowhere vanishing

is required. This implies $L^{\vee 5} \otimes \omega_C^{\log} \cong \mathcal{O}_C$, or equivalently $L^{\otimes 5} \cong \omega_C^{\log}$. Therefore (Σ^C, C, L) is a 5-spin curve and $(\varphi_1, \ldots, \varphi_5)$ gives five fields. We get a moduli space of 5-spin curves with five fields:

$$\overline{M}_{g,\gamma}^{1/5,5p} = \{(\Sigma^C, C, \mathcal{L}, \varphi_1, \cdots, \varphi_5, \rho) \mid \rho \text{ is nowhere zero}\}.$$

Here γ is the monodromy data: if Σ_j is a stacky marking on C, then μ_5 acts on $L|_{\Sigma_j}$ with weight $\gamma_j = \exp(2\pi i r_j/5)$ where $1 \le r_j \le 4$ and we call γ_j narrow. If Σ_j is a scheme marking, we call it broad, and it corresponds to $\gamma_j = 1$.

Similar to the case $\overline{M}_g(\mathbb{P}^4, d)^p$, the moduli stack $\overline{M}_{g,\gamma}^{1/5,5p}$ has a perfect obstruction theory relative to the smooth Artin stack $\mathcal{D} = \{(\Sigma^C, C, L)\}$. There exists a cosection $\sigma \colon \mathcal{O}b \to \mathcal{O}_{\overline{M}_{g,\gamma}^{1/5,5p}}$ whose degeneracy locus is

$$D(\sigma) = \{\xi \in \overline{M}_{g,\gamma}^{1/5,5p} \mid \varphi_i = 0 \text{ for all } i\} = \overline{M}_{g,\gamma}^{1/5} = \{(\Sigma^C, C, L) \mid L^{\otimes 5} \cong \omega_C^{\log}\},$$

which is the moduli space of 5-spin curves.

THEOREM 6.1 (H.L. Chang - J. Li - W. P. Li [**CLL**]). *The (narrow) FJRW invariants can be constructed using cosection localized virtual cycles of* $\overline{M}_{g,\gamma}^{1/5,5p}$:

$$[\overline{M}_{g,\gamma}^{1/5,5p}]^{\mathrm{vir}}_{\mathrm{loc}} \in A_* \overline{M}_{g,\gamma}^{1/5,5p}.$$

The Witten equation mentioned in (6.4), in this case, becomes

$$\bar{\partial} s_i + \overline{\partial_{x_i} W_5(s_1, \ldots, s_5)} = 0, \quad i.e., \quad \bar{\partial} s_i + 5\overline{s_i^4} = 0 \qquad \text{for } i = 1, \cdots, 5. \tag{6.5}$$

This is used to construct Witten's top Chern class to define invariants on the moduli space of 5-spin curves. From Witten's equation (6.5), the term $\bar{\partial} s_i$ gives the obstruction to extending a holomorphic section. Thus the left hand side of (6.5) is a C^∞ section of the obstruction sheaf of the moduli of spin curves with fields. Substituting the complex conjugate in the Witten's equation by the Serre duality, the left hand side of (6.5) becomes a smooth inverse of cosection. The Mathai-Quillen setup in (6.1) and (6.2) generalize naturally here and the Euler class localized near solution of Witten equations would be equal to the Kiem-Li's virtual cycle localized via cosection σ.

We remark here that the form (6.5) indicates the virtual cycle of $([\mathbb{C}^5/\mathbb{Z}_5], W_5)$ is the five self-intersection of the virtual cycle of $([\mathbb{C}/\mathbb{Z}_5], x^5)$ (each defined by using

$\bar{\partial}s + 5\overline{s^5} = 0$ which is (6.3) for the case $r = 5$). This remarkable property is related to self-tensor product of conformal field theories and is discussed in [**FJR1**] or [**CR**].

There is an important subclass of FJRW invariants: those with insertions $-\frac{2}{5}$. Let C have k markings with all $\gamma_j = \zeta^2$ for $1 \le j \le k$ where $\zeta = \exp(2\pi i/5)$. For simplicity we write $\gamma = (\gamma_j)_{j=1}^k = 2^k$. Define

$$\Theta_{g,k} := \int_{[\overline{M}^{1/5,5p}_{g,2^k}]^{\mathrm{vir}}_{\mathrm{loc}}} 1 \in \mathbb{Q}, \quad \text{for } k + 2 - 2g = 0 \bmod 5.$$

It is shown [**CLLL2**] that $\{\Theta_{g,k}\}_{g,k}$ determine all FJRW invariants with descendants for the quintic LG space $([\mathbb{C}^5/\mathbb{Z}_5], W_5)$, where an explicit formula will be given in [**twFJRW**]. For this reason we call $\{\Theta_{g,k}\}_{g,k}$ the primary FJRW invariants.

7. The Puzzle to Link Invariants in Opposite Phases

7.1. Mathematics. We have seen that the three moduli spaces at the bottom of Diagram (5.1) admit virtual fundamental classes, while the moduli space $\mathcal{A}rt := \{(C, L, \varphi, \rho)\}$ at the top of (5.1) does not, because $\mathcal{A}rt$ is not a Deligne Mumford stack. One can introduce all possible stability conditions to define open substacks of $\mathcal{A}rt$ that are Deligne Mumford, just as $\overline{M}_g(\mathbb{P}^4, d)^p$ and $\overline{M}^{1/5,5p}_g$, and then construct virtual classes (path integral measures) for them as we defined $[\overline{M}_g(\mathbb{P}^4, d)^p]^{\mathrm{vir}}_{\mathrm{loc}}$ and $[\overline{M}^{1/5,5p}_g]^{\mathrm{vir}}_{\mathrm{loc}}$ using P-fields and cosection machinery. This is the step that most groups are taking. The theory of ϵ-stability and quasimaps ([**FK1**] [**MOP**]) are developed, for example.

On the other hand, introducing new stability conditions means there are invariants other than the original $N_{g,d}$'s.. Whether these new invariants (defined by new stability conditions) can simplify enumeration of $N_{g,d}$'s or give structures for $N_{g,d}$ predicted by the B-side, is not easy at all. Following Witten's GLSM, we wishfully expect knowing FJRW invariants $\Theta_{g,k}$'s of $(\mathbb{C}^5/\mathbb{Z}_5], W_5)$ would help us to understand/enumerate GW invariants $N_{g,d}$'s of Q. We would like to know whether, and how exactly, the invariants $N^p_{g,d}$'s defined by $[\overline{M}_g(\mathbb{P}^4, d)^p]^{\mathrm{vir}}_{\mathrm{loc}}$ (which are, by Theorem 4.1, $N_{g,d}$'s up to a sign) are related to the FJRW invariants $\Theta_{g,k}$'s defined by $[M^{1/5,5p}_{g,2^k}]^{\mathrm{vir}}_{\mathrm{loc}}$. We understand that the task is to construct theories that quantitatively link all in

(7.1)

$$\text{GW of } Q \quad \overset{\text{Theorem 4.1}}{\Longleftrightarrow} \text{GSW of } (K_{\mathbb{P}^4}, \mathbf{w}) \quad \longleftarrow ? \longrightarrow \quad \text{FJRW of } ([\mathbb{C}^5/\mathbb{Z}_5], W).$$

To pursue this goal, we immediately face a specific problem: "the change of phases' sign". The topological type of fields in $K_{\mathbb{P}^4}$ is labelled by a pair (g, d), where g is the genus of the curve C and $d = \deg L = \deg f^*\mathcal{O}(1)$ is always a non-negative integer; the topological type of fields valued in $([\mathbb{C}^5/\mathbb{Z}_5], W_5)$ is labeled by a pair (g, k), the genus g of the curve and the number k of $2/5$ marking. When g is fixed and when k is general (large) enough, one can show the degree of the line bundle (over the coarse curve) can be arbitrarily negative, namely, in the phase $([\mathbb{C}^5/\mathbb{Z}_5], W_5)$, fields are generally of negative degree. In GLSM this corresponds to the fact that $N_{g,d}$ are invariants near large radius limit point $r \gg 0$ and the LG phase $([\mathbb{C}^5/\mathbb{Z}_5], W_5)$ occurs near the orbifold point $r \ll 0$ (c.f. [**GLSM**, Section 5.1]).

How can a field of positive degree be transformed to a field of negative degree? In which space could this unusual transform happen? How does such transformation – if it exists – change the virtual cycles and counting? We will address these questions in the following sections.

8. Master Space

8.1. Mathematics. If one builds a large moduli space containing $\overline{M}_g(\mathbb{P}^4, d)^p$ and $\overline{M}_g^{1/5,5p}$ as its *disjoint* closed subspace, then intersection theory over the large moduli would give us information relating $\overline{M}_g(\mathbb{P}^4, d)^p$ to $\overline{M}_g^{1/5,5p}$. Recall that the two target spaces $K_{\mathbb{P}^4}$ and $[\mathbb{C}^5/\mathbb{Z}_5]$ are both open subsets of the 5-dimensional stack $[\mathbb{C}^6/\mathbb{C}^*]$, where the two overlap on a large open set $K_{\mathbb{P}^4} - \mathbb{P}^4 = [(\mathbb{C}^5 - \{\vec{0}\})/\mathbb{Z}_5]$. If one embeds these two open substacks as disjoint closed substacks of a higher dimensional stack W, we may consider the space of maps from curves to W as just stated. This higher dimensional stack has a natural construction in various places in "Variation of GIT" (VGIT) before, called the *master space* after M. Thaddeus. Here is a brief introduction.

Consider the following $\mathbb{C}^*$-action on $\mathbb{C}^5 \times \mathbb{C} \times \mathbb{P}^1$: for $t \in \mathbb{C}^*$,

$$(x_1, \dots, x_5, p, [u_1, u_2])^t \colon = (tx_1, \dots, tx_5, t^{-5}p, [tu_1, u_2]).$$

There is a GIT quotient

$$\bar{W} \colon = (\mathbb{C}^5 \times \mathbb{C} \times \mathbb{P}^1 - \mathcal{S})/\mathbb{C}^* \text{ where } \mathcal{S} \colon = \{(x_i = 0 = u_1) \cup (p = 0 = u_2)\}$$

which is a 6-dimensional simplicial toric variety. So $\bar{W}$ has at most orbifold singularities. Indeed, $\bar{W}$ is smooth outside the unique orbifold point given by $x_i = u_2 = 0$. The stacky quotient

$$W = [(\mathbb{C}^5 \times \mathbb{C} \times \mathbb{P}^1 - \mathcal{S})/\mathbb{C}^*]$$

is a 6-dimensional smooth toric Deligne-Mumford (DM) stack with coarse moduli space $\bar{W}$.

Consider a $\mathbb{C}^*$-action on W, and call this action T-action to avoid confusion. For $t \in T = \mathbb{C}^*$,

$$(x_1, \dots, x_5, p, [u_1, u_2])^t = (x_1, \dots, x_5, p, [tu_1, u_2]).$$

The T-fixed locus is a disjoint union of three connected components:

$$W^T = K_{\mathbb{P}^4} \times \{0\} \sqcup \vec{0} \times \left((\mathbb{P}^1 - \{0, \infty\})/\mathbb{C}^*\right) \sqcup [\mathbb{C}^5/\mathbb{Z}_5] \times \{\infty\}$$

where $0 = [1, 0]$ and $\infty = [1, 0]$ in $\mathbb{P}^1$, and the middle term $\left((\mathbb{P}^1 - \{0, \infty\})/\mathbb{C}^*\right)$ is nothing but one single point. The shape of W and its T fixed loci is shown in Figure 2, where $\mathcal{O}_{\mathbb{P}^4}(-5) = K_{\mathbb{P}^4}$ and $[\mathbb{C}^5/\mathbb{Z}_5]$ are disjoint divisors defined by $u_1 = 0$ and $u_2 = 0$ respectively. The single point defined by $x_i = p = 0$ is responsible for the conifold point of complexified Kähler moduli space of the quintic.

9. Mixed Spin Fields: Quantization of the Master Space

9.1. Mixed Spin P-fields. Following the recipe from previous sections, now we consider a field theory valued in W, namely the space of maps from curves to the master space W.

Such an object is called a mixed spin P-field (MSP for short). It consists of

$$\xi = (\Sigma^{\mathcal{C}}, \mathcal{C}, \mathcal{L}, \mathcal{N}, \varphi_1, \dots, \varphi_5, \rho, \nu = [\nu_1, \nu_2])$$

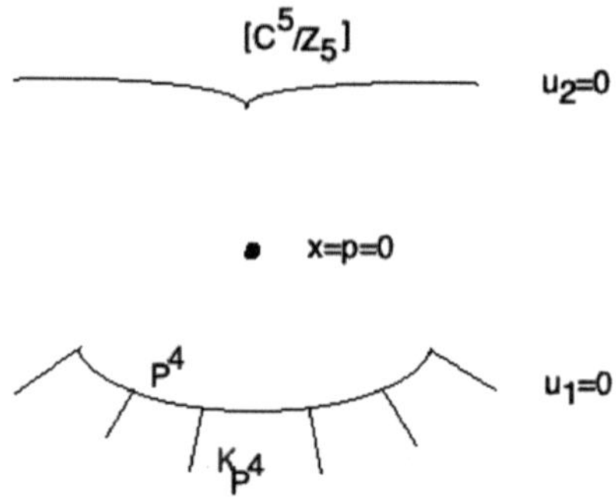

FIGURE 2. picture for the master space W, a six dimensional manifold.

where

(1) $(\Sigma^{\mathcal{C}}, \mathcal{C})$ is a pointed twisted curve,
(2) $\mathcal{L}$ and $\mathcal{N}$ are invertible sheaves on $\mathcal{C}$ ($\mathcal{L}$ is as before but $\mathcal{N}$ is new due to the extra factor $\mathbb{P}^1$ in the master space technique),
(3) $\varphi_i \in H^0(\mathcal{L})$ and $\rho \in H^0(\mathcal{L}^{\vee 5} \otimes \omega_{\mathcal{C}}^{\log})$ (as before),
(4) $\nu = [\nu_1, \nu_2]$ is a new field, where $\nu_1 \in H^0(\mathcal{L} \otimes \mathcal{N})$ and $\nu_2 \in H^0(\mathcal{N})$.

They satisfy the following conditions:

(1) (narrow condition) $\varphi_i|_{\Sigma^{\mathcal{C}}} = 0$,
(2) (combined GIT-like stability conditions)
   (a) $(\varphi_1, \dots, \varphi_5, \nu_1)$ is nowhere vanishing (coming from excluding $\{(x_i = 0 = u_1)\}$),
   (b) (ρ, ν_2) is nowhere vanishing (coming from excluding $\{(p = 0 = u_2)\}$),
   (c) (ν_1, ν_2) is nowhere vanishing (coming from $[u_1, u_2] \in \mathbb{P}^1$).

We say ξ is stable if $\mathrm{Aut}(\xi)$ is finite. For simplicity, we will use φ to represent $(\varphi_1, \dots, \varphi_5)$.

In order to understand why the moduli space of MSP fields geometrically contains the moduli space $\overline{M}_g(\mathbb{P}^4, d)^p$ of stable maps with P-fields and the moduli space $\overline{M}_g^{1/5,5p}$ of 5-spin curves with five P-fields, we examine the moduli space of MSP fields in details.

Let ξ be a MSP field.

(1) When $\nu_1 = 0$, since $(\varphi_1, \dots, \varphi_5, \nu_1)$ is nowhere zero, we must have $(\varphi_1, \dots, \varphi_5)$ is nowhere zero. Since (ν_1, ν_2) is nowhere zero, ν_2 must be nowhere zero. Since ν_2 is a section of $\mathcal{N}$, $\mathcal{N} \cong \mathcal{O}_{\mathcal{C}}$. There is no restriction on ρ. Thus $\xi \in \overline{M}_g(\mathbb{P}^4, d)^p$ and we get GW theory of the quintic threefold Q.
(2) When $\nu_2 = 0$, since (ρ, ν_2) is nowhere zero, ρ must be nowhere vanishing. Since ρ is a section of $\mathcal{L}^{\vee 5} \otimes \omega_{\mathcal{C}}^{\log}$, we must have $\mathcal{L}^5 \cong \omega_{\mathcal{C}}^{\log}$. Also ν_1 must be nowhere zero. Thus $\mathcal{L} \otimes \mathcal{N} \cong \mathcal{O}_{\mathcal{C}}$, i.e., $\mathcal{N} \cong \mathcal{L}^{\vee}$. $\varphi_1, \dots, \varphi_5$ can be arbitrary. Thus $\xi \in \overline{M}_{g,(\gamma_j)}^{1/5,5p}$ and we get the FJRW theory of $([\mathbb{C}^5/\mathbb{Z}_5], W_5)$.
(3) When $\rho = 0$ and $\varphi_i = 0$ for $1 \leq i \leq 5$, ν_1, ν_2 must be nowhere zero. Thus $\mathcal{N} \cong \mathcal{O}_{\mathcal{C}}$ and $\mathcal{L} \cong \mathcal{O}_{\mathcal{C}}$. Hence we get moduli of stable curves $\overline{M}_{g,n}$.

THEOREM 9.1 (H.L. Chang - J. Li - W.P. Li - C.C. Liu [**CLLL**]). *The moduli stack $W_{g,\gamma,\mathbf{d}}$ of stable MSP fields of genus g, monodromy $\gamma = (\gamma_1, \dots, \gamma_\ell)$ of $\mathcal{L}$*

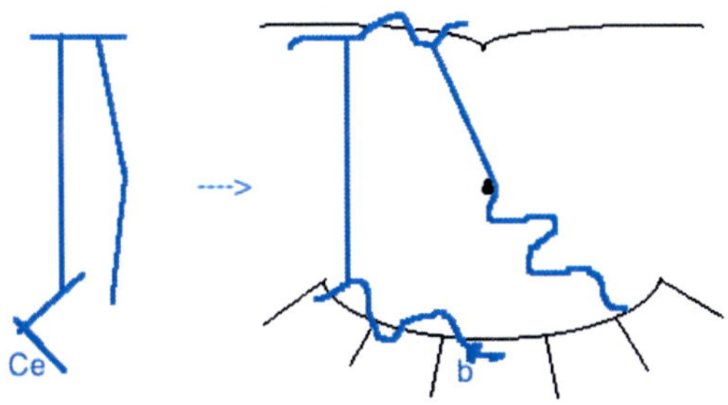

FIGURE 3. graphs for a field valued in master space W.

along $\Sigma^{\mathcal{C}}$ *and degree* $\mathbf{d} = (d_0, d_\infty)$ *of* $\mathcal{L} \otimes \mathcal{N}$ *and* $\mathcal{N}$ *respectively is a separated DM stack of locally finite type.*

The moduli stack $W_{g,\gamma,\mathbf{d}}$ admits a natural $\mathbb{C}^*$-action also called T-action: for $t \in \mathbb{C}^*$,

$$(\Sigma^{\mathcal{C}}, \mathcal{C}, \mathcal{L}, \mathcal{N}, \varphi, \rho, \nu_1, \nu_2)^t \colon = (\Sigma^{\mathcal{C}}, \mathcal{C}, \mathcal{L}, \mathcal{N}, \varphi, \rho, t\nu_1, \nu_2).$$

It is not proper since φ and ρ are sections of invertible sheaves. Thus we cannot do integration on this stack. However, there exists a cosection of its obstruction sheaf. Using the arguments similar to the GW case and LG case, we have the following theorem.

THEOREM 9.2 ([**CLLL**]). *The moduli stack* $W_{g,\gamma,\mathbf{d}}$ *has a* T*-equivariant perfect obstruction theory, a* T*-equivariant cosection* σ *of its obstruction sheaf, and thus carries a* T*-equivariant cosection localized virtual cycle*

$$[W_{g,\gamma,\mathbf{d}}]^{\mathrm{vir}}_{\mathrm{loc}} \in A^T_* W^-_{g,\gamma,\mathbf{d}}$$

where $W^-_{g,\gamma,\mathbf{d}}$ *is the degeneracy locus of* σ, *i.e.,*

$$W^-_{g,\gamma,\mathbf{d}} \colon = (\sigma = 0) = \{\xi \in W_{g,\gamma,\mathbf{d}} \mid \mathcal{C} = (\varphi = 0) \cup (\varphi_1^5 + \ldots + \varphi_5^5 = 0 = \rho)\}.$$

The cycle enumerates "maps to the master space W". Figure 3 is an example where the domain curve is represented as a union of one dimension lines (which is the standard notation in algebraic geometry).

In Figure 3, the component C_e is mapped to single point b, and is what we call a "ghost" (over which ρ can be nonvanishing) in Figure 1. Note that considering ξ as a map is just for easiness of understanding: indeed the map cannot be realized due to the presence of $\omega_{\mathcal{C}}$ in the definition of (ρ in) ξ.

9.2. Properness: Capture Ghost at Infinity. In order to integrate, we need properness of $W^-_{g,\gamma,\mathbf{d}}$. In fact, we have

THEOREM 9.3 ([**CLLL**]). *The degeneracy locus* $W^-_{g,\gamma,\mathbf{d}}$ *is a proper* T*-DM stack of finite type.*

The proof of the properness reveals an important phenomenon transforming fields of different phases in the MSP moduli. Under the transformation, the spin structure of line bundles arises naturally in the LG-phase as a limit of a family of P-fields in CY-phase. We call this phenomenon the "Landau-Ginzburg transition". As mentioned before, the contribution from ghosts is one of the difficulties to approach positive genus Gromov-Witten invariants. The LG transition phenomenon enables the FJRW theory to capture the ghost contribution in the GW theory, inside the MSP moduli space.

9.2.1. *LG-Transition: An Example.* For any positive integer d, we construct a simple example where $g = 1$, $\gamma = \emptyset$, and $\mathbf{d} = (d, 0)$ to illustrate the phenomenon of LG-transition and explain why FJRW theory comes into the picture naturally when we consider GW theory with a P-field. The argument below is also a part of the procedure to prove Theorem 9.3 (properness of the degeneracy locus).

1. A point in the degeneracy locus $W^-_{1,\emptyset,(d,0)}$.

Given a positive integer d, we define an MSP-field

$$\xi = (\mathcal{C}, \mathcal{L}, \mathcal{N}, \varphi, \rho, \nu = [\nu_1, \nu_2]) \tag{9.1}$$

over a point as follows. The curve $\mathcal{C}$ is a union of a smooth elliptic curve C_1 and a smooth rational curve C_0, intersecting at a node p. Under the isomorphism $C_0 \cong \mathbb{P}^1$ we have

$$\mathcal{L}|_{C_0} \cong \mathcal{O}_{\mathbb{P}^1}(d), \quad \mathcal{N}_{C_0} \cong \mathcal{O}_{\mathbb{P}^1}, \quad \varphi|_{C_0} = (x^d, -x^d, 0, 0, 0),$$
$$\rho|_{C_0} = 0, \quad \nu_1|_{C_0} = y^d, \quad \nu_2|_{C_0} = 1$$

where $[x, y]$ are homogeneous coordinates on $C_0 = \mathbb{P}^1$, and $p = [0, 1]$. In particular $\varphi(p) = 0$ and $\nu_1(p) = \nu_2(p) = 1$. On C_1, we have

$$\mathcal{L}|_{C_1} \cong \mathcal{O}_{C_1}, \quad \mathcal{N}|_{C_1} \cong \mathcal{O}_{C_1}, \quad \varphi|_{C_1} = (0, \dots, 0), \quad \nu_1|_{C_1} = 1, \quad \nu_2|_{C_1} = 1.$$

In particular, $\mathcal{N} \cong \mathcal{O}_{\mathcal{C}}$. Finally, we extend $\rho|_{C_0} = 0$ to a non-zero section $\rho \in H^0(\mathcal{L}^{\vee 5} \otimes \omega_{\mathcal{C}}^{\log})$ as follows: $\rho|_{C_1}$ is a non-zero section of $H^0\big((\mathcal{L}^{\vee 5} \otimes \omega_{\mathcal{C}})|_{C_1}\big) = H^0(\omega_{C_1}(p))$ vanishing at p only. The choice of $\rho|_{C_1}$ is unique up to multiplication by a nonzero constant. Then ξ represents a point in the degeneracy locus $W^-_{g=1,\gamma=\emptyset,\mathbf{d}=(d,0)} \subset W_{g=1,\gamma=\emptyset,\mathbf{d}=(d,0)}$.

2. A morphism from $\mathbb{C}^*$ *to* $W^-_{1,\emptyset,(d,0)}$.

Let $S_* = \mathcal{C} \times \mathbb{C}^*$ and let $\pi_1 : S_* \to \mathcal{C}$ be the projection to the first factor. We consider a family of MSP-fields over $\mathbb{C}^*$

$$\xi_* = (S_*, \mathcal{L}_* = \pi_1^* \mathcal{L}, \mathcal{N}_* = \pi_1^* \mathcal{N} = \mathcal{O}_{S_*}, \varphi_* = \pi_1^* \varphi,$$
$$\rho_* = t^{-1} \pi_1^* \rho, [\nu_{1*} = \pi_1^* \nu_1, \nu_{2*} = \pi_1^* \nu_2 = 1])$$

where t is the parameter of $\mathbb{C}^* = \mathrm{Spec}[t, t^{-1}]$. This family over $\mathbb{C}^*$ defines a morphism

$$\phi : \mathbb{C}^* \longrightarrow W^-_{1,\emptyset,(d,0)}. \tag{9.2}$$

By abuse of notation, let π_1 also denote the projection from $S_{*i} := C_i \times \mathbb{C}^*$ to the first factor, where $i = 0, 1$. The restriction of the family ξ_* to S_{0*} is a constant family over $\mathbb{C}^*$:

$$\xi_{0*} = (\Sigma_{0*} = p \times \mathbb{C}^*, S_{0*} = C_0 \times \mathbb{C}^*, \pi_1^* \mathcal{O}_{\mathbb{P}^1}(d), \mathcal{O}_{S_{0*}}, (x^d, -x^d, 0, 0, 0), 0, [y^d, 1])$$

which defines a *constant* map $\phi_0 : \mathbb{C}^* \longrightarrow W^-_{g=0,\gamma=(1),\mathbf{d}=(d,0)}$. The restriction of the family ξ_* to S_{1*} is

$$\xi_{1*} = (\Sigma_{1*} = p \times \mathbb{C}^*, S_{1*} = C_1 \times \mathbb{C}^*, \mathcal{O}_{S_{1*}}, \mathcal{O}_{S_{1*}}, (0, \dots, 0), t^{-1} \pi_1^*(\rho|_{C_1}), [1, 1])$$

which defines a morphism $\phi_1 : \mathbb{C}^* \longrightarrow W_{g=1,\gamma=(1),\mathbf{d}=(0,0)}$.

3. The limits $t \to 0$ and $t \to \infty$
We will see that the morphism (9.2) extends to a morphism

$$\bar{\phi} : \mathbb{P}[1,5] \longrightarrow W^{-}_{1,\emptyset,(d,0)}$$

where the embedding $\mathbb{C}^* \hookrightarrow \mathbb{P}[1,5]$ is given by $t \mapsto [1, t^{-1}]$. The image $\bar{\phi}([0,1])$ (resp. $\bar{\phi}([1,0])$) is the limit in $W^{-}_{1,\emptyset,(d,0)}$ when $t \to 0$ (resp. $t \to \infty$). It is easy to see that

$$\bar{\phi}([1,0]) = (\mathcal{C}, \mathcal{L}, \mathcal{N}, \varphi, 0, \nu = [\nu_1, \nu_2])$$

where $\mathcal{C}, \mathcal{L}, \mathcal{N}, \varphi, \nu_1, \nu_2$ are defined as in Step *1*.

The extension of the constant map $\phi_0 : \mathbb{C}^* \longrightarrow W^{-}_{0,\gamma,\mathbf{d}=(d,0)}$ is the constant map $\bar{\phi}_0 : \mathbb{P}[1,5] \to W^{-}_{0,\gamma,\mathbf{d}=(d,0)}$. To find the limit $\bar{\phi}([0,1])$, it suffices to find $\bar{\phi}_1([0,1])$ where $\bar{\phi}_1$ is the extension of ϕ_1 to $\mathbb{P}[1,5]$.

Let $S_1 = C_1 \times \mathbb{C}^*$ and $\rho_1 = \rho|_{C_1}$. If we naively take $\mathcal{L} = \mathcal{O}_{S_1}$, $\mathcal{N} = \mathcal{O}_{S_1}$, $\varphi_i = 0$, , $\nu_1 = 1$ and $\nu_2 = 1$, the section $t^{-1}\pi_2^*\rho_1$ cannot be extended to a regular section of $\mathcal{L}^{\vee 5} \otimes \omega^{\log}_{S_1/\mathbb{C}}$. Here by abuse of notations, π_i is the projection from S_1 to the i-th factor. One way to solve this problem is to use the equivalence of ξ_{1*} with the following ξ'_{1*}:

$$(S_{1*}, \Sigma_{1*}, \mathcal{L}_* = \mathcal{O}_{S_{1*}}, \mathcal{N}_* = \mathcal{O}_{S_{1*}}, \varphi_{1*} = 0, \ldots, \varphi_{5*} = 0,$$
$$\rho_{S_1*} = \pi_1^*\rho_1, \nu_{1*} = 1, \nu_{2*} = t^{\frac{1}{5}}).$$

Then we can have the extension

$$(S_1, \Sigma_1, \mathcal{L} = \mathcal{O}_{S_1}, \mathcal{N} = \mathcal{O}_{S_1}, \varphi_1 = 0, \ldots, \varphi_5 = 0, \rho_{S_1} = \pi_1^*\rho_1, \nu_1 = 1, \nu_2 = t^{\frac{1}{5}})$$

where $\Sigma_1 = p \times \mathbb{C}$.

The term $t^{\frac{1}{5}}$ may look troublesome. Let's just treat this as indicating that the zero locus of ν_2 is $\frac{1}{5}C_1 \times 0$. The issue of fractional divisor will be resolved once we work in the world of twisted curves.

Let's assume that we can work with fractional divisors. The zero divisor of ρ_{S_1} is Σ_1 and that of ν_2 is $\frac{1}{5}C_1 \times 0$. Since these two divisors intersect, by MSP requirement that ρ_{S_1} and ν_2 cannot be zero simultaneously at any point , we don't get an MSP extension. Thus we need to blow up the intersection of these two divisors to separate them.

Let $\tau\colon S_1' \to S_1$ be the blow up of S_1 at $P = \Sigma_1 \cap (C_1 \times 0) = p \times 0$, E be the exceptional divisor, $\Sigma_1' \subset S'$ be the strict transform of Σ_1, and C_1' be the strict transform of $C_1 \times 0$. The zero divisor of $\tau^*\nu_2$ is $\frac{1}{5}C_1' + \frac{1}{5}E$. The zero divisor of $\tau^*\rho_{S_1}$ is $\Sigma_1' + E$. Now we need to modify $\mathcal{L}$ and $\mathcal{N}$ by replacing $\mathcal{L}$ by $\mathcal{L}' = \mathcal{O}_{S_1'}(E/5)$ and $\mathcal{N}$ by $\mathcal{N}' = \mathcal{O}_{S_1'}(-E/5)$. Here we pretend that $\mathcal{L}'$ and $\mathcal{N}'$ exist. Indeed, they do not exist in the ordinary sense, but their existence again will be resolved once we work with orbifolds. Let ρ' be the section in $H^0(\mathcal{L}'^{\vee 5} \otimes \omega^{\log}_{S_1'/\mathbb{C}}) = H^0(\omega^{\log}_{S_1'/\mathbb{C}}(-E))$ whose image is $\tau^*\rho_{S_1}$ under the natural map $H^0(\omega^{\log}_{S_1'/\mathbb{C}}(-E)) \to H^0(\omega^{\log}_{S_1'/\mathbb{C}})$. The zero divisor of ρ' is Σ_1'. Let ν_2' be the section of $\mathcal{O}_{S_1'}(-E/5)$ whose image under the natural map $\mathcal{O}_{S_1'}(-E/5) \to \mathcal{O}_{S_1'}$ is $\tau^*\nu_2$. Then the zero divisor of ν_2' is $\frac{1}{5}C_1'$. Since C_1' and Σ_1' don't intersect, we can get an MSP extension by taking $\mathcal{L}' = \mathcal{O}_{S_1'}(E/5)$, $\mathcal{N}' \cong \mathcal{L}'^{\vee}$, ν_1' a nonzero constant section $\mathcal{L}' \otimes \mathcal{N}'$, $\varphi_1' = \ldots = \varphi_5' = 0$, $\rho' \in H^0(\mathcal{L}'^{\vee 5} \otimes \omega^{\log}_{S_1'/\mathbb{C}})$ with its zero divisor being Σ_1' which is the marking of $S_1'/\mathbb{C}$, and $\nu_2' \in H^0(\mathcal{N}')$ whose zero divisor is $\frac{1}{5}C_1'$. As we mentioned earlier, to make this

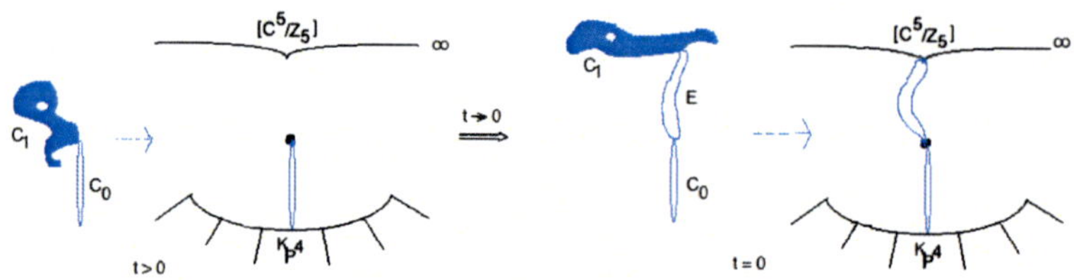

FIGURE 4. via scaling P-field $\rho \mapsto \rho/t$ and making $t \to 0$, the ghost is capture at infinity and become FJRW instantons.

construction rigorous, we have to do base change and introduce stacky structures at nodes of $C_1' \cup E$ (see [**AGV, CLLL**]).

Now we see that the central fiber of S_1' over $\mathbb{C}$ at $0 \in \mathbb{C}$ is set-theoretically $C_1 \cup E$, a union of the elliptic curve C_1 with the smooth rational curve E. The section ν_2' vanishes on $\mathcal{C}_\infty = C_1' \cong C_1$ and ρ' is nowhere vanishing on $\mathcal{C}_\infty$. Hence $\mathcal{C}_\infty$ is a 5-spin twisted curve, and E is a rational smooth twisted curve with a marking where ρ' vanishes.

Then we can glue the MSP field on S_1' with the MSP field on $C_0 \times \mathbb{C}$ by identifying the marking Σ_1' with the marking $p \times \mathbb{C} \in C_0 \times \mathbb{C}$ after possibly a base change. Thus the central fiber of the extension is a union of a smooth rational curve C_0, an elliptic curve C_1 which is a 5-spin curve, and a rational twisted curve E intersecting with C_1 at the stacky point and with C_0 at another point where the nonzero ρ-section vanishes.

We can also deform the MSP field (9.1) to a MSP field in GW sector as follows.

Consider $S = \mathcal{C} \times \mathbb{C}$, Let π_1 be the projection of S to its first factor. Take, for $t \in \mathbb{C}$, we have a family of MSP over $\mathbb{C}$,

$$(S, \quad \pi_1^*\mathcal{L}, \quad \pi_1^*\mathcal{N}, \quad \varphi_S, \quad \pi_1^*\rho, \quad [t\pi_1^*\nu_1, \pi_1^*\nu_2]).$$

Here φ_S is defined as follows.

$$\varphi_S|_{C_0\times\mathbb{C}} = (x^d, -x^d, (1-t)y^d, -(1-t)y^d, 0), \quad \varphi_S|_{C_1\times\mathbb{C}} = (0,0,1-t,t-1,0).$$

When $t=1$, we get the MSP field ξ in (9.1). When $t=0$, we get an MSP lying in GW sector since $t\pi_1^*\nu_1|_{t=0} = 0$.

□

10. Vanishing and Polynomial Relations

How to extract information of GW and/or FJRW invariants from the cycle $[W_{g,\gamma,\mathbf{d}}]^{\mathrm{vir}}_{\mathrm{loc}}$? In this section, we consider a less general case $\gamma = \emptyset$ (i.e. no markings) to illustrate the key ideas. By virtual dimension counting, we have

$$[W_{g,\mathbf{d}}]^{\mathrm{vir}}_{\mathrm{loc}} \in H^{\mathbf{C}^*}_{2(d_0+d_\infty+1-g)}(W^-_{g,\mathbf{d}}, \mathbb{Q}).$$

When $d_0 + d_\infty + 1 - g > 0$, letting $u = c_1(\mathbf{1}|_{wt=1})$, i.e. u is the parameter for $H^*_{\mathbf{C}^*}(pt)$, we have

$$[u^{d_0+d_\infty+1-g} \cdot [W_{g,\mathbf{d}}]^{\mathrm{vir}}_{\mathrm{loc}}]_0 = 0.$$

Here $[\cdot]_0$ is the degree zero term in the variable u.

Let Γ be a graph associated to fixed points of the T-action of $W_{g,\mathbf{d}}$ and F_Γ be a connected component of $W^T_{g,\mathbf{d}}$ of the graph type Γ. Applying the cosection localized version of the virtual localization formula of Graber-Pandaripande [**GP**] proved by Chang-Kiem-J.Li in [**CKL**], we obtain

$$\sum_{\Gamma}\left[u^{d_0+d_\infty+1-g}\frac{[F_\Gamma]^{vir}_{loc}}{e(N_{F_\Gamma})}\right]_0=0. \tag{10.1}$$

To deal with $[F_\Gamma]^{\mathrm{vir}}_{\mathrm{loc}}$, we need a decomposition result to be explained below.

Let $\xi=(\mathcal{C},\mathcal{L},\mathcal{N},\varphi,\rho,\nu_1,\nu_2)\in(W_{g,\mathbf{d}})^T$ be an MSP field fixed by the T-action. We set

(1) $\mathcal{C}_0$ to be the part of $\mathcal{C}$ where $\nu_1=0$;
(2) $\mathcal{C}_1$ to be the part of $\mathcal{C}$ where $\varphi=0=\rho$ and hence $\nu_1=1=\nu_2$, i.e., ν_1 and ν_2 are nowhere zero;
(3) $\mathcal{C}_\infty$ to be the part of $\mathcal{C}$ where $\nu_2=0$.

Thus

- $\xi|_{(\text{connected component of } \mathcal{C}_0)}$ is in $\overline{M}_{g',n'}(\mathbb{P}^4,d')^p$ which gives GW invariants of Q. Here marked points appear coming from some nodes on $\mathcal{C}_0$.
- $\xi|_{(\text{connected component of } \mathcal{C}_1)}$ is in $\overline{M}_{g',n'}$ which gives Hodge integrals.
- $\xi|_{(\text{connected component of } \mathcal{C}_\infty)}$ is in $\overline{M}^{\frac{1}{5},5p}_{g',\gamma'}$ which gives FJRW invariants of $(\mathbb{C}^5/\mathbb{Z}_5,W_5)$ where γ' appears because of some stacky nodes on $\mathcal{C}_\infty$.

We have the following decomposition result:

$$[F_\Gamma]^{vir}_{loc}=c\prod[\text{moduli of }\xi|_{\mathcal{C}_0}]^{\mathrm{vir}}_{\mathrm{loc}}\cdot[\text{moduli of }\xi|_{\mathcal{C}_1}]^{\mathrm{vir}}_{\mathrm{loc}}\cdot[\text{moduli of }\xi|_{\mathcal{C}_\infty}]^{\mathrm{vir}}_{\mathrm{loc}}$$

where c is a constant. The first factor gives GW invariants of stable maps to $\mathbb{P}^4$ with P-fields, i.e. $N_{g',d'}$. The second factor gives Hodge integrals on $\overline{M}_{g',n'}$. The third factor gives FJRW invariants of insertions $-\frac{2}{5}$ (after using a vanishing). After $e(N_{F_\Gamma})$'s are calculated, using the polynomial relations (10.1), we obtain the following results about GW invariants of the quintic.

THEOREM 10.1 ([**CLLL2**]). *Letting $d_\infty=0$, the relations (10.1) provide an effective algorithm to evaluate the GW invariants $N_{g,d}$ provided the following are known*

(1) *$N_{g',d'}$ for (g',d') such that $g'<g$, and $d'\le d$;*
(2) *$N_{g,d'}$ for $d'<g$;*
(3) *$\Theta_{g',k}$ for $g'\le g-1$ and $k\le 2g-4$;*
(4) *$\Theta_{g,k}$ for $k\le 2g-2$.*

Recall that $\Theta_{g,k}$ is the genus g FJRW invariants of insertions $-\frac{2}{5}$ and $\Theta_{g,k}$ may be non-zero only when $k+2-2g\equiv 0(5)$. We can see that when $g=2$ only $\Theta_{2,2}$ is needed, and when $g=3$ only $\Theta_{3,4}$ is needed.

REMARK 10.2. *As we know, on using mathematical induction, upon more numerical datum the induction is, the less effective the computation will be. We can see from Theorem 10.1 that MSP induction for GW invariants is carried out on two numbers, genus and the degree only. Thus this provides a rather effective way to facilitate the induction procedure.*

We can also use the vanishings (10.1) for $d=(0,d_\infty)$, to determine quintic's FJRW invariants up to finite many initial data.

THEOREM 10.3 ([**CLLL2**]). *For a fixed positive genus g, the finite set*

$$\{\Theta_{g,k}\}_{k<7g-2}$$

determine all genus g FJRW invariants $\{\Theta_{g,k}\}_{k=0}^{\infty}$.

These relations are effective in calculating FJRW invariants. For example, for the case of genus 2, $\{\Theta_{2,k}\}_k$ can be inductively derived from only two unknowns $\Theta_{2,2}$ and $\Theta_{2,7}$.

We end this section by some speculations.

Let us look at Theorem 10.1 from a different aspect. Inductively we may suppose all GW/FJRW invariants for genus less than g are known. Then for genus g, Theorem 10.1 reduces the problem of determining the infinitely many GW invariants $\{N_{g,d}\}_{d=1}^{\infty}$ to two finite sets of initial datum

$$\{N_{g,1},\cdots,N_{g,g-1}\} \quad \text{and} \quad \{\Theta_{g,k}\}_{k\leq 2g-2}.$$

We formulate the following speculation:

By suitable choice of positive d_0 and d_∞, the relations (10.1) provide an effective algorithm to determine the first set of initial data $\{N_{g,1},\cdots,N_{g,g-1}\}$.

If this is true, then one is left to determine the second set of initial data $\{\Theta_{g,k}\}_{k\leq 2g-2}$. We propose another conjecture about fully determining all FJRW invariants for the quintic,

CONJECTURE 10.4. *The equations (10.1) using $d_0 = 0$ and nonempty γ's (i.e. with markings) give relations that effectively evaluate all $\Theta_{g,k}$.*

11. Comparison with Physical Theories

11.1. Comparison with Witten's GLSM. In [**GLSM**], Witten introduced a family of theories using path integrals, called the Gauged Linear Sigma Model (GLSM), linking a non-linear sigma models on a Calabi-Yau hypersurface to a sigma model targeting in a Landau-Ginzburg space. The GLSM is parameterized by a complexified Kähler parameter

$$t := r - i\theta$$

where r is "Fayet-Iliopoulos parameter" and θ is called the theta angle[5]. Witten [**GLSM**, Sect 3.1] argued that the GLSM specializes to GW path integral when $r \to +\infty$, and specializes to LG model path integral when $r \to -\infty$. This is known as the Calabi-Yau/Landau-Ginzburg correspondence.

The Mixed-Spin-P fields (MSP fields) introduced in [**CLLL**] is a field theory designed to capture "phase space transition" in one cage[6]. An MSP field can be viewed as an interpolation between fields valued in $K_{\mathbb{P}^4}$ and fields valued in $[\mathbb{C}^5/\mathbb{Z}_5]$, and the interpolation is governed by the "ν field". Over the part of worldsheet (curve) where $\nu = 0$, the MSP field is a pure field taking values in $K_{\mathbb{P}^4}$, and, over $\nu = \infty$, the MSP field is a pure field taking values in $[\mathbb{C}^5/\mathbb{Z}_5]$. In a nutshell, by promoting the phase parameter $r - i\theta$ into a field ν on worldsheet (curve), we transform Witten's family of theories parametrized by $r - i\theta$ into a single new field theory. Also, an advantage of MSP moduli is it works for higher loop (in physics

[5] we use notations in [**Wi2**, Sect 15.2.2];

[6] in the MSP cage (proper) integral does not diverge because the cage is proper and separated by [**CLLL**]

terms) or for higher genus (in math terms), while GLSM in physics literature does not[7].

We recall the question raised by Witten [**GLSM**, Page 28]: "*Are Calabi-Yau and Landau-Ginzburg separated by a true phase transition, at or near r = 0? There is no reason that the answer to this question has to be universal, that is, independent of the path one follows in interpolating from Calabi-Yau to Landau-Ginzburg in a multiparameter space of not necessarily conformally invariant theories. Along a suitable path, there may well be a sharply defined phase transition, while along another path there might not be one. This seems quite plausible.*" Though our ν field does not have a definite value of which we can vary "the theories", ν allows us to introduce a $T = \mathbb{C}^*$ action, and by localizing to T fixed locus, we obtain (possibly as referred to by Witten) a precise phase transition between the quintic Calabi-Yau threefold and the LG of the Fermat quintic. Furthermore, such phase transitions are multi-fold: for each $\mathbf{d} = (d_0, d_\infty)$ that provides a vanishing, we get an interpolation. And when we vary $\mathbf{d}$, we obtain a class of such interpolations. Thus in physics terms, we may say that each $(d_0, d_\infty) \in \mathbb{Z}_{\geq 0} \times \mathbb{Z}$, being a mode describing how worldsheet is wrapped to Kähler moduli spaces, provides a path linking CY point to LG point in the phase spaces.

11.2. Compare with B-model. For the quintic Calabi-Yau threefold, the modularity of the generating function F_g^A is suggested by physicists, but is a mystery in mathematics. In physics literature (cf. [**BCOV, HKQ**]), by assuming all genus mirror symmetry and using "holomorphic anomaly equation", the genus g GW generating function $F_g = \sum N_{g,d} q^d$ (where $g > 0$) can be determined up to $3g - 2$ unknowns. The degree 0 GW invariant $N_{g,0}$ is known, so we are left with $3g-3$ unknowns. The boundary conditions at the orbifold point (which corresponds to Landau-Ginzburg theory of the Fermat quintic polynomial in five variables) impose $\lceil \frac{3}{5}(g-1) \rceil$ constraints on the $3g-3$ unknowns, whereas the "gap condition" at the conifold point imposes $2g-2$ constraints on the $3g-3$ unknowns. In summary, the holomorphic anomaly equation and the boundary conditions determine F_g up to $\lfloor \frac{2}{5}(g-1) \rfloor$ unknowns.

Coincidentally, granting Conjecture A, the number of initial data needed to determine F_g are the FJRW invariants $\Theta_{g,k \leq 2g-2}$, subject to $2g - 2 \equiv k(5)$. Thus $\lfloor \frac{2(g-1)}{5} \rfloor + 1$ many FJRW invariants are needed to determine $N_{g,d}$ via MSP moduli, provided all lower genus invariants $\{F_{g'} : g' < g\}$ are known. We hope there is more geometric explanation then viewing it just as a coincidence.

Acknowledgments

The first author was partially supported by Hong Kong GRF grant 600711 and 6301515. The second author was partially supported by NSF grant DMS-1104553, DMS-1159156, and DMS-1564500. The third author was partially supported by by Hong Kong GRF grant 602512 and 6301515. The fourth author was partially supported by NSF grant DMS-1206667, DMS-1159416, and DMS-1564497. This article is an expansion of J. Li's plenary talk at String Math 2015 in Sanya.

[7] GLSM only treat genus zero worldsheet

References

[ACV] Dan Abramovich, Alessio Corti, and Angelo Vistoli, *Twisted bundles and admissible covers*, Comm. Algebra **31** (2003), no. 8, 3547–3618, DOI 10.1081/AGB-120022434. Special issue in honor of Steven L. Kleiman. MR2007376

[AF] Dan Abramovich and Barbara Fantechi, *Orbifold techniques in degeneration formulas*, Ann. Sc. Norm. Super. Pisa Cl. Sci. (5) **16** (2016), no. 2, 519–579. MR3559610

[AGV] Dan Abramovich, Tom Graber, and Angelo Vistoli, *Gromov-Witten theory of Deligne-Mumford stacks*, Amer. J. Math. **130** (2008), no. 5, 1337–1398, DOI 10.1353/ajm.0.0017. MR2450211

[AJ] Dan Abramovich and Tyler J. Jarvis, *Moduli of twisted spin curves*, Proc. Amer. Math. Soc. **131** (2003), no. 3, 685–699, DOI 10.1090/S0002-9939-02-06562-0. MR1937405

[BF] K. Behrend and B. Fantechi, *The intrinsic normal cone*, Invent. Math. **128** (1997), no. 1, 45–88, DOI 10.1007/s002220050136. MR1437495

[BCOV] M. Bershadsky, S. Cecotti, H. Ooguri, and C. Vafa, *Kodaira-Spencer theory of gravity and exact results for quantum string amplitudes*, Comm. Math. Phys. **165** (1994), no. 2, 311–427. MR1301851

[COGP] Philip Candelas, Xenia C. de la Ossa, Paul S. Green, and Linda Parkes, *A pair of Calabi-Yau manifolds as an exactly soluble superconformal theory*, Nuclear Phys. B **359** (1991), no. 1, 21–74, DOI 10.1016/0550-3213(91)90292-6. MR1115626

[Ce] S. Cecotti, *$N = 2$ Landau-Ginzburg vs. Calabi-Yau σ-models: nonperturbative aspects*, Internat. J. Modern Phys. A **6** (1991), no. 10, 1749–1813, DOI 10.1142/S0217751X91000939. MR1098365

[CKL] Huai-Liang Chang, Young-Hoon Kiem, and Jun Li, *Torus localization and wall crossing for cosection localized virtual cycles*, Adv. Math. **308** (2017), 964–986, DOI 10.1016/j.aim.2016.12.019. MR3600080

[CL1] Huai-Liang Chang and Jun Li, *Gromov-Witten invariants of stable maps with fields*, Int. Math. Res. Not. IMRN **18** (2012), 4163–4217, DOI 10.1093/imrn/rnr186. MR2975379

[CL2] H.-L. Chang and J. Li, "A vanishing associated with irregular MSP fields," arXiv:1708.02902.

[CLL] Huai-Liang Chang, Jun Li, and Wei-Ping Li, *Witten's top Chern class via cosection localization*, Invent. Math. **200** (2015), no. 3, 1015–1063, DOI 10.1007/s00222-014-0549-5. MR3348143

[CLLL] H.-L. Chang, J. Li, W.-P. Li, and C.-C. Melissa Liu, "Mixed-Spin-P fields of Fermat quintic polynomials," (2015) math.AG. arXiv:1505.07532

[CLLL2] H.-L. Chang, J. Li, W.-P. Li, C.-C. Melissa Liu, 'An effective theory of GW and FJRW invariants of quintics Calabi-Yau manifolds," (2016) arXiv:1603.06184.

[twFJRW] H.-L. Chang, S. Guo, and W.-P. Li, "Dual twisted FJRW invariants of quintic singularity," in preparation.

[ChK] J-W. Choi and Y-H. Kiem, "Landau-Ginzburg/Calabi-Yau correspondence via quasi-maps, I," (2011) arXiv:1103.0833.

[Chi] Alessandro Chiodo, *Towards an enumerative geometry of the moduli space of twisted curves and rth roots*, Compos. Math. **144** (2008), no. 6, 1461–1496, DOI 10.1112/S0010437X08003709. MR2474317

[CR] Alessandro Chiodo and Yongbin Ruan, *Landau-Ginzburg/Calabi-Yau correspondence for quintic three-folds via symplectic transformations*, Invent. Math. **182** (2010), no. 1, 117–165, DOI 10.1007/s00222-010-0260-0. MR2672282

[FK1] Ionuţ Ciocan-Fontanine and Bumsig Kim, *Moduli stacks of stable toric quasimaps*, Adv. Math. **225** (2010), no. 6, 3022–3051, DOI 10.1016/j.aim.2010.05.023. MR2729000

[FK2] Ionuţ Ciocan-Fontanine and Bumsig Kim, *Wall-crossing in genus zero quasimap theory and mirror maps*, Algebr. Geom. **1** (2014), no. 4, 400–448, DOI 10.14231/AG-2014-019. MR3272909

[Cad] Charles Cadman, *Using stacks to impose tangency conditions on curves*, Amer. J. Math. **129** (2007), no. 2, 405–427, DOI 10.1353/ajm.2007.0007. MR2306040

[Ch] Alessandro Chiodo, *The Witten top Chern class via K-theory*, J. Algebraic Geom. **15** (2006), no. 4, 681–707, DOI 10.1090/S1056-3911-06-00444-9. MR2237266

[CZ] A. Chiodo and D. Zvonkine, *Twisted r-spin potential and Givental's quantization*, Adv. Theor. Math. Phys. **13** (2009), no. 5, 1335–1369. MR2672465

[FJR1] Huijun Fan, Tyler Jarvis, and Yongbin Ruan, *The Witten equation, mirror symmetry, and quantum singularity theory*, Ann. of Math. (2) **178** (2013), no. 1, 1–106, DOI 10.4007/annals.2013.178.1.1. MR3043578

[FJR2] Huijun Fan, Tyler Jarvis, and Yongbin Ruan, *The Witten equation, mirror symmetry, and quantum singularity theory*, Ann. of Math. (2) **178** (2013), no. 1, 1–106, DOI 10.4007/annals.2013.178.1.1. MR3043578

[FJR3] H. Fan, T. J. Jarvis and Y. Ruan, "A Mathematical Theory of the Gauged Linear Sigma Model," (2015) math.AG. arXiv:1506.02109

[FSZ] Carel Faber, Sergey Shadrin, and Dimitri Zvonkine, *Tautological relations and the r-spin Witten conjecture* (English, with English and French summaries), Ann. Sci. Éc. Norm. Supér. (4) **43** (2010), no. 4, 621–658, DOI 10.24033/asens.2130. MR2722511

[Gath] Andreas Gathmann, *Absolute and relative Gromov-Witten invariants of very ample hypersurfaces*, Duke Math. J. **115** (2002), no. 2, 171–203, DOI 10.1215/S0012-7094-02-11521-X. MR1944571

[Gi] Alexander B. Givental, *Equivariant Gromov-Witten invariants*, Internat. Math. Res. Notices **13** (1996), 613–663, DOI 10.1155/S1073792896000414. MR1408320

[GP] T. Graber and R. Pandharipande, *Localization of virtual classes*, Invent. Math. **135** (1999), no. 2, 487–518, DOI 10.1007/s002220050293. MR1666787

[GS] Josh Guffin and Eric Sharpe, *A-twisted Landau-Ginzburg models*, J. Geom. Phys. **59** (2009), no. 12, 1547–1580, DOI 10.1016/j.geomphys.2009.07.014. MR2583793

[HKQ] M.-x. Huang, A. Klemm, and S. Quackenbush, *Topological string theory on compact Calabi-Yau: modularity and boundary conditions*, Homological mirror symmetry, Lecture Notes in Phys., vol. 757, Springer, Berlin, 2009, pp. 45–102. MR2596635

[Huy] D. Huybrechts, *Fourier-Mukai transforms in algebraic geometry*, Oxford Mathematical Monographs, The Clarendon Press, Oxford University Press, Oxford, 2006. MR2244106

[JK] Tyler J. Jarvis and Takashi Kimura, *Orbifold quantum cohomology of the classifying space of a finite group*, Orbifolds in mathematics and physics (Madison, WI, 2001), Contemp. Math., vol. 310, Amer. Math. Soc., Providence, RI, 2002, pp. 123–134, DOI 10.1090/conm/310/05401. MR1950944

[KKP] Bumsig Kim, Andrew Kresch, and Tony Pantev, *Functoriality in intersection theory and a conjecture of Cox, Katz, and Lee*, J. Pure Appl. Algebra **179** (2003), no. 1-2, 127–136, DOI 10.1016/S0022-4049(02)00293-1. MR1958379

[KKV] Sheldon Katz, Albrecht Klemm, and Cumrun Vafa, *M-theory, topological strings and spinning black holes*, Adv. Theor. Math. Phys. **3** (1999), no. 5, 1445–1537, DOI 10.4310/ATMP.1999.v3.n5.a6. MR1796683

[KL] Young-Hoon Kiem and Jun Li, *Localizing virtual cycles by cosections*, J. Amer. Math. Soc. **26** (2013), no. 4, 1025–1050, DOI 10.1090/S0894-0347-2013-00768-7. MR3073883

[Kon1] Maxim Kontsevich, *Intersection theory on the moduli space of curves and the matrix Airy function*, Comm. Math. Phys. **147** (1992), no. 1, 1–23. MR1171758

[Ko] Maxim Kontsevich, *Enumeration of rational curves via torus actions*, The moduli space of curves (Texel Island, 1994), Progr. Math., vol. 129, Birkhäuser Boston, Boston, MA, 1995, pp. 335–368, DOI 10.1007/978-1-4612-4264-2_12. MR1363062

[Kr2] Andrew Kresch, *Cycle groups for Artin stacks*, Invent. Math. **138** (1999), no. 3, 495–536, DOI 10.1007/s002220050351. MR1719823

[LM] Gérard Laumon and Laurent Moret-Bailly, *Champs algébriques* (French), Ergebnisse der Mathematik und ihrer Grenzgebiete. 3. Folge. A Series of Modern Surveys in Mathematics [Results in Mathematics and Related Areas. 3rd Series. A Series of Modern Surveys in Mathematics], vol. 39, Springer-Verlag, Berlin, 2000. MR1771927

[LRZ] Si-Qi Liu, Yongbin Ruan, and Youjin Zhang, *BCFG Drinfeld-Sokolov hierarchies and FJRW-theory*, Invent. Math. **201** (2015), no. 2, 711–772, DOI 10.1007/s00222-014-0559-3. MR3370624

[KLi] Keke Li, *Recursion relations in topological gravity with minimal matter*, Nuclear Phys. B **354** (1991), no. 2-3, 725–739, DOI 10.1016/0550-3213(91)90374-7. MR1103071

[LT] Jun Li and Gang Tian, *Virtual moduli cycles and Gromov-Witten invariants of algebraic varieties*, J. Amer. Math. Soc. **11** (1998), no. 1, 119–174, DOI 10.1090/S0894-0347-98-00250-1. MR1467172

[LZ] Jun Li and Aleksey Zinger, *On the genus-one Gromov-Witten invariants of complete intersections*, J. Differential Geom. **82** (2009), no. 3, 641–690. MR2534991

[LLY] Bong H. Lian, Kefeng Liu, and Shing-Tung Yau, *Mirror principle. I*, Asian J. Math. **1** (1997), no. 4, 729–763, DOI 10.4310/AJM.1997.v1.n4.a5. MR1621573

[MOP] Alina Marian, Dragos Oprea, and Rahul Pandharipande, *The moduli space of stable quotients*, Geom. Topol. **15** (2011), no. 3, 1651–1706, DOI 10.2140/gt.2011.15.1651. MR2851074

[MP] D. Maulik and R. Pandharipande, *A topological view of Gromov-Witten theory*, Topology **45** (2006), no. 5, 887–918, DOI 10.1016/j.top.2006.06.002. MR2248516

[Mo] Takuro Mochizuki, *The virtual class of the moduli stack of stable r-spin curves*, Comm. Math. Phys. **264** (2006), no. 1, 1–40, DOI 10.1007/s00220-006-1538-3. MR2211733

[PV] Alexander Polishchuk and Arkady Vaintrob, *Algebraic construction of Witten's top Chern class*, Advances in algebraic geometry motivated by physics (Lowell, MA, 2000), Contemp. Math., vol. 276, Amer. Math. Soc., Providence, RI, 2001, pp. 229–249, DOI 10.1090/conm/276/04523. MR1837120

[RR] D. Ross and Y. Ruan, "Wall-crossing in genus zero Landau-Ginzburg theory," (2014) arXiv:1402.6688

[Ito] Kei Ito, *Topological phase of $N = 2$ superconformal field theory and topological Landau-Ginzburg field theory*, Phys. Lett. B **250** (1990), no. 1-2, 91–95, DOI 10.1016/0370-2693(90)91159-9. MR1081501

[Vafa] Cumrun Vafa, *Topological Landau-Ginzburg models*, Modern Phys. Lett. A **6** (1991), no. 4, 337–346, DOI 10.1142/S0217732391000324. MR1093562

[Wi] Edward Witten, *Two-dimensional gravity and intersection theory on moduli space*, Surveys in differential geometry (Cambridge, MA, 1990), Lehigh Univ., Bethlehem, PA, 1991, pp. 243–310. MR1144529

[Wi1] Edward Witten, *The N matrix model and gauged WZW models*, Nuclear Phys. B **371** (1992), no. 1-2, 191–245, DOI 10.1016/0550-3213(92)90235-4. MR1156705

[Wi2] Edward Witten, *Mirror manifolds and topological field theory*, Essays on mirror manifolds, Int. Press, Hong Kong, 1992, pp. 120–158. MR1191422

[GLSM] Edward Witten, *Phases of $N = 2$ theories in two dimensions*, Nuclear Phys. B **403** (1993), no. 1-2, 159–222, DOI 10.1016/0550-3213(93)90033-L. MR1232617

[Wi3] Edward Witten, *Algebraic geometry associated with matrix models of two-dimensional gravity*, Topological methods in modern mathematics (Stony Brook, NY, 1991), Publish or Perish, Houston, TX, 1993, pp. 235–269. MR1215968

[YY] Satoshi Yamaguchi and Shing-Tung Yau, *Topological string partition functions as polynomials*, J. High Energy Phys. **7** (2004), 047, 20, DOI 10.1088/1126-6708/2004/07/047. MR2095047

[Zi] Aleksey Zinger, *Standard versus reduced genus-one Gromov-Witten invariants*, Geom. Topol. **12** (2008), no. 2, 1203–1241, DOI 10.2140/gt.2008.12.1203. MR2403808

[Zi2] Aleksey Zinger, *The reduced genus 1 Gromov-Witten invariants of Calabi-Yau hypersurfaces*, J. Amer. Math. Soc. **22** (2009), no. 3, 691–737, DOI 10.1090/S0894-0347-08-00625-5. MR2505298

Department of Mathematics, Hong Kong University of Science and Technology, Hong Kong
E-mail address: mahlchang@ust.hk

Shanghai Center for Mathematical Sciences, Fudan University, China — and — Department of Mathematics, Stanford University
E-mail address: jli@math.stanford.edu

Department of Mathematics, Hong Kong University of Science and Technology, Hong Kong
E-mail address: mawpli@ust.hk

Department of Mathematics, Columbia University, New York, NY
E-mail address: ccliu@math.columbia.edu

Homological mirror functors via Maurer-Cartan formalism

Cheol-Hyun Cho

ABSTRACT. This is a write-up of the lecture at String Math 2015. We give a survey on recent joint works with Hansol Hong and Siu-Cheong Lau where we develop a Lagrangian Floer theory formalism to explain some of Homological mirror symmetry phenomenons. The formalism does not need a Lagrangian torus fibration, but uses a single or finite collection of Lagrangians to produce a mirror Landau-Ginzburg model, which always comes with a canonical A_∞-functor from Fukaya category to matrix factorization category. We illustrate this by giving many different forms of the mirror of a symplectic torus.

1. Introduction

Homological mirror symmetry (conjectured by **[Kon95]**) claims an equivalence of the derived Fukaya category $\mathrm{CF}(X)$ of a symplectic manifold X (called A-side) and the bounded derived category of coherent sheaves on the mirror $\check{X}$ (called B-side). In another version of the conjecture, on the B-side, we may have derived category of matrix factorizations $MF(W)$ of a Landau-Ginzburg model ($W : \check{X} \to \mathbb{C}$) instead.

HMS has been verified in many interesting cases, generating numerous interesting new ideas in related fields. The most successful approach to explain HMS is Strominger-Yau-Zaslow formalism **[SYZ96]** , which views (homological) mirror symmetry as relations arising between dual torus fibrations. Namely, given a Lagrangian torus fibration $X \to Y$, a point on the mirror space $\check{X}$ is given by a pair $(L_y, \mathscr{L})$ where L_y is a torus fiber and $\mathscr{L}$ is a flat $U(1)$-bundle on L_y. Fiberwise Fourier-Mukai transform can be used to mach symplectic and complex information in simple cases. In this approach, Landau-Ginzburg model W appears from the counting of holomorphic discs whose boundary lies on Lagrangian torus fibers of the SYZ fibration (In fact, this was first verified in **[CO06]** for toric Fano manifolds). The difficulty of SYZ approach lies on the singularity of Lagrangian torus fibration, which is in general rather complicated.

HMS is usually proved via computation (enough of both A, B-sides so that they can be shown to be isomorphic) but recently, based on the idea of Fukaya, Abouzaid **[Abo]** constructed a homological mirror functor for Langrangian torus bundles (without singularity). Roughly, the functor should associate a Lagrangian submanifold L' in X a coherent sheaf on $\check{X}$, and the stalk of the corresponding sheaf at the point $(L_y, \mathscr{L})$ is given by the Floer cohomology $HF((L_y, \mathscr{L}), L')$.

2010 *Mathematics Subject Classification.* Primary 14J33, 53D37.

The seminal work of Seidel [**Sei11**] on genus two curve shows that mirror symmetry does not does not require a Lagrangian torus fibration. This work has been generalized to the case of higher genus by Efimov [**Efi12**], to the case of Calabi-Yau hypersurfaces by Sheridan's [**She11**], [**She15**]. In fact, our formalism was developed while we were trying to understand the work [**Sei11**].

The essence of our formalism is the direct relation of Floer complex and matrix factorization. For a polynomial W, matrix factorization of W is the product of two polynomial entry matrices M_0, M_1 whose products satisfies (for some constant (in fact a critical value) λ)

$$M_0\, M_1 = M_1\, M_0 = (W - \lambda) \cdot Id.$$

Oh[**Oh**] has shown that (in the monotone case, which is generalized by [**FOOO09**]) the Lagrangian Floer differential d between two Lagrangians L_0, L_1 satisfies

$$d \circ (-d) = (-d) \circ (d) = (\lambda_0 - \lambda_1) \cdot Id,$$

where λ_1, λ_0 are counting of J-holomorphic discs of Maslov index two at a generic point of L_1 and L_0 respectively. The above two formulas already look rather similar, but W is a Landau-Ginzburg potential but λ_0 is a constant and they are different.

What we will do is to consider a decoration by formal variables to either flat connection on the Lagrangian L_0 or immersed Floer generators of L_0 using the language of Maurer-Cartan formalism. Then the decorated disc counting gives a function W instead of a constant λ_0. Also, the Floer differential d, which counts J-holomorphic discs, will count decorated strips, hence gives a matrix of polynomial entry. This will provide a direct way to interpret Floer complex as a matrix factorization. The Figure 1 (a) shows the decorated strip by immersed generators on $L_0 = \mathbb{L}$ and (b) indicates how the composition to two strips (corresponding to M_0, M_1) give rise to a decorated disc counting W_0 in the simplest case.

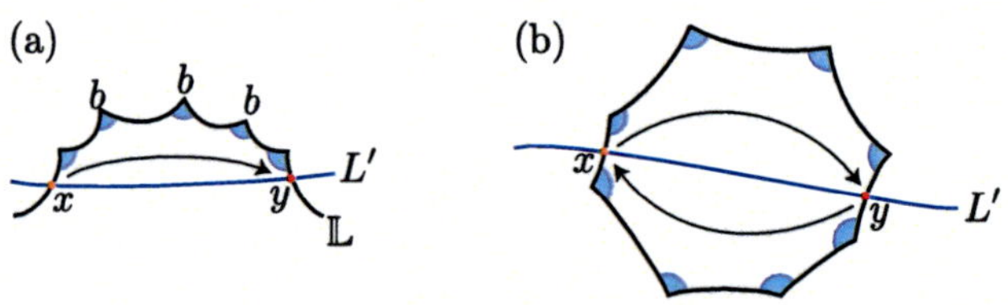

FIGURE 1. Deformed Floer differential

Let us explain how our formalism works in a rigorous way. First, we need to choose a single (or finite) possibly immersed Lagrangian $\mathbb{L}$ in the symplectic manifold X. For this pair $(X, \mathbb{L})$, we will obtain the mirror LG model $W_{\mathbb{L}}$, with an associated A_∞-functor

$$\mathscr{F}^{\mathbb{L}} : F(X) \to MF(W_{\mathbb{L}}).$$

Here, $W_{\mathbb{L}}$ will be a J-holomorphic disc potential function ([**FOOO09**]) defined on the solution space of the Maurer-Cartan equation of $\mathbb{L}$. We will see that it is quite interesting to generalize the Maurer-Cartan formalism of [**FOOO09**] to a non-commutative setting so that we have a non-commutative Maurer-Cartan space (nc algebra $\mathscr{A}$) with a choice of central element $W_{\mathbb{L}}$. Here $F(X)$ is the Fukaya A_∞-category of X and $MF(W_{\mathbb{L}})$ is the dg-category of matrix factorizations of $W_{\mathbb{L}}$. We will give more details later.

It is similar to SYZ approach in the sense that we use Lagrangians to construct the mirror. Note that our mirror construction depends on the choice of $\mathbb{L}$. It is an analogue of a choice of a Lagrangian torus fibration in SYZ approach. But, the notable difference from SYZ is that there is a big freedom, which is the choice of $\mathbb{L}$. In the case of the symplectic torus, we use this to find many different (even non-commutative) forms of the mirror elliptic curve.

There is one issue that we have to deal with. We need a Lagrangian $\mathbb{L}$ with a good Maurer-Cartan solution space. We will see that the mirror functor $\mathscr{F}^{\mathbb{L}}$ has some injectivity property but may not be surjective in general. For an equivalence, we expect that a union of generators of the Fukaya category of X is a good candidate for $\mathbb{L}$. In fact, to have a bigger MC solution space, and for the case that $\mathbb{L}$ is a union of Lagrangians, we introduced non-commutative coefficient system in [**CHL15**].

On the other hand, the most surprising part of our formalism is that the mirror functor $\mathscr{F}^{\mathbb{L}}$ is given automatically. Namely, once LG potential $W_{\mathbb{L}}$ on MC solution space of X is set-up, the functor $\mathscr{F}^{\mathbb{L}}$ is algebraically defined using Fukaya category operations. The functor $\mathscr{F}^{\mathbb{L}}$ in the object level, associates to a Lagrangian submanifold L', a matrix factorization of $W_{\mathbb{L}}$ (factorization of $W_{\mathbb{L}}$ times the identity matrix Id into the product of two matrices M_0, M_1):

$$M_0 \cdot M_1 = M_1 \cdot M_1 = W_{\mathbb{L}} \cdot Id.$$

Naively speaking, we are claiming here that computing Floer complex is almost like computing matrix factorizations! To be more precise, the differential of the standard Floer complex is given by counting J-holomorphic strips, but we will have to decorate the standard Floer complex by a family of Maurer-Cartan elements. These matrices M_0, M_1 are obtained by counting decorated J-holomorphic strips.

In a sense, this formalism explains geometric origin of homological mirror functors (beyond SYZ cases). It uses Fukaya category operations to construct the mirror functor. The algebraic structure of the mirror A_∞-functor is very similar to Yoneda embedding (A_∞-version is developed in [**Fuk02**]), and one may call it a *family of curved Yoneda embedding.* Namely, by taking $\mathrm{Hom}(\mathbb{L}, \bullet)$ on $F(X)$, instead of a chain complex, we get a curved complex with

$$d^2 = \lambda_1 - \lambda_0.$$

and the family of λ_0 over MC solution space of $\mathbb{L}$ is nothing but $W_{\mathbb{L}}$. Hence this gives matrix factorizations

$$d \cdot (-d) = W - \lambda_1.$$

We find that this family version gives the desired A_∞-functor between A_∞-fukaya category and dg-category of matrix factorizations of $W_{\mathbb{L}}$.

To illustrate our formalism in an elementary way, we begin with a simple example.

2. Basic example of $(\mathbb{C}, S^1)$ and $W(x) = x$

This is a rather trivial example, which relates the geometric pair $(\mathbb{C}, S^1)$ and the potential function $W(x) = x$, but still illustrates several important features of our construction, as we will see. We will ignore what happens near infinity of $\mathbb{C}$ for simplicity, and suppress the area contribution T^A of each holomorphic strips from the expressions.

Let us begin with the construction of $W(x)$. We consider $\mathbb{L} = S^1$ the standard unit circle in $\mathbb{C}$. We choose and fix a generator $X \in H^1(\mathbb{L}, \mathbb{C}^*)$, a Poincare dual to the point, say $z_0 \in \mathbb{L}$. In fact, we consider the element $b = xX$, where x is a formal variable (with values in $\mathbb{C}^*$). Recall that H^1 governs the infinitesimal deformation of Lagrangian $\mathbb{L}$. Another way to think about this data is the choice of a flat connection of a trivial complex line bundle over $\mathbb{L}$, whose holonomy is $x \in \mathbb{C}^*$ (the connection may be chosen so that it is nontrivial only in the neighborhood of the point $z_0 \in \mathbb{L}$.

In any case, $W_{\mathbb{L}}(x)$ is called the disc potential of $\mathbb{L}$. Since there is a unique disc (up to $Aut(D^2)$) with boundary on $\mathbb{L}$, We define (precise definition will come later)

$$W_{\mathbb{L}}(x) = 1 \cdot x.$$

Now, we explain the categorical correspondence. Homological mirror symmetry in this case should assign to any curve in $\mathbb{C}$, a matrix factorization of $W_{\mathbb{L}}$. i.e. two square matrices M_0, M_1 with entries in $\mathbb{C}[x, x^{-1}]$ satisfying

$$M_0 \cdot M_1 = M_1 \cdot M_0 = W_{\mathbb{L}} \cdot Id = \mathrm{diag}(x, x, \cdots, x)$$

And such an association should give an A_∞-functor relating the structures in both sides.

For $(\mathbb{C}, S^1)$, an object of Fukaya category is an oriented curve $L \subset \mathbb{C}$ (which can be made transverse to S^1 by a Hamiltonian isotopy). Then the standard Floer complex $CF(\mathbb{L}, L)$ is a (possibly obstructed) chain complex generated by the intersection points $L \cap S^1$, whose differential counts rigid J-holomorphic strips with boundary on $\mathbb{L}$ and L connecting two intersection points.

In our case, we consider the family of Floer complexes $CF((\mathbb{L}, b), L)$ for $x \in \mathbb{C}^*$. The Floer differential in this case is deformed by $b = xX$. This just means that we will deform the differentials from rigid J-holomorphic strips which passes through $z_0 = PD(X)$ by multiplying the holonomy contribution x.

More precisely, given a curve L, we define the corresponding matrix factorizations M_0^L, M_1^L using decorated Floer complex. Namely, by fixing the canonical counter-clockwise orientation of S^1, the intersection points $L \cap S^1$ can be divided into positive or negative intersections. For $R = \mathbb{C}[x, x^{-1}]$, we define

$$P^+ = \bigoplus_{p^+ \in L \cap S^1} R\langle p^+ \rangle, \quad P^- = \bigoplus_{p^- \in L \cap S^1} R\langle p^- \rangle.$$

(Recall that Floer complex between two oriented Lagrangian submanifolds are $\mathbb{Z}/2$-graded in the above sense). Then, M_0^L, M_1^L is given by the Floer differentials (with additional effect of holonomy x) which give R-module maps

$$M_0^L : P^+ \to P^-, M_1^L : P^- \to P^+.$$

In this real two dimensional case, we count bigons with convex corners, and whenever the boundary of the bigon passes through the point z_0, we multiply x (or x^{-1} depending on the choice $X \in H^1(\mathbb{L}, \mathbb{C}^*)$). For simplicity, we will ignore the symplectic area contribution in the differentials in this section.

For example, if $L = \mathbb{R}$ is the real line of $\mathbb{C}$, then we have two intersection points $p = +1, q = -1$, and two bigons, upper bigon as a Floer differential from p to q, and lower bigon from q to p. Let us assume that z_0 the point for the class X, has positive imaginary part. Then, the corresponding matrix factorization is given by

$$R\langle p \rangle \mapsto R\langle q \rangle \mapsto R\langle p \rangle,$$

where the first map (from the bigon containing x) is given by $x \cdot Id$, and the second map (from the bigon without x) is Id. Hence we have the (1×1) factorization $x = x \cdot 1$.

For a more non-trivial example, let us consider another curve L_1, which has more complicated intersection pattern with S^1 as in the Figure 2.

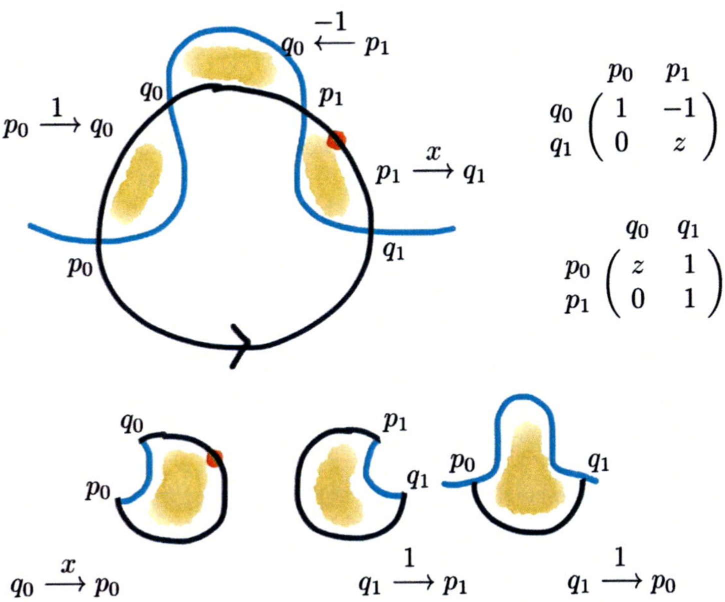

FIGURE 2. (2×2) matrix factorization of x from a curve L_1

We have two positive intersection points p_0, p_1 and two negative intersections q_0, q_1 (so that $P^+ = R^{\oplus 2}, P^- = R^{\oplus 2}$). Decorated Floer differentials by counting bigons provides

$$p_0 \mapsto q_0, \quad p_1 \mapsto -q_0 + xq_1.$$

$$q_0 \mapsto zp_0, \quad q_1 \mapsto p_0 + p_1.$$

First line is given by the counting of three smaller bigons (colored yellow in the original figure), and the second line is given by the counting of three bigger bigons (drawn in the bottom).

These differentials can be written as matrix factorizations

$$M_0 \cdot M_1 = \begin{pmatrix} 1 & -1 \\ 0 & x \end{pmatrix} \cdot \begin{pmatrix} x & 1 \\ 0 & 1 \end{pmatrix} = \begin{pmatrix} x & 0 \\ 0 & x \end{pmatrix} = x \cdot \mathrm{Id}_{2\times 2}.$$

2.1. Why does it work? By considering a curve which intersects the circle in $2n$-points, we can find a $(n \times n)$ matrix factorization of $W_{\mathbb{L}}(x) = x$ in a similar way. One may ask why this procedure always give a matrix factorization of $W_{\mathbb{L}}$. Note that the diagonal entry in the product $M_0 \cdot M_1$ corresponds to d^2, one bigon followed by another one such that it starts and ends at the same intersection point. One can check that the union of two corresponding bigons form the standard disc D^2, and hence produces $W_{\mathbb{L}}$.

This is the typical example where chain complex condition is violated ($d^2 \neq 0$) by the disc bubbling, and in our case the disc is decorated by the parameter $x \in \mathbb{C}^*$

to produce W. A prototype of this argument was first considered by Oh in the case of Lagrangian Floer cohomology of two monotone Lagrangian submanifolds with minimal Maslow index two. Namely, if Φ_i is the one point open Gromov-Witten invariant, counting Maslov index two discs passing through a generic point of L_i for $i = 0, 1$, then we have

$$d^2 = \Phi_1 - \Phi_0.$$

Roughly speaking this identity provides the matrix factorization: d^2 term (with b-contributions) correspond to the matrix factorization $-M_0M_1$ and Φ_1 is a constant (in the sense that it does not involve the deformation parameter $b = xX$), and Φ_0 (with a family of b-contributions) is the potential function for $L_0 = \mathbb{L}$.

For non-diagonal entries, the union of two bigons form a polygon with one non-convex corner. It is not hard to check that such configurations should occur in pairs whose signed counts cancel out.

We remark that the above procedure looks rather simple, but in fact it is already a new construction. This is because that our mirror variable x is different from the standard Strominger-Yau-Zaslow(SYZ) mirror variables. The standard SYZ variable is based on a Lagrangian torus fibration. For the case of $\mathbb{C}$, $\pi : z \to |z|^2$ defines a circle fibration to $\mathbb{R}_{\geq 0}$, and the mirror variable $x = re^{i\theta} \in \mathbb{C}^*$ comes from base parameter $r \in \mathbb{R}_{>0}$ with a holonomy parameter $e^{i\theta}$ of the circle fiber. Hence in the SYZ setting, one cannot interpret the Floer complex directly as matrix factorizations, since the contribution will not be a holomorphic function of x. It is possible in our approach, because our variable is a formal variable in the deformation of the Lagrangian, and in this particular example also because we have chosen the holonomy to be concentrated near a codimension one submanifold so that the effect on each strip is a monomial of x.

This example is generalized to compute the mirror matrix factorizations in toric cases ([**CHL14**]), where we explicitly compute the mirror matrix factorizations mirror to the Clifford torus T^n and also for $\mathbb{R}P^n$ in $\mathbb{C}P^n$, which has not been computed previously.

2.2. Variations. The above construction of matrix factorization involves several choices. For example, we can move the marked point z_0, which corresponds to the choice of a different flat connection. A morphism between two matrix factorizations $(P^\bullet, M_\bullet)$, $(Q^\bullet, N_\bullet)$ is given by R-module maps $\Psi : P^\bullet \to R^\bullet$. The compatibility of Ψ with $M_\bullet, N_\bullet$ gives the differential, $\delta(\Psi) = M_\bullet \circ \Psi - \Psi \circ N_\bullet$. Composition of two morphisms is given by the obvious composition, and these data define the *differential graded category of matrix factorizations*.

If we move the marked point from z_0 to z_1 while passing through a single generator p_i (resp. q_i) then the resulting matrix factorization in the previous computation changes. It can be checked that the corresponding matrix factorizations are isomorphic to each other, and the precise isomorphism between two matrix factorizations turns out to be $\langle p_i \rangle \mapsto x\langle p_i \rangle$ (resp. $\langle q_i \rangle \mapsto x^{-1}\langle q_i \rangle$) and an identity on other generators. We refer readers to Lemma 5.3 of [**CHL14**] for more details.

In this example, we regard $\mathbb{C}$ as a symplectic manifold and L as a Lagrangian submanifold. One may also ask if we choose a different Hamiltonian isotopy of L, say $\phi(L)$. Note that the corresponding matrix factorization is built on $CF(\mathbb{L}, \phi(L))$, which is quasi-isomorphic to $CF(\mathbb{L}, L)$ as Lagrangian Floer complexes. In section 5 of [**CHL14**], it is shown that the matrix factorization for $\phi(L)$ is indeed homotopic to that of L. The proof is a direct translation of Floer theory proof incorporating the family of Maurer-Cartan elements b.

Note that using a compactly supported Hamiltonian isomotpy, one can easily move L away from the unit circle $\mathbb{L}$. Then the corresponding matrix factorization is zero, which shows that the (2×2) matrix factorization that we have computed is indeed homotopic to zero. Thus we find that the homological mirror symmetry correspondence (although trivial after taking derived categories) are very nicely presented in this example.

3. Maurer-Cartan formalism and two different potentials

Given a symplectic manifold X, and a compact oriented Lagrangian (possibly immersed) submanifold L, Fukaya-Oh-Ohta-Ono [**FOOO09**] (Akaho-Joyce [**AJ10**] in the immersed case) constructed a filtered A_∞-algebra $(C^\bullet(L), \{m_k\}_{k\geq 0})$ over the Novikov field Λ. Here

$$\Lambda = \left\{ \sum_{i=0}^{\infty} a_i T^{\lambda_i} \mid a_i \in \mathbb{C}, \lambda_i \in \mathbb{R}, \lim_{i\to\infty} \lambda_i = \infty \right\}.$$

An A_∞-algebra is a generalization of a differential graded associative algebra, where associativity is relaxed to a homotopy associativity. It involves k-ary operation $m_k : (C^\bullet(L))^{\otimes k} \to C^\bullet(L)$ of degree $2-k$ for each k which satisfies

$$\sum_{n_1+n_2=n+1} (-1)^\epsilon m_{n_1}(w_1, \cdots, w_{i-1}, m_{n_2}(w_i, \cdots, w_{i+n_2-1}), w_{i+n_2}, \cdots, w_n) = 0,$$

for any n-tuple $(w_1, \cdots, w_n) \in (C^\bullet(L))^{\otimes n}$ for $n = 0, 1, \cdots$. These operations are constructed via counting J-holomorphic discs(polygons) and we omit details. Note that $(2-k)$ is a natural degree since m_1 is a differential (degree 1 map), and m_2 is a product (degree 0 map) and m_3 is a homotopy for the associativity for the product and hence of degree (-1). If we shift the degree by one to consider $C^\bullet(L)[1]$, m_k is of degree 1 for all k. We denote the shifted degree of an element x by $|x|' = |x| - 1$. Then ϵ is the natural Koszul sign: $\epsilon = |w_1|' + \cdots + |w_{i=1}|'$.

An element $\mathbf{1}_L$ is called a unit if $m_k(\cdots, \mathbf{1}_L, \cdots) = 0$ except $m_2(\mathbf{1}_L, x) = (-1)^{|x|} m_2(x, \mathbf{1}_L) = x$. In Lagrangian Floer theory, (homotopy) unit is given by the Poincare dual of the fundamental class of L. The first few A_∞-equations are given as

$$m_1(m_0) = 0, m_1^2(x) + m_2(m_0, x) + (-1)^{|x|'} m_2(x, m_0) = 0$$

If $m_0 = 0$, then m_1 is a differential ($m_1^2 = 0$) and its homology is called Lagrangian Floer cohomology of L. If m_0 is a non-zero multiple of a unit $\mathbf{1}_L$, then from the unit property, we still have $m_1^2 = 0$. This is what usually happens for monotone Lagrangian submanifolds.

In general Fukaya-Oh-Ohta-Ono studied the obstruction and deformation theory of $(C^\bullet(L), \{m_k\}_{k\geq 0})$ to deal with the general $m_0 \neq 0$ case. An element $b \in C^{\mathrm{odd}}(L)$ is called a weak Maurer-Cartan (MC) element if

$$m(e^b) = m_0(1) + m_1(b) + m_2(b, b) + m_3(b, b, b) + \cdots = c \cdot \mathbf{1}_L.$$

To ensure the convergence of the above sum, one assume that all coefficients of b have positive T-exponents (Λ_+), but in many examples, the sum converges without this assumption also, and in such a case, we can extend the domain of b so that coefficients of b have non-negative T-exponents (Λ_0).

Given a weak MC element b, one can deform the given A_∞-structure $\{m_k\}$ to another A_∞-structure $\{m_k^b\}$:

$$(3.1)\quad m_k^b(x_1,\cdots,x_k)=\sum m(b,\cdots,b,x_1,b,\cdots,x_i,\cdots,b,x_k,b,\cdots,b)= m(e^bx_1e^bx_2e^b\cdots e^bx_ke^b).$$

Note that we have $m_0^b = m(e^b) = c\cdot \mathbf{1}_L$, and hence we have a deformed chain complex $(m_1^b)^2=0$. This is why weak MC elements are introduced in **[FOOO09]** to define Lagrangian Floer cohomology in the general case.

We can introduce a more general deformation

$$(3.2)\qquad m_k^{b_0,\cdots,b_k}=\sum m(b_0,\cdots,b_0,x_1,b_1,\cdots,x_i,\cdots,b_{k-1},x_k,b_k,\cdots,b_k).$$

Let us explain this weak MC condition in a heuristic way. As $m(e^b)$ can be any even degree element, the weak MC condition should be seen as a vanishing condition of the output (except the unit $\mathbf{1}_L$). Let us assume that $C^\bullet(L)$ is finitely generated by $\{f_i\}\cup\{\mathbf{1}_L\}$. By writing $m(e^b)=c\cdot\mathbf{1}_L+\sum_i a_if_i$, Maurer-Cartan equation is equivalent to the vanishing of a_i's.

In general, such a vanishing may not occur, and a generic A_∞-algebra is obstructed in general. In Lagrangian Floer the existence of the solution b indicates certain symmetry of J-holomorphic curves. In **[FOOO09]**, they construct b inductively (from minimal energy level) so that it is a chain bounding the sum of J-holomorphic curve contributions at each energy level (with possible insertion of b's from the lower energy level). This is possible only if the obstruction vanishes at each level. We find in **[CHL15]** that by considering non-commutative coefficients, we get a much bigger MC solution space.

Now, let us explain our construction. Let us choose and fix odd degree elements $X_i\in C^\bullet(L)$ for $i=1,\cdots,k$, and introduce corresponding formal variables $\{x_i\}$ with values in Λ_+. By setting

$$b=x_1X_1+\cdots+x_kX_k,$$

we may consider a family of deformations of the given A_∞-structure $\{m_k\}$. Here, the choice of values of $x_1,\cdots,x_k$ provide the formal family of deformations. In fact, we may choose X_i in any degree, and require that x_i is of degree $1-deg(X_i)$ to consider an extended deformation. But we will restrict ourselves to the former case for simplicity.

For example, if b always satisfies the MC equation $m(e^b)=W_{\mathbb{L}}(b)\cdot\mathbf{1}_L$, then coefficient $W_{\mathbb{L}}(b)$ is called the Lagrangian Floer (or Landau-Ginzburg) potential $W_{\mathbb{L}}:(\Lambda_+)^k\to\Lambda_+$ or sometimes called a worldsheet superpotential). In general, $W_{\mathbb{L}}(b)$ is only defined for b in the MC solution space.

Let us explain why we need the MC equations. The function $W_{\mathbb{L}}(b)$ counts J-holomorphic discs with corners given by X_i's passing through a generic point $p\in L$, and it is decorated by formal variables x_i's so that $W_{\mathbb{L}}(b)$ remembers the labeling of the corners. If the MC equation $m(e^b)=W_{\mathbb{L}}(b)\cdot\mathbf{1}_L$ holds, then since the output is a constant multiple of a fundamental class of L, so the resulting counting is independent of the generic point p. But if MC equation fails, then $W_{\mathbb{L}}(b)$ will not be well-defined.

We may interpret the solution space of MC-equation to be the vanishing locus of all $a_i's$. But in several fundamental examples, such a vanishing condition is related to the partial derivatives of a single function, called space-time superpotential of the A_∞-algebra $\{m_k^b\}$. In fact, in **[CHL15]**, we show that this is the case if $\mathbb{L}$ consists of odd dimensional spheres.

If we assume that the A_∞-algebra is cyclic symmetric (with a pairing $\langle , \rangle$), then we can define the space-time superpotential as

$$\Phi(b) = \sum_k \frac{1}{k+1} \langle m_k(b, b, \cdots, b), b \rangle.$$

$\Phi(b)$ is cyclic expression and its partial derivative with respect to variables x_i gives

$$\frac{\partial \Phi}{\partial x_i} = \sum_k \langle m_k(b, b, \cdots, b), X_i \rangle = \langle m(e^b), X_i \rangle.$$

Partial derivative on one of b turns it into X_i, which can be sent to the last spot (using cyclic symmetry), and there are $k+1$ such terms, which cancels the fraction coefficient. Hence, weak MC equation implies the vanishing of $\frac{\partial \Phi}{\partial x_i}$. In this way, we have two potential functions Φ and W appearing in this story. Previously, these two potentials have not appeared at the same time. The potential W appeared in constructing Landau-Ginzburg mirror usually in Fano cases and Φ appeared in the counting of holomorphic discs in Calabi-Yau cases.

We emphasize that since Φ governs the weak MC equation, the well-definedness of W depends on Φ. Roughly speaking, $W(b)$ is a disc counting function (with all possible b-insertions) passing through a generic point p (we may define it by $W(b) = \langle m(e^b), p \rangle$). MC equation ensures that the output is independent of this choice of generic point p. From now on, let us assume that the weak MC equation is equivalent to the vanishing of partial derivatives of Φ for simplicity.

Now, let us explain the non-commutative generalization of the above story. When we consider the reference Lagrangian consisting of several Lagrangians $\{L_1, \cdots, L_k\}$, we will consider a quiver $Q^{\mathbb{L}}$ whose vertices $\{v_1, \cdots, v_k\}$ corresponding to Lagrangians, and each of arrow x_i of $Q^{\mathbb{L}}$ is for a choice of odd generator $X_i \in CF(L_{i_1}, L_{i_2})$. Later we will consider $b = \sum x_i X_i$, but this b is defined on the tensor product over semi-simple ring, which we now explain. Let $\Lambda Q^{\mathbb{L}}$ be the path algebra of the quiver $Q^{\mathbb{L}}$ which is completed with respect to the energy of Λ. We will use $\Lambda Q^{\mathbb{L}}$ or $\Lambda Q^{\mathbb{L}}/\sim$ for MC relations $\sim$ as non-commutative coefficients. Note that

$$CF(\mathbb{L}, \mathbb{L}) = \bigoplus_{i,j} CF(L_i, L_j)$$

form an A_∞-algebra over a semi-simple ring $\Lambda^{\oplus} = \oplus_{i=1}^k \Lambda$.

First, we can easily extend the coefficient of an A_∞-algebra from the Novikov field Λ to

$$\Lambda Q^{\mathbb{L}} \otimes_{\Lambda^{\oplus}} CF(\mathbb{L}, \mathbb{L}).$$

Tensoring over $\Lambda^{\oplus}$ guarantees that the expression vanishes unless the path and the element in the Floer complex has the same starting Lagrangian L. Such an extention is define by the formula

$$m_k(f_1 w_1, \cdots, f_k w_k) = f_k f_{k-1} \cdots f_1 m_k(w_1, \cdots, w_k).$$

One can easily check that such an extension still defines an A_∞-algebra. The above construction for a single Lagrangian $\mathbb{L} = L$ also generalizes the previous construction since the variables x_i's are not commutative anymore.

Now, let us explain how to define the non-commutative MC solution space or non-commutative algebra $\mathscr{A}$. We start with the coefficient ring $\Lambda Q^{\mathbb{L}}$ and compute $m(e^b)$. As explained before, we have

$$m(e^b) = c \cdot \mathbf{1}_L + \sum_i a_i f_i.$$

Our non-commutative algebra $\mathscr{A}$ is defined from $\Lambda Q^{\mathbb{L}}$ taking a quotient with two-side ideal generated by (a_i). i.e. we set

$$\mathscr{A} = \frac{\Lambda Q^{\mathbb{L}}}{(a_i)_i} = \frac{\Lambda Q^{\mathbb{L}}}{(\partial\Phi)}$$

where the last equality holds as we assumed for simplicity that all MC equation come from partial derivatives of Φ. Thus, we have shown that the non-commutative MC space is given by the quiver algebra with a potential Φ.

Note that we have an additional potential $W_{\mathbb{L}}$, coming from the disc counting through a generic point p. The following is a key theorem that we obtain a non-commutative Landau-Ginzburg model in this way.

THEOREM 3.1 ([**CHL15**] Theorem 1.2). *The disc potential $W_{\mathbb{L}}$ is always a central element of the quiver algeba with potential $\mathscr{A} = \frac{\Lambda Q^{\mathbb{L}}}{(\partial\Phi)}$. In particular we have $x_i W_{\mathbb{L}} = W_{\mathbb{L}} x_i$ in $\mathscr{A}$.*

4. Construction of the canonical A_∞-functor

Let us choose and fix our reference Lagrangian $\mathbb{L}$ together with $b = \sum_i x_i X_i$. We will explain how we obtain a canonical A_∞-functor

$$\mathscr{F}^{\mathbb{L}} : F(X) \to MF(W_{\mathbb{L}}).$$

First, let me explain both sides of the functor. Category of matrix factorizations $MF(W_{\mathbb{L}})$ is the differential graded category as explained in Section 2, and we regard it as an A_∞-category with vanishing $m_{k\geq 3}$. One should note that the sign conventions for dg-categories and A_∞-categories are different, but can be easily related.

The Fukaya A_∞-category $F(X)$ of X has an object which is a spin orientable (weakly) unobstructed Lagrangian submanifold L, together with a choice of complex flat line bundle $\mathcal{L}$, and (weak) Maurer-Cartan element f (here, f is not a family, but just one MC element). We denote by L the triple $(L, \mathcal{L}, f)$ for simplicity. The A_∞ structure m_k on $F(X)$ is defined by counting J-holomorphic polygons.

Let us explain the object level correspondence of the functor $\mathscr{F}^{\mathbb{L}}$, whch sends any Lagrangian L' to a matrix factorization

$$\mathscr{F}^{\mathbb{L}} : L' \to (M(L')_0, M(L')_1)$$

where $M(L')_0 M(L')_1 = M(L')_1 M(L')_0 = W_{\mathbb{L}} \cdot Id$.

As illustrated in Section 2, the matrices $M(L')_0, M(L')_1$ are the differentials of the decorated Floer complex between the reference Lagrangian $\mathbb{L}$ and L' for any object L' in $F(X)$, which is

$$\big(CF\big((\mathbb{L}, b), L'\big), m_1^{b,0}\big).$$

i.e. we count strips, but we allow decorations from b's. If b comes from an immersed generator, this means that the decorated strip could be k-gon for any $k \geq 2$. See Figure 1.

Recall that $b = \sum x_i X_i$ and the A_∞-operation deformed by b is explained in (3.1), (3.2). It is important that the formal parameter x_i's are added, so that we can consider the Lagrangian $(\mathbb{L}, b)$ as a formal family. Namely we use this formal family instead of the Lagrangian torus fibration. The functor $\mathscr{F}^{\mathbb{L}}$ obviously depends on our choice of reference Lagrangian $\mathbb{L}$.

Using the formula (3.2), we find that

$$m_1^{b,0}(m_1^{b,0}(x)) = m(e^b(m(e^b x))) = m(e^b, m(e^b), e^b, x) = m(W_{\mathbb{L}}(b)\mathbf{1}_{\mathbb{L}}, x) = W_{\mathbb{L}}(b) \cdot x$$

Hence this gives the desired matrix factorization

$$(m_1^{b,0})^2 = W_{\mathbb{L}} \cdot Id.$$

We remark that if L' is equipped with a weak Maurer-Cartan element b' (which do not contain any variable) with potential value λ', the matrix factorization that we consider is given by

$$\big(CF((\mathbb{L}, b), (L', b')), m_1^{b,b'}\big).$$

It is not hard to check that we have

$$(m_1^{b,b'})^2 = (W_{\mathbb{L}} - \lambda') \cdot Id.$$

The functor $\mathscr{F}^{\mathbb{L}}$ on the morphism level is defined by $m_2^{b,0,0}$. Namely, given a morphism between two Lagrangians $z \in L_1 \cap L_2$, then

$$\pm m_2^{b,0,0}(\bullet, z)$$

defines a morphism between matrix factorizations. In general, there exist higher morphisms of the A_∞-functor (which relates A_∞-structure with a dg-structure of the other side), $\mathscr{F}_k^{\mathbb{L}}(z_1, \cdots, z_k)$, which is given by

$$\pm m_{k+1}^{b,0,\cdots,0}(\bullet, z_1, \cdots, z_k).$$

These satisfies A_∞-functor equations [**CHL13**].

5. Elliptic curve example

We illustrate our formalism in the case of a symplectic torus T^2. In this section, we explain the work of [**CHL13**] where we find the mirror elliptic curve as the Hesse cubic in $\mathbb{P}^2$. In the next section, we explain how to find a non-commutative mirror of a symplectic torus by modifying this construction following [**CHL15**]. In the last section, we will find another form of mirror elliptic curve as the complete intersection in $\mathbb{P}^3$, as well as its non-commutative versions.

We can choose T^2 with a complex structure so that it has $\mathbb{Z}/3$-symmetry. In Figure 3 (a), by attaching opposite sides, we obtain T^2, and the $\mathbb{Z}/3$-action is given by the rotation by 120-degree. The quotient space is the orbifold sphere $\mathbb{P}^1_{333}$. The immersed Lagrangian $\mathbb{L}$ drawn in $\mathbb{P}^1_{333}$ in Figure 3 can be lifted to T^2 or its universal cover $\mathbb{C}$. Three immersed points X, Y, Z of $\mathbb{L}$ (which are 60 degree angles when used as inputs) are used to form

$$b = xX + yY + zZ.$$

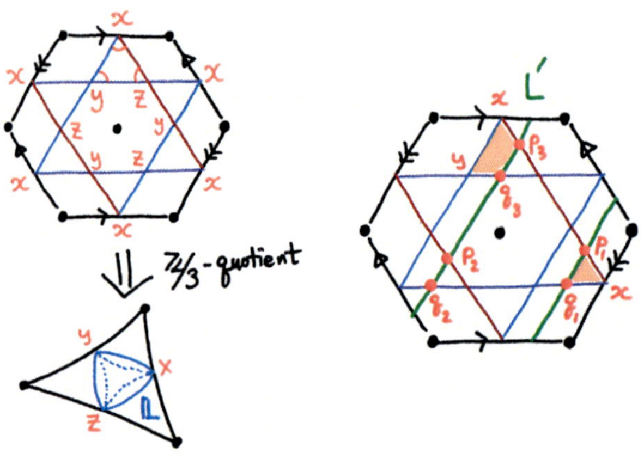

FIGURE 3. Torus quotient and finding matrix factorization of L'

As any smooth holomorphic disc in $\mathbb{P}^1_{333}$ (without twisted sector bulk insertions [**CP12**]) can be lifted to the universal cover $\mathbb{C}$, it is convenient to work in $\mathbb{C}$, by lifting $\mathbb{L}$ as well as the labelings x, y, z. A lifting of $\mathbb{L}$ is given a straight line, and its $\mathbb{Z}/3$-action image (and its translation by a lattice vector) is also a lift of $\mathbb{L}$.

The first thing to check whether the Maurer-Cartan equation $m(e^b) = W_{\mathbb{L}} \cdot \mathbf{1}$ holds. If it holds for any b of the above form, then we can calculate $W_{\mathbb{L}}$ by counting triangles passing through a generic point p. In this case, MC equation is equivalent to the invariance of counting with respect to p.

As illustrated in Figure 4 (a), we may move the green point (generic point p) to the next segment of $\mathbb{L}$, then obviously the collection of triangles passing through p changes. But in this case, there exist an involution symmetry (with respect to the vertical line passing through z) and unsigned counting of triangles (with its areas) do not change. But it turns out that the signed counting is not invariant unless we put a non-trivial spin structure on $\mathbb{L}$. This effect can be enforced by introducing (-1) holonomy flat $U(1)$ bundle on $\mathbb{L}$. It is proved in [**CHL13**] that in such a case, any $b = xX + yY + zZ$ satisfies the weak MC equation.

After some computations, it is shown that the potential function is of the form

$$\psi(q)(x^3 + y^3 + z^3) - \phi(q)(xyz)$$

where $\psi(q), \phi(q)$ are power series in q. It may be re-parametrized as

$$W_{\mathbb{L}} = x^3 + y^3 + z^3 - \sigma(q)xyz$$

As q records symplectic areas, it may be considered as a Kähler parameter, and the ratio $\sigma(q)$ may be considered as the complex parameter of the mirror elliptic curve. It is shown in [**CHL13**] that $\sigma(q)$ is the inverse to the well-known mirror map of Saito, which is obtained from two solutions of Picard-Fuch's equation.

Also, we can use $\mathbb{L}$ to compute matrix factorizations mirror to Lagrangians in T^2. For example, let L' be a Lagrangian in T^2 obtained by a small translation of $\mathbb{L}$. L' and $\mathbb{L}$ has 6 intersection points, which are labeled as $p_1, q_1, \cdots, p_3, q_3$. As illustrated in the right hand side of Figure 3, the corresponding matrix factorization is given by (3×3) matrices, whose entries are given by counting deformed strips. In this case, deformed strips (triangles) from p_i to q_i gives a matrix with linear entries and those (quadrilaterals) from q_i to p_i gives a matrix with quadratic entries.

$$\begin{pmatrix} \alpha_1 x & \alpha_2 z & \alpha_3 y \\ \alpha_3 z & \alpha_1 y & \alpha_2 x \\ \alpha_2 y & \alpha_3 x & \alpha_1 z \end{pmatrix} \cdot \begin{pmatrix} \beta_1 x^2 - \gamma_1 yz & \beta_3 z^2 - \gamma_3 xy & \beta_2 y^2 - \gamma_2 zx \\ \beta_2 z^2 - \gamma_2 xy & \beta_1 y^2 - \gamma_1 zx & \beta_3 x^2 - \gamma_3 yz \\ \beta_3 y^2 - \gamma_3 zx & \beta_2 x^2 - \gamma_2 yz & \beta_1 z^2 - \gamma_1 xy \end{pmatrix} \tag{5.1}$$

The construction generalizes to $\mathbb{P}^1_{a,b,c}$ (which do not have SYZ fibration).

THEOREM 5.1 ([**CHKL**],[**CHL13**]). *(1) The mirror of $W_{a,b,c}$ for $\mathbb{P}^1_{a,b,c}$ can be computed and it is of the form*

$$x^a + y^b + z^c + \sigma xyz + \cdots$$

(2) The A_∞-functor $\mathscr{F}^{\mathbb{L}}$ induces a derived equivalence and hence proves the homological mirror symmetry conjecture for $\frac{1}{a} + \frac{1}{b} + \frac{1}{c} \leq 1$.

The explicit computation of W is the most difficult in the hyperbolic case (the universal cover of $\mathbb{P}^1_{a,b,c}$ is the hyperbolic disc), and in this case, we find a combinatorial algorithm to find all terms step by step, and show that $W_{a,b,c}$ has an isolated singularity at the origin if $\frac{1}{a} + \frac{1}{b} + \frac{1}{c} \leq 1$.

Homological mirror symmetry is proved as follows. Abouzaid's generating criterion for surfaces can be used to show that $\mathbb{L}$ generates the Fukaya category of $\mathbb{P}^1_{a,b,c}$. We explicitly compute the mirror matrix factorization of $\mathbb{L}$ under our functor $\mathscr{F}^{\mathbb{L}}$ and show that it is also a split generator of the matrix factorization category of $W_{a,b,c}$ which corresponds to the skyscraper sheaf at the origin.

In fact, we can use this to prove homological mirror symmetry for the torus T^2. Here is the schematic picture.

$$\begin{array}{ccc} T^2 & \xrightarrow{\ \ mirror\ \ } & (W : \mathbb{C}^3/(\mathbb{Z}/3) \to \mathbb{C}) \\ \Big\downarrow {\scriptstyle \mathbb{Z}/3} & & \Big\uparrow {\scriptstyle \text{dual } \mathbb{Z}/3\text{-action}} \\ \mathbb{P}^1_{333} & \xrightarrow{\ \ mirror\ \ } & (W : \mathbb{C}^3 \to \mathbb{C}) \end{array}$$

By adding the $\mathbb{Z}$-grading information to the $\mathbb{Z}/3$-action, we prove the following

THEOREM 5.2 ([**CHL15**]). *(1) There exist a $\mathbb{Z}$-graded A_∞-functor*

$$\mathscr{F}^{\mathbb{L}} : F^{gr}(T^2) \to MF^{gr}(W_{\mathbb{L}})$$

which gives a derived equivalence

(2) Composing with Orlov isomorphism $DMF^{gr}(W_{\mathbb{L}}) \cong D^b coh(\check{E})$ we obtain the mirror symmetry between symplectic and complex torus.

Recently, Sangwook Lee [**Lee**] has shown that the above homological mirror symmetry is compatible with the work of Polishchuk-Zaslow [**PZ98**] (after composing a certain symplectomorphism). i.e. there is a commuting diagram between our functor and that of Polishchuk-Zaslow.

6. Non-commutative mirrors and deformation quantization

The construction in the previous section has been generalized by allowing non-commutative coefficients [**CHL15**]. Let us write the reference Lagrangian $\mathbb{L}$ as $\mathbb{L}_0$ in this section. And we take a translation of $\mathbb{L}_0$ to $\mathbb{L}_t$ (we have a S^1-worth of possible translations). Or, we could change the holonomy of $\mathbb{L}$ from (-1) to any $\lambda \in U(1)$. This breaks the weak Maurer-Cartan equation of $\mathbb{L}$ since there exist no more involution symmetry between triangles bounded by $\mathbb{L}_t$. Due to $\mathbb{Z}/3$-action, among the two triangles of $\mathbb{L}_0$ with equal area, one of them become smaller, and the other one become larger. This is illustrated in Figure 4 (b).

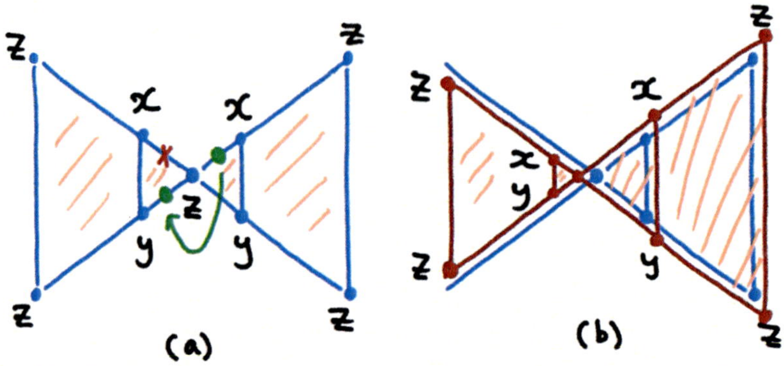

FIGURE 4. MC equation for $\mathbb{L}_0$ and $\mathbb{L}_t$

Hence MC equation is not satisfied in the usual sense. But we find in **[CHL15]** that we can introduce non-commutative coefficients, so that Maurer-Cartan equation is of the form

$$axy + byz + cz^2 = 0, ayz + bzy + cx^2 = 0, azx + bxz + cy^2 = 0.$$

Here, a, b, c depends on two parameters, t, λ, corresponding to the translation of the reference Lagrangian $\mathbb{L}$ (t) and change of the holonomy of $\mathbb{L}$ (λ). The resulting non-commutative algebra with these relations (for each t, λ) is called a Sklyanin algebra. This is one of the fundamental example of non-commutative algebraic geometry, studied by Artin, Tate, Van den Vergh **[ATdB90]** and many others. This also appeared previously in symplectic literatures of Auroux-Katzarkov-Orlov **[AKO08]**, Aldi-Zaslow **[AZ06]**. In both of these works, the relation is found after computing the product m_2. But in our formalism, the relation is systematically derived from Maurer-Cartan equation, and furthermore we have homological mirror functors for each $(t, \lambda) \in T^2$.

THEOREM 6.1. **[CHL15]**

(1) *We have T^2-family of $\mathbb{Z}$-graded A_∞-functors for each $(t, \lambda) \in T^2$*

$$\mathscr{F}^{\mathbb{L}} : F^{gr}(T^2) \to MF^{gr}(A_t, W_t).$$

(2) *Maurer-Cartan coefficients (a, b, c) lies on the mirror cubic*

$$W_0(a, b, c) = a^3 + b^3 + c^3 - \sigma abc = 0.$$

(3) *The quotient of non-commutative algebra $\mathscr{A}_t$ by a central element W_t, $\mathscr{A}_t/W_t$, is a deformation quantization of affine del Pezzo surface $W_0 = 0$ in $\mathbb{C}^3$.*

Let us mention a little bit about deformation quantization in this case, studied by Etingof-Ginzburg **[EG10]**. Given any polynomial of (x, y, z), in this case W_0, there exist a Poisson structure on $\mathbb{C}^3$ defined by

$$\{x, y\} = \frac{\partial W_0}{\partial z}, \{y, z\} = \frac{\partial W_0}{\partial x}, \{z, x\} = \frac{\partial W_0}{\partial y}.$$

One can check that $\{W_0, \bullet\} = 0$, and hence such a Poisson structure descends to the hypersurface $W_0 = 0$ in $\mathbb{C}^3$.

Etingof-Ginzburg found that this Poisson algebra can be deformation quantized, in a two step process. First, they show that the Poisson structure on $\mathbb{C}^3$ can be deformation quantized into a Calabi-Yau algebra $\mathscr{A}_h$, and the Poisson center W_0 then is deformed to a center W_h of non-commuative algebra $\mathscr{A}$. The existence of W_h was proved using deep results of formality theorem of Kontsevich. In our case, Both of them appear naturally in our setting. The Calabi-Yau algebra $\mathscr{A}_h$ is nothing but the non-commutative MC algebra $\mathscr{A}_t$ of the reference Lagrangian $\mathbb{L}_t$, and the central element W_h is nothing but our disc potential function for $\mathbb{L}_t$. To prove the last part of the above theorem, one also has to find the relation between W_t and the derivatives of a, b, c at $t = 0$, and we refer readers to **[CHL15]** for details.

7. Complete intersection mirror of the torus

Let T^2 be a torus, which is given by $\mathbb{C}/(\mathbb{Z} \oplus \mathbb{Z})$. Consider the $\mathbb{Z}/2$-action on T^2, which is the rotation around the origin. The quotient orbifold is the pillowcase orbifold $\mathbb{P}^1_{2222}$, which is a sphere with 4 $\mathbb{Z}/2$-orbifold point.

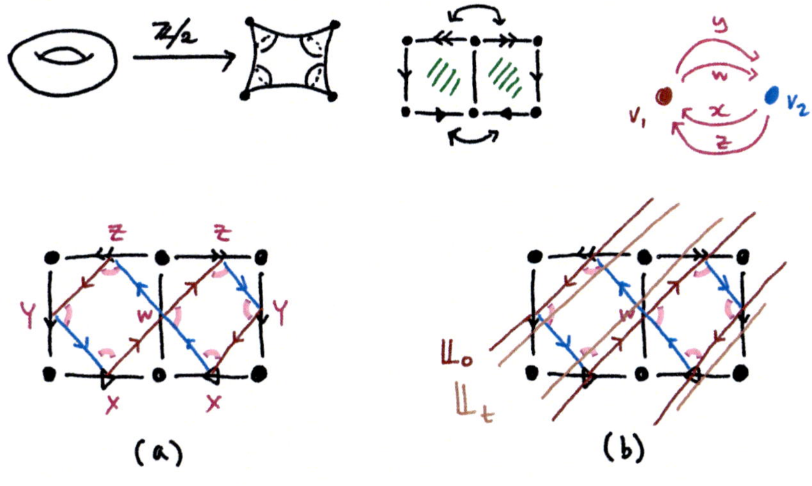

FIGURE 5. $\mathbb{P}^1_{2222}$ Lagrangians, quivers and deformations

In this case, the reference $\mathbb{L} = L_1 \cup L_2$ consists of two Lagrangians, and $L_1 \cap L_2$ has 4 odd degree intersection points X, Y, Z, W. Hence the quiver $Q^{\mathbb{L}}$ has two vertices v_1, v_2 and 4 edges x, y, z, w. The Maurer-Cartan equation is given by the partial derivatives of the spacetime superpotential Φ

$$\Phi = \text{const} \cdot (xyzw - zyxw)$$

For example, the symmetry of polygons at the vertex w corresponds to $\partial_w \Phi$, which gives $xyz = zyx$.

$$\mathbb{C}Q^{\mathbb{L}}/(\partial\Phi) \qquad\qquad\qquad (\widehat{Y} = \mathcal{O}_{\mathbb{P}^1} \bigoplus \mathcal{O}_{\mathbb{P}^1})$$

non comm. resol. $\searrow$ $\swarrow$ crep. resol.

$$\mathbb{C}[x_1, x_2, x_3, x_4]/(x_1x_3 = x_2x_4)$$

In fact, the algebra $\mathbb{C}Q^{\mathbb{L}}/(\partial\Phi)$ is a non-commutative resolution of the conifold. Here, conifold Y is given by the equation $x_1x_3 = x_2x_4$ in $(x_1, x_2, x_3, x_4) \in \mathbb{C}^4$, which has a (commutative) crepant resolution $\widehat{Y} = \mathcal{O}_{\mathbb{P}^1}(-1) \bigoplus \mathcal{O}_{\mathbb{P}^1}(-1)$. Non-commutative and commutative resolutions are related as follows. There exist a tilting bundle $\mathscr{T}$ on $\widehat{Y}$ such that its endomorphism ring $\mathrm{End}(\mathscr{T})$ becomes $\mathscr{A}$. In this case,

$$\mathscr{T} = \mathcal{O}_{\widehat{Y}} \bigoplus \mathcal{O}_{\widehat{Y}}(1).$$

And by considering explicit toric coordinates of $\widehat{Y}$, one can identify y, w as elements of $\mathrm{Hom}(\mathcal{O}_{\widehat{Y}}, \mathcal{O}_{\widehat{Y}}(1))$ and x, z as $\mathrm{Hom}(\mathcal{O}_{\widehat{Y}}(1), \mathcal{O}_{\widehat{Y}})$.

One can recover the conifold Y from $\mathbb{C}Q^{\mathbb{L}}/(\partial\Phi)$ by considering loops at one of the vertex, say v_1. Then, such loops are generated by

$$x_1 = xy, x_2 = xw, x_3 = zw, x_4 = zy$$

and the relation $(\partial\Phi)$ implies that these new variables satisfy the conifold equation $x_1x_3 = x_2x_4$.

Now, the disc potential function $W_{\mathbb{L}} = W_{L_0} + W_{L_1}$ is an element of $\Lambda Q^{\mathbb{L}}/(\partial\Phi)$, which is given by the counting of decorated polygons passing through a generic point $p_i \in L_i$ (which gives W_{L_i}) for $i = 0, 1$. It is not hard to see that the rigid polygons bounded by $\mathbb{L}$ are given by rectangles. We have

$$W_{L_0} = \phi(q)(x_1^2 + x_2^2 + x_3^2 + x_4^2) + \psi(q)x_1x_3.$$

After reparametrization, we get the following complete intersection mirror of the symplectic torus

$$\begin{cases} x_1x_3 - x_2x_4 & = 0 \\ x_1^2 + x_2^2 + x_3^2 + x_4^2 + \sigma x_1x_3 & = 0 \end{cases}$$

Namely, this is the elliptic curve embedded in $\mathbb{P}^3$, which is the same elliptic curve that we have obtained previous as Hesse cubic in $\mathbb{P}^2$.

One can also obtain a deformation quantization in this case too. Namely, we translate L_0 to L_0' by t to obtain $\mathbb{L}_t$. Then, the non-commutative algebra $\Lambda Q^{\mathbb{L}}/(\partial\Phi_t)$ is more complicated where

$$\Phi_t = Axyzw + Bwzyx + \frac{C}{2}((wx)^2 + (yz)^2) + \frac{D}{2}((xy)^2 + (zw)^2).$$

When $t \to 0$, we have $A = -B, C = D = 0$ recovering the previous construction. One can compute A, B, C, D explicitly as certain theta functions. These A, B, C, D are realized as coefficients of certain matrix factorizations for $\mathbb{L}_0$ case, and from this observation, we prove that (A, B, C, D) in fact lies on the mirror complete intersection in $\mathbb{P}^3$. The non-commutative algebra $\mathscr{A}_t = \Lambda Q^{\mathbb{L}}/(\partial\Phi_t)$ is a deformation quantization of the conifold, and $\mathscr{A}_t/(W_t)$ is a deformation quantization of the complete intersection in $\mathbb{C}^4$, whose Poisson structure is given by the Nambu bracket.

References

[Abo] M. Abouzaid, *Family Floer cohomology and mirror symmetry*, preprint, arXiv:1404.2659.

[AJ10] Manabu Akaho and Dominic Joyce, *Immersed Lagrangian Floer theory*, J. Differential Geom. **86** (2010), no. 3, 381–500. MR2785840

[AKO08] Denis Auroux, Ludmil Katzarkov, and Dmitri Orlov, *Mirror symmetry for weighted projective planes and their noncommutative deformations*, Ann. of Math. (2) **167** (2008), no. 3, 867–943, DOI 10.4007/annals.2008.167.867. MR2415388

[ATdB90] M. Artin, J. Tate, and M. Van den Bergh, *Some algebras associated to automorphisms of elliptic curves*, The Grothendieck Festschrift, Vol. I, Progr. Math., vol. 86, Birkhäuser Boston, Boston, MA, 1990, pp. 33–85. MR1086882

[AZ06] Marco Aldi and Eric Zaslow, *Seidel's mirror map for abelian varieties*, Adv. Theor. Math. Phys. **10** (2006), no. 4, 591–602. MR2259690

[CHKL] Cheol-Hyun Cho, Hansol Hong, Sang-hyun Kim, and Siu-Cheong Lau, *Lagrangian Floer potential of orbifold spheres*, Adv. Math. **306** (2017), 344–426, DOI 10.1016/j.aim.2016.10.017. MR3581306

[CHL13] Cheol-Hyun Cho, Hansol Hong, and Siu-Cheong Lau, *Localized mirror functor for Lagrangian immersions, and homological mirror symmetry for* $\mathbb{P}^1_{a,b,c}$, J. Differential Geom. **106** (2017), no. 1, 45–126. MR3640007

[CHL14] Cheol-Hyun Cho, Hansol Hong, and Siu-Cheong Lau, *Localized mirror functor constructed from a Lagrangian torus*, preprint (2014), arXiv:1406.4597.

[CHL15] Cheol-Hyun Cho, Hansol Hong, and Siu-Cheong Lau, *Non-commutative homological mirror functor*, preprint (2015). arXiv:1512.07128

[CO06] Cheol-Hyun Cho and Yong-Geun Oh, *Floer cohomology and disc instantons of Lagrangian torus fibers in Fano toric manifolds*, Asian J. Math. **10** (2006), no. 4, 773–814, DOI 10.4310/AJM.2006.v10.n4.a10. MR2282365

[CP12] Cheol-Hyun Cho and Mainak Poddar, *Holomorphic orbi-discs and Lagrangian Floer cohomology of symplectic toric orbifolds*, J. Differential Geom. **98** (2014), no. 1, 21–116. MR3263515

[Efi12] Alexander I. Efimov, *Homological mirror symmetry for curves of higher genus*, Adv. Math. **230** (2012), no. 2, 493–530, DOI 10.1016/j.aim.2012.02.022. MR2914956

[EG10] Pavel Etingof and Victor Ginzburg, *Noncommutative del Pezzo surfaces and Calabi-Yau algebras*, J. Eur. Math. Soc. (JEMS) **12** (2010), no. 6, 1371–1416, DOI 10.4171/JEMS/235. MR2734346

[FOOO09] Kenji Fukaya, Yong-Geun Oh, Hiroshi Ohta, and Kaoru Ono, *Lagrangian intersection Floer theory: anomaly and obstruction. Part II*, AMS/IP Studies in Advanced Mathematics, vol. 46, American Mathematical Society, Providence, RI; International Press, Somerville, MA, 2009. MR2548482

[Fuk02] Kenji Fukaya, *Floer homology and mirror symmetry. II*, Minimal surfaces, geometric analysis and symplectic geometry (Baltimore, MD, 1999), Adv. Stud. Pure Math., vol. 34, Math. Soc. Japan, Tokyo, 2002, pp. 31–127. MR1925734

[Kon95] Maxim Kontsevich, *Homological algebra of mirror symmetry*, Proceedings of the International Congress of Mathematicians, Vol. 1, 2 (Zürich, 1994), Birkhäuser, Basel, 1995, pp. 120–139. MR1403918

[Lee] Sangwook Lee, *Comparison of mirror functors of the elliptic curve via lg-cy correspondence*, Preprint (2015).

[Oh] Yong-Geun Oh, *Addendum to: "Floer cohomology of Lagrangian intersections and pseudo-holomorphic disks. I." [Comm. Pure Appl. Math.* ***46*** *(1993), no. 7, 949–993; MR1223659 (95d:58029a)]*, Comm. Pure Appl. Math. **48** (1995), no. 11, 1299–1302, DOI 10.1002/cpa.3160481104. MR1367384

[PZ98] Alexander Polishchuk and Eric Zaslow, *Categorical mirror symmetry: the elliptic curve*, Adv. Theor. Math. Phys. **2** (1998), no. 2, 443–470, DOI 10.4310/ATMP.1998.v2.n2.a9. MR1633036

[Sei11] Paul Seidel, *Homological mirror symmetry for the genus two curve*, J. Algebraic Geom. **20** (2011), no. 4, 727–769, DOI 10.1090/S1056-3911-10-00550-3. MR2819674

[She11] Nick Sheridan, *On the homological mirror symmetry conjecture for pairs of pants*, J. Differential Geom. **89** (2011), no. 2, 271–367. MR2863919

[She15] Nick Sheridan, *Homological mirror symmetry for Calabi-Yau hypersurfaces in projective space*, Invent. Math. **199** (2015), no. 1, 1–186, DOI 10.1007/s00222-014-0507-2. MR3294958

[SYZ96] Andrew Strominger, Shing-Tung Yau, and Eric Zaslow, *Mirror symmetry is T-duality*, Nuclear Phys. B **479** (1996), no. 1-2, 243–259, DOI 10.1016/0550-3213(96)00434-8. MR1429831

Department of Mathematical Sciences, Seoul National University, Gwanak-gu, Seoul, South Korea

E-mail address: chocheol@snu.ac.kr

Mirror symmetry, Tyurin degenerations and fibrations on Calabi-Yau manifolds

Charles F. Doran, Andrew Harder, and Alan Thompson

ABSTRACT. We investigate a potential relationship between mirror symmetry for Calabi-Yau manifolds and the mirror duality between quasi-Fano varieties and Landau-Ginzburg models. More precisely, we show that if a Calabi-Yau admits a so-called Tyurin degeneration to a union of two Fano varieties, then one should be able to construct a mirror to that Calabi-Yau by gluing together the Landau-Ginzburg models of those two Fano varieties. We provide evidence for this correspondence in a number of different settings, including Batyrev-Borisov mirror symmetry for K3 surfaces and Calabi-Yau threefolds, Dolgachev-Nikulin mirror symmetry for K3 surfaces, and an explicit family of threefolds that are not realized as complete intersections in toric varieties.

1. Introduction

The aim of this paper is to investigate the relationship between mirror symmetry for Calabi-Yau manifolds and for Fano varieties of the same dimension.

Classically, mirror symmetry is a network of conjectures relating the properties of two mirror dual Calabi-Yau manifolds. For us, unless otherwise stated, a *Calabi-Yau manifold* will always be a smooth compact Kähler manifold V with trivial canonical bundle $\omega_V \cong \mathcal{O}_V$ and vanishing cohomology groups $H^i(V, \mathcal{O}_V)$ for all $0 < i < \dim(V)$.

A similar duality has been proposed for Fano varieties. In physics, Eguchi, Hori and Xiong [**EHX97**] postulated that a d-dimensional manifold X with $c_1(X) > 0$ should be mirror to a *Landau-Ginzburg model* (Y, w), where Y is a d-dimensional Kähler manifold and w is a function $\mathsf{w}\colon Y \to \mathbb{C}$. This correspondence was then incorporated into the framework of homological mirror symmetry, as a correspondence between the directed Fukaya category [**Sei01b**] (resp. the bounded derived

2010 *Mathematics Subject Classification.* Primary 14J33; Secondary 14J32, 14J30, 14J28, 14D06.

Key words and phrases. Mirror symmetry, Calabi-Yau threefold, K3 surface, fibration, degeneration.

The first author would like to acknowledge support from the Natural Sciences and Engineering Research Council of Canada (NSERC) and the Visiting Campobassi Professorship of the University of Maryland.

The second author was supported by an NSERC Post-Graduate Scholarship.

The third author was supported by a Fields-Ontario Postdoctoral Fellowship with funding provided by NSERC and the Ontario Ministry of Training, Colleges and Universities.

category of singularities [**Orl09**]) associated to (Y, w) and the bounded derived category of coherent sheaves on X (resp. the Fukaya category of X). More recently, Katzarkov, Kontsevich and Pantev [**KKP14**] conjectured that if X is a Fano variety, then the Landau-Ginzburg model (Y, w) of X is in fact a quasi-projective variety that satisfies certain specific conditions and, moreover, that there is a mirror relationship between the Hodge numbers of X and certain Hodge-theoretic invariants of (Y, w).

It is expected that this notion of mirror symmetry for Fano varieties is related to classical mirror symmetry for Calabi-Yau manifolds of one dimension lower, as follows. If X is a d-dimensional Fano variety with mirror Landau-Ginzburg model (Y, w), then a general fibre of w is expected to be a $(d-1)$-dimensional Calabi-Yau variety that is mirror, in the classical sense, to a generic anticanonical hypersurface in X.

This raises a natural question: is mirror symmetry for d-dimensional Fano varieties related to classical mirror symmetry for Calabi-Yau manifolds of the same dimension d? In this paper we outline a correspondence that provides a potential answer to this question.

This correspondence may be described as follows. Let V be a d-dimensional Calabi-Yau manifold and suppose that V admits a degeneration to a union $X_1 \cup_Z X_2$ of two quasi-Fano varieties glued along an anticanonical hypersurface Z (such degenerations are called *Tyurin degenerations*). Then we claim that the mirror W of V admits a fibration by $(d-1)$-dimensional Calabi-Yau manifolds, with general fibre S that is mirror to Z. Moreover, W can be constructed topologically by gluing together the Landau-Ginzburg models (Y_1, w_1) and (Y_2, w_2) of X_1 and X_2, in a sense to be made precise in Section 2.2.

The first person to observe traces of such a correspondence was probably Dolgachev [**Dol96**], who noticed that Dolgachev-Nikulin mirror symmetry for K3 surfaces matches Type II degenerations (of which Tyurin degenerations are a special case) with elliptic fibrations on the mirror. After this, the first mention of a higher dimensional version appears to be due to Tyurin, who gave a brief hint of its existence at the very end of [**Tyu04**].

More recently, a variant of the construction presented here was worked out in detail by Auroux [**Aur08**], in the special case where V is a double cover of a Fano variety X ramified over a smooth member of $|-2K_X|$; this V admits a Tyurin degeneration to the union of X with itself.

The structure of this paper is as follows. In Section 2 we describe our construction. We begin with a d-dimensional Calabi-Yau manifold V which admits a Tyurin degeneration to a union $X_1 \cup_Z X_2$ of quasi-Fano varieties glued along an anticanonical hypersurface Z. Then we show that the Landau-Ginzburg models (Y_1, w_1) and (Y_2, w_2) of X_1 and X_2 may be glued together to form a new variety W, which is fibred by Calabi-Yau $(d-1)$-folds topologically mirror to Z, so that the Euler numbers of V and W satisfy the mirror relationship $\chi(V) = (-1)^d \chi(W)$. This suggests that V and W should be thought of as mirror dual. In the threefold case we provide even more evidence for this conjecture: if we make the assumptions that W is Calabi-Yau and that the K3 surface Z is Dolgachev-Nikulin mirror to a general fibre of the fibration on W, then we can show that V and W are in fact topologically mirror.

In the remaining sections of the paper we discuss this correspondence in several special cases. In Section 3 we discuss the case of Batyrev-Borisov mirror symmetry for surfaces and threefolds. Indeed, suppose that V is a K3 surface or Calabi-Yau threefold constructed as an anticanonical hypersurface in a Gorenstein toric Fano 3- or 4-fold, determined by a reflexive polytope Δ. We show that a nef partition Δ_1, Δ_2 of Δ determines both a Tyurin degeneration $X_1 \cup_Z X_2$ of V and a fibration $\pi\colon W \to \mathbb{P}^1$ on a birational model W of its Batyrev mirror, so that the general fibre of π is Batyrev-Borisov mirror dual to the intersection $Z = X_1 \cap X_2$.

Specializing to the threefold case, we further show that the singular fibres of the K3 surface fibration $\pi\colon W \to \mathbb{P}^1$ contain numerical information about X_1 and X_2, and describe a relationship between W and the Landau-Ginzburg models of X_1 and X_2. Unfortunately a corresponding result in the K3 surface case is difficult to prove for combinatorial reasons, but we conjecture the form that such a result should take.

Section 4 is concerned with Dolgachev-Nikulin mirror symmetry for K3 surfaces. We revisit Dolgachev's **[Dol96]** mirror correspondence between Type II degenerations (of which Tyurin degenerations are a special case) and elliptic fibrations, which may be thought of as a generalization of the correspondence described in Section 2. Consideration of several explicit examples suggests a way to enhance our conjectures to cope with more general Type II degenerations, which may contain more than two components.

In Section 5 we discuss how this theory fits with classical mirror symmetry for threefolds. We begin by showing that, if V is a Calabi-Yau threefold that undergoes a Tyurin degeneration (satisfying certain technical conditions), then mirror symmetry predicts the existence of a K3 fibration on the mirror threefold W, with properties consistent with those expected from the theory in Section 2. Following this, we specialize our discussion to the case of threefolds fibred by quartic mirror K3 surfaces, as studied in **[DHNT16]**. In this setting we explicitly construct candidate mirror threefolds, along with Tyurin degenerations of them, and show that they have the properties predicted by Section 2. In particular, this provides an important illustration of our theory using threefolds that are not complete intersections in toric varieties, thereby giving evidence that the ideas of Section 2 apply beyond the toric setting of Section 3.

Finally, Section 6 discusses the limitations of our construction. Indeed, it appears that difficulties arise for Tyurin degenerations of V which occur along loci in moduli that are disjoint from points of maximally unipotent monodromy. In this case, we seem to have no guarantee of the existence of a mirror fibration on W; an example where this occurs is given in Example 6.1. Instead we present evidence that, if W is replaced by its bounded derived category of coherent sheaves $\mathbf{D}^b(W)$, it should be possible to find a non-commutative fibration of $\mathbf{D}^b(W)$ by Calabi-Yau categories, which might be thought of as homologically mirror to the Tyurin degeneration of V.

2. Setup and preliminary results

Our aim is to provide evidence for a mirror correspondence between a certain type of degeneration of Calabi-Yau manifolds, called a Tyurin degeneration, and Calabi-Yau manifolds constructed by gluing Landau-Ginzburg models. We begin by defining these objects.

2.1. Smoothing Tyurin degenerations. A smooth variety X is called a *quasi-Fano* variety if its anticanonical linear system contains a smooth Calabi-Yau member and $H^i(X, \mathcal{O}_X) = 0$ for all $i > 0$. Given this, a *Tyurin degeneration* is a degeneration $\mathcal{V} \to \Delta$ of Calabi-Yau manifolds over the unit disc $\Delta \subset \mathbb{C}$, such that the total space $\mathcal{V}$ is smooth and the central fibre is a union of two quasi-Fano varieties that meet normally along a smooth variety Z, with $Z \in |-K_{X_i}|$ for each $i \in \{1, 2\}$. Degenerations of this type have been studied by Lee [**Lee06**], who coined the name.

This construction can be reversed, and a family of Calabi-Yau manifolds built up from a pair of quasi-Fano varieties X_1 and X_2 as follows. Let Z be a smooth variety which is a member of both $|-K_{X_1}|$ and $|-K_{X_2}|$, and suppose that there are ample classes $D_1 \in \mathrm{Pic}(X_1)$ and $D_2 \in \mathrm{Pic}(X_2)$ which both restrict to the same ample class $D \in \mathrm{Pic}(Z)$ (this last condition is needed to ensure that [**KN94**, Theorem 4.2], which gives the existence of a smoothing, can be applied in our setting). Let $X_1 \cup_Z X_2$ denote the variety which is a normal crossings union of X_1 and X_2 meeting along Z.

With this setup, we say that $X_1 \cup_Z X_2$ is *smoothable* to a Calabi-Yau manifold V if there exists a complex manifold $\mathcal{V}$ equipped with a map $\psi\colon \mathcal{V} \to \Delta$ so that the fibre $\psi^{-1}(0) = X_1 \cup_Z X_2$, the fibre $\psi^{-1}(t)$ is a smooth Calabi-Yau manifold for any $t \in \Delta \setminus \{0\}$, and V is a general fibre of $\mathcal{V}$. It follows from a theorem of Kawamata and Namikawa [**KN94**, Theorem 4.2] that $X_1 \cup_Z X_2$ is smoothable to a Calabi-Yau manifold V if and only if N_{Z/X_1} and N_{Z/X_2} are inverses of one another and, moreover, that the resulting manifold V is unique up to deformation.

2.2. Gluing Landau-Ginzburg models. Let us first define what we mean by Landau-Ginzburg (LG) model in this paper. In [**Har16**], a notion of a LG model is defined which conjecturally encapsulates the LG models of Fano varieties, and even goes further to describe the LG models of many quasi-Fano varieties. For general quasi-Fano varieties, however, we do not believe that this definition is sufficient; in particular, it seems that for general quasi-Fanos we must drop any expectation that our LG model be algebraic.

For this reason, in this paper we adopt a much more general definition.

DEFINITION 2.1. A *Landau-Ginzburg* (*LG*) *model* of a quasi-Fano variety is a pair (Y, w) consisting of a a Kähler manifold Y satisfying $h^1(Y) = 0$ and a proper map $\mathsf{w}\colon Y \to \mathbb{C}$. The map w is called the *superpotential*.

Note that this definition leaves room for the image of w to be an open set in $\mathbb{C}$. If Y is quasi-projective then the Hodge numbers of such LG models (Y, w) are defined in [**KKP14**]; however, in the general case it is unclear how this should be done. Instead, following [**KKP14**], we propose that if (Y, w) is the LG model of a quasi-Fano variety X, then we should have

$$h^i(Y, \mathsf{w}^{-1}(t)) = \sum_j h^{d-i+j,j}(X), \tag{2.1}$$

where $h^i(Y, \mathsf{w}^{-1}(t))$ is the rank of the cohomology group of the pair $H^i(Y, \mathsf{w}^{-1}(t))$ and t is a generic point in the image of w. We also expect that if (Y, w) is the LG model of X, then the smooth fibres of w should be mirror to generic anticanonical hypersurfaces in X.

With notation as in the previous section, it now seems pertinent to ask whether there is any relationship between the LG models of the quasi-Fano varieties X_1 and X_2, and mirror symmetry for V. Indeed, it seems natural to expect that these LG models could be somehow glued together to give a mirror W for V, since we are, in a topological sense, gluing X_1 and X_2 together to form V (see **[Tyu04]** for details on this topological construction).

In more detail, we expect that if Y_i is the LG model of X_i, equipped with superpotential w_i, then the monodromy symplectomorphism on $\mathsf{w}_i^{-1}(t)$ (for t a regular value of w_i) associated to a small loop around ∞ can be identified under mirror symmetry with the restriction of the Serre functor of the bounded derived category of coherent sheaves $\mathbf{D}^b(X_i)$ on X_i to the bounded derived category of coherent sheaves $\mathbf{D}^b(Z)$ on Z **[Sei01a, KKP14]**. This Serre functor is simply $(-) \otimes \omega_{X_i}[d]$ where $[d]$ denotes shift by $d = \dim X_i$. Thus, up to a choice of shift, we see that the action of monodromy on $\mathsf{w}_i^{-1}(t)$ should be identified with the autoequivalence of $\mathbf{D}^b(Z)$ induced by taking the tensor product with $\omega_{X_i}|_Z = N_{Z/X_i}^{-1}$.

Now recall that if $X_1 \cup_Z X_2$ is smoothable to V, then we have $N_{Z/X_1} \otimes N_{Z/X_2} = \mathcal{O}_Z$, so the monodromy symplectomorphism ϕ_1 associated to a clockwise loop around infinity on $\mathsf{w}_1^{-1}(t)$ should be same as the monodromy ϕ_2^{-1} associated to a *counter-clockwise* loop around infinity on $\mathsf{w}_2^{-1}(t)$. It should be noted that, for this to make sense, we must assume that the fibres of w_1 and w_2 are topologically the same Calabi-Yau manifold, which we denote by S; this assumption is stronger than the assumption that both are mirror to Z.

Now we glue these LG models as follows. For each $i \in \{1, 2\}$, choose r_i so that $|\lambda| \leq r_i$ for every λ in the critical locus of w_i. Then choose local trivializations of Y_i over $U_i = \{z \in \mathbb{C} : |z| > r_i\}$ and let $Q_i = \mathsf{w}_i^{-1}(U_i)$. This local trivialization is topologically equivalent to expressing Q_i as a gluing of the ends of $B_i = S \times [-1, 1] \times (-1, 1)$ together via the map

$$\widetilde{\phi}_i \colon p \times \{-1\} \times (z) \longmapsto \phi_i(p) \times \{1\} \times (z),$$

where ϕ_i is the monodromy symplectomorphism, and we identify $S \times \{-1\} \times (-1, 1)$ with $S \times \{1\} \times (-1, 1)$.

Assuming that $\phi_1 = \phi_2^{-1}$ (which, we recall, conjecturally follows from smoothability of $X_1 \cup_Z X_2$), we can identify B_1 with B_2 by the map

$$\tau \colon p \times [x] \times (z) \longmapsto p \times [-x] \times (-z).$$

Under this identification of B_1 and B_2, it is clear that $\tau \cdot \widetilde{\phi}_1 = \widetilde{\phi}_2$. Thus the identification τ gives an isomorphism between Q_1 and Q_2, allowing us to glue Y_1 to Y_2 along Q_1 and Q_2 to produce a C^∞ manifold W. This gluing respects the fibrations w_1 and w_2, so W is equipped with a fibration π over the gluing of $\mathbb{C}$ with $\mathbb{C}$ described above. It is clear that the base of this fibration is just the 2-sphere S^2. This procedure is illustrated in Figure 1.

EXAMPLE 2.2. As a sanity check, we can perform this construction with elliptic curves. Take a degeneration of an elliptic curve to a union of two copies of $\mathbb{P}^1$ meeting in two points (Kodaira type I_2). The LG model of $\mathbb{P}^1$ is the map from $\mathbb{C}^\times$, which is topologically a twice-punctured rational curve, to $\mathbb{C}$ given by

$$\mathsf{w} \colon x \longmapsto x + \frac{1}{x}.$$

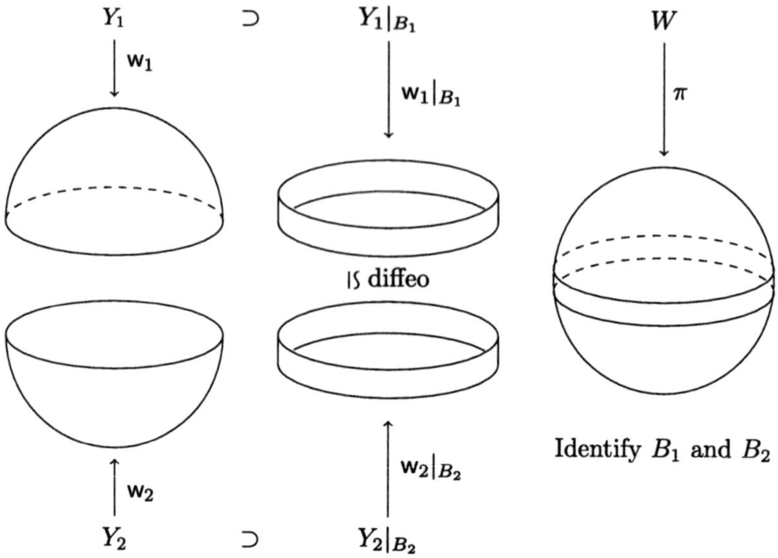

FIGURE 1. Gluing Y_1 and Y_2 to give W

This map w is a double covering of $\mathbb{A}^1$ ramified at two points. One can check that monodromy of this fibration around the point at infinity is trivial. Let Y_1 and Y_2 be copies of this LG model of $\mathbb{P}^1$. Then we may glue Y_1 and Y_2 as described above. The resulting topological space is a double cover of S^2 which is ramified at four points. This is simply the 2-dimensional torus, which is topologically mirror to the original elliptic curve

THEOREM 2.3. *Let X_1 and X_2 be d-dimensional quasi-Fano varieties which contain the same anticanonical Calabi-Yau hypersurface Z, such that $K_{X_1}|_Z + K_{X_2}|_Z = 0$. Let (Y_1, w_1) and (Y_2, w_2) be Landau-Ginzburg models of X_1 and X_2, and suppose that the fibres of w_1 and w_2 are topologically the same Calabi-Yau manifold, which is topologically mirror to Z. Finally, let V be a Calabi-Yau variety obtained from $X_1 \cup_Z X_2$ by smoothing and let W be the variety obtained by gluing Y_1 to Y_2 as above. Then*

$$\chi(V) = (-1)^d \chi(W),$$

where χ denotes the Euler number.

PROOF. Start by recalling the long exact sequence of the pair $(Y_i, \mathsf{w}_i^{-1}(t))$, for t a regular value of w_i,

$$\cdots \to H^n(Y_i, \mathbb{C}) \to H^n(\mathsf{w}_i^{-1}(t), \mathbb{C}) \to H^{n+1}(Y_i, \mathsf{w}_i^{-1}(t); \mathbb{C}) \to H^{n+1}(Y_i, \mathbb{C}) \to \cdots .$$

Since Euler numbers are additive in long exact sequences, we have that $\chi(Y_i) = \chi(Y_i, \mathsf{w}_i^{-1}(t)) + \chi(\mathsf{w}_i^{-1}(t))$. By Equation (2.1), we see that $\chi(Y_i, \mathsf{w}^{-1}(t))$ is equal to $(-1)^d \chi(X_i)$, where d is the dimension of Y_i. Thus $\chi(Y_i) = (-1)^d \chi(X_i) + \chi(\mathsf{w}_i^{-1}(t)))$. Moreover, since $\mathsf{w}^{-1}(t)$ is topologically mirror to Z by assumption, we have that $\chi(Z) = (-1)^{d-1} \chi(\mathsf{w}^{-1}(t))$, which gives $\chi(Y_i) = (-1)^d (\chi(X_i) - \chi(Z))$.

On the other hand, the Mayer-Vietoris exact sequence

$$\cdots \to H^n(W, \mathbb{C}) \to H^n(Y_1, \mathbb{C}) \oplus H^n(Y_2, \mathbb{C}) \to H^n(Y_1 \cap Y_2, \mathbb{C}) \to \cdots$$

gives $\chi(W) = \chi(Y_1)+\chi(Y_2)-\chi(Y_1\cap Y_2)$. Since $Y_1\cap Y_2$ is a fibration over an annulus, we can compute its cohomology using the Wang sequence [**PS08**, Theorem 11.33]

$$\cdots \to H^n(Y_1\cap Y_2,\mathbb{C}) \to H^n(\mathsf{w}^{-1}(t),\mathbb{C}) \xrightarrow{T_n-\mathrm{Id}} H^n(\mathsf{w}^{-1}(t),\mathbb{C}) \to \cdots,$$

where T_n is the action of monodromy on $H^n(\mathsf{w}^{-1}(t),\mathbb{C})$ associated to a small loop around our annulus, to obtain $\chi(Y_1\cap Y_2) = 0$. Putting everything together, we obtain

$$\chi(W) = (-1)^d(\chi(X_1)+\chi(X_2)-2\chi(Z)).$$

Finally, since $X_1\cup_Z X_2$ is smoothable to V, we can compute the Euler characteristic of V by applying [**Lee06**, Proposition IV.6], which states that

$$\chi(V) = \chi(X_1)+\chi(X_2)-2\chi(Z).$$

We therefore have that $\chi(W) = (-1)^d\chi(V)$, as claimed. □

This is precisely the relationship between the Euler characteristics of mirror dual Calabi-Yau varieties. In the next subsection, we will provide more evidence for the hypothesis that W is the mirror dual of the original Calabi-Yau variety V, in the special case where V is a Calabi-Yau threefold.

REMARK 2.4. Note that the requirement that there exist two ample divisors D_1 and D_2, on X_1 and X_2 respectively, which restrict to the same divisor on Z was not used at all in the proof of Theorem 2.3. Moreover, despite the fact that the proof of [**KN94**, Theorem 4.2] uses this assumption in a material way (in order to prove the pro-representability of the log deformation functor), the topological construction of the gluing of X_1 and X_2 can be performed without it.

For instance, let us take a generic K3 surface Z with Picard lattice of rank 2 isomorphic to the lattice with Gram matrix

$$\begin{pmatrix} 4 & 6 \\ 6 & 6 \end{pmatrix}.$$

Such a K3 surface embeds into both $\mathbb{P}^3$ and the intersection of a quadric Q and a cubic C in $\mathbb{P}^5$. Let us blow up $\mathbb{P}^3$ in $Z\cap Z'$ for some generic K3 surface Z' in $\mathbb{P}^3$, calling the result X_1, and blow up $Q\cap C$ in the intersection of Z and a generic hyperplane section in $\mathbb{P}^5$, calling the result X_2. Then the normal crossings variety $X_1\cup_Z X_2$ is not Kähler, so we cannot find D_1 and D_2 as above. However, both V and W can be constructed, as C^∞ manifolds, from $X_1\cup_Z X_2$ by the method we have described. We wonder whether V and W represent a mirror pair of non-Kähler Calabi-Yau manifolds.

2.3. The threefold case. With notation as before, Lee [**Lee10**] has computed the Hodge numbers of V in the case where X_1 and X_2 are smooth threefolds. Let us define $\rho_i\colon H^2(X_i,\mathbb{Q})\to H^2(Z,\mathbb{Q})$ for $i=1,2$ to be the restriction and define $k = \mathrm{rank}(\mathrm{im}(\rho_1)+\mathrm{im}(\rho_2))$.

THEOREM 2.5. [**Lee10**, Corollary 8.2] *Let V be a Calabi-Yau threefold constructed as as smoothing of $X_1\cup_Z X_2$, as above. Then*

$$h^{1,1}(V) = h^2(X_1)+h^2(X_2)-k-1,$$
$$h^{2,1}(V) = 21+h^{2,1}(X_1)+h^{2,1}(X_2)-k.$$

On the other side of the picture, we have a corresponding result for W.

PROPOSITION 2.6. *Let W be as above and let S be a general fibre of the map π. Assume that* $\dim W = \dim S + 1 = 3$. *Then*

$$h^2(W) = 1 + h^2(Y_1, S) + h^2(Y_2, S) + \ell,$$

where ℓ is the rank of the subgroup of $H^2(S, \mathbb{C})$ spanned by the intersection of the images of $H^2(Y_1, \mathbb{C})$ and $H^2(Y_2, \mathbb{C})$ under the natural restriction maps.

PROOF. Let U be the annulus along which B_1 and B_2 are glued, and let $Q = \pi^{-1}(U)$ be its preimage in W. We begin by computing the rank of $H^2(W, \mathbb{C})$ using the Mayer-Vietoris sequence

$$\cdots \to H^1(Q, \mathbb{C}) \to H^2(W, \mathbb{C}) \to H^2(Y_1, \mathbb{C}) \oplus H^2(Y_2, \mathbb{C}) \xrightarrow{r_1^Q - r_2^Q} H^2(Q, \mathbb{C}) \to \cdots,$$

where r_i^Q are the natural restriction maps from $H^2(Y_i, \mathbb{C})$ to $H^2(Q, \mathbb{C})$. From the Wang sequence, we obtain $H^1(Q, \mathbb{C}) = \mathbb{C}$. So, using the assumption that $H^1(Y_1, \mathbb{C}) = H^1(Y_2, \mathbb{C}) = 0$, we see that $H^2(W, \mathbb{C})$ is isomorphic to the direct product of $\mathbb{C}$ and the kernel of the restriction map $r_1^Q - r_2^Q$. We note that this map fits into a commutative triangle

$$\begin{array}{ccc} H^2(Y_1, \mathbb{C}) \oplus H^2(Y_2, \mathbb{C}) & \xrightarrow{r_1^Q - r_2^Q} & H^2(Q, \mathbb{C}) \\ & \searrow^{r_1^S - r_2^S} & \downarrow r_Q^S \\ & & H^2(S, \mathbb{C}). \end{array}$$

Now, since S is a K3 surface, we have $h^1(S) = 0$, and it follows from the Wang sequence that the map r_Q^S is injective. Thus the kernel of $r_1^S - r_2^S$ is the same as the kernel of $r_1^Q - r_2^Q$. Elementary linear algebra gives that the rank of this kernel is $h^2(Y_1) + h^2(Y_2) - \mathrm{rank}(\mathrm{im}(r_1^S) + \mathrm{im}(r_2^S))$. So we obtain

$$h^2(W) = 1 + h^2(Y_1) + h^2(Y_2) - \mathrm{rank}(\mathrm{im}(r_1^S) + \mathrm{im}(r_2^S)).$$

Now, for $i = 1, 2$ we have exact sequences

$$0 \longrightarrow H^2(Y_i, S; \mathbb{C}) \longrightarrow H^2(Y_i, \mathbb{C}) \xrightarrow{r_i^S} H^2(S, \mathbb{C}) \longrightarrow \cdots.$$

Which give

$$h^2(Y_i) = h^2(Y_i, S) + \mathrm{rank}(\mathrm{im}(r_i^S)).$$

Putting together with the previous expression, the proposition follows. □

Therefore, if W admits a complex structure for which it is Calabi-Yau, then we compute

$$\begin{aligned} \chi(W) &= 2h^{1,1}(W) - 2h^{2,1}(W) \\ &= 2(1 + h^2(Y_1, S) + h^2(Y_2, S) + \ell) - 2h^{2,1}(W). \end{aligned}$$

Equation (2.1) then gives that $h^2(Y_i, S) = h^{2,1}(X_i)$, so

$$\chi(W) = 2(h^{2,1}(X_1) + h^{2,1}(X_2) - h^{2,1}(W) + \ell + 1).$$

Furthermore, from Theorems 2.3 and 2.5, we also know that

$$\begin{aligned}\chi(W) &= -\chi(V)\\ &= -2h^{1,1}(V) + 2h^{2,1}(V)\\ &= -2(h^2(X_1) + h^2(X_2) - k - 1) + 2(h^{2,1}(X_1) + h^{2,1}(X_2) + 21 - k)\\ &= 2(h^{2,1}(X_1) + h^{2,1}(X_2) - h^2(X_1) - h^2(X_2) + 22)\end{aligned}$$

Putting this together, we have that $h^{2,1}(W) = \ell - 21 + h^2(X_1) + h^2(X_2)$. So in order for W and V to be topologically mirror to one another, we must have

$$\ell - 21 + h^2(X_1) + h^2(X_2) = h^2(X_1) + h^2(X_2) - k - 1,$$

which is equivalent to $\ell + k = 20$. This is true if S and Z are mirror dual in the sense of Dolgachev-Nikulin, given the lattice polarization on Z (resp. S) coming from the sum of the images of the restriction maps $H^2(X_i,\mathbb{Z}) \to H^2(Z,\mathbb{Z})$ (resp. the intersection of the images of the restriction maps $H^2(Y_i,\mathbb{Z}) \to H^2(S,\mathbb{Z})$). Thus mirror symmetry for V and W is consistent with mirror symmetry for S and Z.

REMARK 2.7. In the case where W is Calabi-Yau and S and Z are Dolgachev-Nikulin mirror dual, the expressions

$$\begin{aligned}h^{2,1}(V) &= 21 + h^{2,1}(X_1) + h^{2,1}(X_2) - k\\ h^{1,1}(W) &= 1 + h^2(Y_1, S) + h^2(Y_2, S) + \ell\end{aligned}$$

could be thought of as mirror dual decompositions of the corresponding Hodge numbers, in the following sense.

The Hodge number $h^{2,1}(X_i)$ may be interpreted as the fibre dimension of the natural map from the moduli space of pairs (X_i, Z) to the moduli space of appropriately polarized K3 surfaces Z (which has dimension $20-k$). Thus the degenerate fibre $X_1 \cup_Z X_2$ should have $h^{2,1}(X_1)+h^{2,1}(X_2)+20-k = h^{2,1}(V)-1$ deformations, and such Tyurin degenerations should appear in codimension 1 in the moduli space of V.

We thus obtain a decomposition of $h^{2,1}(V)$ into contributions $h^{2,1}(X_i)$ coming from deformations of each X_i, a contribution $(20 - k)$ from deformations of the gluing locus Z, and 1 for the codimension in the moduli space.

On the mirror side a similar statement holds: $h^{1,1}(W)$ can be decomposed into contributions $h^2(Y_i, S)$ coming from the LG-models (Y_i, w_i) (these will be interpreted later as counts of components in singular fibres), a contribution ℓ from divisors on the generic fibre S, and 1 for the class of a general fibre (compare [**DHNT16**, Lemma 3.2]). The picture is completed by noting that $h^2(Y_i, S) = h^{2,1}(X_i)$ and $\ell = 20 - k$.

3. Batyrev-Borisov mirror symmetry

In this section we will prove a number of results that illustrate the situation considered in the previous section in the special case of Batyrev-Borisov mirror symmetry. For background on the definitions and concepts used in this section, we refer the reader to [**CK99, CLS11**]. However, since our conventions differ very slightly from those used in the references above, before we proceed we will briefly outline the notation to be used in the remainder of this section.

Let M be a free $\mathbb{Z}$-module of rank d, let Δ be a reflexive polytope in $M \otimes \mathbb{R} = M_{\mathbb{R}}$, and denote the boundary of Δ by $\partial\Delta$. Let $N = \mathrm{Hom}(M, \mathbb{Z})$ be the dual lattice to M and denote by $\langle \cdot, \cdot \rangle$ the natural bilinear pairing from $N \times M$ to $\mathbb{Z}$. Let

$$\Delta^{\circ} = \{u \in N_{\mathbb{R}} : \langle u, v\rangle \geq -1 \text{ for all } v \in \Delta\}$$

denote the polar polytope to Δ.

Let $\mathbb{P}_{\Delta}$ be the d-dimensional toric variety associated to the polytope Δ. The toric variety $\mathbb{P}_{\Delta}$ is Fano and has at worst Gorenstein singularities. Following [**Bat94**, Theorem 2.2.24], one can find a toric variety X_{Δ} which is a toric partial resolution of singularities of $\mathbb{P}_{\Delta}$ and which has at worst Gorenstein terminal singularities. Such X_{Δ} is referred to as a *maximal projective crepant partial (mpcp) resolution of singularities* of $\mathbb{P}_{\Delta}$. In the future, we shall fix one such X_{Δ} for any given $\mathbb{P}_{\Delta}$. The variety X_{Δ} can be presented as a quotient of some Zariski open subset $U \subseteq \mathbb{C}^{|\partial\Delta \cap M|}$ by the torus $(\mathbb{C}^{\times})^{|\partial\Delta|-d}$. There is thus a homogeneous coordinate ring $\mathbb{C}[\{z_{\rho}\}_{\rho \in \partial\Delta \cap M}]$ on X_{Δ}.

The vanishing of each coordinate z_{ρ} determines a divisor on X_{Δ}, invariant under the natural action of the torus $(\mathbb{C}^{\times})^d$, which we call D_{ρ}. The anticanonical divisor $-K_{X_{\Delta}}$ of X_{Δ} is linearly equivalent to $\sum_{\rho \in \partial\Delta \cap M} D_{\rho}$, and the cycle class group $A_1(X_{\Delta})$ is generated by the divisors D_{ρ}. A divisor $\sum_{\rho \in \partial\Delta \cap M} b_{\rho} D_{\rho}$ for $b_{\rho} \in \mathbb{Z}$ is Cartier if and only if there is a piecewise linear function φ on $M_{\mathbb{R}}$, which takes integral values on M and which is linear on the cones of the fan defining X_{Δ}, so that $\varphi(\rho) = b_{\rho}$ for all ρ.

A *nef partition* of Δ is a partition of $\partial\Delta \cap M$ into sets $E_1, \ldots, E_k$, so that for each $i = 1, \ldots, k$, the divisor $\sum_{\rho \in E_i} D_{\rho}$ is nef and Cartier. Let us denote the line bundle thus associated to E_i by $\mathcal{L}_i$. We will let $\Delta_i = \mathrm{Conv}(E_i \cup \{0_M\})$; in a mild abuse of terminology, we also refer to $\Delta_1, \ldots, \Delta_k$ as a nef partition of Δ.

Batyrev's [**Bat94**] toric version of mirror symmetry claims that the generic anticanonical hypersurfaces in X_{Δ} and $X_{\Delta^{\circ}}$ are mirror dual. Moreover, if we have a nef partition of Δ, then the complete intersection V of generic sections of $\mathcal{L}_1, \ldots, \mathcal{L}_k$ is again Calabi-Yau. Borisov [**Bor93**] and Batyrev-Borisov [**BB96**] propose that there is a similar combinatorial construction of the mirror of V. In this case, we define

$$\nabla_i = \left\{ u \in N_{\mathbb{R}} : \begin{array}{l} \langle u, v\rangle \geq 0 \text{ for all } v \in E_j, j \neq i \\ \langle u, v\rangle \geq -1 \text{ for all } v \in E_i \end{array} \right\}$$

and let $\nabla = \mathrm{Conv}(\nabla_1 \cup \cdots \cup \nabla_k)$. This is a reflexive polytope and $\nabla_1, \ldots, \nabla_k$ is a nef partition of ∇. The complete intersection W in X_{∇} cut out by generic sections of the line bundles associated to $\nabla_1, \ldots, \nabla_k$ is a Calabi-Yau variety, which is expected to be mirror dual to V.

Finally, a *refinement* of a nef partition $E_1, \ldots, E_k$ is defined to be another nef partition $F_1, \ldots, F_{k+1}$ so that $F_i = E_i$ for $1 \leq i \leq k-1$ and $E_k = F_k \cup F_{k+1}$.

Now, let X_{Δ} be a d-dimensional toric variety as above. Suppose that V is a Calabi-Yau complete intersection of nef divisors in X_{Δ}, determined by a nef partition $E_1, \ldots, E_k$. Our aim is to show that, if $F_1, \ldots, F_{k+1}$ is a refinement of $E_1, \ldots, E_k$, then this combinatorial data determines

- a Tyurin degeneration of V, and
- a pencil of quasi-smooth varieties birational to Calabi-Yau $(d-k-1)$-folds inside of the Batyrev-Borisov mirror W.

In the case where V is a threefold, we show that this pencil induces a K3 surface fibration on some birational model of W and that the singular fibres of this fibration carry information about the Tyurin degeneration of V. We will then compare this with the LG model picture in the previous section.

3.1. Tyurin degenerations. More precisely, let $\mathcal{L}_i$ be the line bundles on X_Δ associated to the E_i. The refinement $F_1, \ldots, F_{k+1}$ gives rise to a pair of nef line bundles $\mathcal{L}'_k$ and $\mathcal{L}'_{k+1}$ so that $\mathcal{L}'_k \otimes \mathcal{L}'_{k+1} = \mathcal{L}_k$. Let $s_i \in H^0(X_\Delta, \mathcal{L}_i)$ be generic sections determining a quasi-smooth Calabi-Yau complete intersection V in X_Δ. If we let s'_k and s'_{k+1} be sections of $\mathcal{L}'_k$ and $\mathcal{L}'_{k+1}$ respectively, then $s'_k s'_{k+1}$ is a section of $\mathcal{L}_k$.

We can use this to construct a pencil of complete intersections as follows. First, let

$$V' = \cap_{i=1}^{k-1}\{s_i = 0\}$$

and assume that V' is connected and quasi-smooth; it is also clear that V' is quasi-Fano. Then take the pencil

$$\mathcal{Q}\colon \{ts_k - s'_k s'_{k+1} = 0\} \cap V'$$

in $\mathbb{A}^1 \times X_\Delta$, with t a parameter on $\mathbb{A}^1$. If we assume that X_Δ is a smooth resolution of $\mathbb{P}_\Delta$, then the only singularities of $\mathcal{Q}$ in a neighbourhood of $0 \in \mathbb{A}^1$ are along $t = s_k = s'_k = s'_{k+1} = 0$, which we call Σ.

Note that since Σ is the intersection of a set of nef divisors in X_Δ, it has no base locus and its singularities are contained in the singular set of X_Δ. Furthermore, the intersection of Σ with any torus invariant subvariety of X_Δ is irreducible, thus Σ itself is either irreducible or a union of non-intersecting subvarieties of X_Δ. As a result, if X_Δ is smooth, then so is Σ, for general enough choices of sections.

We can thus resolve the singularities of $\mathcal{Q}$ by blowing up $t = s'_k = 0$ inside of $\mathbb{A}^1 \times X_\Delta$ and taking the proper transform of $\mathcal{Q}$. The result is a Tyurin degeneration of V, so that the fibre over 0 of the degeneration is equal to the union of $\hat{X}_1$ and X_2, where $\hat{X}_1$ is a quasi-Fano variety given by blowing up $X_1 := V' \cap \{s'_k = 0\}$ along $V' \cap \{s_k = s'_k = s'_{k+1} = 0\}$ and X_2 is a quasi-Fano variety given by $X_2 := V' \cap \{s'_{k+1} = 0\}$.

In the general situation, where X_Δ is not a smooth resolution of $\mathbb{P}_\Delta$, we can still perform all of the steps above, but we will have singularities occurring at every step in general. The resulting degeneration will not be a Tyurin degeneration in the strict sense, but should still include data corresponding to the quasi-Fano varieties X_1, X_2 and the blown up locus Σ. We note here that a version of the smoothability result of Kawamata and Namikawa that works for mildly singular varieties has been explored in the thesis of Lee [**Lee06**]. The singular case may also be interpreted as equipping the union of X_1 and X_2 with a log structure (see e.g. [**ACG13**] and the references therein), which accounts for the subvariety Σ and determines the smoothing to V.

3.2. Pencils and fibrations on the mirror. Now we will look at how this nef partition is reflected in the mirror. For ease of notation we restrict ourselves to the case where V is a hypersurface; all of the results below generalize in the obvious way to refinements of k-partite nef partitions corresponding to codimension k complete intersections.

Since V is a hypersurface, the nef partition $E_1 = \Delta$ is trivial. Let Δ_1, Δ_2 denote the polytopes corresponding to the refinement F_1, F_2 of E_1; we thus have $\Delta = \mathrm{Conv}(\Delta_1 \cup \Delta_2)$.

Now, the Batyrev dual of V is a Calabi-Yau variety W embedded as an anti-canonical hypersurface in X_{Δ°. By definition, W is cut out by an equation in the homogeneous coordinate ring of X_{Δ°, which may be written as

$$f := \sum_{\rho \in \Delta \cap M} a_\rho \prod_{\sigma \in \partial\Delta^\circ \cap N} z_\sigma^{\langle \sigma, \rho \rangle + 1} = 0,$$

where a_ρ are generically chosen complex coefficients. We will take a pencil $\mathcal{P}$ of hypersurfaces in W, for $[s : t] \in \mathbb{P}^1$, defined by the intersection of W with hypersurfaces of the form

$$s \sum_{\rho \in \Delta_1 \cap M \setminus 0_M} a_\rho \prod_{\sigma \in \partial\Delta^\circ \cap N} z_\sigma^{\langle \sigma, \rho \rangle + 1} = t a_0 \prod_{\sigma \in \partial\Delta^\circ \cap N} z_\sigma.$$

Note that, away from $[s : t] = [0 : 1]$, this pencil may also be defined by the pair of equations

$$s \sum_{\rho \in \Delta_1 \cap M \setminus 0_M} a_\rho \prod_{\sigma \in \partial\Delta^\circ \cap N} z_\sigma^{\langle \sigma, \rho \rangle + 1} = t a_0 \prod_{\sigma \in \partial\Delta^\circ \cap N} z_\sigma$$
$$s \sum_{\rho \in \Delta_2 \cap M \setminus 0_M} a_\rho \prod_{\sigma \in \partial\Delta^\circ \cap N} z_\sigma^{\langle \sigma, \rho \rangle + 1} = (s - t) a_0 \prod_{\sigma \in \partial\Delta^\circ \cap N} z_\sigma.$$

as the sum of these two equations is just sf, which vanishes precisely along W if $s \neq 0$.

We wish to show that the pencil $\mathcal{P}$ induces a fibration on W. If we can show that a general member of this pencil is Calabi-Yau, then this will follow from:

PROPOSITION 3.1. *If $S \subseteq W$ is a smooth Calabi-Yau $(d-1)$-fold in a d-dimensional Calabi-Yau manifold, then the linear system $|S|$ is base-point free and hence there is a map $\pi : W \to \mathbb{P}^1$ with S as a fibre.*

PROOF. By adjunction, $\mathcal{O}_S(S) = N_{S/W} = \omega_S = \mathcal{O}_S$. Let $\iota : S \hookrightarrow W$ be the inclusion map, then we have a short exact sequence of sheaves

$$0 \longrightarrow \mathcal{O}_W \xrightarrow{s} \mathcal{O}_W(S) \longrightarrow \iota^* \mathcal{O}_S \longrightarrow 0,$$

where s is a section of $\mathcal{O}_W(S)$ whose vanishing locus is S. This short exact sequence gives rise to a long exact sequence in cohomology

$$0 \longrightarrow H^0(W, \mathcal{O}_W) \longrightarrow H^0(W, \mathcal{O}_W(S)) \longrightarrow H^0(W, \iota^* \mathcal{O}_S) \longrightarrow 0,$$

where the vanishing of $H^1(W, \mathcal{O}_W)$ follows from the Calabi-Yau property of W. From this sequence, we see that the restriction of a generic section of $\mathcal{O}_W(S)$ to S is a nonzero section of $\iota^* \mathcal{O}_S$, which is *a fortiori* non-vanishing. Thus $|S|$ is base-point free and determines a map $W \to \mathbb{P}^1$, since $h^0(\mathcal{O}_W(S)) = 2$ by the exact sequence above. □

To apply this proposition, we need to show that a general member S of the pencil $\mathcal{P}$ is Calabi-Yau. When $\dim(W) = 2$ or 3, this will follow from the next proposition, which is proved in [**Har16**].

PROPOSITION 3.2. **[Har16]** *If* $\dim(W) = 2$, *then a general member of the pencil* $\mathcal{P}$ *is a smooth elliptic curve. If* $\dim(W) = 3$, *then a general member of the pencil* $\mathcal{P}$ *is a smooth blow-up of a K3 surface* S *and, moreover,* S *is Batyrev-Borisov dual to the intersection of the quasi-Fano varieties* X_1 *and* X_2 *from Section* 3.1.

In the case where $\dim(W) = 2$, it therefore follows immediately from Proposition 3.1 that the pencil $\mathcal{P}$ is an elliptic fibration on W. However, if W is a Calabi-Yau threefold and a general member of $\mathcal{P}$ is a blown up K3 surface $\hat{S}$, then we will need to get rid of the (-1)-curves in S before we can apply Proposition 3.1. We will do this by performing a series of birational transformations.

LEMMA 3.3. *Let* W *be a Calabi-Yau threefold and let* $\hat{S}$ *be a blown up K3 surface in* W. *If* C *is a* (-1)-*curve in* $\hat{S}$, *then* $N_{C/W} \cong \mathcal{O}_{\mathbb{P}^1}(-1) \oplus \mathcal{O}_{\mathbb{P}^1}(-1)$.

PROOF. We have a short exact sequence of sheaves on C,

$$0 \longrightarrow \Theta_C \longrightarrow \Theta_W|_C \longrightarrow N_{C/W} \longrightarrow 0.$$

Since $c_1(\Theta_C) = 2$ and $c_1(\Theta_W|_C) = c_1(\Theta_W)|_C = 0$, it follows that $N_{C/W} \cong \mathcal{O}_{\mathbb{P}^1}(a) \oplus \mathcal{O}_{\mathbb{P}^1}(b)$ for some $a, b \in \mathbb{Z}$ with $a + b = -2$ (see, for example, [**Kat92**, Section 1]).

We may embed the normal bundle $N_{C/\hat{S}}$ into $N_{C/W}$ to get a short exact sequence of line bundles

$$0 \longrightarrow N_{C/\hat{S}} \longrightarrow N_{C/W} \longrightarrow \mathcal{L} \longrightarrow 0$$

for some line bundle $\mathcal{L}$. Since C is a (-1)-curve in $\hat{S}$, we know that $N_{C/\hat{S}} \cong \mathcal{O}_{\mathbb{P}^1}(-1)$. Furthermore, we have that $c_1(\mathcal{L}) = -1$ from the fact that $c_1(N_{C/W}) = -2$, thus $\mathcal{L} \cong \mathcal{O}_{\mathbb{P}^1}(-1)$. The long exact sequence in cohomology coming from the above short exact sequence proves that $H^0(C, N_{C/W}) = 0$ and therefore we must have that $N_{C/W} = \mathcal{O}_{\mathbb{P}^1}(-1) \oplus \mathcal{O}_{\mathbb{P}^1}(-1)$. □

Therefore, any (-1)-curve C in $\hat{S}$ may be blown up to produce a a variety $\widetilde{W}$ with exceptional divisor a copy of $\mathbb{P}^1 \times \mathbb{P}^1$. This copy of $\mathbb{P}^1 \times \mathbb{P}^1$ can be smoothly contracted along either ruling. Contracting along one ruling recovers W; we denote the variety obtained by contracting along the other ruling by W^+. The strict transform of $\hat{S}$ in $\widetilde{W}$ is just $\hat{S}$ itself, but the contraction $\hat{W} \to W^+$ contracts the (-1)-curve C in $\hat{S}$. This is an example of a birational operation called a *flop*.

Repeating this for all (-1) curves in $\hat{S}$, we obtain a birational model of W which is a smooth Calabi-Yau threefold in which $\hat{S}$ has been contracted to its minimal model, which is a K3 surface. Call the Calabi-Yau threefold resulting from this process $\hat{W}$. Applying Proposition 3.1 to $\hat{W}$, we see that we have proved:

THEOREM 3.4. *Let* W *be a Calabi-Yau threefold containing a smooth blown up K3 surface* $\hat{S}$. *By performing a sequence of flops on* W, *we may obtain a birational model* $\hat{W}$ *of* W *which admits a fibration* $\pi\colon \hat{W} \to \mathbb{P}^1$, *so that the minimal model* S *of* $\hat{S}$ *is a general fibre of* π.

Putting everything together, in the case of toric hypersurface Calabi-Yau threefolds we obtain:

COROLLARY 3.5. *If* Δ *is a* 4-*dimensional reflexive polytope that admits a nef partition* Δ_1, Δ_2 *and* V *is a general anticanonical hypersurface in* X_Δ, *then* V *admits a Tyurin degeneration and its Batyrev dual* W *has a birational model which admits a fibration by K3 surfaces. Moreover, the general fibre in this fibration on*

W is Batyrev-Borisov mirror dual to the complete intersection K3 surface in X_Δ determined by Δ_1, Δ_2.

REMARK 3.6. There seems to be an inherent incompatibility between Batyrev (and Batyrev-Borisov) duality and K3 surface fibrations on Calabi-Yau threefolds, since one can show that even in very basic examples, there are exceptional curves in $\hat{S}$ that cannot be avoided by simply changing the birational model of the toric ambient space X_{Δ°. A notable exception occurs when either Δ_1 or Δ_2 is 1-dimensional. In this case, one of the two component quasi-Fano varieties to which V degenerates is itself a toric variety. This is mirrored by the fact that X_{Δ° admits a morphism to $\mathbb{P}^1$ which induces the required K3 fibration on W. These seem to be a subset of the "toric fibrations" which have been studied extensively in the physics literature by a number of authors ([**AKMS97, CCS13, GP13**] to name a few).

REMARK 3.7. One can consider more general refinements of nef partitions, by taking a nef partition $F_1, \dots, F_\ell$ so that for each E_i, there is a subset I_i of $\{1, \dots, \ell\}$ so that $E_i = \cup_{j \in I_i} F_j$. These will give rise to generalized degenerations of the Calabi-Yau V determined by $E_1, \dots, E_k$ to unions of quasi-Fano varieties, and families of Calabi-Yau varieties of codimension $\ell - k$ in W. The issue here, of course, is that it is hard to prove that this gives a fibration on a birational model of W. Despite this, these families of Calabi-Yau varieties surely have properties related to the LG models of the appropriate quasi-Fano varieties.

REMARK 3.8. Generalizing Corollary 3.5 to higher dimensions seems to be a challenge, since we have made use of both the minimal model program for surfaces and a characterization of flops in three dimensions. Of course, both of these objects have analogues in higher dimensions, but they are much more oblique and not likely to be useful in such a general situation. As it stands, the results above suffice to show that we obtain rational maps from W to $\mathbb{P}^1$ corresponding to every bipartite nef partition of Δ.

3.3. Singular fibres and a comparison with LG models. Now we will analyze the singular members of the pencil $\mathcal{P}$, which are birational to the fibres of $\hat{W}$. Using this, we can give a very accurate description of the singular fibres of $\hat{W}$, up to birational transformations. As we shall see, the resulting theory meshes nicely with the LG model picture described in Section 2. The results in this section can be extended quite generally, but for simplicity we will restrict ourselves to the situation of threefold hypersurfaces.

With notation as in the previous section, the nef partition Δ_1, Δ_2 of Δ determines a pair of polytopes $\nabla_1, \nabla_2 \subseteq \Delta^\circ$, which are a nef partition of $\nabla := \mathrm{Conv}(\nabla_1 \cup \nabla_2) \subseteq \Delta^\circ$. Note that this inclusion may be strict; it therefore does not follow that we have a degeneration of W to quasi-Fano varieties dual to X_1 and X_2.

Our first goal is to look at the member of the pencil $\mathcal{P}$ over $[s:t] = [0:1]$ and understand its meaning in terms of the Tyurin degeneration described in Section 3.1.

PROPOSITION 3.9. *The member of the pencil $\mathcal{P}$ corresponding to $[s:t] = [0:1]$ is equal to*

$$\bigcup_{\sigma \in (\Delta^\circ \setminus \nabla) \cap N} (D_\sigma \cap W).$$

In other words, the linear system defining the pencil $\mathcal{P}$ is associated to the line bundle $\mathcal{O}_W(\sum_{\sigma\in(\Delta^\circ\setminus\nabla)\cap N} D_\sigma|_W)$.

PROOF. Recall that the pencil $\mathcal{P}$ is defined as the intersection of W with hypersurfaces of the form

$$s \sum_{\rho\in\Delta_1\cap M\setminus 0_M} a_\rho \prod_{\sigma\in\partial\Delta^\circ\cap N} z_\sigma^{\langle\sigma,\rho\rangle+1} - ta_0 \prod_{\sigma\in\partial\Delta^\circ\cap N} z_\sigma = 0.$$

Thus, at least on an open set of W, we may write this family of hypersurfaces as a rational map from X_{Δ° to $\mathbb{P}^1$, defined by $\phi\colon [z_\sigma] \mapsto [s : t]$ where

$$s = \prod_{\sigma\in(\Delta^\circ\setminus\nabla)\cap N} z_\sigma$$

$$t = \left(\sum_{\rho\in\Delta_1\cap M\setminus 0_M} a_\rho \prod_{\sigma\in\partial\Delta^\circ\cap N} z_\sigma^{\langle\sigma,\rho\rangle+1}\right) \Big/ \left(\prod_{\sigma\in\partial\nabla\cap N} z_\sigma\right)$$

We want to show that this map is defined on W away from the base locus of $\mathcal{P}$. Note that homogeneity away from $\sum_{\sigma\in\partial\nabla\cap N} D_\sigma$ is clear, since both terms are sections of $\mathcal{L} := \mathcal{O}_{X_{\Delta^\circ}}(\sum_{\sigma\in(\Delta^\circ\setminus\nabla)\cap N} D_\sigma)$. Now, if $\sigma\in\nabla_2$, then the numerator in the expression for t above has a factor of z_σ, since $\langle\sigma,\rho\rangle \geq 0$ for all $\rho\in\Delta_1$ and $\sigma\in\nabla_2$. Thus both terms are also sections of $\mathcal{L}$ along D_σ for $\sigma\in\nabla_2$. Moreover, on the restriction of ϕ to W, we notice that the expression for t can also be written as

$$\left(\sum_{\rho\in\Delta_2\cap M} a_\rho \prod_{\sigma\in\partial\Delta^\circ\cap N} z_\sigma^{\langle\sigma,\rho\rangle+1}\right) \Big/ \left(\prod_{\sigma\in\partial\nabla\cap N} z_\sigma\right).$$

and thus, for the same reason as above, both terms are sections of $\mathcal{L}$ along $D_\sigma\cap W$ for $\sigma\in\nabla_1$. Finally, [**BB96**, Proposition 6.3] implies that $\nabla\cap N = (\nabla_1\cap N)\cup(\nabla_2\cap N)$, so we deduce that the expressions for s and t above form honest global sections of $\mathcal{L}|_W$. Thus the map ϕ is well-defined away from the base locus of $\mathcal{L}|_W$ and the fibre of ϕ over $s=0$ is as required. □

It follows from the proof of Proposition 3.9 that the line bundle $\mathcal{O}_W(\hat{S})$ is just $\mathcal{O}_W(\sum_{\sigma\in(\Delta^\circ\setminus\nabla)\cap N} D_\sigma)$. Since W is an anticanonical hypersurface, the intersection of a divisor D_σ with W is empty if and only if σ lies in the relative interior of a facet of Δ°. If σ is in the interior of a codimension 2 face of Δ°, then a calculation analogous to that performed in [**Roh04**, §3.3] shows that $D_\rho\cap W$ has $1+\ell^*(\Gamma(\sigma))\ell^*(\Gamma(\sigma)^\vee)$ irreducible components, where $\Gamma(\sigma)$ is the smallest face of Δ° containing σ, $\Gamma(\sigma)^\vee$ is the face of Δ made up of points ρ satisfying $\langle\sigma,\rho\rangle=-1$, and $\ell^*(\Gamma)$ denotes the number of lattice points in the relative interior of Γ. Finally, if σ lies in a codimension ≥ 3 face of Δ°, then $D_\rho\cap W$ is irreducible for generic W.

PROPOSITION 3.10. *If σ is in $(\Delta^\circ\setminus\nabla)\cap N$, then $D_\rho\cap W$ has a single irreducible component. Therefore, the member of the pencil $\mathcal{P}$ corresponding to $[s:t]=[0:1]$ has*

$$\#(\Delta^\circ\setminus\nabla)\cap N$$

irreducible components.

PROOF. First, if σ is contained on the relative interior of a facet of Δ°, then $\Gamma(\sigma)^\vee$ is a single vertex η of Δ. Without loss of generality, we can assume that $\eta\in\Delta_1$. Therefore, $\langle\sigma,\rho\rangle\geq -1$ for all points $\rho\in\Delta$ and $\langle\sigma,\rho\rangle=-1$ if and only if

$\rho = \eta$, so, by definition, σ is in ∇_1. Since, by [**BB96**, Proposition 6.3], all points of $\nabla \cap N$ are either in ∇_1 or ∇_2, it follows that no point of $(\Delta^\circ \setminus \nabla) \cap N$ is in the interior of a facet of Δ°. Thus for any point $\sigma \in (\Delta^\circ \setminus \nabla) \cap N$, the intersection $D_\sigma \cap W$ is nonempty.

It just remains to treat the case where σ lies in a codimension 2 face of Δ°. Since σ lies in $(\Delta^\circ \setminus \nabla) \cap N$, by definition there must be some $\rho_1 \in \Delta_1$ and $\rho_2 \in \Delta_2$ so that $\langle \rho_1, \sigma \rangle = \langle \rho_2, \sigma \rangle = -1$. Therefore, $\Gamma(\sigma)^\vee$ contains points in both Δ_1 and Δ_2, so is a face of neither. Given this, [**BB96**, Proposition 6.3] implies that $\Gamma(\sigma)^\vee$ does not contain any points in its relative interior. So $\ell^*(\Gamma(\sigma)^\vee) = 0$ and hence $D_\sigma \cap W$ has a single irreducible component. □

REMARK 3.11. For W of arbitrary dimension, the same proof works to find the number of components of the member of the pencil $\mathcal{P}$ corresponding to $[s : t] = [0 : 1]$. However, if $\dim(W) \geq 4$ we do not know whether this may be interpreted as a count of components of a singular fibre in a fibration on some birational model $\hat{W}$ of W (see Remark 3.8).

Next we show that this number also has meaning with respect to the mirror Calabi-Yau variety V and its degeneration to the union of X_1 and X_2.

PROPOSITION 3.12. *If* $\dim(\nabla_1) = \dim(\nabla_2) = 4$, *then* $V \cap X_1 \cap X_2$ *is an irreducible curve* C *of genus*

$$g(C) = \#(\Delta^\circ \setminus \nabla) \cap N - 1.$$

PROOF. By construction, C is a complete intersection of sections of the line bundles $\omega_{X_\Delta}^{-1}$ (which determines V), $\mathcal{L}_1 = \mathcal{O}_{X_\Delta}(\sum_{\rho \in \Delta_1 \cap N} D_\rho)$ (which determines X_1), and $\mathcal{L}_2 = \mathcal{O}_{X_\Delta}(\sum_{\rho \in \Delta_2 \cap N} D_\rho)$ (which determines X_2). The Koszul complex resolving $\mathcal{O}_C$ is thus given by

$$\omega_{X_\Delta}^2 \to (\mathcal{L}_1^{-1} \otimes \omega_{X_\Delta}) \oplus (\mathcal{L}_2^{-1} \otimes \omega_{X_\Delta}) \oplus \omega_{X_\Delta} \to \mathcal{L}_1^{-1} \oplus \mathcal{L}_2^{-1} \oplus \omega_{X_\Delta} \to \mathcal{O}_{X_\Delta}$$

The corresponding second spectral sequence converges to $H^i(C, \mathcal{O}_C[3])$, so

$$\bigoplus_{p+q=i+3} {}''E_\infty^{p,q} \cong H^i(C, \mathcal{O}_C).$$

The relevant portion of ${}''E_1^{p,q}$ is given by

$$\begin{array}{cccccc}
H^4(\omega_{X_\Delta}^2) & \to & H^4(\mathcal{L}_1^{-1} \otimes \omega_{X_\Delta}) \oplus H^4(\mathcal{L}_2^{-1} \otimes \omega_{X_\Delta}) \oplus H^4(\omega_{X_\Delta}) & \to & \mathbb{C} & \to 0 \\
0 & \to & 0 & \to & 0 & \to 0 \\
0 & \to & 0 & \to & 0 & \to 0 \\
0 & \to & 0 & \to & 0 & \to 0 \\
0 & \to & 0 & \to & 0 & \to \mathbb{C}
\end{array}$$

Now, by [**BB96**, Theorem 2.5], we know that

$$\begin{aligned}
h^4(\omega_{X_\Delta}^2) &= \ell^*(2\Delta^\circ), \\
h^4(\omega_{X_\Delta}) &= 1 \\
h^4(\mathcal{L}_1^{-1} \otimes \omega_{X_\Delta}) &= \ell^*(\nabla_1 + \Delta^\circ), \\
h^4(\mathcal{L}_2^{-1} \otimes \omega_{X_\Delta}) &= \ell^*(\nabla_2 + \Delta^\circ).
\end{aligned}$$

It is not then hard to see that this spectral sequence degenerates at the ${}''E_2$ term and $h^0(\mathcal{O}_C) = 1$, hence C is irreducible. Since $h^i(\mathcal{O}_C) = 0$ for $i > 1$, we have that the top row of $''E_1^{p,q}$ above is exact except at the left-most term. Thus we can compute that

$$g(C) = \ell^*(2\Delta^\circ) - (\ell^*(\nabla_1 + \Delta^\circ) + \ell^*(\nabla_2 + \Delta^\circ)).$$

It remains to show that this is precisely the number of points in $(\Delta^\circ \setminus \nabla) \cap N$. For this we need a small lemma.

LEMMA 3.13. *If Q is either ∇_i or Δ°, the number $\ell^*(Q + \Delta^\circ)$ is equal to $\ell(Q)$, where $\ell(Q)$ denotes the number of lattice points in Q.*

PROOF. The polytope Δ° is defined by the inequalities $\langle \sigma, \rho \rangle \geq -1$ for all points $\rho \in \Delta$. Similarly, ∇_1 is defined by the inequalities $\langle \sigma, \rho \rangle \geq -1$ for all points $\rho \in \Delta_1$ and $\langle \sigma, \rho \rangle \geq 0$ for all points $\rho \in \Delta_2$. We shall prove the lemma for $Q = \nabla_1$; the other cases are analogous.

Now, the polytope $\nabla_1 + \Delta^\circ$ is defined by the inequalities $\langle \sigma, \rho \rangle \geq -2$ for $\rho \in \Delta_1$ and $\langle \sigma, \rho \rangle \geq -1$ for $\rho \in \Delta_2$. Therefore a point in the interior of $\nabla_1 + \Delta^\circ$ satisfies these inequalities strictly, and thus any lattice point in the interior of $(\nabla_1 + \Delta^\circ)$ has $\langle \sigma, \rho \rangle \geq 0$ for all $\rho \in \Delta_2$ and $\langle \sigma, \rho \rangle \geq -1$ for all $\rho \in \Delta_1$. But this is just the set of all lattice points in ∇_1. □

From this lemma, we see that

$$\ell^*(2\Delta^\circ) - (\ell^*(\nabla_1 + \Delta^\circ) + \ell^*(\nabla_2 + \Delta^\circ)) = \ell(\Delta^\circ) - \ell(\nabla_1) - \ell(\nabla_1).$$

Moreover, [**BB96**, Proposition 6.3] shows that all lattice points of ∇ are lattice points of either ∇_1 or ∇_2, so this is equal to

$$\ell(\Delta^\circ) - \ell(\nabla) + 1 = \#(\Delta^\circ \setminus \nabla) \cap N + 1;$$

here the extra $(+1)$ term corresponds to the fact that we have over-counted the origin, which is the intersection of ∇_1 and ∇_2. This completes the proof of Proposition 3.12. □

REMARK 3.14. A very minor modification of this proof shows that, in the case where V has dimension $d \geq 3$, we have $h^{d-2,0}(V \cap X_1 \cap X_2) = \#(\Delta^\circ \setminus \nabla) \cap N - 1$. If $\dim(V) = 2$, then $h^0(V \cap X_1 \cap X_2) = \#(\Delta^\circ \setminus \nabla) \cap N$.

Putting everything together, we obtain the following theorem.

THEOREM 3.15. *If $\dim(V) = \dim(W) = d \geq 3$, then the member of the pencil of hypersurfaces $\mathcal{P}$ corresponding to $[s : t] = [0 : 1]$ has exactly $h^{d-2,0}(V \cap X_1 \cap X_2) + 1$ components. If $\dim(V) = \dim(W) = 2$, then $V \cap X_1 \cap X_2$ is a set of points and the member of the pencil of hypersurfaces $\mathcal{P}$ corresponding to $[s : t] = [0 : 1]$ has exactly $\#(V \cap X_1 \cap X_2)$ components.*

Next we analyze the rest of the members of the pencil $\mathcal{P}$ on W. Our goal is to show that the members corresponding to $[1 : 0]$ and $[1 : 1]$ are essentially the singular fibres of the LG models of X_1 and X_2. Thus there is a very real sense in which the pencil $\mathcal{P}$ on W is collecting information about the LG models of X_1 and X_2.

First, however, we describe how these LG models are constructed. In [**Har16**], it is shown that the naïve compactification of Givental's [**Giv98**] Landau-Ginzburg model for a complete intersection X in a toric variety X_Δ is smooth if X has dimension less than or equal to 3, and otherwise has only mild singularities.

This compactification is defined as follows. Assume that we have a polytope Δ and a nef partition Δ_1, Δ_2 of Δ, so that Δ_1 and Δ_2 contain no interior points. In this setting, a general enough global section of the line bundle $\mathcal{L}_1$ associated to Δ_1 determines a quasi-Fano hypersurface X in X_Δ. The compactified version of Givental's LG model for X is then the complete intersection $Y \subset X_\nabla \times \mathbb{A}^1$ cut out by the equations

$$\sum_{\rho \in \Delta_1 \cap M} a_\rho \prod_{\sigma \in \nabla \cap N} z_\sigma^{\langle \sigma, \rho \rangle - \sigma^1_{\min}} = 0$$

$$t a_0 \prod_{\sigma \in \nabla_2 \cap N \setminus 0_N} z_\sigma - \sum_{\rho \in \Delta_2 \cap M \setminus 0_M} a_\rho \prod_{\sigma \in \partial \nabla \cap N} z_\sigma^{\langle \sigma, \rho \rangle - \sigma^2_{\min}} = 0$$

where a_ρ are complex constants, t is the coordinate on $\mathbb{A}^1$, and $\sigma^i_{\min}$ is -1 if σ is in ∇_i and 0 otherwise. The superpotential w is just projection of this complete intersection onto $\mathbb{A}^1$. It is shown in [**Har16**] that (Y, w) has the expected properties for an LG model of X.

With this in place, we find:

THEOREM 3.16. *The members of the pencil $\mathcal{P}$ corresponding to $[s : t] = [1 : 1]$ and $[1 : 0]$ are birational to the fibres over 0 of the LG models (Y_1, w_1) and (Y_2, w_2) of X_1 and X_2 respectively. In fact, for any choice of W and hypersurface $\mathcal{P}([1 : t])$ with $t \in \mathbb{C}$, there is a choice of LG model (Y, w) of either X_1 or X_2 so that $\mathcal{P}([1 : t])$ is birational to a fibre of (Y, w).*

PROOF. Recall from Section 3.2 that we have an expression for a birational model of W as the complete intersection in $\mathbb{P}^1[s, t] \times X^\circ_\Delta$ given by the vanishing of

$$f_1 := s \sum_{\rho \in \Delta_1 \cap M \setminus 0_M} a_\rho \prod_{\sigma \in \partial \Delta^\circ \cap N} z_\sigma^{\langle \sigma, \rho \rangle + 1} - t a_0 \prod_{\sigma \in \partial \Delta^\circ \cap N} z_\sigma$$

$$f_2 := s \sum_{\rho \in \Delta_2 \cap M \setminus 0_M} a_\rho \prod_{\sigma \in \partial \Delta^\circ \cap N} z_\sigma^{\langle \sigma, \rho \rangle + 1} - (s - t) a_0 \prod_{\sigma \in \partial \Delta^\circ \cap N} z_\sigma .$$

Note that f_1 has a factor of $\prod_{\sigma \in \nabla_2 \cap N \setminus 0_N} z_\rho$ by the definition of ∇_2, and an analogous statement holds for f_2, by the definition of ∇_2.

Now if we let $[s : t] = [1 : 1]$, then we obtain the complete intersection of

$$f_1|_{[1:1]} = \sum_{\rho \in \Delta_1 \cap M} a_\rho \prod_{\sigma \in \partial \Delta^\circ \cap N} z_\sigma^{\langle \sigma, \rho \rangle + 1}$$

$$f_2|_{[1:1]} = \sum_{\rho \in \Delta_2 \cap M \setminus 0_M} a_\rho \prod_{\sigma \in \partial \Delta^\circ \cap N} z_\sigma^{\langle \sigma, \rho \rangle + 1} .$$

Note that this is precisely the complete intersection determining the fibre over 0 of the LG model of X_1, except compactified in X_{Δ° instead of X_∇.

To compare these compactifications, define a birational map φ from X_{Δ° to X_∇, which sends z_σ to z_σ if $\sigma \in \partial \nabla \cap N$. The restriction of φ to the complement of the torus invariant loci of codimension ≥ 2 simply has the effect of removing those codimension 1 tori corresponding to points $\sigma \in (\Delta^\circ \setminus \nabla) \cap N$. Thus we see that φ induces a birational map between $\mathcal{P}([1 : 1])$ and the fibre over 0 of (Y_1, w_1) if no components of $\mathcal{P}([1 : 1])$ are contained in torus invariant loci of X_{Δ° of codimension ≥ 2, and no component of $\mathcal{P}([1 : 1])$ is contained in a divisor D_σ for $\sigma \in (\Delta^\circ \setminus \nabla) \cap N$.

The first of these two claims is trivial: since each component of $\mathcal{P}([1:1])$ is of codimension 2 in X_{Δ°, it is contained in a codimension ≥ 2 torus invariant subvariety of X_{Δ° if and only if it is the closure of such a torus invariant subvariety. Since W contains no torus invariant subvarieties of X_{Δ°, this cannot happen. The second claim follows from the fact that if $\sigma \in (\Delta^\circ \setminus \nabla) \cap N$, then $D_\sigma \cap W$ is in $\mathcal{P}([0:1])$. Thus $\mathcal{P}([1:1])$ can only intersect D_σ in at most a codimension 2 subvariety of X_{Δ°.

An identical argument suffices to show that $\mathcal{P}([1:0])$ is birational to the fibre over 0 of (Y_2, w_2), and in fact this shows that for any $p \in \mathbb{P}^1 \setminus \{[0:1]\}$, for a generic choice of W there is a choice of $i \in \{1,2\}$ and an LG model (Y_i, w_i) so that $\mathcal{P}(p)$ is birational to a fibre of (Y_i, w_i). □

As a philosophical remark, this proves that all of the interesting data surrounding the pencil $\mathcal{P}$ on W is related to either the LG models of X_1 and X_2, or the variety $V \cap X_1 \cap X_2$. Indeed, this is the same data as was used to determine the Tyurin degeneration of V: X_1 and X_2 were the quasi-Fano hypersurfaces in X_Δ, and $V \cap X_1 \cap X_2$ was the locus that needed to be blown up in X_1 to obtain a smooth degeneration.

In the case where both X_1 and X_2 are pullbacks to X_Δ of ample hypersurfaces in $\mathbb{P}_\Delta$ along the mpcp resolution map, then we can say even more using the following theorem, proved in [**Har16**].

THEOREM 3.17. [**Har16**] *With notation as above, suppose in addition that X_i is the pull-back to X_Δ of an ample hypersurface in $\mathbb{P}_\Delta$ and that* $\dim \Delta = d \geq 4$. *Let (Y_i, w_i) denote the LG model of X_i and let ρ_0 denote the number of components in its singular fibre over* 0. *Then*

$$\rho_0 = h^{d-2,1}(X_i) + 1.$$

From this, we immediately obtain:

COROLLARY 3.18. *Suppose that* $\dim(V) = \dim(W) = 3$ *and let ρ_p be the number of irreducible components in the member of the pencil $\mathcal{P}$ corresponding to $p \in \mathbb{P}^1$. If X_1 and X_2 are pullbacks of ample hypersurfaces in $\mathbb{P}_\Delta$, then*

- $\rho_{[1:0]} = h^{2,1}(X_1) + 1$,
- $\rho_{[1:1]} = h^{2,1}(X_2) + 1$,
- $\rho_{[0:1]} = h^{1,0}(C) + 1$.

REMARK 3.19. Note that the same result is true for $\hat{W}$, as the birational transformation from W to $\hat{W}$ is an isomorphism in codimension 1. Thus the preceding corollary can be interpreted as a count of components in singular fibres of the K3 surface fibration $\pi \colon \hat{W} \to \mathbb{P}^1$ on $\hat{W}$.

Now let ℓ be the rank of the image of the restriction map $H^2(\hat{W}, \mathbb{C}) \to H^2(S, \mathbb{C})$, for S a smooth fibre of π. Using [**DHNT16**, Lemma 3.2], one can easily show that

$$h^{1,1}(\hat{W}) = \sum_{p \in \mathbb{P}^1} (\rho_p - 1) + \ell + 1$$

Moreover, by [**Bat94**] we see that $h^{2,1}(V) = h^{1,1}(W) = h^{1,1}(\hat{W})$. So, noting that $h^{2,1}(\hat{X}_1) = h^{2,1}(X_1) + g(C)$ (see, for instance, [**Voi07**, Theorem 7.31]),

Theorem 2.5 gives

$$\sum_{p \in \mathbb{P}^1 \setminus \{[1:0],[0:1],[1:1]\}} (\rho_p - 1) + \ell + k = 20.$$

This implies that if Dolgachev-Nikulin mirror symmetry does not hold (in a precise sense) for the K3 surfaces associated to the nef partition Δ_1, Δ_2 and their Batyrev-Borisov duals, then this failure is seen by the fibres of the fibration $\pi \colon \hat{W} \to \mathbb{P}^1$ away from the points in the set $\{[1:0], [0:1], [1:1]\}$.

Finally, one may ask whether an analogue of Corollary 3.18 holds when V and W are K3 surfaces. The difficulty here is in proving an analogue of Theorem 3.17: for subtle combinatorial reasons, the proof given in [**Har16**] does not easily generalize to the K3 surface case. However, we expect the following conjecture to hold in this case:

CONJECTURE 3.20. *Suppose that* $\dim(V) = \dim(W) = 2$ *and let* ρ_p *be the number of irreducible components in the fibre of the elliptic fibration* $\pi \colon W \to \mathbb{P}^1$ *over* $p \in \mathbb{P}^1$. *If* X_1 *and* X_2 *are pullbacks of ample hypersurfaces in* $\mathbb{P}_\Delta$, *then*

- $\rho_{[1:0]} = h^{1,1}(X_1) - h^{1,1}(X_\Delta) + 1$,
- $\rho_{[1:1]} = h^{1,1}(X_2) - h^{1,1}(X_\Delta) + 1$,
- $\rho_{[0:1]}$ *is the number of points in* $V \cap X_1 \cap X_2$ *and the corresponding singular fibre is semistable (Kodaira type* I_n*), and*
- *all other fibres of* π *are irreducible.*

We will illustrate this conjecture with an example.

EXAMPLE 3.21 (Anticanonical hypersurfaces in $(\mathbb{P}^1)^3$). Let us take V to be an anticanonical hypersurface in $(\mathbb{P}^1)^3$. This is a K3 surface with Picard lattice of rank 3, isomorphic to the lattice with Gram matrix

$$\begin{pmatrix} 0 & 2 & 2 \\ 2 & 0 & 2 \\ 2 & 2 & 0 \end{pmatrix}.$$

There is a Tyurin degeneration of V to a union of two $(1,1,1)$ divisors X_1, X_2 in $(\mathbb{P}^1)^3$. The intersection $V \cap X_1 \cap X_2$ is 12 points. On the other side, we see that there is an elliptic fibration on the mirror dual K3 surface W, which has three reducible fibres of types I_{12}, I_2 and I_2.

The polytope Δ determining $(\mathbb{P}^1)^3$ has vertices $\sigma_0, \dots, \sigma_5$ given by the columns of the matrix

$$\begin{pmatrix} 1 & -1 & 0 & 0 & 0 & 0 \\ 0 & 0 & 1 & -1 & 0 & 0 \\ 0 & 0 & 0 & 0 & 1 & -1 \end{pmatrix}$$

The appropriate nef partition is $E_1 = \{\sigma_0, \sigma_2, \sigma_4\}$ and $E_2 = \{\sigma_1, \sigma_3, \sigma_5\}$, which has dual nef partition with

$$\nabla_1 = \mathrm{Conv}(\{(1,0,0), (0,1,0), (0,0,1), (1,1,1), (1,1,0), (0,1,1), (0,1,1), (0,0,0)\})$$
$$\nabla_2 = -\nabla_1.$$

We draw the polytopes Δ°, ∇_1 and ∇_2 in Figure 2. The leftmost picture in Figure 2 is just the polytope Δ°, the middle picture denotes ∇_1 and ∇_2 and the picture on the right shows $(\Delta^\circ \setminus \nabla) \cap N$. It is clear from the description of the fibre over $[0:1]$ that it is actually semi-stable, so it follows from Kodaira's classification

FIGURE 2. Polytopes related to Example 3.21

of singular fibres of elliptic fibrations that the resulting fibre is necessarily of type I_{12}. The same cannot be said for the fibres over $[1 : 0]$ and $[1 : 1]$, which have two components each, since it is not necessarily true that these fibres have normal crossings. Kodaira's classification can only be used to determine that these fibres are either of type I_2 or of type III.

4. Dolgachev-Nikulin mirror symmetry

Next we'll turn our attention to K3 surfaces. As noted in the previous section, in the setting of Batyrev-Borisov mirror symmetry, a refinement of the nef partition defining our K3 surface V gives rise to both a Tyurin degeneration of V and an elliptic fibration on the mirror K3 surface W. In the K3 surface case this appears to be part of a wider correspondence, which seems to have first been noticed by Dolgachev [**Dol96**], between Type II degenerations (of which Tyurin degenerations are an example) and elliptic fibrations on the Dolgachev-Nikulin mirror.

We begin by giving a precise statement of this correspondence. Suppose we have a (pseudo-ample) L-polarized K3 surface V, for some lattice L. To define the Dolgachev-Nikulin mirror of V, we first fix a primitive isotropic vector f in the orthogonal complement $L^\perp$ of L in the K3 lattice $\Lambda_{K3} \cong H^{\oplus 3} \oplus E_8^{\oplus 2}$ (where H denotes the hyperbolic plane lattice and E_8 is the negative definite root lattice E_8). With this in place, the Dolgachev-Nikulin mirror W of V is defined to be an $\check{L}$-polarized K3 surface, where

$$\check{L} := (\mathbb{Z}f)^\perp_{L^\perp}/\mathbb{Z}f.$$

Note that this depends upon the choice of isotropic vector f.

By the discussion in [**Dol96**, Section 6] (see also [**Sca87**, Section 2.1]), fixing f is equivalent to fixing a 0-dimensional cusp (Type III point) in the Baily-Borel compactification $\overline{D}_L$ of the period domain D_L of L-polarized K3 surfaces. Call this cusp P.

Then we have the following result, which is essentially contained in [**Dol96**, Remark 7.11]:

PROPOSITION 4.1. *With notation as above, there is a bijective correspondence between* 1*-dimensional cusps in* $\overline{D}_L$ *that pass through* P*, and primitive isotropic vectors* $e \in \check{L}$.

REMARK 4.2. Since 1-dimensional cusps in $\overline{D}_L$ correspond to Type II degenerations of V and isotropic vectors $e \in \check{L}$ correspond to elliptic fibrations on W, this gives rise to a correspondence between Type II degenerations of V and elliptic fibrations on W (up to automorphism).

PROOF OF PROPOSITION 4.1. Suppose first that we have a 1-dimensional cusp $C \subset \overline{D}_L$ that passes through P. By [**Sca87**, Section 2.1], such cusps are in bijection with rank two primitive isotropic sublattices of $L^\perp$ that contain the vector f. So C gives rise to a sublattice E of $L^\perp$ and, since E is a primitive sublattice of $(\mathbb{Z}f)^\perp_{L^\perp}$ and E contains f, we see that $E \cap \check{L}$ is a primitive isotropic sublattice of $\check{L}$ of rank 1. Let e denote a generator of this sublattice; then e is a primitive isotropic vector in $\check{L}$.

Conversely, suppose we have a primitive isotropic vector $e \in \check{L}$. By definition of $\check{L}$, we have $\langle e, f\rangle = 0$. So the lattice E spanned by e and f is a rank two primitive isotropic sublattice of $L^\perp$. But this gives rise to a 1-dimensional cusp passing through P. □

We will now illustrate this correspondence in some explicit examples, which will provide a glimpse of some possible deeper structure.

4.1. H-polarized K3's. We begin by looking at the Type II degenerations of H-polarized K3's. An H-polarized K3 surface may be constructed as an anticanonical hypersurface in an mpcp resolution of the weighted projective space $\mathbb{WP}(1,1,4,6)$. However, the defining polytope of this weighted projective space does not admit any nef partitions, so we cannot apply the theory of Section 3 to study it.

Instead, we will try a different approach to comparing Type II degenerations and elliptic fibrations on the mirror. Note first that an H-polarized K3 surface V naturally corresponds to a double cover of the Hirzebruch surface $\mathbb{F}_4$, ramified over a divisor in the linear system $|4s+12f|$, where s is the class of the (-4)-section and f is the class of a fibre (when there is no risk of confusion, we will always denote the $(-n)$-section in a Hirzebruch surface $\mathbb{F}_n$ by s and a fibre of the ruling by f). So H-polarized K3 surfaces are equivalent to pairs $(\mathbb{F}_4, B)$, where $B \in |4s+12f|$ is the branch divisor.

The moduli space of such pairs admits a *KSBA compactification* (see [**Ale96a, Ale96b, Ale06, KSB88**]), which has been studied in detail by Brunyate [**Bru15**]. This compactification admits a birational morphism to the usual Baily-Borel compactification of the moduli space of H-polarized K3 surfaces. The Type II degenerations occur along two boundary components in this KSBA compactification, which map to the two 1-dimensional cusps in the Baily-Borel compactification. These two kinds of Type II degenerations may be described as follows:

(1) Degenerate V to a union $X_1^1 \cup_Z X_2^1$ defined as follows. X_1^1 is a double cover of $\mathbb{F}_4$ ramified over a smooth divisor in the linear system $|2s+12f|$, and $Z \subset X_1^1$ is an elliptic curve given by the pull-back of the (-4)-section. X_2^1 is also a double cover of $\mathbb{F}_4$, this time ramified over a smooth divisor in the linear system $|s+4f|$, and $Z \subset X_2^1$ is the pull-back of a second smooth divisor in the linear system $|s+4f|$.

    Riemann-Roch easily yields that there is a 17-dimensional space of deformations of (X_1^1, Z) preserving the double covering structure, and a 1-dimensional space of deformations of (X_2^1, Z). To glue these components together along Z, we need to ensure that the elliptic curves Z in each component are isomorphic; this imposes a single gluing condition. The total configuration therefore has $17+1-1=17$ moduli. It is easy to see that any variety $X_1^1 \cup_Z X_2^1$ defined in this way is d-semistable (in the sense of [**Fri83**]), so admits a smoothing to a K3 surface by [**Fri83**, Theorem 5.10]. Thus, such degenerations lie along a 17-dimensional boundary component (i.e. a boundary divisor) in the KSBA moduli space.

(2) Degenerate V to a union $X_1^2 \cup_Z X_2^2$, where both X_1^2 and X_2^2 are double covers of $\mathbb{F}_2$ ramified over smooth divisors in the linear system $|4s+6f|$ and Z is the pull-back of a fibre of the ruling on $\mathbb{F}_2$. Riemann-Roch yields that each component (X_i^2, Z) has 9 deformations that preserve the double covering structure and, as before, there is a single gluing condition along Z. The total configuration therefore has $9+9-1=17$ moduli and is d-semistable, so gives another boundary divisor of the moduli space.

Now we look at the mirror. Up to isometry, there are two 1-dimensional cusps in $\overline{D}_H$, meeting in a unique 0-dimensional cusp. So we only have one choice of mirror W: a K3 surface polarized by the lattice $M := H \oplus E_8 \oplus E_8$.

Now we match the Type II degenerations above with elliptic fibrations on the mirror. From [**CD07**, Section 3.3], we know that an M-polarized K3 surface admits two elliptic fibrations:

(1) The alternate fibration, which has an I_{12}^* and six I_1's. Note that the I_{12}^* fibre has 17 components, corresponding to the 17 moduli of the component X_1^1 in degeneration (1) above.

(2) The standard fibration, which has two II^*'s and four I_1's. Note that each II^* fibre has 9 components, corresponding to the 9 moduli of each of the components X_i^2 in the Type II degeneration (2) above.

4.2. K3 surfaces of degree two. Now consider K3 surfaces of degree two (i.e. polarized by the rank one lattice $\langle 2\rangle$). A K3 surface V of degree 2 may be constructed as an anticanonical hypersurface in the weighted projective space $\mathbb{WP}(1,1,1,3)$. This toric variety is determined by the polytope $\Delta \subset M_{\mathbb{R}} \cong \mathbb{R}^3$ with vertices given by the columns $\sigma_1, \sigma_2, \sigma_3, \sigma_4$ of the matrix

$$\begin{pmatrix} 1 & 0 & 0 & -1 \\ 0 & 1 & 0 & -1 \\ 0 & 0 & 1 & -3 \end{pmatrix}.$$

Up to automorphism, there is only one bipartite nef partition of this polytope, given by $E_1 = \{\sigma_1, \sigma_2, \sigma_4\}$ and $E_2 = \{\sigma_3\}$. The dual nef partition in $N_{\mathbb{R}}$ has

$$\nabla_1 = \mathrm{Conv}(-e_1^* - e_2^*, -e_1^* - e_2^* + e_3^*, -e_1^* + 2e_2^*, 2e_1^* - e_2^*)$$
$$\nabla_2 = \mathrm{Conv}(0_N, -e_3^*, 3e_1^* - e_3^*, 3e_2^* - e_3^*)$$

where e_1, e_2 and e_3 are basis vectors of $M_{\mathbb{R}}$ and e_i^* are their duals.

As in Example 3.21, we may look at the polytopes ∇_1 and ∇_2 in comparison to Δ° and note that the integral points in $\Delta^\circ \setminus \nabla$ form a cycle of length 18. Thus the fibre over $[0:1]$ in the elliptic fibration on the Batyrev dual W described in Section 3.2 is of type I_{18}. One can also check, using techniques described in [**Har16**], that the fibres over $[1:0]$ and $[1:1]$ are irreducible.

Finally, in the degree two case it is well known that the Batyrev dual family of K3 surfaces is actually Dolgachev-Nikulin dual (see, for instance, [**Roh04**]), so a generic Batyrev dual K3 surface W has Picard lattice $M_2 := H \oplus E_8 \oplus E_8 \oplus A_1$. The elliptic fibrations on such an M_2-polarized K3 surface were computed by Dolgachev [**Dol96**, Remark 7.11]; we see that one of them has an I_{18} fibre and six I_1's, as expected.

However, it is known (see, for instance, [**Sca87**, Section 6]) that the Baily-Borel compactification of the moduli space of K3 surfaces of degree two has four 1-dimensional cusps, corresponding to four types of Type II degenerations, yet the example above only gives one. We can analyse the others using the same techniques that we used in the H-polarized case above.

Indeed, it is well known that a K3 surface of degree two V naturally corresponds to a double cover of $\mathbb{P}^2$ ramified over a smooth sextic curve. The moduli space of K3 surfaces of degree two is therefore the same as the moduli space of pairs $(\mathbb{P}^2, B)$ where B is a sextic curve.

Alexeev and Thompson have studied a KSBA compactification [**AT16**] for the moduli space of such pairs. The Type II degenerations occur along four boundary components, which map to the four cusps in the Baily-Borel compactification. The four corresponding kinds of Type II degenerations may be described as follows:

(1) Degenerate V to a union $X_1^1 \cup X_2^1$, defined as follows. X_1^1 is a double cover of $\mathbb{F}_4$ ramified over a smooth divisor in the linear system $|2s + 12f|$, and $Z \subset X_1^1$ is an elliptic curve given by the pull-back of the (-4)-section. X_2^1 is a double cover of $\mathbb{P}^2$ ramified over a smooth conic, and $Z \subset X_2^1$ is the pull-back of a second smooth conic. This degeneration is in many ways analogous to degeneration (1) in the H-polarized case.

    In this case the pair (X_1^1, Z) admits 16 deformations preserving the double covering structure, and (X_2^1, Z) has two. As in the H-polarized case, there is a single gluing condition corresponding to choice of the elliptic curve Z, so the total configuration has $17 + 2 - 1 = 18$ moduli. It is easy to see that all such varieties are d-semistable, so they give rise to an 18-dimensional boundary component (i.e. a boundary divisor) in the KSBA moduli space.

(2) Degenerate V to a union $X_1^2 \cup_Z X_2^2$, defined as follows. X_1^2 is a double cover of $\mathbb{F}_1$ ramified over a smooth divisor in the linear system $|2s + 6f|$ and $Z \subset X_1^1$ is the pull-back of the (-1)-section. X_2^2 is a double cover of $\mathbb{P}^2$ ramified over a smooth quartic and $Z \subset X_2^2$ is the pull-back of a line.

    Counting deformations that preserve the double cover structure, we see that (X_1^2, Z) has 11 deformations and the (X_2^2, Z) component has 8. However, as usual there is a gluing condition along Z, so the total configuration therefore has $11 + 8 - 1 = 18$ moduli. Since such varieties are all d-semistable, they correspond to a boundary divisor in the moduli space.

(3) The last two cases are more interesting. In the first, we degenerate V to a single component X_1^3, defined to be the double cover of a cubic cone ramified over a smooth anticanonical divisor. X_1^3 thus contains two log canonical singularities, corresponding to the pull-backs of the vertex of the cone. This double cover has 18 moduli, and corresponds to a boundary divisor in the KSBA moduli space.

    Interestingly, this degeneration may be seen to be equivalent to the degeneration given by the nef partition above. Indeed, if $X_1 \cup_Z X_2$ is the degeneration corresponding to this nef partition, we may obtain the degeneration X_1^3 by blowing up along Z, then contracting X_1 and X_2. These contractions give rise to the two log canonical singularities present in X_1^3.

(4) The final case is most difficult. In [**AT16**], the corresponding degeneration of V is a union $X_1^4 \cup_Z X_2^4$, where each X_i^4 is a double cover of $\mathbb{F}_2^0 \cong \mathbb{WP}(1,1,2)$ (obtained by taking a copy of $\mathbb{F}_2$ and contracting the (-2)-section) ramified over a divisor in the linear system $\mathcal{O}(6)$ and cyclically over the A_1 singularity, and Z is the pull-back of a divisor in the linear system $\mathcal{O}(1)$. This degeneration is in many ways analogous to degeneration (2) in the H-polarized case.

Counting moduli, we see that each component (X_i^4, Z) has 9 deformations that preserve the double covering structure, but there is, as always, one gluing condition along Z. We thus have $9+9-1=17$ deformations, which does not give the expected divisor on the boundary. However, there is also something wrong with this degeneration: one can easily check that it is not d-semistable, so does not appear as the central fibre in a smooth semistable degeneration of K3 surfaces.

There are two ways to resolve this. The first may be thought of as analogous to the blow-up of $V \cap X_1 \cap X_2$ in X_1 from Section 3: we simply blow up an arbitrary point on the image of Z in one copy of $\mathbb{F}_2^0$. This corresponds to blowing up a pair of points on Z in X_1^4, under the condition that the two points blown up are exchanged by the involution defining the double covering. After blowing up, $X_1^4 \cup_Z X_2^4$ becomes d-semistable and acquires an extra modulus, corresponding to the choice of point to blow up, making the corresponding boundary component into a divisor.

However, there is also a second way to resolve this problem. Inspired by (3), above, we could try introducing a third component between X_1^4 and X_2^4; this has the appeal of maintaining the symmetry of the degenerate fibre. Such a central component X_3^4 can be constructed as follows. Take a copy of $\mathbb{P}^2$ and blow up three points in general position, to get a (-1)-hexagon. Blow up one of the vertices of this hexagon again, then contract the two (-2)-curves that result. One obtains a surface that has two A_1 singularities, and the (-1)-hexagon becomes a pentagon whose sides have self-intersections $(-1,-1,-\frac{1}{2},0,-\frac{1}{2})$. The sum of the two (-1)-curves in this pentagon gives a ruling and X_3^4 is a double cover of this surface ramified over three of its fibres, as well as cyclically over both of the A_1 singularities. X_3^4 glues to X_1^4 and X_2^4 along two isomorphic elliptic curves Z_1 and Z_2, which are the pull-backs of the two $(-\frac{1}{2})$-curves. We thus obtain a degenerate fibre $X_1^4 \cup_{Z_1} X_3^4 \cup_{Z_2} X_2^4$.

Accounting for automorphisms, (X_3^4, Z_1, Z_2) has 2 deformations. So, with this component included, the total moduli count is $9+9+2-1-1=18$ (the two (-1)'s appear because there is a gluing condition associated to each double curve). Moreover, the fibre $X_1^4 \cup_{Z_1} X_3^4 \cup_{Z_2} X_2^4$ is d-semistable, so we get a boundary divisor in moduli.

Now we again match with elliptic fibrations on the mirror. These are computed by Dolgachev [**Dol96**, Remark 7.11] as:

(1) An elliptic fibration which has one I_{12}^*, one I_2, and four I_1's. Note that the I_{12}^* fibre has 17 components, corresponding to the 17 moduli of (X_1^1, Z) in degeneration (1) above, and the I_2 fibre has two components, corresponding to the two moduli of (X_2^1, Z).

(2) An elliptic fibration with one fibre of type I_6^*, one fibre of type III^*, and three I_1's. As above, the I_6^* fibre has 11 components, corresponding to the 11 moduli of (X_1^2, Z) in degeneration (4) above, and the III^* fibre has eight components, corresponding to the eight moduli of (X_2^2, Z).
(3) An elliptic fibration with one fibre of type I_{18} and six I_1's. Once again, the I_{18} fibre has 18 components, corresponding to the 18 moduli of the single component X_3^1 in degeneration (3), above.
(4) An elliptic fibration which has two II^*'s, one I_2, and two I_1's. Note that each II^* fibre has 9 components, corresponding to the 9 moduli of (X_1^4, Z) and (X_2^4, Z) in degeneration (2) above, and the I_2 fibre has two components. These two components can be thought of as corresponding to the two points on Z which are blown up to make $X_1 \cup_Z X_2$ d-semistable (c.f. Conjecture 3.20), or as corresponding to the two moduli of the "extra" central component (X_3^4, Z_1, Z_2).

4.3. Discussion. These results are highly suggestive of the idea that the correspondence between Tyurin degenerations and elliptic fibrations on Batyrev dual K3 surfaces, explored in Section 3, may extend to a broader correspondence between Type II degenerations of an L-polarized K3 surface V and elliptic fibrations on its Dolgachev-Nikulin mirror W. Such a correspondence should have properties that generalize those described in Section 3; the aim of this section is to discuss the form that such properties may take.

Firstly, cases (3) and (4) of the considerations in Section 4.2 suggest that there may be some correspondence between the choice of model for the Type II degeneration of V and some properties of the elliptic fibration. Thinking about this in the context of the threefold philosophy outlined in Remark 2.7, one might conjecture the following.

Suppose first that we have a Type II degeneration of V to a configuration $X_1 \cup_Z X_2 \cup_Z \cdots \cup_Z X_k$, where each X_i meets X_{i-1} and X_{i+1}. Let n_i be the number of deformations of (X_i, Z) (where Z denotes the double locus on X_i and may have more than one component) that preserve some notion of lattice polarization – this last condition is to ensure that the deformed Type II fibres still smooth to L-polarized K3 surfaces, and was arranged in the preceding examples by requiring that deformations preserve the double covering structure; it also accounts for the appearance of the $h^{1,1}(X_\Delta)$ term in Conjecture 3.20. Then the fibre dimension of the natural map from the moduli space of such pairs (X_i, Z) to the moduli space of elliptic curves Z is equal to $n_i - 1$.

Thus, noting that all elliptic double curves in the Type II degeneration are isomorphic, we see that such a Type II fibre should have $\sum_{i=1}^{k}(n_i-1)+1$ deformations. We first conjecture that this number should equal $19-\ell$, where $\ell = \text{rank}(L)$. Then, since the moduli space of V (as an L-polarized K3 surface) is $(20-\ell)$-dimensional, such Type II degenerations should lie along codimension 1 loci in an appropriate compactification. We would thus get a decomposition of the $20-\ell$ deformations of V into contributions $(n_i - 1)$ from deformations of each X_i, a contribution 1 from deformations of Z, and 1 for the codimension in the moduli space.

Now we look at the mirror picture. As proved in Proposition 4.1, the Type II degeneration of V given above should correspond to an elliptic fibration $\pi\colon W \to \mathbb{P}^1$ on the mirror W. We suggest that the decomposition $X_1 \cup_Z X_2 \cup_Z \cdots \cup_Z X_k$ of V corresponds to a "slicing" of the $\mathbb{P}^1$ base of π, so that each X_i corresponds to a slice S_i, as follows. S_1 is a disc, which is glued along its boundary to one of the boundaries of an annulus S_2. The other boundary of the annulus S_2 is then glued to one of the boundaries of an annulus S_3, and so on, until the remaining boundary of the annulus S_{k-1} is glued to a disc S_k.

The singular fibres of π should then be apportioned amoungst the slices as follows. If ρ_p denotes the number of components in the fibre of π over p, then the slicing should satisfy

$$n_i - 1 = \sum_{p \in S_i} (\rho_p - 1).$$

This gives a decomposition of the Picard rank $(20 - \ell)$ of W into contributions $n_i - 1$ from the singular fibres lying on the slice S_i, a contribution 1 from the class of a section, and a contribution 1 from the class of a fibre.

Note that this is completely compatible with the Tyurin degeneration picture presented in Section 2 and the Batyrev-Borisov picture of Conjecture 3.20. Indeed, in the Tyurin degeneration picture the slicing has two pieces S_1 and S_2, which are the LG models of the two components X_1 and X_2 of the degenerate fibre $X_1 \cup_Z X_2$. Furthermore, in the setting of Conjecture 3.20 we also have a slicing into two pieces S_1 and S_2, one of which contains the point $[1:0]$ and the other of which contains $[1:1]$. Which slice the point $[0:1]$ falls into depends upon the choice of blow-up of $V \cap X_1 \cap X_2$: if we blow up $V \cap X_1 \cap X_2$ in X_i (for $i \in \{1,2\}$), then the corresponding I_n fibre appears in the slice S_i.

Finally, we describe how this slicing picture works in the H-polarized and degree 2 cases considered above. In the H-polarized case we have the following two possibilities, numbered compatibly with Section 4.1. The labelling of the slices is chosen so that the slice corresponding to the component X_i^j is labelled S_i^j.

(1) $\mathbb{P}^1$ is sliced into two pieces S_1^1 and S_2^1, such that S_1^1 contains the I_{12}^* fibre.
(2) $\mathbb{P}^1$ is sliced into two pieces S_1^2 and S_2^2, each of which contains a II^* fibre.

Moreover, in the degree two case we have the following four possibilities, numbered compatibly with Section 4.2.

(1) $\mathbb{P}^1$ is sliced into two pieces S_1^1 and S_2^1, such that S_1^1 contains the I_{12}^* fibre and S_2^1 contains the I_2.
(2) $\mathbb{P}^1$ is sliced into two pieces S_1^2 and S_2^2, such that S_1^2 contains the I_6^* fibre and S_2^2 contains the III^*.
(3) In this case we have two choices of slicing, corresponding to the two types of degeneration. For the degeneration given by the nef partition and Batyrev mirror symmetry, $\mathbb{P}^1$ is sliced into two pieces S_1 and S_2, one of which contains the I_{18} fibre. For the KSBA degeneration, we have a "degenerate" slicing of $\mathbb{P}^1$ into a single piece, which contains all singular fibres.
(4) In the final case we also have two choices of slicing, corresponding to the two types of degeneration. In the case where we blow up a pair of points, $\mathbb{P}^1$ is sliced into two pieces S_1^4 and S_2^4, such that S_1^4 contains the II^* fibre and the I_2 fibre, and S_2^4 contains the other II^*. In the case where we have three components in the degeneration, $\mathbb{P}^1$ is sliced into three pieces S_1^4, S_3^4 and S_2^4, such that S_1^4 and S_2^4 are discs containing one II^* fibre each, and S_4^3 is an annulus containing the I_2 fibre.

5. Beyond Batyrev-Borisov mirror symmetry for threefolds

The aim of this section is to provide some evidence that the ideas presented in Section 2 also hold for threefolds outside the toric setting considered in Section 3. We begin by showing that classical mirror symmetry suggests a correspondence between Tyurin degenerations and K3 fibrations on mirror dual pairs of Calabi-Yau threefolds. This should be thought of as a threefold analogue of Proposition 4.1. Then we specialize our discussion to the case of threefolds fibred by mirror quartics, as studied in **[DHNT16]**, and show that, in that setting, the correspondence predicted by classical mirror symmetry is consistent with the construction presented in Section 2.

5.1. Classical mirror symmetry for threefolds. Classical mirror symmetry predicts that if V and W are mirror dual Calabi-Yau threefolds, then there is a relation between monodromy operators acting on $H^3(V,\mathbb{Q})$ and divisors in the closure of the Kähler cone of W. We will briefly sketch some of the details of this relationship here, the interested reader may find more details in [**CK99**, Chapter 6].

Suppose that $\mathcal{V} \to (\Delta^*)^n$ is a family of Calabi-Yau threefolds over the punctured polydisc, with fibre $\mathcal{V}_t = V$ above some $t \in (\Delta^*)^n$. For each $i \in \{1,\ldots,n\}$, let T_i be the unipotent monodromy operator acting on $H^3(V,\mathbb{Q})$ coming from the loop $(t_1,\ldots,t_{i-1},e^{2\pi i t},t_{i+1},\ldots,t_n)$, where $(t_1,\ldots,t_{i-1},t_{i+1},\ldots,t_n)$ are fixed constants, and let $N_i = \log(T_i)$. The family $\mathcal{V}$ is said to have *maximally unipotent monodromy* at $(0,\ldots,0)$ if

(1) for any n-tuple $(a_1,\ldots,a_n)$ of positive integers, the weight filtration $W_\bullet$ on $H^3(V,\mathbb{Q})$ induced by $\sum_{i=1}^n a_iN_i$ has $\dim W_0 = \dim W_1 = 1$ and $\dim W_2 = n+1$, and
(2) if $g_0,\ldots,g_n$ is a basis of W_2 chosen so that g_0 spans W_0, and m_{ij} are defined by $N_ig_j = m_{ij}g_0$, then the matrix (m_{ij}) is invertible.

If $\mathcal{V}$ has maximally unipotent monodromy, then mirror symmetry should produce a map which assigns to each N_i a divisor D_i in the closure of the Kähler cone of W. Moreover, there should be an identification under mirror symmetry which gives an isomorphism $H^{3-i,i}(V) \cong H^{i,i}(W)$, and hence an isomorphism $H^3(V,\mathbb{C}) \cong \bigoplus_{i=0}^3 H^{i,i}(W)$, so that the action of N_i on $H^3(V,\mathbb{C})$ agrees with the action of the cup product operator $J_i(-) = (-)\cup[-D_i]$ under this correspondence. Thus, for any n-tuple $(a_1,\ldots,a_n)$ of non-negative integers, the weight filtration on $H^3(V,\mathbb{C})$ induced by $N := \sum_{i=1}^n a_iN_i$ should be mirrored by the filtration on $\bigoplus_{i=0}^3 H^{i,i}(W)$ induced by $J := \sum_{i=1}^n a_iJ_i$, and the limit Hodge decomposition should correspond to the decomposition $\bigoplus_{i=0}^3 H^{i,i}(W)$.

Now we specialize this discussion to the case of a Tyurin degeneration of Calabi-Yau threefolds $\mathcal{V} \to \Delta$. As in the previous sections, we write the central fibre of $\mathcal{V}$ as $X_1 \cup_Z X_2$ and let V denote a general fibre. Let T be the monodromy operator acting on $H^3(V,\mathbb{Q})$ associated to a counterclockwise loop around 0. In order to apply the predictions of mirror symmetry, we assume that T may be identified with a loop $\prod_{i=1}^n T_i^{a_i}$ around a point of maximally unipotent monodromy in the complex moduli space of V, where T_i are as above and a_i are non-negative integers. Define $N := \log(T) = \sum_{i=1}^n a_iN_i$. We will use the Clemens-Schmid exact sequence associated to N to compute the limit mixed Hodge structure on $H^3(V)$, then see what this allows us to deduce about the mirror threefold W.

REMARK 5.1. We note that the Tyurin degeneration $\mathcal{V}$ cannot have maximally unipotent monodromy, for purely topological reasons (see, for instance, [**Mor84**, Corollary 2]), so T must correspond to a loop around some positive-dimensional boundary component of the compactified complex moduli space of V. In particular, this implies that some of the a_i must be zero.

We begin by looking at the mixed Hodge structure on $H^3(\mathcal{V})$ given by Griffiths and Schmid [**GS75**, Section 4]. The weight filtration $W_\bullet$ on $H^3(\mathcal{V},\mathbb{Q})$ has

$$\mathrm{Gr}_3^W = H^3(X_1,\mathbb{Q}) \oplus H^3(X_2,\mathbb{Q})$$

and, if r_1 and r_2 are the restriction maps $r_i\colon H^2(X_i,\mathbb{Q}) \to H^2(Z,\mathbb{Q})$, then

$$\mathrm{Gr}_2^W = H^2(Z,\mathbb{Q})/(\mathrm{im}(r_1)+\mathrm{im}(r_2)).$$

These weight graded pieces are then equipped with the appropriate Hodge filtrations. Define integers $u := \mathrm{rank}(\mathrm{Gr}_2^W) - 2$ and $v := \frac{1}{2}\mathrm{rank}(\mathrm{Gr}_3^W)$. Noting that K_{X_i} is anti-effective, so that $h^{3,0}(X_i) = 0$, we see that $v = h^{2,1}(X_1) + h^{2,1}(X_2)$.

The Clemens-Schmid exact sequence gives us an exact sequence of mixed Hodge structures

$$\cdots \longrightarrow H_5(\mathcal{V}) \longrightarrow H^3(\mathcal{V}) \xrightarrow{i^*} H^3_{\lim}(V) \xrightarrow{N} H^3_{\lim}(V) \longrightarrow H_3(\mathcal{V}) \longrightarrow \cdots$$

where i^* is the pull-back on cohomology induced by the inclusion $i\colon V \hookrightarrow \mathcal{V}$.

LEMMA 5.2. *$H_5(\mathcal{V}) = 0$, so the map i^* is an injection.*

PROOF. The Mayer-Vietoris sequence for $X_1 \cup_Z X_2$ gives

$$\cdots \longrightarrow H_5(X_1)\oplus H_5(X_2) \longrightarrow H_5(\mathcal{V}) \longrightarrow H_4(Z) \xrightarrow{\alpha} H_4(X_1)\oplus H_4(X_2) \longrightarrow \cdots,$$

where the map α is induced by the inclusions $Z \hookrightarrow X_i$.

Now, $H_5(X_1)\oplus H_5(X_2)$ vanishes by Poincaré duality and the assumption that $h^{0,1}(X_i) = 0$. Moreover, as Z is an effective anticanonical divisor in both X_1 and X_2, the image of the class $[Z] \in H_4(Z)$ of Z under α is non-trivial. But $[Z]$ generates $H_4(Z)$, so α must be injective. Thus the sequence above gives $H_5(\mathcal{V}) = 0$. □

Applying this lemma and some standard results on the Clemens-Schmid exact sequence (see, for instance, [**Mor84**]) we obtain the following limit mixed Hodge structure on $H^3(V)$

	Gr_F^3	Gr_F^2	Gr_F^1	Gr_F^0
Gr_4^M	$\mathbb{C}$	$\mathbb{C}^u$	$\mathbb{C}$	0
Gr_3^M	0	$\mathbb{C}^v$	$\mathbb{C}^v$	0
Gr_2^M	0	$\mathbb{C}$	$\mathbb{C}^u$	$\mathbb{C}$

where $M_\bullet$ is the monodromy weight filtration induced by N and $F^\bullet$ is the limit Hodge filtration.

Therefore, the divisor $D = \sum_{i=1}^n a_iD_i$ on W which corresponds to N under mirror symmetry should have

	$H^{0,0}(Y^\vee)$	$H^{1,1}(W)$	$H^{2,2}(W)$	$H^{3,3}(W)$
coimage(J)	$\mathbb{C}$	$\mathbb{C}^u$	$\mathbb{C}$	0
$\ker(J)/\mathrm{im}(J)$	0	$\mathbb{C}^v$	$\mathbb{C}^v$	0
$\mathrm{im}(J)$	0	$\mathbb{C}$	$\mathbb{C}^u$	$\mathbb{C}$

where, as before, $J(-) = (-)\cup[-D]$ denotes the cup-product operator. In particular, we see that $J^2 = 0$. Since D is in the closure of the Kähler cone of W, results of Oguiso [**Ogu01**, Example 2.3] show that mD is the class of a fibre in a fibration of W by K3 or abelian surfaces, for some positive integer m.

REMARK 5.3. Based on the ideas in the previous sections, we conjecture that mD will always be the class of fibre in a K3 fibration on W. Oguiso [**Ogu01**, Example 2.3] gives a simple criterion to test for this: mD defines a K3 fibration on W if and only if $c_2(W) \cdot D > 0$.

In light of this remark, we will assume throughout the remainder of this section that mD defines a K3 fibration on W. Then the calculation above also shows that the classes in $\mathrm{Pic}(W)$ supported on fibres span a $v+1$ dimensional subspace, where one of these classes is mD itself. Moreover, there is a rank u subspace of $\mathrm{Pic}(W)$ with $J(\eta) \neq 0$ for each class $\eta \neq 0$ in this subspace. By the global invariant cycles theorem, classes in this second subspace come from monodromy invariant cycles on fibres of the K3-fibration on W. Thus the K3 surface fibration on W induced by mD is $\check{L}$-polarized (in the sense of [**DHNT15**, Definition 2.1]), for some lattice $\check{L}$ of rank u.

Therefore we see that, if V admits a Tyurin degeneration to a union of threefolds $X_1 \cup_Z X_2$, and if restriction of divisors from X_1 and X_2 induces a lattice polarization of Z by a lattice L of rank $20-u$, then we expect the mirror W to admit an $\check{L}$-polarized K3 surface fibration, for some lattice $\check{L}$ of rank u. Moreover, the space of divisors in W that are supported on fibres of the fibration should have rank $v+1 = h^{2,1}(Y_1) + h^{2,1}(Y_2) + 1$. Note that this is completely consistent with the predictions of Section 2.3.

5.2. Threefolds fibred by mirror quartics. Our next aim is to demonstrate how this works in a special case: that of threefolds fibred by mirror quartic K3 surfaces. As we will see, in this setting the predictions of classical mirror symmetry, described above, mesh perfectly with the construction presented in Section 2.

A detailed study of threefolds fibred by mirror quartic K3 surfaces was conducted in [**DHNT16**]. We begin by briefly recapping the main construction and results of that paper, before describing how it fits into our picture.

The goal of [**DHNT16**] is to answer the following question: let W be a Calabi-Yau threefold and assume that W admits a fibration over $\mathbb{P}^1$ by K3 surfaces, $\pi\colon W \to \mathbb{P}^1$. Assume that the general fibre of π is a K3 surface S with $\mathrm{Pic}(S) = M_2$, for $M_2 := H \oplus E_8 \oplus E_8 \oplus \langle -4 \rangle$ (i.e. S is Dolgachev-Nikulin mirror to a quartic hypersurface in $\mathbb{P}^3$), and that the monodromy representation acts trivially on M_2. Such a structure is called an *M_2-polarized K3 fibration* on W. In [**DHNT16**] we attempted to classify Calabi-Yau threefolds admitting M_2-polarized K3 fibrations.

In order to describe this classification, we start by taking a basic family, called $\mathcal{X}$ in [**DHNT16**], which is a smooth resolution of the family of hypersurfaces

$$\{\lambda w^4 + xyz(x+y+z-w) = 0\} \subset \mathbb{P}^3,$$

for $\lambda \neq 1/256, 0$. This defines a smooth family of K3 surfaces over $\mathbb{P}^1 \setminus \{0, 1/256, \infty\}$. The classification of M_2-polarized fibrations from [**DHNT16**] can then be stated as:

THEOREM 5.4. [**DHNT16**, Section 2] *If $\pi\colon W \to \mathbb{P}^1$ is an M_2-polarized K3 fibration on a Calabi-Yau threefold W, then there is a map*

$$g\colon \mathbb{P}^1 \to \mathbb{P}^1$$

so that W is birational to $g^*\mathcal{X}$ where

$$\begin{array}{ccc} g^*\mathcal{X} & \longrightarrow & \mathcal{X} \\ \downarrow & & \downarrow \\ U & \xrightarrow{g|_U} & \mathbb{P}^1 \setminus \{0, 1/256, \infty\} \end{array}$$

and $U = g^{-1}(\mathbb{P}^1 \setminus \{0, 1/256, \infty\})$.

Moreover, the preimage of 0 under g consists of either 1 or 2 points. If $g^{-1}(0)$ is 2 points, then g is ramified to order $1, 2$ or 4 at each point in $g^{-1}(0)$. If $g^{-1}(0)$ is a single point, then g is ramified to order 8 at that point. If g is unramified over $1/256$ then W is smooth, but if g ramifies over $1/256$ then W may have isolated singularities.

Using this, in [**DHNT16**, Section 4] we obtained a classification of smooth deformation equivalent families of such Calabi-Yau threefolds, under the assumption that g is unramified over $1/256$ (which, by the theorem above, ensures that our Calabi-Yau threefolds will be smooth). This classification is determined by two pieces of data: a pair of numbers i and j in $\{1, 2, 4\}$, denoting the orders of ramification of g over 0, and a choice of partition $\mu := [x_1, \ldots, x_k]$ of $\deg(g) = i + j$, denoting the ramification profile of g over ∞. Note here that the case where $g^{-1}(0)$ is a single point is a deformation of the case where $i = j = 4$, so may be ignored; see [**DHNT16**, Remark 3.1]. We call a general member of this family $W_{i,j}^{\mu}$.

THEOREM 5.5. [**DHNT16**, Propositions 3.5, 3.8] *If $W_{i,j}^{\mu}$ is as above, then*

(1) $h^{2,1}(W_{i,j}^{\mu}) = k$,

(2) $h^{1,1}(W_{i,j}^{\mu}) = 20 + \sum_{s=1}^{k}(2x_s^2 + 1) + c_i + c_j$.

where $c_1 = 30$, $c_2 = 10$ and $c_4 = 0$.

Finally, the proof of [**DHNT16**, Proposition 2.5] classifies the singular fibres of these threefolds, up to small birational transformations. We give names to each possibility and a description of one member of each equivalence class.

(1) I_0: A smooth K3 surface. Along with the generic fibre, which clearly has type I_0, if $t \in g^{-1}(0)$ and g ramifies to order 4 at t, then the fibre over t is of type I_0.

(2) $\mathrm{I}_{\mathrm{odp}}$: A K3 surface with a single node. If $t \in g^{-1}(1/256)$, then the fibre over t is of type $\mathrm{I}_{\mathrm{odp}}$.

(3) I_n^{Δ} for $n \in \mathbb{N}$: this is a semistable singular fibre whose dual graph is a triangulation of the faces of a 3-dimensional simplex with sides of length n. Such a fibre has $2n^2 + 2$ irreducible components. If $t \in g^{-1}(\infty)$ and g ramifies to order n at t, then the fibre over t is of type I_n^{Δ}.

(4) II: A union of 11 smooth rational surfaces, one of which, E, has multiplicity 2, and the others, $F_1, \ldots, F_{10}$, have multiplicity 1. Each F_i intersects E in a smooth rational curve, but $F_i \cap F_j$ is empty for $i \neq j$. If $t \in g^{-1}(0)$ and g ramifies to order 2 at t, then the fibre over t is of type II.

(5) IV: A normal crossings union of 31 smooth rational surfaces. One has multiplicity 4, and the rest have multiplicities 3, 2 or 1. If $t \in g^{-1}(0)$ and g is unramified at t, then the fibre over t is of type IV.

5.3. Degenerations in the mirror. Now we look at the mirror picture. The Dolgachev-Nikulin mirror to an M_2-polarized K3 surface is a K3 surface of degree 4 ($\langle 4\rangle$-polarized); generically such K3 surfaces are just smooth quartics in $\mathbb{P}^3$. There are three Fano threefolds which contain a K3 surface of degree 4 as their anticanonical hypersurface; these are X_1, the quartic threefold in $\mathbb{P}^4$, X_2, the double cover of $\mathbb{P}^3$ ramified along a smooth generic quartic surface, and X_4, which is just $\mathbb{P}^3$ itself. Note that the integer i assigned to each Fano X_i is just the index of that Fano threefold.

Let Z be a generic smooth anticanonical K3 surface in both X_i and X_j, for some choice of integers i,j in $\{1,2,4\}$. Then $N_{Z/X_i} \cong \omega_{X_i}^{-1}|_Z \cong \mathcal{O}_Z(i)$, where $\mathcal{O}_Z(1)$ is the restriction of a hyperplane section in $\mathbb{P}^3$ to Z. Thus if we take a normal crossings union of X_i and X_j meeting along Z then, in accordance with the discussion in Section 2.1, we cannot construct a Calabi-Yau threefold by smoothing $X_i \cup_Z X_j$, since $N_{Z/X_i} \otimes N_{Z/X_j} = \mathcal{O}_Z(i+j) \neq \mathcal{O}_Z$.

This is similar to the situation in Section 3.1, where we could not smooth $X_1 \cup_Z X_2$ without first blowing up the locus $V \cap X_1 \cap X_2$, and we will solve it in the same way. Let $C_1, \ldots, C_k$ be smooth curves in S cut out by sections of $\mathcal{O}_Z(x_s)$, for positive integers $x_1, \ldots, x_k$, so that $\sum_{s=1}^k x_s = i+j$. Then let $f\colon \widetilde{X}_i \to X_i$ be the blow up of X_i in $C_1, \ldots, C_k$ sequentially and let E_i be the exceptional divisor over C_i, for $i = 1, \ldots, k$.

The canonical divisor of $\widetilde{X}_i$ is given by $f^*K_{X_i} - \sum_{s=1}^k E_s$, so $\omega_{\widetilde{X}_i}^{-1} \cong \mathcal{O}_Z(-j)$. Therefore, according to [**KN94**, Theorem 4.2], we may smooth $\widetilde{X}_i \cup_Z X_j$ to a Calabi-Yau threefold. We denote this threefold by $V_{i,j}^\mu$, where, as before, μ denotes the partition $[x_1, \ldots, x_k]$ of $(i+j)$.

We claim that $V_{i,j}^\mu$ and $W_{i,j}^\mu$ are mirror dual, in the classical sense. As a first piece of evidence for this, we compute the Hodge numbers of $V_{i,j}^\mu$.

PROPOSITION 5.6. *Let $i,j \in \{1,2,4\}$ and let $\mu = [x_1, \ldots, x_k]$ be a partition of $i+j$. Then the Hodge numbers of the threefold $V_{i,j}^\mu$ are given by*

$$h^{1,1}(V_{i,j}^\mu) = k,$$

$$h^{2,1}(V_{i,j}^\mu) = 20 + \sum_{s=1}^k (2x_s^2 + 1) + h^{2,1}(X_i) + h^{2,1}(X_j),$$

where $h^{2,1}(X_s) = 30$ (resp. 10, 0) for $s = 1$ (resp. 2, 4).

PROOF. By definition, $V_{i,j}^\mu$ is a smoothing of $\widetilde{X}_i \cup_Z X_j$. Define

$$q := \operatorname{rank}(\operatorname{im}(H^2(\widetilde{X}_i, \mathbb{Z}) \oplus H^2(X_j, \mathbb{Z}) \to H^2(Z, \mathbb{Z}))).$$

Then Lee [**Lee10**, Corollary 8.2] shows that the Hodge numbers of $V_{i,j}^\mu$ are given by

$$h^{1,1}(V_{i,j}^\mu) = h^2(\widetilde{X}_i) + h^2(X_j) - q - 1,$$
$$h^{2,1}(V_{i,j}^\mu) = 21 + h^{2,1}(\widetilde{X}_i) + h^{2,1}(X_j) - q.$$

Now, since the Néron-Severi group of Z is generated by the restriction of a hyperplane section from X_i, we must have $q = 1$. Moreover, since we blew up X_i a total of k times to obtain $\widetilde{X}_i$, we have $h^2(\widetilde{X_i}) = k+1$ and $h^2(X_j) = 1$. Thus $h^{1,1}(V_{i,j}^\mu) = k$.

To compute $h^{2,1}(V^{\mu}_{i,j})$, we begin by noting that a smooth curve C_s defined by a section of $\mathcal{O}_Z(x_s)$ has self-intersection $4x_s^2$ in Z. So the genus formula for curves on a surface gives $g(C_s) = 2x_s^2 + 1$. Thus, by standard results on the cohomology of a blow-up (see, for instance, [**Voi07**, Theorem 7.31]), we find

$$h^{2,1}(\widetilde{X}_i) = h^{2,1}(X_i) + \sum_{s=1}^{k}(2x_s^2 + 1),$$

giving the claimed result for $h^{2,1}(V^{\mu}_{i,j})$. Finally, the values of $h^{2,1}(X_s)$ are easy to compute explicitly. □

Putting this proposition together with the result of Theorem 5.5, we obtain:

COROLLARY 5.7. *Let $i, j \in \{1, 2, 4\}$ be a pair of integers and let $\mu = [x_1, \ldots, x_k]$ be a partition of $i + j$. Then there is a mirror duality between the Hodge numbers of the Calabi-Yau threefolds $V^{\mu}_{i,j}$ and $W^{\mu}_{i,j}$.*

We expect that $V^{\mu}_{i,j}$ is actually mirror to $W^{\mu}_{i,j}$, but of course this is not a proof. As further evidence, however, we can also compare filtrations as in Section 5.1. For the threefolds $V^{\mu}_{i,j}$, we may compute the limit mixed Hodge structure associated to the degeneration to $\widetilde{X}_i \cup_Z X_j$, to obtain

	Gr_F^3	Gr_F^2	Gr_F^1	Gr_F^0
Gr_4^M	$\mathbb{C}$	$\mathbb{C}^{19}$	$\mathbb{C}$	0
Gr_3^M	0	$\mathbb{C}^v$	$\mathbb{C}^v$	0
Gr_2^M	0	$\mathbb{C}$	$\mathbb{C}^{19}$	$\mathbb{C}$

for $v = h^{2,1}(X_i) + h^{2,1}(X_j) + \sum_{s=1}^{k}(2x_s^2 + 1)$.

Now, for the threefolds $W^{\mu}_{i,j}$, let J be the cup product operator with the negative of the class of a fibre of the M_2-polarized K3 surface fibration on $W^{\mu}_{i,j}$. Then, by the proof of [**DHNT16**, Proposition 3.5] and Proposition 5.6, we see that the rank of the space of divisors in $H^{1,1}(W^{\mu}_{i,j})$ that are supported on fibres is

$$\mathrm{rank}(\ker(J)) = 1 + \sum_{s=1}^{k}(2x_s^2 + 1) + h^{2,1}(X_i) + h^{2,1}(X_j) = v + 1.$$

Moreover, the image of J in $H^{1,1}(W^{\mu}_{i,j})$ is the span of the class of a fibre and the image of J in $H^{3,3}(W^{\mu}_{i,j})$ spans $H^{3,3}(W^{\mu}_{i,j})$. Finally, the image of J in $H^{2,2}(W^{\mu}_{i,j})$ is the space of classes dual to divisors in $H^{1,1}(W^{\mu}_{i,j})$ swept out by monodromy invariant divisors on a general fibre, which has rank 19. Thus, we obtain

	$H^{0,0}(X^{\mu}_{i,j})$	$H^{1,1}(W^{\mu}_{i,j})$	$H^{2,2}(W^{\mu}_{i,j})$	$H^{3,3}(W^{\mu}_{i,j})$
$\mathrm{coimage}(J)$	$\mathbb{C}$	$\mathbb{C}^{19}$	$\mathbb{C}$	0
$\ker(J)/\mathrm{im}(J)$	0	$\mathbb{C}^v$	$\mathbb{C}^v$	0
$\mathrm{im}(J)$	0	$\mathbb{C}$	$\mathbb{C}^{19}$	$\mathbb{C}$

and the duality of bifiltered vector spaces discussed in Section 5.1 is satisfied in this case.

Finally, we note that restriction of divisors from X_i and X_j induces a lattice polarization of Z by the lattice $\langle 4 \rangle$, whilst the K3 surface fibration on $W^{\mu}_{i,j}$ is M_2-polarized. As expected from our previous calculations, these two lattices are Dolgachev-Nikulin mirror dual.

5.4. Relationship to LG models. Now we will see how this fits with the results of Section 2. We begin by describing the LG models of X_1, X_2 and X_4. Again, we note that these threefolds can be constructed from the basic K3 fibred threefold $\mathcal{X}$.

THEOREM 5.8. [**Gol07, DHK16**] *Define $g_\ell\colon \mathbb{A}^1 \to \mathbb{A}^1$ by $g_\ell(t) = t^\ell$. Then the LG model $(Y_\ell, \mathsf{w}_\ell)$ of X_ℓ is a partial compactification of $g_\ell^*\mathcal{X}$.*

In fact, it follows from our classification of singular fibres in Section 5.2 that

(1) (Y_1, w_1) has two singular fibres in $\mathbb{A}^1$, located at 0 and 1/256, of types IV and $\mathrm{I}_{\mathrm{odp}}$ respectively,
(2) (Y_2, w_2) has three singular fibres in $\mathbb{A}^1$, located at 0 and $\pm\sqrt{1/256}$, of types II and $\mathrm{I}_{\mathrm{odp}}$ respectively,
(3) (Y_4, w_4) has four singular fibres in $\mathbb{A}^1$, located at $\sqrt[4]{1/256}$, all of which have type $\mathrm{I}_{\mathrm{odp}}$. Its fibre over 0 has type I_0.

The following ansatz has been suggested by Katzarkov [**Kat09**]:

ANSATZ 5.9. *Blowing up a Fano threefold X in a smooth curve of genus $g \geq 2$ has the effect of deforming the LG model (Y, w) of X, so that a semistable fibre with $g+1$ irreducible components moves from infinity to a point in $\mathbb{A}^1$.*

By this logic, we can induce that if $\widetilde{X}_\ell$ is as above, then its LG model $(\widetilde{Y}_\ell, \widetilde{\mathsf{w}}_\ell)$ has singular fibres of the types listed above for $(Y_\ell, \mathsf{w}_\ell)$, along with fibres of type $\mathrm{I}^\Delta_{x_s}$, for $s = 1, \ldots, k$, since a smooth curve C_s determined by a section of $\mathcal{O}_Z(x_s)$ has genus $2x_s^2 + 1$.

By the results of [**DHNT16**, Section 4], generically we may choose the map g determining $W^\mu_{i,j}$ so that its ramification points not lying over 0 and ∞ are all of order 2, and so that all nonzero branch points λ of g have $|\lambda| \geq R$ for some real number $R > 1/256$. Then there are two components to the preimage of the disc $U_R = \{z \in \mathbb{C} : |z| < R\}$, corresponding to the two preimages of 0. Let U_R^ℓ be the component of $g^{-1}(U_R)$ containing the ramification point of g of order ℓ over 0 (for $\ell \in \{i, j\}$). Then the restriction of the fibration π on $W^\mu_{i,j}$ to U^ℓ_R is a fibration over a disc with a fibre of type IV, II or I_0 over $g^{-1}(0)$, depending on whether $\ell = 1$, 2 or 4 respectively, and ℓ fibres of type $\mathrm{I}_{\mathrm{odp}}$. In fact, this fibration over U^ℓ_R is deformation equivalent to the LG model of X_ℓ, in the sense that as $R \to \infty$, the map g degenerates to a stable map from a pair of rational curves C_i and C_j meeting at a point, such that the restriction of g to C_ℓ is g_ℓ. We also see that monodromy around the boundary of U^ℓ_R is equal to the monodromy around ∞ of the the LG model $(Y_\ell, \mathsf{w}_\ell)$.

Away from 0 and ∞, we have that the restriction of π to $g^{-1}(\mathbb{P}^1 \setminus U_R)$ has singular fibres of type $\mathrm{I}^\Delta_{x_i}$, for $i = 1, \ldots, k$, and these account for all of the singular fibres of π restricted to $g^{-1}(\mathbb{P}^1 \setminus U_R)$. We thus obtain the following theorem; which is highly reminiscent of our philosophy in the K3 surface case (see Section 4.3).

THEOREM 5.10. *The threefold $W^\mu_{i,j}$ is topologically equivalent to the gluing of the LG model of X_j to the LG model of $\widetilde{X}_i$, as described in Section 2.2.*

6. Non-commutative fibrations

Now we turn our attention to ways in which this construction can fail. Suppose that V and W are mirror dual Calabi-Yau threefolds. The computation in

Section 5.1 suggests that, if a Tyurin degeneration of V occurs along a locus in the moduli space of V that contains a maximally unipotent monodromy point, then there should exist a K3 fibration on W whose properties are consistent with the results in Section 2.3.

However, it is also possible for Tyurin degenerations to occur along loci in the moduli space of V that are disjoint from points of maximally unipotent monodromy. For such Tyurin degenerations the argument in Section 5.1 will break down and, in particular, in such cases we will have no guarantee of the existence of K3 fibrations on W. For an example where this occurs, consider:

EXAMPLE 6.1. Let W be the complete intersection of two cubics in $\mathbb{P}^5$ and let V be its Batyrev-Borisov mirror. Since $h^{2,1}(V) = h^{1,1}(W) = 1$, the dimension of the moduli space of V is 1 and, indeed, it can be shown that the moduli space of Calabi-Yau varieties deformation equivalent to V is $\mathbb{P}^1 \setminus \{0, 1, \infty\}$. In **[KP09]** Katzarkov and Przyjalkowski show that, after semi-stabilization, the degeneration of V over ∞ is a Tyurin degeneration. The degenerate fibre is a union of two quasi-Fano threefolds X_1, X_2 meeting in a K3 surface Z, which has Picard rank 20 and transcendental lattice isomorphic to

$$\begin{pmatrix} 2 & 1 \\ 1 & 2 \end{pmatrix}.$$

There are several problems with this example. First of all, we find that this Tyurin degeneration is not connected to a point of maximally unipotent monodromy in the moduli space of V, so the discussion above does not hold. An even more grave issue is the fact that the K3 surface Z has no Dolgachev-Nikulin mirror. We would therefore not expect a K3 surface fibration corresponding to this Tyurin degeneration to appear on the mirror Calabi-Yau threefold W and, indeed, the non-existence of such a fibration is easily verified, since $h^{1,1}(W) = 1$.

However, if we are willing to replace the mirror Calabi-Yau threefold W with its bounded derived category of coherent sheaves $\mathbf{D}^b(W)$, we *do* know how to apply mirror symmetry to monodromy in non-maximally unipotent families. Indeed, if $\mathcal{V}$ is a family of Calabi-Yau threefolds over a punctured disc U, such that there is a symplectic form ω on $\mathcal{V}$ which restricts to a symplectic form on each fibre, then the action of monodromy around 0 induces a symplectomorphism on a smooth fibre V. According to **[Sze04]**, this symplectomorphism induces an autoequivalence on the Fukaya category $\mathcal{F}\mathrm{uk}(V, \omega)$, which passes through mirror symmetry to produce an element of $\mathrm{AutEq}(\mathbf{D}^b(W))$.

Our goal is to understand what sort of autoequivalence this is. If $\pi\colon W \to \mathbb{P}^1$ is a fibration of W by K3 surfaces, then there is a right derived pullback functor $\mathbf{R}\pi^*\colon \mathbf{D}^b(\mathbb{P}^1) \to \mathbf{D}^b(W)$. We conjecture that the autoequivalence of $\mathbf{D}^b(W)$ obtained above is an autoequivalence related to the spherical functor $\mathbf{R}\pi^*$. Moreover, if p is a point in $\mathbb{P}^1$, then the inclusion gives a pullback $\mathbf{R}i^*\colon \mathbf{D}^b(\mathbb{P}^1) \to \mathbf{D}^b(p)$. We can take the fibre product of these categories, denoted $\mathbf{D}^b(W) \otimes_{\mathbf{D}^b(\mathbb{P}^1)} \mathbf{D}^b(p)$, to give a category associated to a general fibre of W. We want this category to be the bounded derived category of coherent sheaves on a K3 surface.

To state this idea more precisely, we need a definition. A triangulated category $\mathbf{T}$ with Serre functor S is called *d-Calabi-Yau* if S is equivalent to $[d]$, where $[d]$ indicates the natural "shift by d" functor. This is a useful notion because the bounded derived category $\mathbf{D}^b(X)$ of coherent sheaves on a smooth projective variety X is d-Calabi-Yau if and only if X is a Calabi-Yau variety of dimension d.

CONJECTURE 6.2. *Let V be a Calabi-Yau variety of dimension d that admits a Tyurin degeneration, and let W be a homological mirror of V. Then there is a functor $F\colon \mathbf{D}^b(\mathbb{P}^1) \to \mathbf{D}^b(W)$ so that for a generic point $p \in \mathbb{P}^1$, the category $\mathbf{D}^b(V) \otimes_{\mathbf{D}^b(\mathbb{P}^1)} \mathbf{D}^b(p)$ is a $(d-1)$-Calabi-Yau category.*

Moreover, we expect that the fibres of such a categorical fibration are equivalent, in an appropriate sense, to the Fukaya category of the Calabi-Yau $Z = X_1 \cap X_2$, where $X_1 \cup_Z X_2$ is the Tyurin degeneration of V.

EXAMPLE 6.3. Calabrese and Thomas **[CT16]** showed how this should work in the setting of Example 6.1. Let f_1 and f_2 be cubics so that $W = \{f_1 = f_2 = 0\} \subset \mathbb{P}^5$. Blowing up $\mathbb{P}^5$ along W, one obtains a five-fold $\widetilde{\mathbb{P}}^5$ which is fibred over $\mathbb{P}^1$ by cubics $sf_1 + tf_2 = 0$, for $[s:t] \in \mathbb{P}^1$. This variety has derived category

$$\mathbf{D}^b(\widetilde{\mathbb{P}}^5) = \langle \mathbf{D}^b(W), \widetilde{\pi}^*\mathbf{D}^b(\mathbb{P}^1)(i,0) : i = 3,4,5 \rangle$$

where $\widetilde{\pi}\colon \widetilde{\mathbb{P}}^5 \to \mathbb{P}^1$ is the natural map.

Using this, Calabrese and Thomas show that there is a fibration structure on $\mathbf{D}^b(W)$ satisfying the conditions of Conjecture 6.2 above. In this case the derived categories $\mathbf{D}^b(W) \otimes_{\mathbf{D}^b(\mathbb{P}^1)} \mathbf{D}^b(p)$ are 2-Calabi-Yau categories, but they are not the bounded derived category of coherent sheaves on any K3 or abelian surface; instead they are semi-orthogonal summands of the derived categories of the cubic fourfolds cut out by the equations $sf_1 + tf_2 = 0$.

Moreover, according to Hassett **[Has99]**, the transcendental lattice of a generic cubic fourfold is isomorphic to

$$\mathrm{E}_8^2 \oplus \mathrm{U}^2 \oplus \begin{pmatrix} -2 & -1 \\ -1 & -2 \end{pmatrix},$$

which is precisely the Dolgachev-Nikulin dual lattice associated to the transcendental lattice of the K3 surface Z in Example 6.1. We thus conjecture that the non-commutative fibration found by Calabrese and Thomas **[CT16]** is mirror dual to the Tyurin degeneration of Katzarkov and Przyjalkowski **[KP09]**.

It may be possible to generate other examples like this one as follows. Suppose that $\mathcal{V}$ is a family of Calabi-Yau threefolds over a small disc U, so that the general fibre V has $h^{2,1}(V) = 1$ and the fibre over 0 is a Tyurin degeneration. Then the matrix describing the monodromy action on $H^3(V,\mathbb{Q})$ associated to a loop around 0 has two Jordan blocks of rank 1.

The families of Calabi-Yau variations of Hodge structure lying over $\mathbb{P}^1 \setminus \{0,1,\infty\}$ with appropriate monodromy properties have been classified by Doran and Morgan **[DM06]**. One can check that there are precisely three families in this classification for which the monodromy matrix associated to ∞ has two Jordan blocks of rank 1. These families are mirrors to

(1) The $(3,3)$ complete intersection in $\mathbb{P}^5$,
(2) The $(4,4)$ complete intersection in $\mathbb{WP}(1,1,1,1,2,2)$,
(3) The $(6,6)$ complete intersection in $\mathbb{WP}(1,1,2,2,3,3)$.

This raises the following natural question.

QUESTION 6.4. Are there categorical fibrations on $\mathbf{D}^b(W)$ satisfying Conjecture 6.2, for W a $(4,4)$ complete intersection in $\mathbb{WP}(1,1,1,1,2,2)$ or a $(6,6)$ complete intersection in $\mathbb{WP}(1,1,2,2,3,3)$?

Neither of these Calabi-Yau threefolds W can admit commutative K3 fibrations, since $h^{1,1}(W) = 1$, so if mirror dual fibrations exist then they are necessarily non-commutative, as in Example 6.3 above.

References

[ACG13] Dan Abramovich, Qile Chen, Danny Gillam, Yuhao Huang, Martin Olsson, Matthew Satriano, and Shenghao Sun, *Logarithmic geometry and moduli*, Handbook of moduli. Vol. I, Adv. Lect. Math. (ALM), vol. 24, Int. Press, Somerville, MA, 2013, pp. 1–61. MR3184161

[AKMS97] A. C. Avram, M. Kreuzer, M. Mandelberg, and H. Skarke, *Searching for K3 fibrations*, Nuclear Phys. B **494** (1997), no. 3, 567–589, DOI 10.1016/S0550-3213(97)00214-9. MR1453773

[Ale96a] V. Alexeev, *Log canonical singularities and complete moduli of stable pairs*, Preprint, August 1996, arXiv:alg-geom/9608013.

[Ale96b] ———, *Moduli spaces $M_{g,n}(W)$ for surfaces*, Higher-Dimensional Complex Varieties (Trento, 1994) (Berlin), de Gruyter, 1996, pp. 1–22.

[Ale06] ———, *Higher-dimensional analogues of stable curves*, International Congress of Mathematicians. Vol. II, Eur. Math. Soc., Zürich, 2006, pp. 515–536.

[AT16] V. Alexeev and A. Thompson, *Modular compactification of moduli of K3 surfaces of degree 2*, Manuscript in preparation, 2016, preliminary version available at http://alpha.math.uga.edu/~valery/.

[Aur08] Denis Auroux, *Special Lagrangian fibrations, mirror symmetry and Calabi-Yau double covers* (English, with English and French summaries), Astérisque **321** (2008), 99–128. Géométrie différentielle, physique mathématique, mathématiques et société. I. MR2521645

[Bat94] Victor V. Batyrev, *Dual polyhedra and mirror symmetry for Calabi-Yau hypersurfaces in toric varieties*, J. Algebraic Geom. **3** (1994), no. 3, 493–535. MR1269718

[BB96] Victor V. Batyrev and Lev A. Borisov, *On Calabi-Yau complete intersections in toric varieties*, Higher-dimensional complex varieties (Trento, 1994), de Gruyter, Berlin, 1996, pp. 39–65. MR1463173

[Bor93] L. A. Borisov, *Towards the mirror symmetry for Calabi-Yau complete intersections in Gorenstein toric Fano varieties*, Preprint, October 1993, arXiv:alg-geom/9310001.

[Bru15] A. Brunyate, *A compact moduli space of elliptic K3 surfaces*, Ph.D. thesis, University of Georgia, 2015.

[CCS13] Philip Candelas, Andrei Constantin, and Harald Skarke, *An abundance of K3 fibrations from polyhedra with interchangeable parts*, Comm. Math. Phys. **324** (2013), no. 3, 937–959, DOI 10.1007/s00220-013-1802-2. MR3123541

[CD07] Adrian Clingher and Charles F. Doran, *Modular invariants for lattice polarized K3 surfaces*, Michigan Math. J. **55** (2007), no. 2, 355–393, DOI 10.1307/mmj/1187646999. MR2369941

[CK99] David A. Cox and Sheldon Katz, *Mirror symmetry and algebraic geometry*, Mathematical Surveys and Monographs, vol. 68, American Mathematical Society, Providence, RI, 1999. MR1677117

[CLS11] David A. Cox, John B. Little, and Henry K. Schenck, *Toric varieties*, Graduate Studies in Mathematics, vol. 124, American Mathematical Society, Providence, RI, 2011. MR2810322

[CT16] John R. Calabrese and Richard P. Thomas, *Derived equivalent Calabi-Yau threefolds from cubic fourfolds*, Math. Ann. **365** (2016), no. 1-2, 155–172, DOI 10.1007/s00208-015-1260-6. MR3498907

[DHK16] C. F. Doran, A. Harder, L. Katzarkov, J. Lewis, and V. Przyjalkowski, *Moduli spaces of K3 surfaces and mirror symmetry for Fano threefolds*, Manuscript in preparation, 2016.

[DHNT15] Charles F. Doran, Andrew Harder, Andrey Y. Novoseltsev, and Alan Thompson, *Families of lattice polarized K3 surfaces with monodromy*, Int. Math. Res. Not. IMRN **23** (2015), 12265–12318, DOI 10.1093/imrn/rnv071. MR3431621

[DHNT16] C. F. Doran, A. Harder, A. Y. Novoseltsev, and A. Thompson, *Calabi-Yau threefolds fibred by mirror quartic K3 surfaces*, Adv. Math. **298** (2016), 369–392, DOI 10.1016/j.aim.2016.03.045. MR3505744

[DM06] Charles F. Doran and John W. Morgan, *Mirror symmetry and integral variations of Hodge structure underlying one-parameter families of Calabi-Yau threefolds*, Mirror symmetry. V, AMS/IP Stud. Adv. Math., vol. 38, Amer. Math. Soc., Providence, RI, 2006, pp. 517–537. MR2282973

[Dol96] I. V. Dolgachev, *Mirror symmetry for lattice polarized $K3$ surfaces*, J. Math. Sci. **81** (1996), no. 3, 2599–2630, DOI 10.1007/BF02362332. Algebraic geometry, 4. MR1420220

[EHX97] Tohru Eguchi, Kentaro Hori, and Chuan-Sheng Xiong, *Gravitational quantum cohomology*, Internat. J. Modern Phys. A **12** (1997), no. 9, 1743–1782, DOI 10.1142/S0217751X97001146. MR1439892

[Fri83] Robert Friedman, *Global smoothings of varieties with normal crossings*, Ann. of Math. (2) **118** (1983), no. 1, 75–114, DOI 10.2307/2006955. MR707162

[Giv98] Alexander Givental, *A mirror theorem for toric complete intersections*, Topological field theory, primitive forms and related topics (Kyoto, 1996), Progr. Math., vol. 160, Birkhäuser Boston, Boston, MA, 1998, pp. 141–175. MR1653024

[Gol07] Vasily V. Golyshev, *Classification problems and mirror duality*, Surveys in geometry and number theory: reports on contemporary Russian mathematics, London Math. Soc. Lecture Note Ser., vol. 338, Cambridge Univ. Press, Cambridge, 2007, pp. 88–121, DOI 10.1017/CBO9780511721472.004. MR2306141

[GP13] Antonella Grassi and Vittorio Perduca, *Weierstrass models of elliptic toric $K3$ hypersurfaces and symplectic cuts*, Adv. Theor. Math. Phys. **17** (2013), no. 4, 741–770. MR3262515

[GS75] Phillip Griffiths and Wilfried Schmid, *Recent developments in Hodge theory: a discussion of techniques and results*, Discrete subgroups of Lie groups and applicatons to moduli (Internat. Colloq., Bombay, 1973), Oxford Univ. Press, Bombay, 1975, pp. 31–127. MR0419850

[Har16] A. Harder, *The geometry of Landau-Ginzburg models*, Ph.D. thesis, University of Alberta, 2016.

[Has99] Brendan Hassett, *Some rational cubic fourfolds*, J. Algebraic Geom. **8** (1999), no. 1, 103–114. MR1658216

[Kat92] Sheldon Katz, *Rational curves on Calabi-Yau threefolds*, Essays on mirror manifolds, Int. Press, Hong Kong, 1992, pp. 168–180. MR1191424

[Kat09] L. Katzarkov, *Homological mirror symmetry and algebraic cycles*, Homological mirror symmetry, Lecture Notes in Phys., vol. 757, Springer, Berlin, 2009, pp. 125–152. MR2596637

[KKP14] Ludmil Katzarkov, Maxim Kontsevich, and Tony Pantev, *Bogomolov-Tian-Todorov theorems for Landau-Ginzburg models*, J. Differential Geom. **105** (2017), no. 1, 55–117. MR3592695

[KN94] Yujiro Kawamata and Yoshinori Namikawa, *Logarithmic deformations of normal crossing varieties and smoothing of degenerate Calabi-Yau varieties*, Invent. Math. **118** (1994), no. 3, 395–409, DOI 10.1007/BF01231538. MR1296351

[KP09] L. Katzarkov and V. Przyjalkowski, *Generalized homological mirror symmetry and cubics*, Tr. Mat. Inst. Steklova **264** (2009), no. Mnogomernaya Algebraicheskaya Geometriya, 94–102, DOI 10.1134/S0081543809010118; English transl., Proc. Steklov Inst. Math. **264** (2009), no. 1, 87–95. MR2590839

[KSB88] J. Kollár and N. I. Shepherd-Barron, *Threefolds and deformations of surface singularities*, Invent. Math. **91** (1988), no. 2, 299–338, DOI 10.1007/BF01389370. MR922803

[Lee06] Nam-Hoon Lee, *Constructive Calabi-Yau manifolds*, ProQuest LLC, Ann Arbor, MI, 2006. Thesis (Ph.D.)–University of Michigan. MR2708981

[Lee10] Nam-Hoon Lee, *Calabi-Yau construction by smoothing normal crossing varieties*, Internat. J. Math. **21** (2010), no. 6, 701–725, DOI 10.1142/S0129167X10006173. MR2658406

[Mor84] David R. Morrison, *The Clemens-Schmid exact sequence and applications*, Topics in transcendental algebraic geometry (Princeton, N.J., 1981/1982), Ann. of Math. Stud., vol. 106, Princeton Univ. Press, Princeton, NJ, 1984, pp. 101–119. MR756848

[Ogu01] Keiji Oguiso, *On the finiteness of fiber-space structures on a Calabi-Yau 3-fold*, J. Math. Sci. (New York) **106** (2001), no. 5, 3320–3335, DOI 10.1023/A:1017959510524. Algebraic geometry, 11. MR1878052

[Orl09] Dmitri Orlov, *Derived categories of coherent sheaves and triangulated categories of singularities*, Algebra, arithmetic, and geometry: in honor of Yu. I. Manin. Vol. II, Progr. Math., vol. 270, Birkhäuser Boston, Inc., Boston, MA, 2009, pp. 503–531, DOI 10.1007/978-0-8176-4747-6_16. MR2641200

[PS08] Chris A. M. Peters and Joseph H. M. Steenbrink, *Mixed Hodge structures*, Ergebnisse der Mathematik und ihrer Grenzgebiete. 3. Folge. A Series of Modern Surveys in Mathematics [Results in Mathematics and Related Areas. 3rd Series. A Series of Modern Surveys in Mathematics], vol. 52, Springer-Verlag, Berlin, 2008. MR2393625

[Roh04] F. Rohsiepe, *Lattice polarized toric K3 surfaces*, Preprint, September 2004, arXiv:hep-th/0409290.

[Sca87] Francesco Scattone, *On the compactification of moduli spaces for algebraic K3 surfaces*, Mem. Amer. Math. Soc. **70** (1987), no. 374, x+86, DOI 10.1090/memo/0374. MR912636

[Sei01a] Paul Seidel, *More about vanishing cycles and mutation*, Symplectic geometry and mirror symmetry (Seoul, 2000), World Sci. Publ., River Edge, NJ, 2001, pp. 429–465, DOI 10.1142/9789812799821_0012. MR1882336

[Sei01b] Paul Seidel, *Vanishing cycles and mutation*, European Congress of Mathematics, Vol. II (Barcelona, 2000), Progr. Math., vol. 202, Birkhäuser, Basel, 2001, pp. 65–85. MR1905352

[Sze04] Balázs Szendrői, *Contractions and monodromy in homological mirror symmetry*, Strings and geometry, Clay Math. Proc., vol. 3, Amer. Math. Soc., Providence, RI, 2004, pp. 301–314. MR2103728

[Tyu04] Andrei N. Tyurin, *Fano versus Calabi-Yau*, The Fano Conference, Univ. Torino, Turin, 2004, pp. 701–734. MR2112600

[Voi07] Claire Voisin, *Hodge theory and complex algebraic geometry. II*, Reprint of the 2003 English edition, Cambridge Studies in Advanced Mathematics, vol. 77, Cambridge University Press, Cambridge, 2007. Translated from the French by Leila Schneps. MR2449178

Department of Mathematical and Statistical Sciences, 632 CAB, University of Alberta, Edmonton, AB, T6G 2G1, Canada
E-mail address: charles.doran@ualberta.ca

Department of Mathematical and Statistical Sciences, 632 CAB, University of Alberta, Edmonton, AB, T6G 2G1, Canada
E-mail address: aharder@ualberta.ca

Department of Pure Mathematics, University of Waterloo, 200 University Ave West, Waterloo, ON, N2L 3G1, Canada
E-mail address: am6thomp@uwaterloo.ca

SL(2,ℂ) Chern-Simons theory and four-dimensional quantum geometry

Muxin Han

ABSTRACT. This paper presents SL(2,ℂ) Chern-Simons theory on a class of graph complement 3-manifolds, and its relation with classical and quantum geometries on 4-dimensional manifolds. In classical theory, we explain the correspondence between a class of SL(2,ℂ) flat connections on 3-manifold and the Lorentzian simplicial geometries in 4 dimensions. The class of flat connections on the graph complement 3-manifold is specified by a certain boundary condition. The corresponding simplicial 4-dimensional geometries are made by constant curvature 4-simplices. The quantization of 4d simplicial geometry can be carried out via the quantization of flat connection on 3-manifold in Chern-Simons theory. In quantum SL(2,ℂ) Chern-Simons theory, a basis of physical wave functions is the class of (holomorphic) 3d block, defined by analytically continued Chern-Simons path integral over Lefschetz thimbles. Here we propose that the (holomorphic) 3d block with the proper boundary condition imposed gives the quantization of simplicial 4-dimensional geometry. Interestingly in the semiclassical asymptotic expansion of (holomorphic) 3d block, the leading contribution gives the classical action of simplicial Einstein-Hilbert gravity in 4 dimensions. Such a result suggests a relation between SL(2,ℂ) Chern-Simons theory on a class of 3-manifolds and simplicial quantum gravity on 4-dimensional manifolds.

1. Introduction

Chern-Simons theory in 3-dimensions is the quintessential topological quantum field theory (TQFT) and has been studied extensively since the 1980's (see e.g. [**1**]). In addition to its importance in the formulation of TQFT [**2**], Chern-Simons theory has applications in many branches of modern mathematics and physics. The celebrated work of Witten [**3**], exposed the remarkable relation between Chern-Simons theory with compact gauge group and knot theory. Chern-Simons theory plays an important role in the formulation of the Volume Conjecture, which relates knot polynomials to the hyperbolic geometry of 3-manifolds [**4**–**7**]. It also has been applied to quantum gravity in 3-dimensions [**8**, **9**]. Moreover, many aspects of String/M-theory and Supersymmetric Gauge Theory have close relationship with Chern-Simons theory (e.g. [**10**–**15**]). Chern-Simons theory has also the interesting applications to Loop Quantum Gravity (LQG) in 4 dimensions, in the covariant formulation and black hole physics (e.g. [**16**–**20**]).

2010 *Mathematics Subject Classification.* Primary 57M25, 83C27.

Chern-Simons theory with compact gauge group and its quantization are well understood after the intensive investigations of the last 20 years. However, quantum Chern-Simons theory with complex gauge group $G_{\mathbb{C}}$ (G being a compact Lie group) is a rather open subject. The Chern-Simons theories with gauge groups $G_{\mathbb{C}}$ being noncompact are qualitatively different from Chern-Simons theory with a compact group. In general the Hilbert space of Chern-Simons theory with a noncompact group is infinite-dimensional [**7, 21–23**], while the Hilbert space of the theory with compact group is finite-dimensional. Recently, there has been substantial progress in the understanding of the theory with complex gauge group [**5–7, 23, 24**]. an active area of research.

The present paper focuses on Chern-Simons theory with the complex gauge group SL(2, $\mathbb{C}$) on a compact oriented 3-manifold $\mathscr{M}_3$:

$$CS\left[\mathscr{M}_3 \,|\, A, \bar{A}\right] = \frac{t}{8\pi}\int_{\mathscr{M}_3} \mathrm{tr}\left(A\wedge \mathrm{d}A + \frac{2}{3}A\wedge A\wedge A\right) + \frac{\bar{t}}{8\pi}\int_{\mathscr{M}_3} \mathrm{tr}\left(\bar{A}\wedge \mathrm{d}\bar{A} + \frac{2}{3}\bar{A}\wedge \bar{A}\wedge \bar{A}\right). \tag{1.1}$$

with the possible boundary terms when M_3 has a boundary. $t = k + is$ is Chern-Simons coupling with $k, s \in \mathbb{R}$, and $\bar{t}$ is assumed to be the complex conjugate of t. $A = A^j\tau_j$ with $\tau_j = -\frac{i}{2}\sigma_j$ ($\sigma_{1,2,3}$ denote the Pauli matrices) is a connection 1-form taking value in the complex Lie algebra $\mathfrak{sl}_2\mathbb{C}$. Here in this work, we focus on a class of 3-manifolds $\mathscr{M}_3$, in which the simplest example is a complement 3-manifold of Γ_5-graph FIG.1. If a graph is embedded in S^3, the graph complement 3-manifold is obtained by removing the graph as well as the interior of its tubular neighborhood from S^3. The Γ_5-graph complement 3-manifold is often denoted by $S^3 \setminus \Gamma_5$. Obviously $S^3 \setminus \Gamma_5$ has a boundary, which is a genus-6 closed 2-surface Σ_6.

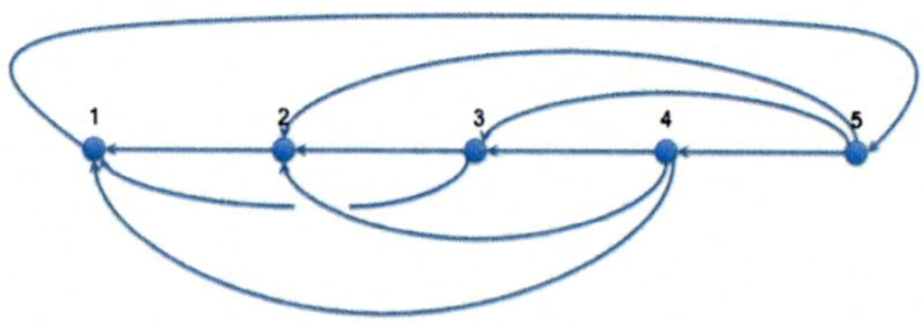

FIGURE 1. The Γ_5 graph can be drawn with five 4-valent vertices, ten edges ℓ_{ab} with the curve ℓ_{24} over-crossing ℓ_{13}. Γ_5 may also be drawn as a graph with all vertices being 3-valent, by simply slip each 4-valent vertex into two connected 3-valent vertices, which results in 10 vertices and 15 edges. However both ways to draw Γ_5 lead to the same 3-manifold $S^3 \setminus \Gamma_5$.

Chern-Simons theory with graph defects have been considered in [**25**] of compact gauge group; and the volume conjecture has been generalized to quantum spin-networks with knotted graphs in [**26, 29**]. From the mathematical point of view, the space of knotted graphs may be more interesting than the space of knots – due to the fact that the space of trivalent knotted graphs are finitely generated. This means that there is a finite (small) set of trivalent knotted graphs such that all trivalent knotted graphs can be generated by a few algebraic operations, while the space of knots is a proper subset of the space of trivalent graphs [**27**]. A recent study of trivalent knotted graphs, from the perspective of perturbative Vassiliev-Kontsevich invariants, gives the right algebraic operations, [**28**].

Classically, the equation of motion from SL(2, ℂ) Chern-Simons theory gives

$$F = \mathrm{d}A + A \wedge A = 0, \qquad \bar{F} = \mathrm{d}\bar{A} + \bar{A} \wedge \bar{A} = 0. \tag{1.2}$$

Namely A is a flat connection on the 3-manifold $\mathscr{M}_3$. The moduli space of flat connection $\mathcal{M}_{\text{flat}}(\mathscr{M}_3, \mathrm{SL}(2,\mathbb{C}))$ is the space of solutions. When $\mathscr{M}_3$ has a boundary being a closed 2-surface $\Sigma_g = \partial\mathscr{M}_3$ of genus-g, the space of boundary values of $A \in \mathcal{M}_{\text{flat}}(\mathscr{M}_3, \mathrm{SL}(2,\mathbb{C}))$ is a subvariety inside $\mathcal{M}_{\text{flat}}(\Sigma_g, \mathrm{SL}(2,\mathbb{C}))$, the moduli space of SL(2, ℂ) flat connections on Σ. In general, $\mathcal{M}_{\text{flat}}(\Sigma_g, \mathrm{SL}(2,\mathbb{C}))$, known as Hitchin moduli space, is a hyper-Kähler variety of $\dim_{\mathbb{C}} = 6g - 6$, which has 3 distinct complex structures I, J, K[1] [**30**]. The three corresponding Kähler forms is denoted by $\omega_I, \omega_J, \omega_K$. When we consider $\mathcal{M}_{\text{flat}}(\Sigma_g, \mathrm{SL}(2,\mathbb{C}))$ as the phase space of SL(2, ℂ) Chern-Simons theory, the holomorphic Chern-Simons (Atiyah-Bott-Goldman) symplectic structure ω_{CS} is given by

$$\omega_{CS} = \frac{t}{4\pi}\int_{\Sigma_g} \mathrm{tr}\left[\delta_1 A \wedge \delta_2 A\right] = \frac{t}{\pi}\left[\omega_I - i\omega_K\right], \tag{1.3}$$

which comes from the holomorphic part of $CS\left[\mathscr{M}_3 \mid A, \bar{A}\right]$. $\mathcal{M}_{\text{flat}}(\mathscr{M}_3, \mathrm{SL}(2,\mathbb{C}))$ is embedded as a subvariety $\mathcal{L}_{\mathbf{A}}$ of $\dim_{\mathbb{C}} = 3g-3$ in $\mathcal{M}_{\text{flat}}(\Sigma_g, \mathrm{SL}(2,\mathbb{C}))$ by considering the boundary values.

$$\mathcal{L}_{\mathbf{A}} \simeq \mathcal{M}_{\text{flat}}(\mathscr{M}_3, \mathrm{SL}(2,\mathbb{C})) \tag{1.4}$$

$\mathcal{L}_{\mathbf{A}}$ is holomorphic with respect to the complex structure J, and is Lagrangian with respect to I and K, i.e. ω_I, ω_K thus ω_{CS} vanish on $\mathcal{L}_{\mathbf{A}}$ [**31, 32**].

The complex Fenchel-Nielsen (FN) coordinate $x_m, y_m \in \mathbb{C}$, $m = 1\cdots 3g-3$ can be used to locally parametrize the connections of $\mathcal{M}_{\text{flat}}(\Sigma_g, \mathrm{SL}(2,\mathbb{C}))$ [**33, 34**], using a pants decomposition of the closed 2-surface Σ_g. Here the complex FN "length variable" x_m is the eigenvalue of the holonomy along a closed pant curve c_m for the pants decomposition. The complex FN "twist variable" y_m is the conjugate variable such that ω_{CS} is written as

$$\omega_{CS} = \left(-\frac{t}{2\pi}\right)\sum_{m=1}^{3g-3} \frac{\mathrm{d}y_m}{y_m} \wedge \frac{\mathrm{d}x_m}{x_m}. \tag{1.5}$$

The explicit relation between y_m and holonomies is given in e.g. [**31, 34**]. In terms of $\{x_m, y_m\}_{m=1}^{3g-3}$, the holomorphic Lagrangian subvariety $\mathcal{L}_{\mathbf{A}} \simeq \mathcal{M}_{\text{flat}}(\mathscr{M}_3, \mathrm{SL}(2,\mathbb{C}))$ can be expressed locally as a set of holomorphic polynomial equations

$$\mathbf{A}_m(x, y) = 0, \qquad m = 1, \cdots, 3g-3. \tag{1.6}$$

When $\mathscr{M}_3$ is the complement of a knot thus $\partial\mathscr{M}_3 = T^2$, we have $\mathcal{M}_{\text{flat}}(T^2, \mathrm{SL}(2,\mathbb{C})) \simeq \mathbb{C}^* \times \mathbb{C}^*/\mathbb{Z}_2$, and $\mathcal{L}_{\mathbf{A}}$ is specified as by the zero-locus of a single holomorphic polynomial $\mathbf{A}(x, y)$, known as A-polynomial [**6, 35**].

[1] I is induced from the complex structure of Σ_g, J is from the complex structure of the complex group SL(2, ℂ), and K is obtained by the relation $K = IJ$.

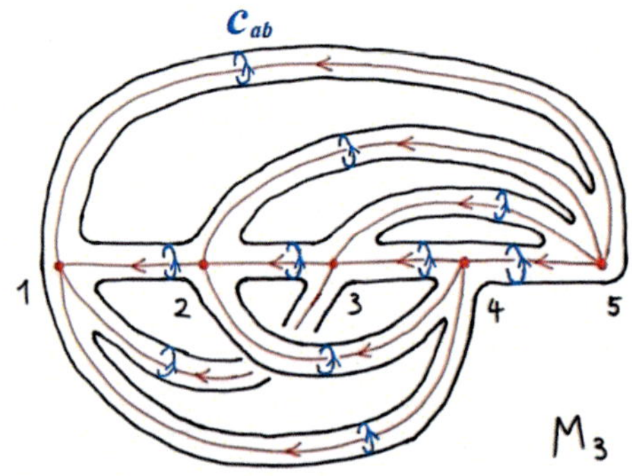

FIGURE 2. The graph complement 3-manifold M_3 after removing the thickened Γ_5-graph from S^3. The 2d boundary $\partial M_3 = \Sigma_6$ of graph complement M_3 is a genus-6 closed 2-surface. A set of meridian closed curves c_{ab} are defined on Σ_6 such that $\Sigma_6 \setminus \{c_{ab}\}$ is a set of 4-holed spheres. $a, b = 1, \cdots, 5$ label the vertices of the graph.

2. Classical Correspondence

Let us focus on the 3-manifold $S^3 \backslash \Gamma_5$ whose boundary is a genus-6 closed surface Σ_6 (see FIG.2). We are interested in a subspace of $\mathcal{L}_{\mathbf{A}} \simeq \mathcal{M}_{\text{flat}}(S^3 \setminus \Gamma_5, \text{SL}(2,\mathbb{C}))$, in which the $\text{SL}(2,\mathbb{C})$ flat connections have 4-geometry interpretations. More precisely, a flat connection in the subspace determines the geometry of a convex 4-simplex in 4-dimensional Lorentzian constant curvature spacetime (de-Sitter or Anti-de-Sitter)[2]. Here we consider 4-dimensional Lorentzian geometry with signature $(-,+,+,+)$. We have that all the convex constant curvature 4-simplex geometries with both $\Lambda > 0$ and $\Lambda < 0$ can be described by a class of $\text{SL}(2,\mathbb{C})$ flat connections on the graph complement 3-manifold $S^3 \setminus \Gamma_5$.[3] [**19, 36**]

$$\text{(2.1)} \quad \text{A class of } (A, \bar{A}) \text{ in } \mathcal{M}_{\text{flat}}\left(S^3 \setminus \Gamma_5, \text{SL}(2,\mathbb{C})\right) = \text{Constant Curvature 4-Simplex Geometries.}$$

The subspace of flat connections on $S^3 \setminus \Gamma_5$ is specified by certain boundary condition imposed on their boundary values on Σ_6. The boundary condition is introduced in [**36**], which can be described briefly in the following way: Σ_6 can be decomposed into five 4-holed spheres $\mathcal{S}_{a=1,\cdots,5}$ by cutting through 10 meridian closed curves in FIG.2. The boundary condition requires that the boundary value of $A \in \mathcal{M}_{\text{flat}}(S^3 \setminus \Gamma_5, \text{SL}(2,\mathbb{C}))$ reduces to an $\text{SU}(2)$ flat connection up to gauge transformation, when it is restricted onto one of the 4-holed spheres $\mathcal{S}_a$[4].

The boundary condition is motivated by the geometrical interpretation of $\text{SU}(2)$ flat connection on a 4-holed sphere $\mathcal{S}_a$. Namely, an $\text{SU}(2)$ flat connection on a 4-holed sphere determines uniquely a convex tetrahedron geometry in constant

[2] A 4-simplex is the elementary building block of the simplicial decomposition of 4-dimensional manifold. It is analogous to the tetrahedron in 3 dimensions and the triangle in 2 dimensions. See FIG.3 for a projection diagram.

[3] We do not consider degenerate 4-simplex geometries here.

[4] It doesn't imply A becomes an $\text{SU}(2)$ flat connection on entire Σ_6, since the different 4-holed spheres may correspond to different $\text{SU}(2)$ subgroups in $\text{SL}(2,\mathbb{C})$.

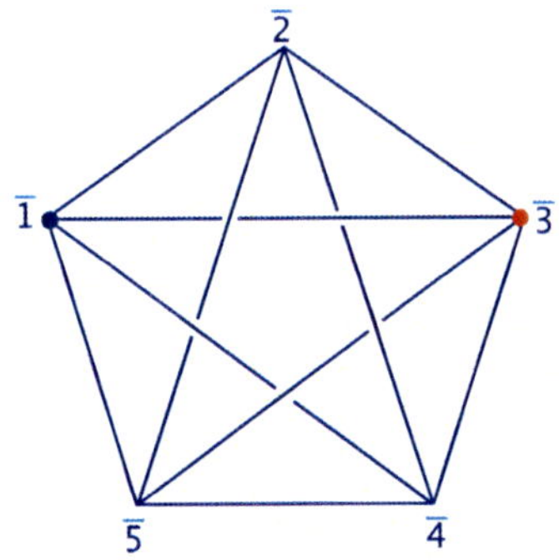

FIGURE 3. An abstract 4-simplex, whose vertices are labeled by $\bar{1}, \cdots, \bar{5}$. The tetrahedron tetra$_a$ denotes the tetrahedron which doesn't have the vertex $\bar{a}$. The triange Δ_{ab} (resp. Δ_{ba}) denotes the triangle belonging to tetra$_a$ (resp. tetra$_b$) shared by tetra$_a$ and tetra$_b$. The edges are denoted by $(\bar{a}, \bar{b})$ oriented from $\bar{b}$ to $\bar{a}$.

curvature 3d space (spherical or hyperbolic 3d space). The statement holds for a dense subset of $\mathcal{M}_{\text{flat}}(\mathcal{S}_a, \text{SU}(2))$ up to the flat connections corresponding to degenerate geometries. If we consider PSU(2) flat connections instead of SU(2), the correspondence becomes 1-to-1 between a flat connection on 4-holed sphere and a convex constant curvature tetrahedron. Such an interpretation of SU(2) flat connections on 4-holed sphere has been introduced in **[19]** (see also **[37]** for a thorough exploration).

Given a flat connection $A \in \mathcal{M}_{\text{flat}}(S^3 \setminus \Gamma_5, \text{SL}(2,\mathbb{C}))$ on the Γ_5 graph complement 3-manifold, satisfying the above boundary condition, it is shown in **[36]** that the flat connection A determines uniquely a convex 4-simplex geometry in 4-dimensional Lorentzian spacetime with constant curvature Λ. The closed boundary of the 4-simplex determined by A is formed by 5 constant curvature tetrahedra, which are congruent to the tetrahedron geometries determined by the boundary data of A on 4-holed spheres $\mathcal{S}_a$. Again the statement holds up to the flat connections corresponding to degenerate 4-simplex geometries. If we consider PSL$(2,\mathbb{C})$ flat connections instead of SL$(2,\mathbb{C})$, the correspondence becomes 1-to-1 between $A \in \mathcal{M}_{\text{flat}}(S^3 \setminus \Gamma_5, \text{PSL}(2,\mathbb{C}))$ satisfying the boundary condition and a convex 4-simplex geometry of constant curvature.

A simple intuition lies behind the above correspondence between flat connections on a 3-manifold and the geometry of a 4-manifold. The 1-skeleton of a 4-simplex gives a triangulation of 3-sphere S^3 as the boundary of 4-simplex. The Γ_5 graph can be viewed as an "dual" graph of the 4-simplex skeleton, in the sense that the fundamental group of $S^3 \setminus \Gamma_5$ is isomorphic to the fundamental group of the 4-simplex skeleton π_1(simplex). On one side, an SL$(2,\mathbb{C})$ flat connection on $S^3 \setminus \Gamma_5$ is a representation of fundamental group $\pi_1(S^3 \setminus \Gamma_5)$ up to conjugation. On the other side, if the 4-simplex is embedded in a geometrical 4d spacetime $(\mathfrak{M}_4, g_{\alpha\beta})$, the spin connection on $\mathfrak{M}_4$ gives a representation up to conjugation of π_1(simplex) by the holonomies. The isomorphism between $\pi_1(S^3 \setminus \Gamma_5)$ and π_1(simplex) identifies the flat connection on $S^3 \setminus \Gamma_5$ and spin connection on 4-simplex. More precisely, it identifies the holonomies of flat connection along the loops in $\pi_1(S^3 \setminus \Gamma_5)$ and the holonomies of spin connection along the closed paths on π_1(simplex). In terms of a commutative diagram,

$$
\begin{array}{ccc}
\pi_1(S^3\setminus\Gamma_5) & \xleftarrow{X} & \pi_1(\text{simplex}) \\
 & \omega_{\text{flat}} \searrow \qquad \swarrow \omega_{\text{spin}} & \\
\multicolumn{3}{c}{\Big\langle \{\tilde{H}_{ab}\in \mathrm{SL}(2,\mathbb{C})\}_{a<b} \;\Big|\; \text{algebraic relations Eqs.(2.3)} - \text{(2.8)} \Big\rangle \Big/ \text{conjugation}}
\end{array}
\tag{2.2}
$$

where X denotes the isomorphism between $\pi_1(S^3\setminus\Gamma_5)$ and $\pi_1(\text{simplex})$ and ω_{flat}, ω_{spin} denote the representation from flat connection on $S^3\setminus\Gamma_5$ and spin connection on $\mathfrak{M}_4$. The algebraic relations Eq.(2.3) - (2.8) represent the group structure of $\pi_1(S^3\setminus\Gamma_5)$ using holonomies along the close loop around defects:

$$\text{vertex } 1: \quad \tilde{H}_{14}\tilde{H}_{13}^{(1)}\tilde{H}_{12}\tilde{H}_{15}=1, \tag{2.3}$$

$$\text{vertex } 2: \quad \tilde{H}_{12}^{-1}\tilde{H}_{24}\tilde{H}_{23}\tilde{H}_{25}=1, \tag{2.4}$$

$$\text{vertex } 3: \quad \tilde{H}_{23}^{-1}(\tilde{H}_{13}^{(2)})^{-1}\tilde{H}_{34}\tilde{H}_{35}=1, \tag{2.5}$$

$$\text{vertex } 4: \quad \tilde{H}_{34}^{-1}\tilde{H}_{24}^{-1}\tilde{H}_{14}^{-1}\tilde{H}_{45}=1, \tag{2.6}$$

$$\text{vertex } 5: \quad \tilde{H}_{25}^{-1}\tilde{H}_{35}^{-1}\tilde{H}_{45}^{-1}\tilde{H}_{15}^{-1}=1, \tag{2.7}$$

$$\text{crossing}: \quad \tilde{H}_{13}^{(1)}=\tilde{H}_{24}\tilde{H}_{13}^{(2)}\tilde{H}_{24}^{-1}. \tag{2.8}$$

In this way, the $\mathrm{SL}(2,\mathbb{C})$ flat connections on $S^3\setminus\Gamma_5$ relates to the spin connections on a spacetime $(\mathfrak{M}_4, g_{\alpha\beta})$. If we consider $(\mathfrak{M}_4, g_{\alpha\beta})$ is a Lorentzian spacetime with constant curvature Λ, and all the 10 triangles of the 4-simplex are flatly embedded in $(\mathfrak{M}_4, g_{\alpha\beta})$ (i.e. vanishing extrinsic curvature), the holonomy of spin connection along a closed path in $\pi_1(\text{simplex})$ enclosing a single triangle determines the area of the triangle, as well as the embedding property of the triangle, i.e. the 2 normal directions of the triangle embedded in $\mathfrak{M}_4$ [**19**]. The above relation between ω_{flat} and ω_{spin}, as well as the geometrical properties from spin connections, result in the correspondence between the $\mathrm{SL}(2,\mathbb{C})$ flat connections on $S^3\setminus\Gamma_5$ and 4d geometry on a constant curvature 4-simplex.

Given $A\in\mathcal{M}_{\text{flat}}(S^3\setminus\Gamma_5,\mathrm{SL}(2,\mathbb{C}))$ corresponding to a constant curvature 4-simplex, it is naturally accompanied by $\tilde{A}\in\mathcal{M}_{\text{flat}}(S^3\setminus\Gamma_5,\mathrm{SL}(2,\mathbb{C}))$, which is the complex conjugate of A with respect to the complex structure J induced from the complex group $\mathrm{SL}(2,\mathbb{C})$. $A,\tilde{A}$ determine the same 4-simplex geometry but result in 2 opposite 4d orientation for the 4-simplex. Here we call $A,\tilde{A}$ a "parity pair", because the complex conjugate with respect to J naturally relates to a parity inversion in 4d spacetime [**19**]. Such a complex conjugate leaves $\mathrm{SU}(2)$ flat connection invariant. So $A,\tilde{A}$ induce the same $\mathrm{SU}(2)$ flat connections on the 4-holed spheres $\mathcal{S}_{a=1,\cdots,5}$, consistent with the fact that the 4-simplex geometries determined by $A,\tilde{A}$ are the same, and give the same set of geometrical tetrahedra on the boundary.

We consider $\mathcal{M}_{\text{flat}}(S^3\setminus\Gamma_5,\mathrm{SL}(2,\mathbb{C}))\simeq\mathcal{L}_{\mathbf{A}}$ as the holomorphic Lagrangian subvariety in $\mathcal{M}_{\text{flat}}(\Sigma_6,\mathrm{SL}(2,\mathbb{C}))$. Given $A\in\mathcal{M}_{\text{flat}}(S^3\setminus\Gamma_5,\mathrm{SL}(2,\mathbb{C}))$ corresponding to a constant curvature 4-simplex, the complex FN variables of A have the interpretations in terms of 4-simplex geometry. Namely, the 10 length variables x_{ab} for the closed curves c_{ab} in FIG.2 relate respectively to the 10 areas $\mathbf{a}_{ab}$ of the triangles Δ_{ab} in the 4-simplex. The 10 conjugate twist variable y_{ab} relate respectively to the 10 hyper-dihedral angles Θ_{ab}. Each hyper-dihedral angle Θ_{ab} between a pair of boundary tetrahedra is hinged by the triangle Δ_{ab} shared by the tetrahedra. Interestingly the canonical conjugation relation between $\mathbf{a}_{ab}$ and Θ_{ab}, resulting from flat connections and their geometrical correspondence, relates to the canonical structure induced from 4-dimensional Einstein-Hilbert action in General Relativity (GR)[5]. Such a fact suggests the relation between the flat connections on 3-manifold and (simplicial) gravity on 4-dimensional manifold.

The phase space $\mathcal{M}_{\text{flat}}(\Sigma_6, \text{SL}(2,\mathbb{C}))$ has $\dim_{\mathbb{C}} = 30$. In addition to $\{x_{ab}, y_{ab}\}_{a<b}$, there are 5 pairs of variables $\{x_a, y_a\}_{a=1}^5$ parametrize the SU(2) flat connections on $\mathcal{S}_{a=1\cdots,5}$. Geometrically they parametrize the shapes of 5 constant curvature tetrahedra on the boundary of the geometrical 4-simplex.

3. Quantum Correspondence

Our correspondence between $\text{SL}(2,\mathbb{C})$ flat connections on $S^3 \setminus \Gamma_5$ and constant curvature geometry of 4-simplex motivates a new understanding of (simplicial) quantum 4-dimensional geometry in terms of the quantization of flat connections on 3-manifold. For any 3-manifold $\mathscr{M}_3$ with boundary Σ_g, the quantization of $\mathcal{M}_{\text{flat}}(\Sigma_g, \text{SL}(2,\mathbb{C}))$ with the symplectic structure ω_{CS} gives the operator algebra for the canonical conjugate variables, e.g. the operators for the complex FN variables $\hat{x}_m, \hat{y}_m$ satisfying $\hat{x}_m\hat{y}_m = e^{-\frac{2\pi i\hbar}{t}}\hat{y}_m\hat{x}_m$ ($\hbar \in \mathbb{R}$) and $\hat{x}_m\hat{y}_n = \hat{y}_n\hat{x}_m$ for $n \neq m$. The states are represented as the wave functions $Z(u)$, where u is the logarithmic coordinate $u_m = \ln x_m$. The readers are referred to e.g **[38, 39]** for details of quantizing $\mathcal{M}_{\text{flat}}(\Sigma_g, \text{SL}(2,\mathbb{C}))$. The quantization of the holomorphic Lagrangian subvariety $\mathcal{M}_{\text{flat}}(\mathscr{M}_3, \text{SL}(2,\mathbb{C})) \simeq \mathcal{L}_{\mathbf{A}}$ gives a set of operator constraint equations:

$$\hat{\mathbf{A}}_m(\hat{x}, \hat{y}, \hbar)\, Z(u) = 0, \qquad m = 1, \cdots, 3g-3. \tag{3.1}$$

The solutions $Z(u)$ of the above operator constraint equations are the physical states if $\text{SL}(2,\mathbb{C})$ Chern-Simons theory on $\mathscr{M}_3$. A basis of solutions $Z^{(\alpha)}_{CS}(u)$ can be found by the semiclassical WKB method **[5, 7, 24]**:

$$Z^{(\alpha)}_{CS}(u) = \exp\left[\frac{i}{\hbar}\int_{\mathfrak{C}\subset\mathcal{L}_{\mathbf{A}}}^{u,v^{(\alpha)}} \vartheta + \cdots\right] \tag{3.2}$$

The leading term is completely determined by the classical information of phase space and Lagrangian subvariety. ϑ is the Liouville 1-form satisfying $\mathrm{d}\vartheta = \omega_{CS}$. ϑ is integrated along a contour $\mathfrak{C}$ in Lagrangian subvariety $\mathcal{L}_{\mathbf{A}}$. The logarithmic coordinates u_m, v_m relate to x_m, y_m by $x_m = e^{u_m}$ and $y_m = e^{-\frac{2\pi}{t}v_m}$. The label α denotes the branches of $\mathcal{L}_{\mathbf{A}}$ on which a unique set of v_m can be solved from u_m by $\mathbf{A}_m(x,y) = 0$. The end point of the contour $\mathfrak{C}$, which labels $Z^{(\alpha)}_{CS}(u)$, is a flat connection determined by $u, v^{(\alpha)}$ in $\mathcal{M}_{\text{flat}}(\mathscr{M}_3, \text{SL}(2,\mathbb{C})) \simeq \mathcal{L}_{\mathbf{A}}$ (or more precisely, in the cover space of $\mathcal{M}_{\text{flat}}(\mathscr{M}_3, \text{SL}(2,\mathbb{C})) \simeq \mathcal{L}_{\mathbf{A}}$). Thus $Z^{(\alpha)}_{CS}(u)$ associates with a single flat connection $A \in \mathcal{M}_{\text{flat}}(\mathscr{M}_3, \text{SL}(2,\mathbb{C}))$. The starting point of $\mathfrak{C}$ is ambiguous corresponding to the overall phase ambiguity of $Z^{(\alpha)}_{CS}(u)$. Here "$\cdots$" stands for the quantum corrections which in principle can be obtained recursively from the operator constraint equations. The semiclassical wave function $Z^{(\alpha)}_{CS}(u)$, usually called *holomorphic 3d block*, can also be formulated nonperturbatively as a "state-integral model" **[7, 40]**.

The holomorphic 3d block $Z^{(\alpha)}_{CS}(u)$ can also be defined by a functional integral of the holomorphic part of $CS\left[\mathscr{M}_3 \,|\, A, \bar{A}\right]$ over certain integration cycle known as

[5]See **[41]** for the canonical conjugation relation between $\mathbf{a}_{ab}$ and Θ_{ab} derived from Einstein-Hilbert action in GR

Lefschetz thimble [**5**]. Lefschetz thimble is an integration cycle which only contains a single critical point of the action, as another way to understand $Z_{CS}^{(\alpha)}(u)$ associates with a single flat connection on $\mathscr{M}_3$.

The holomorphic 3d block $Z_{CS}^{(\alpha)}(u)$ plays a central role in the quantum part of this work. We now focus on the 3-manifold $S^3 \setminus \Gamma_5$ with boundary Σ_6. We impose the boundary condition to Σ_6 to pick out the flat connections in $\mathcal{M}_{\text{flat}}(S^3 \setminus \Gamma_5, \mathrm{SL}(2,\mathbb{C}))$ corresponding to the 4d geometry of a constant curvature 4-simplex. Given $A \in \mathcal{M}_{\text{flat}}(S^3 \setminus \Gamma_5, \mathrm{SL}(2,\mathbb{C}))$ corresponding to a geometrical 4-simplex, as well as its parity partner $\tilde{A}$, we construct the holomorphic 3d block $Z_{CS}^{(\alpha)}(u)$ that associates with A and with reference $\tilde{A}$. Namely, we let A to be the end point of the contour $\mathfrak{C}$, and let $\tilde{A}$ to be the starting point of $\mathfrak{C}$ to remove the overall phase ambiguity. Our classical correspondence between flat connection on $S^3 \setminus \Gamma_5$ and constant curvature 4-simplex geometry suggests that $Z_{CS}^{(\alpha)}(u)$ from the construction is a wave function for *quantum 4d geometry* of a constant curvature 4-simplex:

$$Z_{CS}^{(\alpha)}(u) \text{ with Boundary Condition } = \text{ Quantum Constant Curvature 4-Simplex Geometry.} \tag{3.3}$$

As a consequence from the above quantum correspondence, in the asymptotic expansion of $Z_{CS}^{(\alpha)}(u)$ in Eq.(3.2), the leading term should give a classical action for the simplicial 4d geometry. The previously mentioned relation between the symplectic structures of flat connections and 4d simplicial gravity suggests that the leading term should give an action of 4d gravity in the simplicial context.

The above argument is confirmed by the semiclassical analysis. Indeed on the branch of Lagrangian sub-variety, the Liouville 1-form relates to the geometrical quantities

$$\vartheta = v \cdot \mathrm{d}u = \sum_{a<b} \Theta_{ab} \mathrm{d}\mathbf{a}_{ab} \tag{3.4}$$

by the relation between FN coordinates and the geometries quantities of 4-simplex. The resulting 1-form is actually exact by the Schläfli identity of constant curvature 4-simplex

$$\sum_{a<b} \mathbf{a}_{ab} \mathrm{d}\Theta_{ab} = \Lambda \mathrm{Vol}_4^{\Lambda} \tag{3.5}$$

where Vol_4^{Λ} is the 4-volume of the 4-simplex. Therefore

$$\sum_{a<b} \Theta_{ab} \mathrm{d}\mathbf{a}_{ab} = \mathrm{d}\left(\sum_{a<b} \mathbf{a}_{ab}\Theta_{ab} - \Lambda \mathrm{Vol}_4^{\Lambda} \right). \tag{3.6}$$

The quantity inside the differential is precisely the 4-dimensional Einstein-Hilbert action of a constant curvature 4-simplex. Therefore the leading contribution (as $\hbar \to 0$) from $Z_{CS}^{(\alpha)}(u)$ on the exponential is a simplicial discretization of 4-dimensional Einstein-Hilbert action, namely Regge action, on a constant curvature 4-simplex

$$S_{Regge}^{\Lambda} = \sum_{a<b} \mathbf{a}_{ab}\Theta_{ab} - \Lambda \mathrm{Vol}_4^{\Lambda}, \tag{3.7}$$

up to an integration constant. Here Λ is the constant curvature of the 4-simplex, being also the cosmological constant. We refer the readers to e.g. [**42–45**] for the derivation of 4d Regge action by a discretization of Einstein-Hilbert action (see also [**19**] for a summary).

The leading asymptotic behavior of $Z_{CS}^{(\alpha)}(u)$ is not necessarily an oscillatory phase by the holomorphicity. If we consider the full $\mathrm{SL}(2,\mathbb{C})$ Chern-Simons action $CS\left[\mathscr{M}_3 \,|\, A, \bar{A}\right]$ including both holomorphic and anti-holomorphic parts, we are interested in the 3d block $Z_{CS}^{(\alpha)}(u)Z_{CS}^{(\overline{\alpha})}(\bar{u})$, where $Z_{CS}^{(\overline{\alpha})}(\bar{u})$ associates with $\bar{A}$. For a flat connection having 4-simplex geometry correspondence, the leading asymptotic behavior of $Z_{CS}^{(\alpha)}(u)Z_{CS}^{(\overline{\alpha})}(\bar{u})$ is an oscillatory phase:

$$(3.8)\qquad \begin{aligned} &Z_{CS}^{(\alpha)}(u)\, Z_{CS}^{(\overline{\alpha})}(\bar{u}) \\ &= \exp\left[\frac{i}{\hbar}2\mathrm{Re}\left(\frac{\Lambda t}{12\pi i}\right) S^{\Lambda}_{Regge} + \frac{i}{\hbar}2\mathrm{Re}\left(\frac{\Lambda t}{12}\right)\sum_{a<b}\mathbf{K}_{ab}\mathbf{a}_{ab} + i\mathrm{C}_{\mathrm{int}} + \cdots\right]. \end{aligned}$$

Thus $Z_{CS}^{(\alpha)}(u)Z_{CS}^{(\overline{\alpha})}(\bar{u})$ is an analog of the functional integration quantization for Einstein-Hilbert action in the simplicial context,

$$(3.9)\qquad \exp\left[\frac{i}{\ell_P^2}\int_{\mathfrak{M}_4} R + \text{“Quantum Corrections”}\right].$$

By such an analog, we identify the gravitational constant G_{N} in terms of Chern-Simons coupling t and cosmological constant Λ:

$$(3.10)\qquad G_{\mathrm{N}} = \left|\frac{3}{2\mathrm{Im}(t)\Lambda}\right|.$$

$\mathrm{C}_{\mathrm{int}} \in \mathbb{R}$ is an (integration) constant which is independent of 4-simplex geometry. The additional term $\frac{i}{\hbar}2\mathrm{Re}\left(\frac{\Lambda t}{12}\right)\sum_{a<b}\mathbf{K}_{ab}\mathbf{a}_{ab}$ ($\mathbf{K}_{ab} \in \mathbb{Z}$) in the leading asymptotics comes from the choice of lift from FN variables x_m, y_m to the logarithmic variables u_m, v_m. This term disappears trivially when $t \in i\mathbb{R}$. However for general complex t, the additional term disappears when a quantization condition of triangle areas $\mathbf{a}_{ab}$ is satisfied:

$$(3.11)\qquad 2\mathrm{Re}\left(\frac{\Lambda t}{12}\right)\sum_{a<b}\mathbf{K}_{ab}\delta\mathbf{a}_{ab} \in 2\pi\hbar\mathbb{Z}$$

The quantization condition is indeed natural. When the boundary condition of $A \in \mathcal{M}_{\mathrm{flat}}(S^3\setminus\Gamma_5, \mathrm{SL}(2,\mathbb{C}))$ is imposed by the Wilson graph operator, the quantization condition is automatically satisfied. The quantization condition is also consistent with the discrete area spectrum in Loop Quantum Gravity (LQG) [**46**].

The analysis is generalized in [**47**] to the level of simplicial complex with an arbitrary number of 4-simplices. A class of 3-manifolds $\mathscr{M}_3$ are constructed by gluing a number of $S^3\setminus\Gamma_5$. The flat connections $A \in \mathcal{M}_{\mathrm{flat}}(\mathscr{M}_3, \mathrm{SL}(2,\mathbb{C}))$ are identified, which determine all possible (nondegenerate) 4-dimensional simplicial geometries on a simplicial complex $\mathcal{K}_4$. In each resulting simplicial geometry, the 4-simplices are of constant curvature Λ, while the large simplicial geometry built by many 4-simplices can approximate arbitrary smooth geometry on a 4-manifold. However a generic $A \in \mathcal{M}_{\mathrm{flat}}(\mathscr{M}_3, \mathrm{SL}(2,\mathbb{C}))$ having 4d simplicial geometry correspondence may result in a non-uniform 4d orientation on the simplicial complex, i.e. different 4-simplex may obtain different 4d orientations. For an orientable simplicial complex, we can find the class of flat connections on $\mathscr{M}_3$, which not only determine all possible (nondegenerate) 4-dimensional simplicial geometries on $\mathcal{K}_4$, but also induce the global 4d orientations. Each flat connection A in the class is accompanied by its *global parity* partner $\tilde{A}$. We construct the Chern-Simons 3d block $Z^{(\alpha)}_{CS}(\mathscr{M}_3|\,u)\, Z^{(\overline{\alpha})}_{CS}(\mathscr{M}_3|\,\bar{u})$ that associates with A and with reference $\tilde{A}$ in the same way as above. The asymptotic expansion in $\hbar$ of the resulting 3d block generalize Eq.(3.9) to the level of simplicial complex:

$$
\begin{aligned}
(3.12)\qquad & Z^{(\alpha)}_{CS}(\mathscr{M}_3|\,u)\, Z^{(\overline{\alpha})}_{CS}(\mathscr{M}_3|\,\bar{u}) \\
= & \exp\left[\frac{i}{\hbar}2\mathrm{Re}\left(\frac{\Lambda t}{12\pi i}\right)\mathbf{S}^{\Lambda}_{Regge} + \frac{i}{\hbar}2\mathrm{Re}\left(\frac{\Lambda t}{12}\right)\sum_{\Delta}\mathbf{K}_\Delta \mathbf{a}_\Delta + i\mathbf{C}_{\mathrm{int}} + \cdots\right].
\end{aligned}
$$

where $\mathbf{S}^{\Lambda}_{Regge}$ is the 4-dimensional Lorentzian Regge action on the entire simplicial complex $\mathcal{K}_4$:

$$
(3.13)\qquad \mathbf{S}^{\Lambda}_{Regge} = \sum_{\Delta\ \text{internal}} \mathbf{a}(\Delta)\varepsilon(\Delta) + \sum_{\Delta\ \text{boundary}} \mathbf{a}(\Delta)\Theta(\Delta) - \Lambda\sum_{\sigma}\mathrm{Vol}^{\Lambda}_4(\sigma)
$$

Here Δ denotes a triangle in $\mathcal{K}_4$ and σ denotes a 4-simplex. If we denote $\Theta_\Delta(\sigma)$ is the hyper-dihedral boost angle of Δ in the 4-simplex σ (the same as Θ_{ab} above), $\varepsilon(\Delta)$ is the Lorentzian deficit angle defined by $\varepsilon(\Delta) := \sum_{\sigma,\Delta\subset\sigma}\Theta_\Delta(\sigma)$ for internal Δ, and $\Theta(\Delta)$ is the Lorentzian boundary hyper-dihedral angle defined by $\Theta(\Delta) := \sum_{\sigma,\Delta\subset\sigma}\Theta_\Delta(\sigma)$ for boundary Δ. In Eq.(3.13) the additional term

$$
\frac{i}{\hbar}2\mathrm{Re}\left(\frac{\Lambda t}{12}\right)\sum_{\Delta}\mathbf{K}_\Delta\mathbf{a}_\Delta(\mathbf{K}_\Delta\in\mathbb{Z})
$$

is again disappear when $t\in i\mathbb{R}$, or when the above quantization condition is satisfied for general t.

The above asymptotics expansion in $\hbar$ suggests that the Chern-Simons 3d block

$$
Z^{(\alpha)}_{CS}(u)\, Z^{(\overline{\alpha})}_{CS}(\bar{u}),
$$

that associates with a flat connection on $\mathscr{M}_3$ corresponding to 4d simplicial geometry on $\mathcal{K}_4$, is a wave function of 4-dimensional simplicial quantum gravity, where its subleading terms in $\hbar$ gives the quantum correction to the classical Einstein-Hilbert action.

4. Wilson Graph Operator and Loop Quantum Gravity

In [**19**], a class of Wilson graph operators are studied in SL(2, $\mathbb{C}$) Chern-Simons theory on S^3. The Wilson graph operators are defined by Γ_5 graph embedded in S^3 colored by certain principle unitary irreducible representations of SL(2, $\mathbb{C}$). we have studied the Chern-Simons expectation value $\mathscr{A}$ of the Wilson graph operators on S^3, and in particular the asymptotic behavior of $\mathscr{A}$ in the "double-scaling limit", i.e. scaling both Chern-Simons coupling t and Wilson-graph representation labels to infinity, and keeping their ratio fixed. Under the double-scaling limit, the Chern-Simons expectation value $\mathscr{A}$ of the Wilson graph operator produces 4d Regge action S^{Λ}_{Regge} of a constant curvature 4-simplex as the leading asymptotics:

$$
(4.1)\qquad \mathscr{A} = e^{\frac{i}{\ell_P^2}S^{\Lambda}_{Regge}+\cdots} + e^{-\frac{i}{\ell_P^2}S^{\Lambda}_{Regge}+\cdots}
$$

up to an possible overall phase factor. Here $\cdots$ stands for the subleading terms in the double-scaling limit. In the sense of asymptotics and relation with simplicial gravity, $\mathscr{A}$ can be viewed as a 4d gravity analog of quantum 6j symbol in Turaev-Viro model of 3d quantum gravity [**57, 58**][6].

As a motivation for the analysis in [**19**], the Chern-Simons expectation value $\mathscr{A}$ has a close relationship with Loop Quantum Gravity (LQG). LQG is an attempt to make a background independent, non-perturbative quantization of 4-dimensional gravity, for reviews, see [**48–50**]. *Spinfoam amplitude* is one of the central object for the covariant dynamics of LQG, which adapts the idea of path integral quantization to the framework of LQG. A spinfoam amplitude is defined on a 4-dimensional simplicial complex $\mathcal{K}_4$ and encodes the transition amplitude for the boundary quantum 3-geometry. In LQG, the quantum 3-dimensional geometry is described by *spin-network states.* The spinfoam amplitude sums over the history of spin-networks, and suggests a foam-like quantum spacetime structure. An important building block of the spinfoam amplitude is Engle-Pereira-Rovelli-Livine (EPRL) 4-simplex amplitude $\mathscr{A}_{EPRL}$ associated to a 4-simplex σ in $\mathcal{K}_4$ [**51**][7]. The Chern-Simons expectation value $\mathscr{A}$ of the Wilson graph is an deformation of EPRL 4-simplex amplitude, in the sense that $\mathscr{A}$ approaches $\mathscr{A}_{EPRL}$ asymptotically in the decoupling limit, i.e. Chern-Simons coupling t is scaled to infinity, keeping Wilson graph representations fixed. Such an deformation has been motivated from mainly two perspectives: (1) the studies of relation between LQG and Topological Quantum Field Theory (TQFT) [**16, 17, 53**], (2) the quantum group deformation of spinfoam amplitude and inclusion of cosmological constant [**18, 54**]. We have the following commutative diagram for the relations among $\mathscr{A}$, $\mathscr{A}_{EPRL}$, and Regge action S^{Λ}_{Regge} or S_{Regge} with or without cosmological constant term:

$$\begin{array}{ccc} \mathscr{A} & \overset{\text{double-scaling limit}}{\longrightarrow} & e^{\frac{i}{\ell_P^2}S^{\Lambda}_{Regge}} + e^{-\frac{i}{\ell_P^2}S^{\Lambda}_{Regge}} \\ \Big\downarrow {\scriptstyle\text{decoupling}} & & \Big\downarrow {\scriptstyle\Lambda\to 0} \\ \mathscr{A}_{EPRL} & \overset{\text{large-j limit}}{\longrightarrow} & e^{\frac{i}{\ell_P^2}S_{Regge}} + e^{-\frac{i}{\ell_P^2}S_{Regge}} \end{array} \tag{4.2}$$

The asymptotic behavior downstairs is shown for EPRL 4-simplex amplitude $\mathscr{A}_{EPRL}$ in [**55, 56**]. S_{Regge} from the asymptotics of $\mathscr{A}_{EPRL}$ is the Regge action without cosmological constant for a *flat* 4-simplex, while S^{Λ}_{Regge} from the Chern-Simons expectation value $\mathscr{A}$ is the Regge action with cosmological constant Λ for a *constant curvature* 4-simplex, the same as Eq.(3.7). Therefore $\mathscr{A}$ is a deformation of $\mathscr{A}_{EPRL}$ in the spinfoam amplitude, which includes the cosmological constant in the framework of LQG.

Probably one has noticed that the 4-dimensional Lorentzian Regge action S^{Λ}_{Regge} appears in the leading asymptotics of both Chern-Simons expectation value $\mathscr{A}$ of Wilson graph operator and Chern-Simons 3d block $Z^{(\alpha)}_{CS}(u)\,Z^{(\overline{\alpha})}_{CS}(\bar{u})$. It is not a coincidence. Firstly it turns out that the double-scaling limit in Chern-Simons theory on S^3 with Wilson graph is the same as the semiclassical limit $\hbar \to 0$ of Chern-Simons theory on the graph complement, keeping the boundary data fixed.

[6]The double-scaling limit of quantum 6j symbol gives the 3d Regge action on a constant curvature tetrahedron [**58**].

[7]It is sometimes also called EPRL/FK, for Freidel and Krasnov, when referring to the version for Euclidean gravity [**52**]

Secondly the Chern-Simons expectation value $\mathscr{A}$ can be understood as an inner product

$$\mathscr{A} = \langle N(\Gamma_5) \,|\, S^3 \setminus \Gamma_5 \rangle \tag{4.3}$$

where $|N(\Gamma_5)\rangle$ is the Chern-Simons state on the tubular neighborhood of Γ_5 excited by the Wilson graph operator, and $|S^3 \setminus \Gamma_5\rangle$ is the Chern-Simons ground state on $S^3 \setminus \Gamma_5$. In the double-scaling limit, the Wilson graph operators in [**19**] defining $\mathscr{A}$ impose the right boundary condition to the boundary Σ_6 of $S^3 \setminus \Gamma_5$ (including the quantization condition Eq.(3.11)), so that the boundary condition picks the parity pair of flat connections $A, \tilde{A}$ on $S^3 \setminus \Gamma_5$, which determines a constant curvature 4-simplex geometry. In other words, the state $|N(\Gamma_5)\rangle$ is a "semiclassical state" picked at the right phase space point in $\mathcal{M}_{\text{flat}}(\Sigma_6, \mathrm{SL}(2,\mathbb{C}))$. The state $|S^3 \setminus \Gamma_5\rangle$ is a linear combination of Chern-Simons 3d blocks $Z^{(\alpha)}_{CS}(u)\, Z^{(\overline{\alpha})}_{CS}(\bar{u})$ on $S^3 \setminus \Gamma_5$. The peakedness of $|N(\Gamma_5)\rangle$ selects the right pair of 3d blocks that associates with A and $\tilde{A}$ respectively, which have respectively $e^{\frac{i}{\ell_P^2} S^{\Lambda}_{Regge}}$ and $e^{-\frac{i}{\ell_P^2} S^{\Lambda}_{Regge}}$ in their leading asymptotics. Note that in the study of Chern-Simons 3d block, we only obtain a single phase because the 3d block is defined to associate with A and with reference $\tilde{A}$, which is the ratio of two 3d blocks with A and $\tilde{A}$ respectively (their overall phase ambiguities cancel).

The separate studies of Chern-Simons 3d block and Wilson graph operator clarify their different roles played in the asymptotics of $\mathscr{A}$. The Regge-action asymptotic behavior of $\mathscr{A}$ crucially depends on the peakedness of $|N(\Gamma_5)\rangle$ created by the Wilson graph operator. However different Wilson graph operators may produce the same peakedness in phase space[8], thus leads to the same asymptotics of $\mathscr{A}$. The close relationship with EPRL 4-simplex amplitude motivates us to study the particular type of Wilson graph operators in [**19**]. But in principle other types of Wilson graph operators are possible to work equally well, when they produce the same peakedness[9]. However independent of the choice of Wilson graphs, the essential role played behind the Regge-action asymptotics of $\mathscr{A}$ is Chern-Simons 3d block on $S^3 \setminus \Gamma_5$ with the right boundary condition imposed. Therefore the Chern-Simons 3d block $Z^{(\alpha)}_{CS}(u)\, Z^{(\overline{\alpha})}_{CS}(\bar{u})$ studied here plays an important role in the covariant formulation of LQG. The classical and quantum correspondences between flat connection on 3-manifold and simplicial geometry on 4-manifold studied here may be viewed as a re-formulation of covariant LQG, by emphasizing the relation with $\mathrm{SL}(2,\mathbb{C})$ Chern-Simons theory.

Our result for the quantum correspondence suggests the Chern-Simons 3d block

$$Z^{(\alpha)}_{CS}(u)\, Z^{(\overline{\alpha})}_{CS}(\bar{u})$$

to be the wave function of simplicial quantum gravity in 4 dimensions. Given its relation with LQG, $Z^{(\alpha)}_{CS}(u)\, Z^{(\overline{\alpha})}_{CS}(\bar{u})$ may be understood as the physical wave function for LQG in 4 dimensions, at least in the level of simplicial geometry. As a future research, it is interesting to find the behavior of $Z^{(\alpha)}_{CS}(u)\, Z^{(\overline{\alpha})}_{CS}(\bar{u})$ under the procedure which corresponds to refining the simplicial complex $\mathcal{K}_4$. it should shed light on the understanding of continuum limit in the covariant formulation of LQG.

---

[8]For instance, for harmonic oscillator, different squeezed coherent state can have the same peakedness.

[9]The different types of Wilson graphs having the same pickedness may relate to the spinfoam amplitudes defined in [DingYou]

Acknowledgements

The author acknowledges the scientific committee of String-Math 2016 for the opportunity of the talk. He also acknowledge the helpful discussion with Pei Du, Sergei Gukov, and Jian Qiu. The research receives support from the US National Science Foundation through grant PHY-1602867, and the Start-up Grant at Florida Atlantic University, USA.

References

[1] J. E. Andersen, et al. *Chern-Simons Gauge Theory: 20 Years After.* AMS/IP Studies in Advanced Mathematics 2011

[2] M. Atiyah. Topological quantum field theories. Publications Mathématiques de l'IHÉS 68 (1989) 175
E. Witten. Topological quantum field theory. Commun. Math. Phys. 117 (1988) 353

[3] Edward Witten, *Quantum field theory and the Jones polynomial*, Comm. Math. Phys. **121** (1989), no. 3, 351–399. MR990772

[4] Hitoshi Murakami and Jun Murakami, *The colored Jones polynomials and the simplicial volume of a knot*, Acta Math. **186** (2001), no. 1, 85–104, DOI 10.1007/BF02392716. MR1828373

[5] Edward Witten, *Analytic continuation of Chern-Simons theory*, Chern-Simons gauge theory: 20 years after, AMS/IP Stud. Adv. Math., vol. 50, Amer. Math. Soc., Providence, RI, 2011, pp. 347–446. MR2809462

[6] Sergei Gukov, *Three-dimensional quantum gravity, Chern-Simons theory, and the A-polynomial*, Comm. Math. Phys. **255** (2005), no. 3, 577–627, DOI 10.1007/s00220-005-1312-y. MR2134725

[7] Tudor Dimofte and Sergei Gukov, *Quantum field theory and the volume conjecture*, Interactions between hyperbolic geometry, quantum topology and number theory, Contemp. Math., vol. 541, Amer. Math. Soc., Providence, RI, 2011, pp. 41–67, DOI 10.1090/conm/541/10678. MR2796627

[8] A. Achúcarro and P. K. Townsend, *A Chern-Simons action for three-dimensional anti-de Sitter supergravity theories*, Phys. Lett. B **180** (1986), no. 1-2, 89–92, DOI 10.1016/0370-2693(86)90140-1. MR865090

[9] Steven Carlip, *Quantum gravity in* $2+1$ *dimensions*, Cambridge Monographs on Mathematical Physics, Cambridge University Press, Cambridge, 1998. MR1637718

[10] E. Witten, *Chern-Simons gauge theory as a string theory*, The Floer memorial volume, Progr. Math., vol. 133, Birkhäuser, Basel, 1995, pp. 637–678. MR1362846

[11] Sergei Gukov and Edward Witten, *Branes and quantization*, Adv. Theor. Math. Phys. **13** (2009), no. 5, 1445–1518. MR2672467

[12] Marcos Mariño, *Chern-Simons theory, matrix models and topological strings*, International Series of Monographs on Physics, vol. 131, The Clarendon Press, Oxford University Press, Oxford, 2005. MR2177747

[13] Sungjay Lee and Masahito Yamazaki, *3d Chern-Simons theory from M5-branes*, J. High Energy Phys. **12** (2013), 035, front matter + 22. MR3158956

[14] Ofer Aharony, Oren Bergman, Daniel Louis Jafferis, and Juan Maldacena, $\mathcal{N}=6$ *superconformal Chern-Simons-matter theories, M2-branes and their gravity duals*, J. High Energy Phys. **10** (2008), 091, 38, DOI 10.1088/1126-6708/2008/10/091. MR2452954

[15] Simone Giombi and Xi Yin, *The higher spin/vector model duality*, J. Phys. A **46** (2013), no. 21, 214003, 41, DOI 10.1088/1751-8113/46/21/214003. MR3062129

[16] L. Smolin, linking topological quantum field theory and nonperturbative quantum gravity. J. Math. Phys. 36 (1995) 6417-6455
S. Major, L. Smolin, Quantum deformation of quantum gravity. Nucl.Phys. B473 (1996) 267-290
L. Smolin, Quantum gravity with a positive cosmological constant. [arXiv:hep-th/0209079]

[17] W. Wieland. Complex Ashtekar variables, the Kodama state and spinfoam gravity. [arXiv:1105.2330]

[18] M. Han. Cosmological constant in LQG vertex amplitude. Phys. Rev. D 84 (2-11) 064010 [arXiv:1105.2212]

[19] Hal M. Haggard, Muxin Han, Wojciech Kamiński, and Aldo Riello, SL(2, C) *Chern-Simons theory, a non-planar graph operator, and 4D quantum gravity with a cosmological constant: semiclassical geometry*, Nuclear Phys. B **900** (2015), 1–79, DOI 10.1016/j.nuclphysb.2015.08.023. MR3414420

[20] Jonathan Engle, Karim Noui, and Alejandro Perez, *Black hole entropy and* SU(2) *Chern-Simons theory*, Phys. Rev. Lett. **105** (2010), no. 3, 031302, 4, DOI 10.1103/PhysRevLett.105.031302. MR2670985

[21] Edward Witten, *Quantization of Chern-Simons gauge theory with complex gauge group*, Comm. Math. Phys. **137** (1991), no. 1, 29–66. MR1099255

[22] E. Buffenoir, K. Noui, and Ph. Roche, *Hamiltonian quantization of Chern-Simons theory with* SL(2, ℂ) *group*, Classical Quantum Gravity **19** (2002), no. 19, 4953–5015, DOI 10.1088/0264-9381/19/19/313. MR1933315

[23] Tudor Dan Dimofte, *Refined BPS Invariants, Chern-Simons Theory, and the Quantum Dilogarithm*, ProQuest LLC, Ann Arbor, MI, 2010. Thesis (Ph.D.)–California Institute of Technology. MR3054865

[24] Tudor Dimofte, Sergei Gukov, Jonatan Lenells, and Don Zagier, *Exact results for perturbative Chern-Simons theory with complex gauge group*, Commun. Number Theory Phys. **3** (2009), no. 2, 363–443, DOI 10.4310/CNTP.2009.v3.n2.a4. MR2551896

[25] Edward Witten, *Gauge theories and integrable lattice models*, Nuclear Phys. B **322** (1989), no. 3, 629–697, DOI 10.1016/0550-3213(89)90232-0. MR1010205

[26] Roland van der Veen, *The volume conjecture for augmented knotted trivalent graphs*, Algebr. Geom. Topol. **9** (2009), no. 2, 691–722, DOI 10.2140/agt.2009.9.691. MR2496886

[27] Dylan P. Thurston, *The algebra of knotted trivalent graphs and Turaev's shadow world*, Invariants of knots and 3-manifolds (Kyoto, 2001), Geom. Topol. Monogr., vol. 4, Geom. Topol. Publ., Coventry, 2002, pp. 337–362. MR2048108

[28] Dror Bar-Natan and Zsuzsanna Dancso, *Homomorphic expansions for knotted trivalent graphs*, J. Knot Theory Ramifications **22** (2013), no. 1, 1250137, 33, DOI 10.1142/S0218216512501374. MR3024021

[29] S. Nawata, P. Ramadevi, Zodinmawia. Trivalent graphs, volume conjectures and character varieties. [arXiv:1404.5119]

[30] N. J. Hitchin, *The self-duality equations on a Riemann surface*, Proc. London Math. Soc. (3) **55** (1987), no. 1, 59–126, DOI 10.1112/plms/s3-55.1.59. MR887284

[31] Tudor Dimofte, Davide Gaiotto, and Roland van der Veen, *RG domain walls and hybrid triangulations*, Adv. Theor. Math. Phys. **19** (2015), no. 1, 137–276, DOI 10.4310/ATMP.2015.v19.n1.a2. MR3418500

[32] Tudor Dimofte, Maxime Gabella, and Alexander B. Goncharov, *K-decompositions and 3d gauge theories*, J. High Energy Phys. **11** (2016), 151, front matter+144. MR3594814

[33] Werner Fenchel and Jakob Nielsen, *Discontinuous groups of isometries in the hyperbolic plane*, De Gruyter Studies in Mathematics, vol. 29, Walter de Gruyter & Co., Berlin, 2003. Edited and with a preface by Asmus L. Schmidt; Biography of the authors by Bent Fuglede. MR1958350

[34] Yuichi Kabaya, *Parametrization of* PSL(2, ℂ)*-representations of surface groups*, Geom. Dedicata **170** (2014), 9–62, DOI 10.1007/s10711-013-9866-x. MR3199475

[35] D. Cooper, M. Culler, H. Gillet, D. D. Long, and P. B. Shalen, *Plane curves associated to character varieties of 3-manifolds*, Invent. Math. **118** (1994), no. 1, 47–84, DOI 10.1007/BF01231526. MR1288467

[36] H. M. Haggard, M. Han, W. Kaminski, A. Riello. SL(2,C) Chern-Simons Theory, Flat Connections, and Four-dimensional Quantum Geometry. [arXiv:1512.07690]

[37] Hal M. Haggard, Muxin Han, and Aldo Riello, *Encoding curved tetrahedra in face holonomies: phase space of shapes from group-valued moment maps*, Ann. Henri Poincaré **17** (2016), no. 8, 2001–2048, DOI 10.1007/s00023-015-0455-4. MR3522023

[38] J. Teschner, *Quantization of the Hitchin moduli spaces, Liouville theory and the geometric Langlands correspondence I*, Adv. Theor. Math. Phys. **15** (2011), no. 2, 471–564. MR2924236

[39] V. V. Fok and L. O. Chekhov, *Quantum Teichmüller spaces* (Russian, with Russian summary), Teoret. Mat. Fiz. **120** (1999), no. 3, 511–528, DOI 10.1007/BF02557246; English transl., Theoret. and Math. Phys. **120** (1999), no. 3, 1245–1259. MR1737362

[40] J. E. Andersen and R. Kashaev. Complex quantum Chern-Simons. [arXiv:1409.1208]

[41] Bianca Dittrich and James P. Ryan, *Phase space descriptions for simplicial 4D geometries*, Classical Quantum Gravity **28** (2011), no. 6, 065006, 34, DOI 10.1088/0264-9381/28/6/065006. MR2773466
[42] Benjamin Bahr and Bianca Dittrich, *Regge calculus from a new angle*, New J. Phys. **12** (2010), no. March, 033010, 10, DOI 10.1088/1367-2630/12/3/033010. MR2606604
[43] T. Regge, *General relativity without coordinates* (English, with Italian summary), Nuovo Cimento (10) **19** (1961), 558–571. MR0127372
[44] J. W. Barrett and T. J. Foxon, *Semiclassical limits of simplicial quantum gravity*, Classical Quantum Gravity **11** (1994), no. 3, 543–556. MR1265378
[45] G. Feinberg, R. Friedberg, T. D. Lee, and H. C. Ren, *Lattice gravity near the continuum limit*, Nuclear Phys. B **245** (1984), no. 2, 343–368, DOI 10.1016/0550-3213(84)90436-X. MR763854
[46] Abhay Ashtekar and Jerzy Lewandowski, *Quantum theory of geometry. I. Area operators*, Classical Quantum Gravity **14** (1997), no. 1A, A55–A81, DOI 10.1088/0264-9381/14/1A/006. Geometry and physics. MR1691887
[47] Muxin Han, *4d quantum geometry from 3d supersymmetric gauge theory and holomorphic block*, J. High Energy Phys. **1** (2016), 065, front matter+62. MR3471515
[48] Thomas Thiemann, *Modern canonical quantum general relativity*, Cambridge Monographs on Mathematical Physics, Cambridge University Press, Cambridge, 2007. With a foreword by Chris Isham. MR2374859
[49] Abhay Ashtekar and Jerzy Lewandowski, *Background independent quantum gravity: a status report*, Classical Quantum Gravity **21** (2004), no. 15, R53–R152, DOI 10.1088/0264-9381/21/15/R01. MR2079936
[50] Jonathan Engle, Etera Livine, Roberto Pereira, and Carlo Rovelli, *LQG vertex with finite Immirzi parameter*, Nuclear Phys. B **799** (2008), no. 1-2, 136–149, DOI 10.1016/j.nuclphysb.2008.02.018. MR2401503
[51] Jonathan Engle, Etera Livine, Roberto Pereira, and Carlo Rovelli, *LQG vertex with finite Immirzi parameter*, Nuclear Phys. B **799** (2008), no. 1-2, 136–149, DOI 10.1016/j.nuclphysb.2008.02.018. MR2401503
[52] Laurent Freidel and Kirill Krasnov, *A new spin foam model for 4D gravity*, Classical Quantum Gravity **25** (2008), no. 12, 125018, 36, DOI 10.1088/0264-9381/25/12/125018. MR2425405
[53] Hideo Kodama, *Specialization of Ashtekar's formalism to Bianchi cosmology*, Progr. Theoret. Phys. **80** (1988), no. 6, 1024–1040, DOI 10.1143/PTP.80.1024. MR988605
[54] K. Noui and Ph. Roche, *Cosmological deformation of Lorentzian spin foam models*, Classical Quantum Gravity **20** (2003), no. 14, 3175–3213, DOI 10.1088/0264-9381/20/14/318. MR1992001
[55] John W. Barrett, R. J. Dowdall, Winston J. Fairbairn, Frank Hellmann, and Roberto Pereira, *Lorentzian spin foam amplitudes: graphical calculus and asymptotics*, Classical Quantum Gravity **27** (2010), no. 16, 165009, 34, DOI 10.1088/0264-9381/27/16/165009. MR2660965
[56] Muxin Han and Mingyi Zhang, *Asymptotics of spinfoam amplitude on simplicial manifold: Lorentzian theory*, Classical Quantum Gravity **30** (2013), no. 16, 165012, 57, DOI 10.1088/0264-9381/30/16/165012. MR3094877
[57] V. G. Turaev and O. Ya. Viro, *State sum invariants of* 3*-manifolds and quantum* $6j$*-symbols*, Topology **31** (1992), no. 4, 865–902, DOI 10.1016/0040-9383(92)90015-A. MR1191386
[58] Shun'ya Mizoguchi and Tsukasa Tada, *Three-dimensional gravity from the Turaev-Viro invariant*, Phys. Rev. Lett. **68** (1992), no. 12, 1795–1798, DOI 10.1103/PhysRevLett.68.1795. MR1153379

DEPARTMENT OF PHYSICS, FLORIDA ATLANTIC UNIVERSITY, 777 GLADES ROAD, BOCA RATON, FLORIDA 33431-0991

E-mail address: `hanm@fau.edu`

Quantum cohomology under birational maps and transitions

Yuan-Pin Lee, Hui-Wen Lin, and Chin-Lung Wang

ABSTRACT. This is an expanded version of the third author's lecture in String-Math 2015 at Sanya. It summarizes some of our works in quantum cohomology.

After reviewing the quantum Lefschetz and quantum Leray–Hirsch, we discuss their applications to the functoriality properties under special smooth flops, flips and blow-ups. Finally, for conifold transitions of Calabi–Yau 3-folds, formulations for small resolutions (blow-ups along Weil divisors) are sketched.

0. Introduction

0.1. Classical aspects on algebraic geometry. In the study of algebraic geometry, we usually encounter *projective morphisms* $f : Y \longrightarrow X$, e.g. blow-ups, bundle morphisms etc., and it is a basic question to study relations of geometric quantities under such morphisms.

By its very definition of being projective, there are *factorizations* of f into compositions of simpler morphisms in the following form. There are vector bundles $\mathscr{E} \to X$ and associated projective bundles $\pi : P = P_X(\mathscr{E}) \to X$ such that $f = \pi \circ \iota$:

$$\begin{array}{ccc} Y & \xrightarrow{\ \iota\ } & P \\ & {\scriptstyle f}\searrow & \downarrow{\scriptstyle \pi} \\ & & X \end{array}$$

where $\iota : Y \hookrightarrow P$ is an imbedding. The choice of P is by no means unique. In fact $\mathscr{E}$ can be taken to be a trivial bundle, say of rank r, and then $P = P^{r-1} \times X$ is a product. However, a good choice of P is usually important so that the induced imbedding ι has good structures.

If $\iota(Y) \subset P$ is a complete intersection, namely that there is a split vector bundle $V = \bigoplus L_i \to P$ and a section $\sigma \in \Gamma(P, V)$ such that $\iota(Y) = \sigma^{-1}(0)$ is its zero loci, then one develops

Lefschetz Hyperplane Theorem

to study relations between Y and P. In most cases one does not obtain complete intersection imbedding automatically. Nevertheless sometimes one may employ the technique of *deformations to the normal cone* to reduce the problem under study to such a situation. The most famous one is the proof of *Grothendieck–Riemann–Roch theorem* [**5**]. Historically it is the proof of GRR which lays the foundation of these techniques. Among many later applications, it is notable that *deformations*

2010 *Mathematics Subject Classification.* Primary 14N35, 14J32, 14D07.

to the normal cone is now a standard method to reduce a problem to its local models (e.g., the proof of *invariance of complex elliptic genera* under K-equivalent birational maps [**23**]).

For the (projective) bundle map π, one develops

Leray–Hirsch Theorem

to study relations between P and X. The techniques in this package include direct images, spectral sequences etc., and usually can be applied to more general bundle maps such that the fiber manifold is well understood. A combination of results for ι and π then gives the desired result for f.

0.2. Quantum aspects. In this article, following the above flow-chart, we survey the related developments on the quantum cohomology ring $QH(X)$, or equivalently the genus zero Gromov–Witten theory. More precisely we consider the Dubrovin (flat) connection ∇ on $TH(X)$ and analyze its behavior under various maps including complete intersection imbedding and projective bundle maps. The essential mathematical tools are the corresponding *Quantum Lefschetz Hyperplane Theorem* and the *Quantum Leray–Hirsch Theorem.*

The first version of quantum Lefschetz was proved around 1996 by Lian, Liu, and Yau [**19**] and by Givental [**6**] independently (cf. [**3**]). It is also known as the *mirror theorem* since its major motivation and application is to prove the counting formula of rational curves on quintic Calabi–Yau 3-folds predicted by Candelas et. al.. In that setup P could be a semi-Fano toric manifold and $V = \bigoplus L_i$ is a sum of convex line bundles such that $c_1(Y) \geq 0$ (semi-Fano or sub Calabi–Yau). There are several improvements of quantum Lefschetz afterwards (e.g. [**8**]). The most general version which allows the background manifold P to be general and without the condition on $c_1(Y)$ was obtained by Coates and Givental [**2**]. We will review this latest theory in terms of Dubrovin connections and introduce the notion of *Birkhoff factorizations* and *generalized mirror transforms.*

For quantum Leray–Hirsch, a version for $P = P_X(\mathscr{E})$ with $\mathscr{E} = \mathscr{O} \oplus L$ a rank two split bundle first appeared in the work of Maulik and Pandharipande [**20**]. A version on the more general case of toric bundle $P \to X$ build on a *split vector bundle* $\mathscr{E} \to X$ was proved by Brown [**1**]. It had been formulated in the framework of Dubrovin connections and used to prove the *invariance of quantum cohomology rings under ordinary flops* by us in 2011 [**11, 12**]. More recently, together with F. Qu, we proved a *quantum splitting principle* and remove the splitting assumption in the quantum Leray–Hirsch [**9**]. For the applications to be discussed here, we will only focus on the Dubrovin connection in the split case. The ideas of *näive quantization basis* and *admissible lift of Mori cone* are the essential ingredients to formulate the quantum Leray–Hirsch.

By combining quantum Lefachetz and quantum Leray–Hirsch, we discuss several applications on birational maps including

(i) the analytic continuations, i.e. invariance, of QH under ordinary flops,
(ii) smooth blow-ups along *complete intersection* centers, as well as
(iii) *decomposition theorem* of projective local models of simple ordinary flips.

The latter two applications are new, and both address the issue of the *functoriality* of QH under non K-equivalent transformations.

Indeed for simple (r, r') flips $f : X \dashrightarrow X'$ with $r > r'$, there is an orthogonal decomposition

$$H(X) = \mathscr{T}^{-1} H(X') \oplus K$$

where $\mathscr{T} = [\Gamma_f]_* : H(X) \to H(X')$ is the map induced from the graph correspondence with kernel (vanishing cycles) $K \cong \mathbb{C}^{r-r'}$, and $\mathscr{T}^{-1} := [\Gamma_f^t]$ is the transpose correspondence. Under a suitable choice of quantum frame $\{\tilde{T}_i\}$ which deforms the classical cohomology basis $\{T_i\}$, we have a ring isomorphism (decomposition)

$$QH(X) \cong \langle \tilde{T}_1, \ldots, \tilde{T}_{\dim H(X')} \rangle \times \mathbb{C}^{r-r'}$$

such that $\langle \tilde{T}_1, \ldots, \tilde{T}_{\dim H(X')} \rangle \cong QH(X')$ as $\mathscr{D}^z$-modules, but not as rings. In this survey we illustrate this for the case of $(2,1)$ flips.

Comparing with the current developments on string-math related topics, the subjects discussed here are more classical in flavor as there is essentially no higher genus theory nor modern mirror symmetry involved. Mirror symmetry phenomenon happens in *the large complex/Kähler structure limit* while birational maps are essentially located at *finite distance Kähler degenerations*. Also we treat only smooth varieties. However, in higher dimensional birational geometry, namely the *minimal model theory*, it is indispensable to include singularities in the variety under consideration.

It is clear that most of the subjects discussed here can be extended to orbifolds since orbifold Gromov–Witten theory is now well developed. However MMP requires more general singularities then orbifold ones and it is still a long way towards a useful *quantum minimal model program.* Of course the smooth case is the first step, and it is our hope that further progress in the general case can be made in the near future.

0.3. Towards a QMMP. As the next step, we are led to consider *birational maps up to complex deformations*, i.e. *transitions*. There are many technical issues from classic algebraic geometry on this regard. Nevertheless it has become clear in recent years that it is indispensable to allow certain transitions in the classification of higher dimensional varieties.

For Calabi–Yau 3-folds the famous Ried's fantasy [**21**] on connecting CY with different topology through transitions is still one of the major research problems in this area. We give a very brief sketch on our recent work [**13**] in understanding the transition of quantum $\mathscr{A}$ model and $\mathscr{B}$ model in *projective* conifold transitions $X \nearrow Y$ through a conifold degeneration $\mathfrak{X} \to \Delta$ from $X = \mathfrak{X}_t$, $t \neq 0$, to $\bar{X} = \mathfrak{X}_0$, followed by a small resolution $Y \to \bar{X}$. In particular the notion of linked GW invariants with respect to a set of vanishing spheres $S_i \subset X$ is introduced which corresponds to the non-exceptional GW invariants on Y. For GW invariants supported on exceptional curves, a *basic exact sequence* shows that there is a local transition between them and the Yukawa couplings of the vanishing periods.

This article ends with discussions on issues of *effective computations* on the quantum transitions in terms of blow-up formula of GW invariants when the blow-up is along certain non-complete intersection (Weil) divisors.

1. Quantum cohomology

In this article, unless stated otherwise, all varieties $X, Y, P, \ldots$ under considerations are assumed to be smooth and projective over $\mathbb{C}$. The cone of effective one cycles (Mori cone) in X is denoted by $NE(X)$.

1.1. Dubrovin connection. A general reference is [**3**]. We fix a cohomology basis $T_i \in H = H(X) := H^*(X, \mathbb{C})$, with dual basis $\{T^i\}$. A general element in cohomology is denoted by $t = \sum t^i T_i$. The genus zero GW theory is encoded by its generating function (pre-potential) $F(t) = \langle\langle\rangle\rangle$, where for $a_i \in H(X)$,

$$\langle\langle a_1, \ldots, a_m \rangle\rangle := \sum_{\beta \in NE(X)} \sum_{n=0}^{\infty} \frac{q^\beta}{n!} \langle a_1, \ldots, a_m, t, \ldots, t \rangle_{g=0, m+n, \beta}.$$

The (formal) Novikov variables q^β's are inserted to avoid the issue of convergence. It also keeps track on the natural grading arising from the virtual dimension (i.e., the *conformal* structure on the *Frobenius manifold* $H(X)$, though we do not make use of this language in this article). In general t is also treated as a formal variable (cf. (1.2) below for $t \in H^0 \oplus H^2$).

Denote by $F_{ijk} = \partial^3_{ijk} F = \langle\langle T_i, T_j, T_k \rangle\rangle$ the 3-point generating functions, and set $A_{ij}^k := \sum_l F_{ijl}\, g^{lk}$. Then the big quantum product at t is defined by

$$T_i *_t T_j = \sum_k A_{ij}^k(t)\, T_k.$$

The product is associative due to the WDVV equations. Equivalently it corresponds to the flatness of the Dubrovin connection on $TH \otimes \mathbb{C}[\![q^\bullet]\!]$:

$$\nabla := d - \frac{1}{z} A \equiv d - \frac{1}{z} \sum_i dt^i \otimes A_i. \tag{1.1}$$

The special role played by the z parameter shows that ∇ is flat if and only if $dA = 0 = A \wedge A$, which is equivalent to WDVV.

1.2. J function and cyclic $\mathscr{D}^z$-modules. Write $t = t_0 + t_1 + t_2$ with $t_0 \in H^0$ and $t_1 \in H^2$. The generating function of all genus zero GW invariants with at most one descendent insertion is organized as

$$\begin{aligned} J(t, z^{-1}) &:= 1 + \frac{t}{z} + \sum_{\beta, n, i} \frac{q^\beta}{n!} T_i \left\langle \frac{T^i}{z(z-\psi)}, (t)^n \right\rangle_\beta \\ &= e^{\frac{t}{z}} + \sum_{\beta \neq 0, n, i} \frac{q^\beta}{n!} e^{\frac{t_0 + t_1}{z} + (t_1 . \beta)} T_i \left\langle \frac{T^i}{z(z-\psi)}, (t_2)^n \right\rangle_\beta \end{aligned} \tag{1.2}$$

where the fundamental class axiom (string equation) and the divisor axiom are applied to get the second equality. The important role played by the J function comes from the following quantum differential equation (QDE)

$$z\partial_i\, z\partial_j\, J = \sum_k A_{ij}^k\, z\partial_k\, J, \tag{1.3}$$

which follows from the topological recursion relation (TRR). It implies that the quantum cohomology $QH(X)$ can be regarded as the cyclic $\mathscr{D}^z$-module $\mathscr{D}^z J$ with base (frame) given by

$$z\partial_i\, J \equiv e^{t/z}\, T_i \pmod{q^\bullet} = T_i + \ldots.$$

It is clear that for $T_1 = \mathbf{1}$ being the fundamental class, $z\partial_1\, J = J$.

The ring of differential operators $\mathscr{D}^z = \mathbb{C}[\![t, q^\bullet]\!]\{z, z\partial_\bullet\}$ is defined so that $p = \sum_\beta q^\beta p_\beta \in \mathscr{D}^z$ implies that p_β is a polynomial in z and $z\partial_\bullet$.

In practice it is sometimes easier to study the GW theory or the J function on the small parameter space $H^0 \oplus H^2$. The expression of J with $t_2 = 0$ is known as the small J function. Using the divisorial reconstruction theorem in [**16**], the small J function determines the sub-algebra of $QH(X)$ generated by $H^2(X)$.

2. Review on quantum Lefschetz

2.1. Quantum Lefschetz for toric base and concavex bundle with $c_1 \geq 0$. Let P be a projective manifold, $L_i \to P$, $1 \leq i \leq r$ be convex line bundles, and $\sigma \in \Gamma(P, \bigoplus_{i=1}^r L_i)$ be a section such that $Y = \sigma^{-1}(0) \hookrightarrow P$ is a smooth submanifold. Given $QH(P)$, the problem is to compute $QH(Y)$.

The method of localizations on stable map moduli spaces leads to the so called *factorial trick* or *hypergeometric modifications.* To state it, we start with the cohomology valued factorial

$$(L)_\beta := \prod_{m=1}^{L.\beta} (L + mz)$$

whenever the intersection number $L.\beta \geq 0$. Then we set

$$I^Y(t, z, z^{-1}) := \sum_{\beta \in NE(P)} q^\beta J^P_\beta(t, z^{-1}) \times \prod_{i=1}^r (L_i)_\beta \tag{2.1}$$

as an approximation of J^Y.

When the ambient space P is a semi-Fano toric manifold (i.e. $c_1(P) \geq 0$), Lian–Liu–Yau and Givental used $\mathbb{C}^\times$ localizations to determine J^P. (This in turn determines $QH(P)$ since $H(P)$ is generated by divisors.) Over such a semi-Fano toric base, they further proved the

THEOREM 2.1 (Mirror Theorem). [**6, 19**] *For $c_1(Y) \geq 0$, $t \in H^0 \oplus H^2$, we have*

$$(I^Y / I^Y_0)(t, z^{-1}) = J^Y(\tau, z^{-1})$$

up to the mirror map $t \mapsto \tau(t)$ which matches $1/z$ coefficients on both sides.

Here I^Y_0 is the component of z^0 terms. Notice that the assumption $c_1(Y) \geq 0$ implies that I^Y is still an expression in z^{-1}.

In [**19**], the line bundles L_i's are also allowed to be concave. In that case their *mirror principle* determines a certain type of twisted GW invariants.

2.2. Quantum Lefachetz over general base and split bundles. Without the condition that $c_1(Y) \geq 0$, the approximation I^Y might contain terms with positive z powers. But J^Y, by definition, contains only terms in powers of z^{-1}. Hence a more sophisticated transformation is needed in order to relate I^Y to J^Y.

Coates and Givental considered the following situation. Let P be a general projective manifold whose big quantum cohomology ring $QH(P)$ is given. Let $L_i \to P$, $1 \leq i \leq r$, be line bundles. They defined twisted GW invariants in this setup. When L_i's are base-point free the twisted invariants are the GW invariants of the complete intersection sub-manifold $Y = \sigma^{-1}(0)$ for a generic section $\sigma \in \Gamma(P, \bigoplus_{i=1}^r L_i)$. They proved

THEOREM 2.2. [2] *Given $J^P(t)$ in $t \in H(P)$, we have $I^Y \in \mathscr{D}^z J^Y$. More precisely, there exists a linear differential operator $b = b(\tau, z, q^\bullet, z\partial_\bullet)$ which is polynomial in z on each finite truncation of the Novikov variables $q^\bullet$ such that*

$$I^Y(t, z, z^{-1}) = b(\tau, z, q^\bullet, z\partial_\bullet)\, J^Y(\tau, z^{-1}).$$

Here $t \mapsto \tau(t) \in H(P)$ is a transformation determined by this property.

To get a better understanding of the statement, we notice that b can be taken to be linear in $z\partial_i$'s because of the QDE (1.3) on J^Y. In a similar fashion, the differentiation $z\partial_i I^Y$ can also be represented by $\sum_j B_{ij}\, z\partial_j J^Y$ for some formal functions B_{ij}'s which are polynomials in z in any $\beta \in NE(P)$. This leads to the so called *Birkhoff factorization*

$$(z\vec{\partial} I)(t, z, z^{-1}) = (z\vec{\partial} J)(\tau, z^{-1}) B(\tau, z) \tag{2.2}$$

of the square matrix $(z\vec{\partial} I) = (z\partial_1 I, \ldots, z\partial_{\dim H(P)} I)$.

Since $z\partial_i I \equiv e^{t/z} T_i \equiv z\partial_i J \pmod{q^\bullet}$, we have $B \equiv Id \pmod{q^\bullet}$ and this implies the isomorphism on $\mathscr{D}^z$ modules

$$\mathscr{D}^z\, I(t) \cong \mathscr{D}^z\, J(\tau)$$

up to a *generalized mirror transform* $\tau(t)$ on $H(X)$. Now it is clear that

$$p(t, z, q^\bullet, z\partial_\bullet)\, I^Y(t, z, z^{-1}) = J^Y(\tau, z^{-1})$$

for some linear operator p. In fact this operator plays the role to remove the $z^{\geq 0}$ terms in I^Y and it can be effectively constructed by induction on $NE(P)$ (cf. [12, Theorem 1.10] for a related construction). The map $t \mapsto \tau(t)$ is then determined by matching the $1/z$ coefficients on both sides.

We also notice that from (2.2) the matrix B^{-1} is the gauge transformation to bring the wrong frame $z\partial_i I$'s back to the preferred frame $z\partial_i J$ so that the connection matrix takes the form expected in (1.1).

3. Quantum Leray–Hirsch

3.1. Factorial trick for split projective (toric) bundles. Let

$$\pi : P = P_X(V) \to X$$

be a projective bundle. The classical Leray–Hirsch theorem computes the cohomology of the total space P in terms of the base X and the fibers. Let $h = c_1(\mathscr{O}_P(1))$, then

$$H(P) \cong \pi^* H(X)[h]/(f_V(h))$$

where $f_V(h)$ is the Chern polynomial of the vector bundle $V \to X$.

It is natural to ask for a similar description on quantum cohomology. For this purpose we assume that $QH(X)$ is given, and $V = \bigoplus_{i=1}^r L_i$ is a sum of line bundles. We seek for an analogous factorial trick as in the case of quantum Lefschetz. However the formulation must be different since now $QH(P)$ contains additional variables.

Let $\bar{t} \in H(X)$ be a general element from the base, $D = t^h h$ be the fiber divisor class with coordinate t^h, and we consider the mixed variable

$$\hat{t} = \bar{t} + D.$$

Then a hypergeometric modification of J^X is defined by

$$I^P(\hat{t}, z, z^{-1}) = \sum_{\beta \in NE(P)} q^\beta J^X_{\pi_*\beta}(\bar{t}) \times e^{\frac{D}{z} + (D.\beta)} \prod_{i=1}^{r} \frac{1}{(h + L_i)_\beta}. \tag{3.1}$$

Here the convention on factorial $\prod_1^s := \prod_\infty^s / \prod_{-\infty}^0$ is used so that $1/(L)_\beta$ makes sense even if $L.\beta < 0$.

When $X = \mathrm{pt}$, this is the I function of P^{r-1} coming from localizations. In general (3.1) arises from fiber localization of a $\mathbb{C}^\times$-action, which exists by the split assumption on V. The formulation works for other fiber bundles as long as the fiber localization is well understood. In that case $D = \sum_{i=1}^{\rho} t^i D_i$ for $D_1, \ldots, D_\rho$ being a basis of $H^2(P/X)$. The following is due to Brown:

THEOREM 3.1. [1] *Given $J^X(\bar{t})$ in $\bar{t} \in H(X)$, we have $I^P \in \mathscr{D}^z J^P$. More precisely, there is a linear differential operator $b = \sum_{\beta \in NE(P)} q^\beta b_\beta$ with $\deg_z b_\beta < \infty$, and a graph $\hat{t} \mapsto \tau(\hat{t}) : H(X) \oplus \mathbb{C}h \to H(P)$, such that*

$$I^P(\hat{t}, z, z^{-1}) = b(\tau, z, q^\bullet, z\partial_\bullet)\, J^P(\tau, z^{-1}).$$

REMARK 3.2. The result was proved in [1] for split toric bundles, and stated in the language of *Lagrangian cones.* We have presented it in an equivalent form to avoid introducing this machinery.

3.2. The Dubrovin connection. Based on Theorem 3.1, we had developed a method to compute the Dubrovin connection on the bundle space P [12]. It is roughly represented by following implication:

$$\mathrm{PF}^{P/X} + \nabla^X \Longrightarrow \nabla^P.$$

To be precise, we need to introduce a system of equations controlling both the fiber directions and the base directions.

For the fibers, we introduce the *Picard–Fuchs ideal.* For the primitive fiber curve class $\ell \in NE(P/X)$, it is easily checked that $\Box_\ell I = 0$ where

$$\Box_\ell = \prod\nolimits_{i=1}^{r} z\partial_{h+L_i} - q^\ell e^{t^h} \tag{3.2}$$

is the Picard–Fuchs operator. (Here ∂_L is the directional derivative in direction L.) The Picard–Fuchs ideal is the left ideal of $\mathscr{D}^z$ generated by the Picard–Fuchs operators.

For differentiations in the variables corresponding to the base, i.e. $H(X)$, directions, we introduce *the lifting of QDE* from X to P. For each $\bar{\beta} \in NE(X)$, a lift of $\bar{\beta}$ is a curve class $\beta \in NE(P)$ such that $\pi_*\beta = \bar{\beta}$. Moreover, β is called *admissible* if

$$-(h + L_i).\beta \geq 0, \quad \text{for all} \quad i = 1, \ldots, r.$$

Admissible lift exists. In fact a minimal lift in effective classes is admissible.

Let $\bar{\beta}^* \in NE(P)$ be an admissible lift of $\bar{\beta} \in NE(X)$. We define

$$D_{\bar{\beta}^*}(z) := \prod_{i=1}^{r} \prod_{m=0}^{-(h+L_i).\bar{\beta}^*-1} (z\partial_{h+L_i} - mz).$$

Then it can be shown by direct computations that (cf. [12, Theorem 3.6])

$$z\partial_i\, z\partial_j\, I = \sum\nolimits_{k,\bar{\beta}} q^{\bar{\beta}^*} e^{D.\bar{\beta}^*} \bar{A}^k_{ij,\bar{\beta}}(\bar{t})\, D_{\beta^*}(z)\, z\partial_k\, I. \tag{3.3}$$

REMARK 3.3. For toric bundles with $\rho(P/X) > 1$, the admissible lift still exists, though not unique, and the lifting of QDE is independent of the choices of $\bar{\beta}^*$ modulo the Picard–Fuchs ideal. A case of double projective bundle (hence $\rho(P/X) = 2$) will be discussed in Lemma 4.2.

Let $\bar{t} = \sum \bar{t}^i \bar{T}_i \in H(X)$, and $e = h^l \bar{T}_i \in H(P)$. The näive quantization of e is defined to be the operator

$$\hat{e} = \partial^{ze} := (z\partial_h)^l \, z\partial_{\bar{T}_i} = (z\partial_{t^h})^l \, z\partial_{\bar{t}^i}.$$

The idea for doing so is clear: since we do not have a variable corresponding to e, we simply use l-th derivatives in t^h to approximate the directional derivative in h^l. On the base direction we keep the first order derivative in direction $\bar{T}_i$ since the variable $\bar{t}^i$ is available.

Using (3.2) and (3.3) we get the first order system on the frame $\partial^{ze} I$ (with e runs through a basis of $H(P)$) over the variables $t^a = \bar{t}^i, t^h$:

$$z\partial_a \left(\partial^{ze} I\right) = \left(\partial^{ze} I\right) C_a(\hat{t}, z).$$

As in §2.2, we have the Birkhoff factorization matrix B such that

$$(\partial^{ze} I)(\hat{t}, z, z^{-1}) = (z\vec{\partial} J)(\tau, z^{-1}) B(\tau, z).$$

In fact B^{-1} is the gauge transformation to remove $z^{\geq 0}$ in $C_a(\hat{t}, z)$. Furthermore, the map $\hat{t} \mapsto \tau(\hat{t})$ is uniquely determined by matching the $1/z$ coefficients of the first column of $(\partial^{ze} I)B^{-1}$ with J:

$$J(\tau, z^{-1}) = z\partial_1 J = p(\hat{t}, z, \partial^{z\bullet}) \, I(\hat{t}, z, z^{-1}).$$

Set $z = 0$ in the gauge transformation we find $-(z\partial_a B)B^{-1} \mapsto 0$ and

$$B_0 \, C_{a;0} \, B_0^{-1}(\hat{t}) = \sum_{i=1}^{\dim H(P)} A_i(\tau(\hat{t})) \frac{\partial \tau^i}{\partial t^a}(\hat{t}), \tag{3.4}$$

where $B_0 = B(z = 0)$ and $C_{a;0} = C_a(z = 0)$.

Since $\tau \equiv \hat{t} \pmod{q^\bullet}$, by the Mori cone induction and divisorial reconstruction we may then determine all the Dubrovin connection matrices $A_i(t)$ from (3.4). Of course the computations involved are necessarily complicated and very demanding. In applying these results special attention is paid to maintain the structural information. We will demonstrate on this through a few applications.

4. Application I: Ordinary flops

4.1. The statement. Let $f : X \dashrightarrow X'$ be a P^r flop. That is, there are two vector bundles $F, F' \to S$ of the same rank $r + 1$, such that the exceptional loci $Z \subset X$ has the following projective bundle structure:

$$\operatorname{Exc} f = Z = P_S(F) \xrightarrow{\bar{\psi}} S.$$

Moreover, the normal bundle of Z in X is given by

$$N = N_{Z/X} \cong \bar{\psi}^* F' \otimes \mathscr{O}_Z(-1).$$

It was shown in [10] that the graph correspondence $\mathscr{T} = [\Gamma_f]_*$ induces an isomorphism on cohomology spaces $H(X) \cong H(X')$, but it does not preserve the effectivity of one cycles. Indeed if ℓ (resp. ℓ') is the class of primitive extremal rational curve in X (resp. X') then

$$\mathscr{T}\ell = -\ell'.$$

Moreover, $\mathscr{T}$ preserves the Poincaré pairing, but not the product structure. It turns out that the topological defects are corrected by the extremal ray GW invariants and the following is true:

THEOREM 4.1. **[9–12]** *The graph correspondence $\mathscr{T}$ induces isomorphism of big quantum cohomology rings $QH(X) \cong QH(X')$ under the analytic continuations induced from $q^\beta \mapsto q^{\mathscr{T}\beta}$.*

Here is a brief history on this problem. For $\dim X = 3$, the multiple cover formula for $\mathscr{O}_{P^1}(-1)^2 \to P^1$ gives the quantum corrections of $(\mathscr{T}D)^3 - D^3$ (cf. Witten **[24]**). The global case was treated by Li–Ruan **[18]** around 2000. They proved a degeneration formula of GW invariants in the *symplectic category*, and used it to show that in fact no degeneration might occur in the threefold case. The problem was then reduced to the case of extremal rays which had already been solved.

In higher dimensions the statement was conjectured to hold for general *birational K-equivalent manifolds* by Ruan and Wang (cf. **[22]**). The case of *simple* ordinary flops, namely $S = \mathrm{pt}$, was solved in 2006 **[10]**. We worked in the algebraic category and reduced the problem to local models by way of the deformations to the normal cone and the degeneration formula of Li **[17]**. In this case non-trivial degenerations do arise and the relations between relative GW invariants, descendent invariants, and absolute GW invariants are carefully studied through degenerations (inspired by a method of Maulik and Pandharipande in **[20]**). For local models, $X = P_{P^r}(\mathscr{O}(-1)^{r+1} \oplus \mathscr{O})$ is a semi-Fano toric variety whose GW theory is well studied (cf. Theorem 2.1). In fact $I^X = J^X$ on small parameters, and the analytic continuation can be solved. The result was further extended to the higher genus GW theory in **[7]** by studying *ancestor invariants* and *quantization*.

For general base S with split bundles F, F', the analytic continuation was later solved in 2011. Indeed, the problem was reduced to the local models in **[11]**, and the case of local models was solved through the quantum Leary–Hirsch theorem in **[12]**. More recently, a *quantum splitting principle* was proved in a joint work with Qu in **[9]**, which reduced the problem for general vector bundles F, F' to the case of split bundles, hence proved Theorem 4.1 completely.

4.2. A sketch of proof. Now we sketch how the Quantum Leray–Hirsch is applied to solve the case of local split flops.

The flop is achieved by first blowing up $Z \subset X$ to get $Y = \mathrm{Bl}_Z X \to X$ and then contracting the exceptional divisor $E = P_S(F) \times_S P_S(F') \subset Y$ in another fiber direction to get $Z' \subset X'$. We have $\bar\psi' : Z' = P_S(F') \to S$ being a projective bundle and $N' = N_{Z'/X'} = \bar\psi'^* F \otimes \mathscr{O}_{Z'}(-1)$:

$$\begin{array}{ccc} X = P_Z(N \oplus \mathscr{O}) & \overset{f}{\dashrightarrow} & X' = P_{Z'}(N' \oplus \mathscr{O}) \\ & {\scriptstyle p}\searrow \quad \swarrow{\scriptstyle p'} & \\ & S & \end{array},$$

where $p = \bar\psi \circ \pi$ in $X \xrightarrow{\pi} Z \xrightarrow{\bar\psi} S$ and similarly $p' = \bar\psi' \circ \pi'$. As a double projective bundle, we have $NE(X/S) = \langle \ell, \gamma \rangle$, where ℓ (resp. γ) is the $\bar\psi$ (resp. π) fiber line classes. Let $\xi = \mathscr{O}_X(1)$ and $h = \mathscr{O}_Z(1)$, and $D = t^h h + t^\xi \xi \in H^2(X/S)$ a general fiber divisor. Then

$$H(X) = p^* H(S)[h, \xi]/(f_F, f_{N\oplus\mathscr{O}}),$$

where f_V is the Chern polynomial of a bundle V.

When $F = \bigoplus_{i=1}^r L_i$, $F' = \bigoplus_{i=1}^r L_i'$ are split bundles, we have

$$f_F(h) = \prod (h + L_i),$$
$$f_{N\oplus\mathscr{O}}(h, \xi) = \xi \prod (\xi - h + L_i').$$

By symmetry we have similar formulae on the X' side. However, it is not compatible with that on X: since $\mathscr{T} h = \xi' - h'$, $\mathscr{T}\xi = \xi'$, it is easy to see that the cup product structure is not preserved under $\mathscr{T}$. (See [**11**, Theorem 1.8] for the explicit computations on the topological defects.)

Now comes the key point: to remedy the topological defects we replace the cohomology class by its "quantized" version, namely we consider differential operators instead. This gives rise to the Picard–Fuchs operators

$$\Box_\ell = \prod z\,\partial_{h+L_i} - q^\ell e^{t^h} \prod z\,\partial_{\xi - h + L_i'},$$
$$\Box_\gamma = z\,\partial_\xi \prod z\,\partial_{\xi - h + L_i'} - q^\gamma e^{t^\xi}$$

which are regarded as the "quantized version" of the Chern polynomials. Similarly we have on the X' side

$$\Box_{\ell'} = \prod z\,\partial_{h'+L_i'} - q^{\ell'} e^{t^{h'}} \prod z\,\partial_{\xi' - h' + L_i},$$
$$\Box_{\gamma'} = z\,\partial_{\xi'} \prod z\,\partial_{\xi' - h' + L_i} - q^{\gamma'} e^{t^{\xi'}}.$$

The coordinates are related by requiring $t^{h'} h' + t^{\xi'} \xi' = \mathscr{T}(t^h h + t^\xi \xi) = t^h(\xi' - h') + t^\xi \xi'$. That is, $t^{h'} = -t^h$ and $t^{\xi'} = t^\xi + t^h$.

Now it is a simple exercise to check that

LEMMA 4.2. *$\mathscr{T}$ induces an isomorphism on Picard–Fuchs ideals*

$$\mathscr{T}\langle \Box_\ell, \Box_\gamma \rangle \cong \langle \Box_{\ell'}, \Box_{\gamma'} \rangle.$$

REMARK 4.3. Lemma 4.2 can be extended to split toric bundle flops.

For the base directions, the lift of QDE in (3.3) is independent of the choice of admissible lift $\bar{\beta}^*$ modulo the Picard–Fuchs ideal. The admissible condition for β in this case is given by $-\beta.(h + L_i) \geq 0$, $-\beta.(\xi - h + L_i') \geq 0$ and $-\beta.\xi \geq 0$. It is readily seen that β is admissible in X if and only if $\mathscr{T}\beta$ is admissible in X'. This implies that the lifting of QDE from S to X and the one from S to X' are indeed equivalent under $\mathscr{T}$ modulo the Picard–Fuchs ideal. Thus, the quantum Leray–Hirsch theorem implies that X and X' have compatible first order PDE systems up to analytic continuations.

To achieve $\mathscr{T} : QH(X) \cong QH(X')$, we still need to show that the Birkhoff factorization B and the generalized mirror map $\tau(\hat{t})$, as appeared in (3.4), are compatible on both sides. We refer the details to [**12**, §3.3].

5. Application II: Blow-ups along complete intersection centers

Let $L_i = \mathscr{O}_X(D_i)$, $1 \le i \le r$, and $Z = D_1 \cap \cdots \cap D_r$ be a smooth complete intersection subvariety of X with codimenison r. Let $\mathscr{E} = \bigoplus_{i=1}^r L_i$ with a given section $s = (s_i)$ such that $D_i = (s_i)$, $Z = s^{-1}(0)$. Consider the blow-up $\phi : Y \to X$ along Z:

$$\begin{array}{ccc} E & \hookrightarrow & Y = \mathrm{Bl}_Z X\,. \\ \downarrow & & \downarrow \phi \\ Z & \hookrightarrow & X \end{array}$$

Given $QH(X)$ and $(\mathscr{E}, s)$, the problem is to determine $QH(Y)$.

By construction, we have a surjective morphism $\mathscr{E}^* \twoheadrightarrow \mathscr{I}_Z$. This leads to the imbedding

$$\iota : Y := \mathrm{Proj}_X \bigoplus_{d=0}^{\infty} \mathscr{I}_Z^d \hookrightarrow \mathrm{Proj}_X \mathrm{Sym}\, \mathscr{E}^* = P_X(\mathscr{E}).$$

Let $\pi : P := P_X(\mathscr{E}) \to X$ be the bundle map with associated Euler sequence $0 \to S \to \pi^*\mathscr{E} \to Q \to 0$ over P. It is shown in [**14**] that there is a canonical section $\sigma \in \Gamma(P, Q)$ of the universal quotient bundle such that $Y = \sigma^{-1}(0) \subset P$. We emphasize that this is not true if Z is not a complete intersection. The situation is summarized in the following diagram:

$$(5.1)\qquad \begin{array}{ccccccccc} 0 & \longrightarrow & S & \longrightarrow & \pi^*\mathscr{E} & \longrightarrow & Q & \longrightarrow & 0\,. \\ & & & \searrow & \downarrow & \swarrow\nearrow \sigma & & & \\ & & Y & \overset{\iota}{\hookrightarrow} & P & & & & \\ & & & \underset{\phi}{\searrow} & \downarrow \pi & & & & \\ & & & & X & & & & \end{array}$$

Let $\eta = c_1(\mathscr{O}_P(1))$. It follows that $S = \mathscr{O}_P(-\eta)$ and $-\eta|_Y = E$ (cf. [**5**]).

Notice that the quotient bundle Q is in general not a split bundle over P and the quantum Lefschetz can not be applied directly. However, a suitable extension of it to short exact sequences allows us to apply it to the Euler sequence (5.1) where both S and $\pi^*\mathscr{E}$ are split bundles.

With the above understood, the (extended) quantum Lefschetz together with the quantum Leray–Hirsch lead to the following factorial trick on the β component of I^Y:

$$(5.2)\qquad \begin{aligned} I^Y_\beta &= J^P_\beta \frac{\prod_{i=1}^r (D_i)_\beta}{(-\eta)_\beta} \\ &\sim J^X_{\pi_*\beta}\, e^{\frac{t^\eta \eta}{z} + t^\eta(\eta.\beta)} \frac{\prod_{i=1}^r (D_i)_\beta}{(-\eta)_\beta \prod_{i=1}^r (\eta + D_i)_\beta}. \end{aligned}$$

Here t^η denotes a suitable dual coordinate of η.

THEOREM 5.1. [**14**] *For the smooth blow-up $\phi : Y \to X$ along a complete intersection center $Z = \bigcap_{i=1}^r D_i$, the relative I factor is given by*

$$I^{Y/X}_\beta = e^{s(\frac{E}{z} + E.\beta)} \left(\prod_{i=1}^r \frac{(D_i)_\beta}{(D_i - E)_\beta (E)_\beta} \right) (E)^{r-1}_\beta,$$

where $E \subset Y$ is the exceptional divisor and s is a suitable dual coordinate.

REMARK 5.2. The formula suggests nice structures of the relative factor. Indeed, $K_Y = \phi^* K_X + (r-1)E$ and $(E)_\beta^{r-1}$ is responsible for the Jacobian of ϕ. For each i, $(D_i)_\beta/((D_i - E)_\beta (E)_\beta)$ is a *K-trivial factor* which describes the decomposition of the linear system $|D_i - Z|$ into moving part and the fixed part (on Y). We expect that this intrinsic formulation will be useful for general blow-ups.

As in quantum Leray–Hirsch described in §3.2, to get the Dubrovin connection on Y (or $QH(Y)$) we proceed by (1) choosing the corresponding näive quantization basis (2) determining the Picard–Fuchs ideal on fibers (3) finding the lifting of QDE on the base X to Y. Then (1) + (2) + (3) determines $\mathscr{D}^z J^Y$, and hence $QH(Y)$. The details will appear in [**14**].

6. Application III: Simple flips

Let $r, r' \in \mathbb{N}$ and $r > r'$. In the definition of ordinary flops, if the underlying vector bundles F and F' have different rank $r+1$ and $r'+1$ respectively then in exactly the same construction as §4.2 we get ordinary (r, r') flips. The effect on quantum cohomology under flips are discussed in [**15**]. In contrast to analytic continuations in the flops case, the situation for flips is more subtle and complex and new phenomena appear. We give a sketch in the simplest case of local models of simple $(2,1)$ flips.

6.1. $H(X)$ **vs** $H(X')$ **and the Picard–Fuchs systems.** The local model of $(2,1)$ flips has the following data: $Z = P^2$, $Z' = P^1$, and

$$f : X = P_{P^2}(\mathscr{O}(-1)^2 \oplus \mathscr{O}) \dashrightarrow X' = P_{P^1}(\mathscr{O}(-1)^3 \oplus \mathscr{O}).$$

The cohomology rings are given by

$$H(X) = \mathbb{C}[h, \xi]/(h^3, \xi(\xi - h)^2),$$
$$H(X') = \mathbb{C}[h', \xi']/(h'^2, \xi'(\xi' - h')^3),$$

where $\dim H(X) = 9$ and $\dim H(X') = 8$.

The graph correspondence $\mathscr{T} = [\Gamma_f]_*$ induces a short exact sequence

$$0 \longrightarrow K \longrightarrow H(X) \xrightarrow{\mathscr{T}} H(X') \longrightarrow 0$$

where $K = \ker \mathscr{T} = \mathbb{C}\mathbf{k}_1$ with $\mathbf{k}_1 = (\xi - h)^2 = [Z]$.

The transpose correspondence $\mathscr{T}^{-1} := [\Gamma_f^t]_*$ preserves the Poincaré pairing and induces an imbedding $\mathscr{T}^{-1} : H(X') \hookrightarrow H(X)$ (indeed, of motives) which leads to an orthogonal decomposition (cf. [**10**, §2.3])

$$H(X) = \mathscr{T}^{-1} H(X') \overset{\perp}{\oplus} K.$$

However, $\mathscr{T}^{-1}$ does not preserves the cup product. In fact $K^\perp$ is not closed under cup product. As in the case of flops we have curve classes ℓ, γ in X and ℓ', γ' in X'. They are related by $\mathscr{T}\ell = -\ell'$ and $\mathscr{T}\gamma = \ell' + \gamma'$. Also $\mathscr{T}h = \xi' - h'$ and $\mathscr{T}\xi = \xi'$.

The divisor variable takes the form $D = t^h h + t^\xi \xi$. To simplify notations in our discussion, we will use variables

$$q_1 = q^\ell e^{t^h}, \quad q_2 = q^\gamma e^{t^\xi}; \qquad q_1' = q^{\ell'} e^{-t^h}, \quad q_2' = q^{\gamma'} e^{t^h + t^\xi}.$$

From the computational point of view of quantum cohomology, X' is bad since $c_1(X') = -h' + 4\xi'$ which contains both K-positive and K-negative directions. In other words, the expression I^X is complicated and contains positive z powers. Its Picard–Fuchs equations are given by

$$\Box_{\ell'} = (z\partial_{h'})^2 - q_1'(z\partial_{\xi'-h'})^3, \qquad \Box_{\gamma'} = z\partial_{\xi'}(z\partial_{\xi'-h'})^3 - q_2'.$$

The first equation $\Box_{\ell'} I^{X'} = 0$ shows that in order to reduce $(z\partial^2_{h'})I^{X'}$ we will receive derivatives of even higher power. It does give the correct reduction algorithm in the Mori cone topology since there is also a q_1' multiplied. It is difficult to compute $\nabla^{X'}$ or to get any structure of it from this approach.

On the other hand, X is toric Fano with $c_1(X) = h + 3\xi$. The computation of $QH(X)$ is in principle easy since $I^X = J^X$ along small parameters. Assume that this has been done, then a natural question is

Can we get $QH(X')$ from $QH(X)$ in a canonical manner?

Now we restrict ourselves to the small parameters $t \in H^0 \oplus H^2$ so that we can work with variables q_1, q_2 and q_1', q_2' directly. The Picard–Fuchs equations on $J = J^X$ can be easily determined to be

$$\Box_\ell = (z\partial_h)^3 - q_1(z\partial_{\xi-h})^2, \qquad \Box_\gamma = z\partial_\xi(z\partial_{\xi-h})^2 - q_2.$$

It is closely related to the one for X':

LEMMA 6.1. *Along the partially compactified two dimensional Kähler moduli* $\mathscr{K} := \{(q_1, q_2)\} \bigcup \{(q_1' = 1/q_1, q_2' = q_1 q_2)\} \cong \mathscr{O}_{P^1}(1)$, *the Hopf-Mobius stripe, we have*

$$\mathscr{T} : \langle \Box_\ell, \Box_\gamma \rangle \cong \langle \Box_{\ell'}, \Box_{\gamma'} \rangle$$

outside the divisors $D_0 = \{q_1 = 0\}$ *and* $D_\infty = \{q_1' = 0\}$.

6.2. Exact formula for ∇^X [15]. The following frame (recall that $I = J$)

$$\begin{aligned}
&v_1 = \hat{\mathbf{1}}J = J,\\
&v_2 = \hat{h}J, \quad v_3 = (\hat{\xi} - \hat{h})J,\\
&v_4 = \hat{h}^2 J - (\hat{\xi} - \hat{h})^2 J, \quad v_5 = \hat{h}(\hat{\xi} - \hat{h})J + (\hat{\xi} - \hat{h})^2 J,\\
&v_6 = \hat{h}^3 J - \hat{h}(\hat{\xi} - \hat{h})^2 J, \quad v_7 = \hat{h}^2(\hat{\xi} - \hat{h})J + \hat{h}(\hat{\xi} - \hat{h})^2 J,\\
&v_8 = \hat{h}^3(\hat{\xi} - \hat{h})J + \hat{h}^2(\hat{\xi} - \hat{h})^2 J,\\
&v_9 = \hat{\mathbf{k}}_1 J = (\hat{\xi} - \hat{h})^2 J,
\end{aligned}$$

respects $H(X) = \mathscr{T}^{-1}H(X') \oplus^\perp K$ when modulo q_1, q_2. They are precisely

$$z\partial_i J \quad \text{at } t \in H^0 \oplus H^2, \quad 1 \le i \le 9,$$

and we get the Dubrovin connection matrices

$$A_1 = h*_{small} = \begin{bmatrix} & & & & & q_1q_2 & & & \\ 1 & & & & & & & & \\ & & & & & & & q_1q_2 & \\ & 1 & & & & & & & \\ & & 1 & & & & & & \\ & & & 1 & & & & & -1 \\ & & & & 1 & & & & \\ & & & & & -1 & 1 & & \\ & 1 & -1 & & & & & & q_1 \end{bmatrix},$$

$$A_2 = \xi*_{small} = \begin{bmatrix} & & & -q_2 & q_2 & q_1q_2 & & & q_2 \\ 1 & & & & & -q_2 & q_2 & & \\ 1 & & & & & & & q_1q_2 & \\ & 1 & & & & & & q_2 & \\ & 1 & 1 & & & & & & \\ & & & 1 & & & & & \\ & & & 1 & 1 & & & & \\ & & & & & & 1 & & \\ & & & & & & & q_2 & \end{bmatrix}.$$

The crucial observation is that there is a single appearance of q_1 in $A^9_{1,9}$. This shows that the system has irregular singularity along $D_\infty = (q_1 = \infty)$ in the K direction. Let $x = q_1'$, $y = q_2'$. After a constant change of basis from v_i's to w_i's such that the Poincaré pairing

$$(w_i, w_j) = \delta_{9,i+j}, \qquad 1 \le i, j \le 8$$

and $w_9 := v_9$ with $(w_9, w_i) = \delta_{9,i}$, the fundamental solution matrix S satisfies

$$z(x\,\partial_x)S = \begin{bmatrix} & & & -\frac{1}{2}xy & xy & & & & xy \\ & & & & & -\frac{1}{2}xy & xy & & \\ 1 & & & & & \frac{1}{4}xy & -\frac{1}{2}xy & & \\ & & & & & & & xy & \\ & 1 & & & & & & -\frac{1}{2}xy & \\ & & & & & & & & 1 \\ & & & 1 & & & & & -\frac{1}{2} \\ & & & & & 1 & & & \\ & -\frac{1}{2} & 1 & & & & & xy & -1/x \end{bmatrix} S,$$

which is irregular in the K-block, i.e. the (9, 9) entry, of Poincaré rank one.

6.3. Block diagonalization. The classical theory of ODE and the flatness of ∇^X imply that there exists a unique formal gauge transformation $S = PZ$:

$$P(x,y,z) = \begin{bmatrix} 1 & & & g_1 \\ & \ddots & & \vdots \\ & & 1 & g_8 \\ f_1 & \cdots & f_8 & 1 \end{bmatrix}, \tag{6.1}$$

such that

$$z(x\,\partial_x)Z = B_1\,Z, \qquad z(y\,\partial_y)Z = B_2\,Z,$$

with B_1, B_2 being block-diagonalized. Moreover, each $f_i(x,y,z) = -g_{9-i}(x,y,-z)$ is a formal series expansion of certain special function. The claim is that we may relate the first 8×8 blocks of $B_1(x,y,z)$ and $B_2(x,y,z)$ with the Dubrovin connection $\nabla^{X'}$.

Under the new z-dependent frame $\tilde{w}_1,\ldots,\tilde{w}_8,\widetilde{\mathbf{k}}_1 J$ from (6.1), namely

$$\tilde{w}_i = w_i + f_i\,\hat{\mathbf{k}}_1 J, \qquad \widetilde{\mathbf{k}}_1 J = \hat{\mathbf{k}}_1 J + \sum_{i=1}^{8} g_i\, w_i, \tag{6.2}$$

we have (a special case of [**15**] for $(r,r') = (2,1)$):

THEOREM 6.2. *For a simple* $(2,1)$ *flip* $f : X \dashrightarrow X'$, *under the frame* (6.2) *at* $z=0$, *we have a ring isomorphism*

$$QH(X) \cong \langle \tilde{w}_1(0),\ldots,\tilde{w}_8(0)\rangle \times \mathbb{C}.$$

Moreover, $\langle \tilde{w}_1(0),\ldots,\tilde{w}_8(0)\rangle \cong QH(X')$ *as* $\mathscr{D}^z$*-modules, but not as rings.*

The proof is based on Lemma 6.1 and we refer to [**15**] for the details.

7. Conifold transitions of Calabi–Yau 3-folds

7.1. Relations on vanishing A and B cycles. A Calabi–Yau variety is a $\mathbb{Q}$-Gorenstein variety with $K \sim 0$ and $h^1(\mathscr{O}) = 0$.

Let $X \nearrow Y$ be a *projective* conifold transition of Calabi–Yau 3-folds X, Y through a singular Calabi–Yau variety $\bar{X}$ with k ordinary double points (ODPs) $p_1,\ldots,p_k \in \bar{X}$. During the complex degeneration $\pi : \mathfrak{X} \to \Delta$ with $\mathfrak{X}_0 = \bar{X}$, there are k vanishing 3-spheres $S_1,\ldots,S_k$ with $N_{S_i/X} = T^*S^3$. During the Kähler degeneration (small contraction) $\psi : Y \to \bar{X}$, there are k vanishing 2-spheres (exceptional curves) $C_1,\ldots,C_k$ with $N_{C_i/Y} = \mathscr{O}_{\mathbb{P}^1}(-1)^{\oplus 2}$:

$$\begin{array}{ccc} & & C_i \subset Y \\ & & \big\downarrow{\scriptstyle\psi} \\ S_i \subset X & \overset{\pi}{\rightsquigarrow} & p_i \in \bar{X}\,. \end{array}$$

Let $\mu := h^{2,1}(X) - h^{2,1}(Y) > 0$ be the lose of complex moduli and $\rho := h^{1,1}(Y) - h^{1,1}(X) > 0$ be the gain of Kähler moduli. From

$$\chi(X) - k\chi(S^3) = \chi(Y) - k\chi(S^2),$$

we get the following well-known elementary relation

$$\mu + \rho = k.$$

This implies that the ψ-exceptional curve classes $[C_i] \in NE(Y/\bar{X})$ admit μ independent relations, and the π vanishing cycles $[S_i] \in V \hookrightarrow H_3(X) \to H_3(\bar{X})$ admit ρ independent relations. (The vanishing cycle space V has $\dim V = \mu$.) Let A, B be the corresponding relation matrices:

$$A = (a_{ij}) \in M_{k\times\mu}, \qquad \sum\nolimits_{i=1}^{k} a_{ij}[C_i] = 0,$$
$$B = (b_{ij}) \in M_{k\times\rho}, \qquad \sum\nolimits_{i=1}^{k} b_{ij}[S_i] = 0.$$

THEOREM 7.1 (Basic exact sequence). [**13**, Theorem 1.14] *The Hodge realization of $\mu + \rho = k$ is represented by an exact sequence*

$$0 \to H^2(Y)/H^2(X) \xrightarrow{B} \mathbb{C}^k \xrightarrow{A^t} V \to 0$$

of weight two Hodge structures.

Indeed $V \cong H^{1,1}_{\infty} H^3(X)$ in the limiting Hodge diamond for π:

$$\begin{array}{ccccc} & & H^{2,2}_{\infty} H^3 & & \\ \mathbb{C} \cong H^{3,0}_{\infty} H^3 & H^{2,1}_{\infty} H^3 & \sim \downarrow N & H^{1,2}_{\infty} H^3 & H^{0,3}_{\infty} H^3 \\ & & H^{1,1}_{\infty} H^3 & & \end{array}$$

and the *invariant subsystem* is $Gr^W_3 H^3(X) \cong H^3(Y)$.

7.2. Local quantum transition. By the Bogomolov–Tian–Todorov theorem and its extension to Calabi–Yau conifolds by Ran and Kawamata, the moduli spaces $\mathcal{M}_Y$ and $\mathcal{M}_{\bar{X}}$ are smooth of dimension $h^{2,1}(Y)$ and $h^{2,1}(X)$ respectively. Also the contraction $\psi : Y \to \bar{X}$ deforms in projective families. This then identifies $\mathcal{M}_Y$ as a codimenison μ boundary strata in $\mathcal{M}_{\bar{X}}$ and *locally* near $[\bar{X}] \in \mathcal{M}_{\bar{X}}$ we have $\mathcal{M}_{\bar{X}} \cong \Delta^{\mu} \times \mathcal{M}_Y$.

We represent $V = \mathbb{C}\langle \Gamma_1, \ldots, \Gamma_\mu \rangle$ in terms of a basis Γ_j's. It was shown in [**13**, Proposition 3.15] that the α-periods

$$r_j = \int_{\Gamma_j} \Omega, \qquad 1 \le j \le \mu$$

form the *degeneration coordinates* around $[\bar{X}] \in \mathcal{M}_{\bar{X}} \cong \Delta^{\mu} \times \mathcal{M}_Y$.

In order to describe the discriminant loci of $\mathcal{M}_{\bar{X}}$ near $[\bar{X}]$, we recall Friedman's result on (partial) smoothing of ODPs:

PROPOSITION 7.2. [**4**] *Let $w_i = a_{i1} r_1 + \ldots + a_{i\mu} r_\mu$, then the divisor $D_i :=$ $\{w_i = 0\} \subset \mathcal{M}_{\bar{X}}$ is the loci where the sphere S_i shrinks to an ODP p_i.*

It is clear that the discriminant loci $D_B = \bigcup_{i=1}^k D_i$ is not a normal crossing divisor. Rather it is a *central hyperplane arrangement.*

Under a suitable choice of homology symplectic basis, the β-periods in the transversal directions are given by

$$u_p = \partial_p u = \int_{\beta_p} \Omega$$

for some function u. The Bryant–Griffiths–Yukawa couplings are then extended over the boundary D_B and satisfy

$$u_{pmn} := \partial^3_{pmn} u = O(1) + \sum_{i=1}^{k} \frac{1}{2\pi\sqrt{-1}} \frac{a_{ip} a_{im} a_{in}}{w_i}$$

for $1 \le p, m, n \le \mu$. It is holomorphic if one of the indices is outside this range.

The collection $\{u_{pmn}\}$ is the essential part of the Gauss–Manin connection ∇^{GM} on $\mathcal{M}_X$ which has regular singular extension over D_B.

Similarly, let $u = \sum_{p=1}^{\rho} u^p T_p \in H^2(Y)/H^2(X)$, $D^i := \{\sum_{p=1}^{\rho} b_{ip} u^p = 0\}$, $i = 1, \ldots, k$. By the multiple cover formula of GW invariants we know that $QH(Y)$, or its Dubrovin connection, is regular singular along $D^A = \bigcup D^i$.

Let $y = \sum_{i=1}^{k} y_i e_i \in \mathbb{C}^k$, with $e^1, \ldots, e^k$ being the dual basis on $(\mathbb{C}^k)^\vee$. The trivial logarithmic connection on $\underline{\mathbb{C}}^k \oplus (\underline{\mathbb{C}}^k)^\vee \longrightarrow \mathbb{C}^k$ is defined by

$$\nabla^k = d + \frac{1}{z} \sum_{i=1}^{k} \frac{dy_i}{y_i} \otimes (e^i \otimes e_i^*).$$

The statement $A^t B = 0$ in Theorem 7.1 leads to an orthogonal sum

$$\mathbb{C}^k = \text{image}\, A \overset{\perp}{\oplus} \text{image}\, B \cong V^* \oplus H^2(Y)/H^2(X). \tag{7.1}$$

THEOREM 7.3. [**13**, Theorem 4.1] *Under the identification* (7.1),

(1) *when restricted to* V^*, ∇^k *is naturally identified with the logarithmic (regular singular) part of* ∇^{GM};
(2) *when restricted to* $H^2(Y)/H^2(X)$, ∇^k *is naturally identified with to the logarithmic part of* ∇^{Dubrovin}.

7.3. Global aspects. Denote by $\mathscr{A}(-)$ the GW theory and $\mathscr{B}(-)$ the variations of Hodge structure. Theorem 7.3 provides evidence to

$$\text{“excess } \mathscr{A} \text{ theory”} + \text{“excess } \mathscr{B} \text{ theory”} = \text{“trivial”}$$

through the partial exchange of quantum information attached to *vanishing cycles* on both the $\mathscr{A}$ and $\mathscr{B}$ theories. For the full information on quantum $\mathscr{A}$, $\mathscr{B}$ theories, we proved the following result:

THEOREM 7.4. [**13**, Theorem 0.3] *Let* $[X]$ *be a nearby point of* $[\bar{X}]$ *in* $\mathscr{M}_{\bar{X}}$.

(1) *The theory* $\mathscr{A}(X)$ *is a sub-theory of* $\mathscr{A}(Y)$ *(e.g. quantum sub-ring in genus 0).*
(2) *The theory* $\mathscr{B}(Y)$ *is a sub-theory of* $\mathscr{B}(X)$ *(invariant sub-VHS).*
(3) *The theory* $\mathscr{A}(Y)$ *can be reconstructed from a “refined* $\mathscr{A}$ *theory” on*
$$X^\circ := X \setminus \bigcup\nolimits_{i=1}^{k} S_i$$
“linked” by the vanishing spheres in $\mathscr{B}(X)$.
(4) *The theory* $\mathscr{B}(X)$ *can be reconstructed from the VMHS on* $H^3(Y^\circ)$,
$$Y^\circ := Y \setminus \bigcup\nolimits_{i=1}^{k} C_i,$$
“linked” by the exceptional curves in $\mathscr{A}(Y)$.

The definition of the linked GW invariant in (3) is really a reformulation of the *discreteness of components* appearing in the *virtual cycle* form of the degeneration formula for conifold transitions of Calab–Yau 3-folds:

$$\langle - \rangle_{g,\beta}^{X} = \sum\nolimits_{\gamma \mapsto \beta} \langle - \rangle_{g,\gamma}^{Y}.$$

The sum is a *finite sum.* However, no method is known to single out the individual term in it. To get $QH(Y)$ from $QH(X)$, it requires a blow-up formula of GW invariants where the blow-up center is a Weil divisor.

Weil divisors on $\bar{X}$ can be constructed from the relation matrix B on $\mathfrak{X}_t$. Indeed, $\sum_{i=1}^k b_{ij}[S_i] = 0$ implies that there are real 4-chains $W_{j,t}$ such that

$$\sum\nolimits_{i=1}^k b_{ij} S_i = \partial W_{j,t}.$$

When $t \to 0$, we get homology cycles $W_j := W_{j,0}$, $j = 1, \dots, \rho$, since now ∂W_j is supported at the ODPs. From our definition of Calabi–Yau varieties it can be shown that W_j's are represented by algebraic cycles, hence they give rise to Weil divisors on $\bar{X}$.

The projective small resolution $\psi : Y \to \bar{X}$ transforms all these W_j's into Cartier divisors. Indeed, let W be the sum of supports of all W_j's. Then $Y = \mathrm{Bl}_W \bar{X}$, i.e. the blow-up of the ideal sheaf $\mathscr{I}_W \subset \mathscr{O}_{\bar{X}}$. A non-Cartier Weil divisor is simply a *non-complete intersection divisor*. Thus the problem is essentially a problem on finding *a blow-up formula of GW theory with non-complete intersection center*. Notice that the GW theory on $\bar{X}$ is so far undefined in the literature. However from the deformation invariance of (log) GW theory one may in practice identify it with the GW theory on X.

EXAMPLE 7.5 (Determinantal transitions [**14**]). Let $Y \subset S \times P^n$ be the zero loci of sections $s_i \in \Gamma(S \times P^n, \mathscr{L}_i)$ where $\mathscr{L}_i \to S \times P^n$ are line bundles of the form $\mathscr{L}_i = L_i \boxtimes \mathscr{O}_{P^n}(1)$ with L_i being semi-ample on S.

Let $[x_0 : \cdots : x_n]$ be the homogeneous coordinates on P^n. We write

$$s_i = \sum\nolimits_{j=0}^n s_{ij}\, x_j, \qquad i = 0, \dots, n, \tag{7.2}$$

where $s_{ij} \in \Gamma(S, L_i)$. We are interested in studying the restriction of the projection map $\pi : S \times P^n \to S$ to Y. Define $\bar{X} = \pi(Y) \subset S$ and

$$\psi = \pi|_Y : Y \to \bar{X}.$$

The variety $\bar{X}$ has defining equation

$$\Delta := \det s_{ij} = 0.$$

For $p \in \bar{X}$, since $s_{ij}(p)$'s are fixed, $\psi^{-1}(p)$ is not unique if and only if equation (7.2) has more than one dimensional solutions in P^n, i.e. p is a singular point of $\bar{X}$. The contraction $\psi : Y \to \bar{X}$ is called a determinantal contraction. Notice that

$$\Delta \in \Gamma(S, \bigotimes\nolimits_{i=0}^n L_i).$$

If for general sections $\tau \in \Gamma(S, \bigotimes_{i=0}^n L_i)$ the variety X_τ defined by $\tau = 0$ is smooth, then it gives rise to a transition $Y \searrow X$. If furthermore $\bar{X}$ has only ODPs, then we get a conifold transition. These properties hold for CICY 3-folds transitions which have been studied extensively in the literature.

Given a determinantal transition, we proceed to determine the GW theory on Y in terms of the one on X and the data L_i's. Our goal is to replace the *extrinsic data* L_i's by the *intrinsic data* associated to ψ, namely the Weil divisor W which gives $Y = \mathrm{Bl}_W \bar{X}$. Here we consider the $g = 0$ case:

Let $h = c_1(\mathscr{O}_{P^n}(1))$. As in (5.2), the quantum Lefschetz gives

$$I^Y = J^S \frac{\prod_{i=0}^n (L_i + h)}{(h)^{n+1}},$$
$$I^X = J^S (\sum\nolimits_{i=0}^n L_i),$$

where we omit the curve class β in the subscript. Then

$$I^{Y/X} = \frac{\prod_{i=0}^{n}(L_i + h)}{(h)^{n+1}(\sum_{i=0}^{n} L_i)}. \tag{7.3}$$

We note that the divisor h on $S \times P^n$ coming from $\mathscr{O}_{P^n}(1)$ restricts to a divisor, still called h, on Y. If $p \in \bar{X}$ is a point with positive dimensional fiber $\psi^{-1}(p) \subset \{p\} \times P^n$ then h intersects $\psi^{-1}(p)$ non-trivially since h comes from a hyperplane in P^n. When ψ is a small contraction, this effective divisor $W = \psi_*(h) \subset \bar{X}$ is thus the Weil divisor we are seeking for, and then Y is the blow-up of $\bar{X}$ along W.

It remains to interpret the *K-trivial factor* (7.3) in term of the linear system $|W|$. As in the case of smooth blow-ups along complete intersection centers (cf. Remark 5.2), we will give intrinsic meaning of $I^{Y/X}$ in terms of decompositions of linear systems. The details will appear in [**14**].

References

[1] Jeff Brown, *Gromov-Witten invariants of toric fibrations*, Int. Math. Res. Not. IMRN **19** (2014), 5437–5482. MR3267376

[2] Tom Coates and Alexander Givental, *Quantum Riemann-Roch, Lefschetz and Serre*, Ann. of Math. (2) **165** (2007), no. 1, 15–53, DOI 10.4007/annals.2007.165.15. MR2276766

[3] David A. Cox and Sheldon Katz, *Mirror symmetry and algebraic geometry*, Mathematical Surveys and Monographs, vol. 68, American Mathematical Society, Providence, RI, 1999. MR1677117

[4] Robert Friedman, *Simultaneous resolution of threefold double points*, Math. Ann. **274** (1986), no. 4, 671–689, DOI 10.1007/BF01458602. MR848512

[5] William Fulton, *Intersection theory*, Ergebnisse der Mathematik und ihrer Grenzgebiete (3) [Results in Mathematics and Related Areas (3)], vol. 2, Springer-Verlag, Berlin, 1984. MR732620

[6] Alexander B. Givental, *Equivariant Gromov-Witten invariants*, Internat. Math. Res. Notices **13** (1996), 613–663, DOI 10.1155/S1073792896000414. MR1408320

[7] Y. Iwao, Y.-P. Lee, H.-W. Lin, and C.-L. Wang, *Invariance of Gromov-Witten theory under a simple flop*, J. Reine Angew. Math. **663** (2012), 67–90, DOI 10.1515/CRELLE.2011.097. MR2889709

[8] Y.-P. Lee, *Quantum Lefschetz hyperplane theorem*, Invent. Math. **145** (2001), no. 1, 121–149, DOI 10.1007/s002220100145. MR1839288

[9] Yuan-Pin Lee, Hui-Wen Lin, Feng Qu, and Chin-Lung Wang, *Invariance of quantum rings under ordinary flops III: A quantum splitting principle*, Camb. J. Math. **4** (2016), no. 3, 333–401. MR3550285

[10] Yuan-Pin Lee, Hui-Wen Lin, and Chin-Lung Wang, *Flops, motives, and invariance of quantum rings*, Ann. of Math. (2) **172** (2010), no. 1, 243–290, DOI 10.4007/annals.2010.172.243. MR2680420

[11] Yuan-Pin Lee, Hui-Wen Lin, and Chin-Lung Wang, *Invariance of quantum rings under ordinary flops I: Quantum corrections and reduction to local models*, Algebr. Geom. **3** (2016), no. 5, 578–614, DOI 10.14231/AG-2016-026. MR3568338

[12] Yuan-Pin Lee, Hui-Wen Lin, and Chin-Lung Wang, *Invariance of quantum rings under ordinary flops II: A quantum Leray-Hirsch theorem*, Algebr. Geom. **3** (2016), no. 5, 615–653, DOI 10.14231/AG-2016-027. MR3568339

[13] ——; *Towards $A + B$ theory in conifold transitions for Calabi–Yau threefolds*, to appear in J. Diff. Geom. arXiv:1502.03277.

[14] ——; *A blowup formula in Gromov–Witten theory*, work in progress.

[15] ——; *Quantum flips*, work in progress.

[16] Y.-P. Lee and R. Pandharipande, *A reconstruction theorem in quantum cohomology and quantum K-theory*, Amer. J. Math. **126** (2004), no. 6, 1367–1379. MR2102400

[17] Jun Li, *A degeneration formula of GW-invariants*, J. Differential Geom. **60** (2002), no. 2, 199–293. MR1938113

[18] An-Min Li and Yongbin Ruan, *Symplectic surgery and Gromov-Witten invariants of Calabi-Yau 3-folds*, Invent. Math. **145** (2001), no. 1, 151–218, DOI 10.1007/s002220100146. MR1839289

[19] Bong H. Lian, Kefeng Liu, and Shing-Tung Yau, *Mirror principle. I*, Asian J. Math. **1** (1997), no. 4, 729–763, DOI 10.4310/AJM.1997.v1.n4.a5. MR1621573

[20] D. Maulik and R. Pandharipande, *A topological view of Gromov-Witten theory*, Topology **45** (2006), no. 5, 887–918, DOI 10.1016/j.top.2006.06.002. MR2248516

[21] Miles Reid, *The moduli space of 3-folds with $K = 0$ may nevertheless be irreducible*, Math. Ann. **278** (1987), no. 1-4, 329–334, DOI 10.1007/BF01458074. MR909231

[22] Chin-Lung Wang, *K-equivalence in birational geometry*, Second International Congress of Chinese Mathematicians, New Stud. Adv. Math., vol. 4, Int. Press, Somerville, MA, 2004, pp. 199–216. MR2497984

[23] Chin-Lung Wang, *K-equivalence in birational geometry and characterizations of complex elliptic genera*, J. Algebraic Geom. **12** (2003), no. 2, 285–306, DOI 10.1090/S1056-3911-02-00312-0. MR1949645

[24] Edward Witten, *Phases of $N = 2$ theories in two dimensions*, Nuclear Phys. B **403** (1993), no. 1-2, 159–222, DOI 10.1016/0550-3213(93)90033-L. MR1232617

Department of Mathematics, University of Utah, Salt Lake City, Utah 84112-0090
E-mail address: `yplee@math.utah.edu`

Department of Mathematics and Taida Institute for Mathematical Sciences (TIMS), National Taiwan University, Taipei 10617, Taiwan
E-mail address: `linhw@math.ntu.edu.tw`

Department of Mathematics and Taida Institute for Mathematical Sciences (TIMS), National Taiwan University, Taipei 10617, Taiwan
E-mail address: `dragon@math.ntu.edu.tw`

L^2-kernels of Dirac-type operators on monopole moduli spaces

Gregory W. Moore, Andrew B. Royston, and Dieter Van den Bleeken

ABSTRACT. We state some mathematical predictions concerning the kernels of Dirac-type operators on moduli spaces of (singular) monopoles in $\mathbb{R}^3$. These predictions follow from the semiclassical interpretation of physical results on spaces of (framed) BPS states in d=4, N=2 gauge theories.

1. Introduction and Conclusion

The purpose of this note is to state some mathematical predictions for the differential geometry of moduli spaces of monopoles that follow from a semiclassical interpretation of some of the recent results on the BPS spectrum of four-dimensional gauge theories with N=2 supersymmetry. These predictions concern the nature of the L^2-kernel of a family of Dirac-type operators on moduli spaces of both singular and non-singular monopoles. We state three conjectures in Section §4. The first is a generalization of the famous conjecture of Ashoke Sen concerning the L^2-harmonic forms on monopole moduli space [**27**,**29**]. The second and third conjectures describe the way in which the kernel is expected to jump as parameters in the family are continuously varied. We illustrate the conjectures with two examples in Section 5. In this note we only consider "pure" vectormultiplet theories and "pure" 't Hooft line defects. The generalizations to theories with matter and general Wilson-'t Hooft line defects will appear elsewhere [**2**].

This note, aimed primarily at mathematicians, is a brief summary of the much more comprehensive paper [**25**] where the reader can find extended material explaining conventions, notation and background, as well as fuller explanations of the reasoning leading to the statements we make here. The papers [**23**–**25**] also contain a more extensive list of references.

2010 *Mathematics Subject Classification.* Primary 47A53, 53C26, 53Z05, 58J20.

Key words and phrases. Monopoles, Moduli Spaces, Dirac operators.

The first author thanks N. Hitchin, C. LeBrun, N. Manton and E. Witten for useful remarks, and D. Brennan for collaboration on related matters. GM is supported by the U.S. Department of Energy under grant DOE-SC0010008, and also gratefully acknowledges the hospitality the Isaac Newton Institute.

The second author was supported by the Mitchell Family Foundation and thanks the NHETC at Rutgers for hospitality.

The third author was partially supported by TUBITAK grant 113F164 and by the Boğaziçi University Research Fund under grant number 13B03SUP7.

2. (Singular) Monopole Moduli Space

Let G be a compact simple Lie group. Consider Yang-Mills-Higgs theory on $\mathbb{R}^{1,3}$ with action proportional to

$$\int_{\mathbb{R}^{1,3}} (F, *F) + (DX, *DX). \tag{2.1}$$

Here $(\cdot,\cdot)$ is a Killing form, normalized so the long-root has square-length 2 and we have omitted the theta-term. (The latter is fully taken into account in [**25**].) Magnetic monopoles are static solutions of the Yang-Mills-Higgs equations defined by solutions of the Bogomolnyi equations on $\mathbb{R}^3$

$$F = *_3 DX \tag{2.2}$$

satisfying the following boundary conditions: We choose a regular element X_∞ of the Lie algebra $\mathfrak{g}$ of G. This selects a Cartan subalgebra $\mathfrak{t}$ as well as a system of simple roots α_I, and coroots H_I, where I runs from one to the rank of the group. In a suitable gauge we impose the asymptotic conditions as the radial distance r from a choice of spatial origin goes to infinity:

$$\begin{aligned} X &= X_\infty - \frac{\gamma_{\mathrm{m}}}{2r} + \cdots \\ F &= \frac{1}{2}\gamma_{\mathrm{m}}\omega + \cdots \end{aligned} \tag{2.3}$$

Here γ_{m} is the magnetic charge and the gauge bundle, restricted to spheres of constant radius S_r^2 for sufficiently large r, has a transition function on the equator $e^{i\phi} \to e^{\gamma_{\mathrm{m}}\phi}$, so γ_{m} is in the coroot lattice[1] Λ_{cr} of $\mathfrak{g}$, while $\omega = \sin\theta d\theta d\phi$ is a volume form on S_r^2. The moduli space of such solutions, identified by the group of gauge transformations that approach 1 as $r \to \infty$ is denoted $\mathcal{M}(\gamma_{\mathrm{m}}, X_\infty)$ and is is a smooth hyperkähler manifold [**1**]. Expanding $\gamma_{\mathrm{m}} = \sum_I n_{\mathrm{m}}^I H_I$, with $n_{\mathrm{m}}^I \in \mathbb{Z}$, the moduli space is nonempty iff all n_{m}^I are nonnegative and at least one is positive [**31**], in which case the real dimension of the moduli space is $4\sum_I n_{\mathrm{m}}^I$ [**32, 33**].

There is a Lie algebra of Killing vector fields on $\mathcal{M}(\gamma_{\mathrm{m}}, X_\infty)$ isomorphic to $\mathbb{R}^3 \oplus \mathfrak{so}(3) \oplus \mathfrak{t}$, the summands corresponding to symmetries of translation, rotation about the origin, and global gauge transformations, respectively. We define the Killing vector associated with an element $h \in \mathfrak{t}$ as follows. In a gauge $A_0 = 0$ let $\hat{A} = A_i dx^i + X dx^4$ be a four-dimensional gauge field so that solutions of (2.2) are equivalent to self-dual gauge fields $\hat{A}$ that are translation invariant in x^4. Given such a solution solve $\hat{D}^2\epsilon = 0$ for a smooth map $\epsilon : \mathbb{R}^4 \to \mathfrak{g}$ that is translation invariant in x^4 and satisfies $\epsilon \to h$ for $r \to \infty$. Then the Killing vector $\mathrm{G}(h)$ has directional derivative at $\hat{A}$ given by $\frac{d}{ds}\hat{A} = -\hat{D}\epsilon$. The map $h \to \mathrm{G}(h)$ is a Lie algebra homomorphism from $\mathfrak{t}$ to the Lie algebra of triholomorphic Killing vectors of the universal cover $\widetilde{\mathcal{M}}(\gamma_{\mathrm{m}}, X_\infty)$ of $\mathcal{M}(\gamma_{\mathrm{m}}, X_\infty)$. In this paper we make the simplifying assumption that $n_{\mathrm{m}}^I > 0$ for all I so that the action is effective. (The

[1] We denote the coroot lattice by $\Lambda_{\mathrm{cr}} \subset \mathfrak{t}$. It is the integral dual of the weight lattice $\Lambda_{\mathrm{wt}} \subset \mathfrak{t}^\vee$, and can be identified with the elements $h \in \mathfrak{t}$ such that $\exp[2\pi h] = 1$. Similarly, the magnetic weight lattice $\Lambda_{\mathrm{mw}} \subset \mathfrak{t}$ is the dual of the root lattice of $\Lambda_{\mathrm{rt}} \subset \mathfrak{t}^\vee$. In this paper we only consider adjoint-valued matter fields and hence we take the gauge group to be the adjoint group. Thus the cocharacter lattice of G is Λ_{mw} and the character lattice of G is Λ_{rt}. Since we have chosen a Killing form on $\mathfrak{t}$, for any $h \in \mathfrak{t}$ we may define a dual vector $h^\vee \in \mathfrak{t}^\vee$ by $\langle h^\vee, h'\rangle := (h, h')$ for all $h' \in \mathfrak{t}$.

more general situation is fully treated in [25].) The universal cover is metrically a product

$$\widetilde{\mathcal{M}}(\gamma_{\mathrm{m}}, X_\infty) = \mathbb{R}^4 \times \mathcal{M}_0(\gamma_{\mathrm{m}}, X_\infty) \tag{2.4}$$

where the $\mathbb{R}^4$ factor has the standard Euclidean metric and is generated by the translation Killing vectors along with $\mathrm{G}(X_\infty)$. The space $\mathcal{M}_0(\gamma_{\mathrm{m}}, X_\infty)$ is the "strongly centered" moduli space in the terminology of [16]. The moduli space is a quotient $\mathcal{M}(\gamma_{\mathrm{m}}, X_\infty) = \widetilde{\mathcal{M}}(\gamma_{\mathrm{m}}, X_\infty)/\mathbb{D}$ where the group of Deck transformations $\mathbb{D} \cong \mathbb{Z}$, and is generated by an isometry ϕ. An important point for us is that the group of hyperholomorphic isometries acting on the universal cover and defined by $\mathbb{D}_{\mathrm{g}} := \{\exp[2\pi \mathrm{G}(h)] | h \in \Lambda_{\mathrm{mw}}\}$, is in general a proper subgroup of $\mathbb{D}$. Through any point $m_0 \in \mathcal{M}(\gamma_{\mathrm{m}}, X_\infty)$ there is an orbit of the maximal torus defining a homomorphism $\pi_1(T,1) \to \pi_1(\mathcal{M}, m_0)$. We may view this as a homomorphism $\mu : \Lambda_{\mathrm{mw}} \to \mathbb{Z}$. Based on the rational map formulation of monopole moduli spaces [1, 9, 17, 18] we prove that $\mu(\lambda) = (\gamma_{\mathrm{m}}, \lambda)$ and for any $\lambda \in \Lambda_{\mathrm{mw}}$

$$\exp[2\pi \mathrm{G}(\lambda)] = \phi^{\mu(\lambda)}. \tag{2.5}$$

The image of μ is a subgroup $\ell\mathbb{Z} \subset \mathbb{Z}$ where ℓ is the positive integer such that $\ell^{-1}\gamma_{\mathrm{m}}^\vee \in \Lambda_{\mathrm{rt}}$ is primitive. (Equivalently, it is the gcd of the components of $\gamma_{\mathrm{m}}^\vee$ along a basis of simple roots.)

We now allow singularities in the monopole field corresponding, physically, to the introduction of an 't Hooft line defect at the origin. The data of the singularity is given by (the Weyl orbit of) an element P of the cocharacter lattice $\Lambda_G = \mathrm{Hom}(U(1), T)$ where T is the Cartan torus of G with Lie algebra $\mathfrak{t}$. Singular monopoles satisfy equation (2.2), together with boundary conditions (2.3) for $r \to \infty$ along with the boundary conditions that read, in a suitable gauge,

$$\begin{aligned} X &= -\frac{P}{2r} + \mathcal{O}(r^{-1/2}) \\ F &= \frac{1}{2} P\omega + \mathcal{O}(r^{-3/2}) \end{aligned} \tag{2.6}$$

for $r \to 0$. The space of solutions to (2.2), (2.3), and (2.6), quotiented by gauge transformations approaching 1 at $r \to \infty$ and the normalizer of P for $r \to 0$, is the moduli space $\overline{\mathcal{M}}(P, \gamma_{\mathrm{m}}, X_\infty)$. It is a hyperkähler manifold with singularities on codimension four loci. The singularities are due to the phenomenon of monopole bubbling [19, 24] and in some cases are loci of orbifold singularities. Let

$$\tilde{\gamma}_{\mathrm{m}} := \gamma_{\mathrm{m}} - P^- = \sum_I \tilde{n}_{\mathrm{m}}^I H_I \tag{2.7}$$

where P^- is the Weyl image of P in the chamber with $\langle \alpha_I, P \rangle \leq 0$ for all I. It is conjectured in [23] (with some supporting evidence given in [24]) that the moduli space is nonempty iff $\tilde{n}_{\mathrm{m}}^I \geq 0$. The presence of the singularity at the spatial origin reduces the Lie algebra of Killing vectors to $\mathfrak{so}(3) \oplus \mathfrak{t}$. The space $\overline{\mathcal{M}}(P, \gamma_{\mathrm{m}}, X_\infty)$ only depends on the Weyl group orbit $[P]$ of P, up to hyperkähler isometry.

3. Consequences of $N = 2$ Supersymmetry

3.1. General Remarks. We now consider the above Yang-Mills-Higgs theory in the context of N=2 supersymmetric gauge theory. The requirement of N=2 supersymmetry brings in a number of new features. (See [**34**] for a comprehensive review.)

The first new feature is the presence of a second adjoint Higgs field Y. The Hamiltonian contains a potential energy term proportional to $\int_{\mathbb{R}^3}([X,Y],*[X,Y])$ and this term forces finite energy field configurations to obey $Y \to Y_\infty$ for $r \to \infty$ with $Y_\infty \in \mathfrak{t}$.

A second new feature is the existence in the quantum theory of distinguished quantum mechanical states known as "BPS states" [**28, 35**]. These are quantum mechanical analogs of the classical dyon field configurations satisfying the Bogomolnyi bound, and indeed the latter configurations are semiclassical approximations to BPS states at weak coupling. On the Coulomb branch of vacua of the N=2 quantum field theory the gauge group is broken down to the maximal torus T and the BPS states carry electromagnetic charges with respect to this unbroken gauge symmetry. In the semiclassical limit there is a distinguished duality frame and the electromagnetic charge γ is valued in the lattice[2]

$$\Gamma := \Lambda_{\mathrm{mw}} \oplus \Lambda_{\mathrm{wt}} \subset \mathfrak{t} \oplus \mathfrak{t}^\vee. \tag{3.1}$$

The lattice Γ has a natural symplectic form defined by

$$\langle\!\langle \gamma_1, \gamma_2 \rangle\!\rangle = \langle\!\langle \gamma_{1,\mathrm{m}} \oplus \gamma_1^{\mathrm{e}}, \gamma_{2,\mathrm{m}} \oplus \gamma_2^{\mathrm{e}} \rangle\!\rangle := \langle \gamma_1^{\mathrm{e}}, \gamma_{2,\mathrm{m}} \rangle - \langle \gamma_2^{\mathrm{e}}, \gamma_{1,\mathrm{m}} \rangle. \tag{3.2}$$

It is valued in $\frac{1}{z}\mathbb{Z}$, where z is the order of the center of the universal cover $\widetilde{G}$.

A third new feature is that the 't Hooft line defects can be generalized to line defects preserving four out of the eight supersymmetries. In the physical theory the line defect depends on a choice of a further parameter, namely a phase ζ that determines which half-dimensional subspace of supersymmetries is preserved by the line defect. Since this parameter will not be visible in our mathematical predictions we will simply denote the line defect by $L([P])$. The insertion of a line defect changes the Hilbert space, but, when singularities at the origin are properly taken into account one can still define a Bogomolnyi bound. The so-called *framed BPS states* are the BPS states in the presence of a supersymmetric line defect [**10**]. They have electromagnetic charge γ in the Γ-torsor $(P + \Lambda_{\mathrm{cr}}) \oplus \Lambda_{\mathrm{rt}}$.

A fourth new feature arising from N=2 supersymmetry is that low energy excitations of (singular) monopoles (the dynamics of "collective coordinates") are described by supersymmetric quantum mechanics on moduli space [**11–13, 21, 34**]. The paper [**25**] has an in-depth review of these results, including generalizations accounting for the effects of a theta angle, the presence of singular monopoles, and the inclusion of one-loop effects.

3.2. Geometrical Definitions Of BPS States. The supersymmetric dynamics of collective coordinates leads to geometrical definitions of the spaces of BPS states. For brevity we henceforth denote $\mathcal{X} := X_\infty$ and $\mathcal{Y} := Y_\infty$.

We begin with the framed BPS states. Let $D\!\!\!\!/^{\mathcal{Y}}$ be the Dirac-type operator on the bundle of L^2-normalizable Dirac spinors on $\overline{\mathcal{M}}(P, \gamma_{\mathrm{m}}; \mathcal{X})$ defined by $D\!\!\!\!/^{\mathcal{Y}} = D\!\!\!\!/ +$

[2] Actually, the semiclassical regime is not simply connected, and Γ is really a local system. This well-known complication is thoroughly discussed in [**25**].

$\not{G}(\mathcal{Y})$, where $\not{D}$ is the standard Dirac operator and $\not{G}(\mathcal{Y})$ is Clifford multiplication by the one-form derived from the triholomorphic Killing vector $\mathrm{G}(\mathcal{Y})$. For brevity, in what follows we will simply refer to $\not{D}^{\mathcal{Y}}$ as a "Dirac operator."

Definition: *The space of semiclassical framed BPS states, in the vacuum $(\mathcal{X}, \mathcal{Y})$, in the presence of the supersymmetric defect $L([P])$, and which have magnetic charge $\gamma_{\mathrm{m}} \in P + \Lambda_{\mathrm{cr}}$, is:*

$$\ker_{L^2}(\not{D}^{\mathcal{Y}}). \tag{3.3}$$

The reason for the phrase "in the vacuum $(\mathcal{X}, \mathcal{Y})$" is explained in Remark 2 below. In the semiclassical description the magnetic charge of the BPS states is encoded in the choice of moduli space $\overline{\underline{\mathcal{M}}}(P, \gamma_{\mathrm{m}}; \mathcal{X})$ whereas the electric charge is the character of the triholomorphic action of $\mathfrak{t}$ on $\ker_{L^2}(\not{D}^{\mathcal{Y}})$. Indeed, we can define the spaces of framed BPS states of fixed electromagnetic charge $\gamma = \gamma_{\mathrm{m}} \oplus \gamma^{\mathrm{e}}$ from the isotypical decomposition with respect to the action of $\mathfrak{t}$:

$$\ker_{L^2}(\not{D}^{\mathcal{Y}}) = \oplus_{\gamma^{\mathrm{e}} \in \Lambda_{\mathrm{wt}}} \overline{\underline{\mathcal{H}}}([P]; \gamma; \mathcal{X}, \mathcal{Y}). \tag{3.4}$$

A parallel definition can be given for the ordinary ("unframed" or "vanilla") BPS states, but there are two extra complications since one must "factor out the center of mass degrees of freedom" and then impose the proper equivariance condition with respect to the Deck group $\mathbb{D}$.

As preparation introduce the rank one projection operator $\mathbf{P} : \mathfrak{t} \to \mathfrak{t}$ onto the line spanned by $\mathcal{X}$ and defined by $\mathbf{P}(h) := \frac{(h, \gamma_{\mathrm{m}})}{(\mathcal{X}, \gamma_{\mathrm{m}})} \mathcal{X}$. Since $\mathbf{P}$ is a projection operator we have a direct sum decomposition:

$$\mathfrak{t} \cong \mathrm{Im}(\mathbf{P}) \oplus \mathrm{Im}(1 - \mathbf{P}) := \mathfrak{t}_{\mathrm{com}} \oplus \mathfrak{t}_0 \tag{3.5}$$

of vector spaces. The subspace $\mathfrak{t}_0$ is the subspace orthogonal to γ_{m} in the Killing metric (but $\mathbf{P}$ is <u>not</u> the orthogonal projection in the Killing metric). In **[25]** we prove that for all $h \in \mathfrak{t}$,

$$(\mathrm{G}(h), \mathrm{G}(\mathcal{X})) = (h, \gamma_{\mathrm{m}}). \tag{3.6}$$

On the left hand side of this equation we use the hyperkähler metric on moduli space, and on the right hand side we use the Killing metric on $\mathfrak{t}$. Therefore, the orthogonal decomposition, using the hyperkähler metric, along $T\mathbb{R}^4 \oplus T\mathcal{M}_0$ for the Killing vectors $\mathrm{G}(h)$ takes the form

$$\mathrm{G}(h) = \mathrm{G}(h_{\mathrm{com}}) + \mathrm{G}(h_0) \tag{3.7}$$

where, for any $h \in \mathfrak{t}$ we define $h_{\mathrm{com}} := \mathbf{P}(h)$ and $h_0 := (1 - \mathbf{P})(h)$. Now we can choose orthogonal coordinates (x^i, x^4) for the $\mathbb{R}^4$ factor so that the three translation Killing vectors give translations in x^i (holding all else fixed) and $\frac{\partial}{\partial x^4}$ is a vector field parallel to $\mathrm{G}(\mathcal{X})$, normalized so that

$$(\frac{\partial}{\partial x^4}, \mathrm{G}(\mathcal{X})) := 1 \tag{3.8}$$

It follows from (2.5) that $\phi^*(x^4) = x^4 + 2\pi$.

We now split the Dirac operator $\not{D}^{\mathcal{Y}}$ on $\widetilde{\mathcal{M}}(\gamma_{\mathrm{m}}, X_\infty)$ using the splitting of the tangent bundle following from (2.4):

$$\not{D}^{\mathcal{Y}} = \not{D}^{\mathcal{Y}}_{\mathrm{com}} + \not{D}^{\mathcal{Y}}_0. \tag{3.9}$$

Explicitly, in terms of the coordinates on $\mathbb{R}^4$ defined above we have

$$\not{D}^{\mathcal{Y}}_{\rm com} = \Gamma^i \frac{\partial}{\partial x^i} + \Gamma^4 \left(\frac{\partial}{\partial x^4} - {\rm i} \frac{(\mathcal{Y}, \gamma_{\rm m})}{(\mathcal{X}, \gamma_{\rm m})} \right) \tag{3.10}$$

where Γ^i, Γ^4 are 4×4 gamma matrices. We now define subspaces of sections of spinor bundles:

$$\mathcal{K}_{\rm com} := \ker \not{D}^{\rm Y'}_{\rm com} \subset \Gamma[\mathcal{S} \to \mathbb{R}^4] \tag{3.11}$$

$$\mathcal{K}_0 := \ker_{L^2} \not{D}^{\rm Y}_0 \subset \Gamma_{L^2}[\mathcal{S} \to \mathcal{M}_0] \tag{3.12}$$

with $\mathrm{Y}' \in \mathfrak{t}_{\rm com}$ and $\mathrm{Y} \in \mathfrak{t}_0$. By pulling back from the two factors and multiplying we get a subspace of sections of the spinor bundle on $\widetilde{\mathcal{M}}$:

$$\mathcal{K} := \pi^*_{\rm com}(\mathcal{K}_{\rm com}) \otimes \pi^*_0(\mathcal{K}_0) \tag{3.13}$$

and $\mathcal{K}$ is annihilated by $\not{D}^{\mathcal{Y}}$ for $\mathcal{Y} = \mathrm{Y}' + \mathrm{Y}$. There are no L^2-harmonic spinors on $\mathbb{R}^4$ for $\not{D}^{\mathcal{Y}}_{\rm com}$. This is physically reasonable since the monopoles are particles and definite momentum eigenstates are plane waves. On the other hand, we do not wish to have such continuous degrees of freedom "internally" so it is important to impose the L^2 condition in the definition of $\mathcal{K}_0$.

Now, $\mathcal{K}$ is a representation of $\mathfrak{t}$ and we can decompose it into eigenspaces of charge $\gamma^{\rm e} \in \mathfrak{t}^\vee$. Let $\mathcal{G}(h)$ denote the lift of the diffeomorphism $\exp[2\pi \mathrm{G}(h)]$ to the spinor bundle. (It exists.) By definition, a "spinor of electric charge $\gamma^{\rm e} \in \mathfrak{t}^\vee$" is one such that for all $h \in \mathfrak{t}$:

$$\mathcal{G}(h)^* \Psi = \exp[-2\pi {\rm i} \langle \gamma^{\rm e}, h \rangle] \Psi. \tag{3.14}$$

The general element $\Psi_{\rm com} \in \mathcal{K}_{\rm com}$ is of the form $\Psi_{\rm com} = e^{-{\rm i} q x^4} s$, where $s \in \mathbb{C}^4$ is a constant Dirac spinor and $q = -(\mathcal{Y}, \gamma_{\rm m})/(\mathcal{X}, \gamma_{\rm m})$. On the other hand, the $\mathfrak{t}$-action on $\mathcal{K}$ factors through a product of a $\mathfrak{t}_{\rm com}$ action on $\mathcal{K}_{\rm com}$ and a $\mathfrak{t}_0$-action on $\mathcal{K}_0$. It follows that we must also have

$$q = \frac{\langle \gamma^{\rm e}, \mathcal{X} \rangle}{(\gamma_{\rm m}, \mathcal{X})} \tag{3.15}$$

and hence

$$(\mathcal{Y}, \gamma_{\rm m}) + \langle \gamma^{\rm e}, \mathcal{X} \rangle = 0 \tag{3.16}$$

is a necessary condition for the existence of a spinor in $\mathcal{K}$ with the data $(\mathcal{X}, \mathcal{Y}, \gamma_{\rm m}, \gamma^{\rm e})$.

A spinor $\Psi \in \mathcal{K}$ will only descend to $\mathcal{M}(\gamma_{\rm m}, \mathcal{X})$ if $\gamma^{\rm e}$ is properly quantized, that is, if $\gamma^{\rm e} \in \Lambda^\vee_{\rm mw} = \Lambda_{\rm rt}$. However, because of equation (2.5) we must impose a further invariance condition under a cyclic group $\mathbb{Z}/\ell\mathbb{Z}$. Thus the physically relevant kernel of the Dirac operator is

$$\left(\oplus_{\gamma^{\rm e} \in \Lambda_{\rm rt}} \mathcal{K}^{\gamma^{\rm e}} \right)^{\mathbb{Z}/\ell\mathbb{Z}} \tag{3.17}$$

and since ϕ commutes with the $\mathfrak{t}$ action we may impose the cyclic group invariance on each isotypical summand. We define (3.17) to be *the space of semiclassical BPS states in the vacuum* $(\mathcal{X}, \mathcal{Y})$*, with magnetic charge* $\gamma_{\rm m}$. Since $\mathcal{K}_{\rm com} \cong \mathbb{C}^4$ we may factor it out of (3.17) and define both the space of BPS states and the "internal BPS states" by[3]

$$(\mathcal{K}^{\gamma^e})^{\mathbb{Z}/\ell\mathbb{Z}} := \mathcal{H}(\gamma;\mathcal{X},\mathcal{Y}) := \mathcal{K}_{\text{com}} \otimes \mathcal{H}_0(\gamma;\mathcal{X},\mathcal{Y}). \tag{3.18}$$

We conclude this section with three remarks

(1) The Killing vector fields arising from rotational symmetry around the origin are not triholomorphic. Nevertheless, they preserve the L^2-kernel of $\not{D}^{\mathcal{Y}}$ and $\not{D}_0^{\mathcal{Y}}$ on $\overline{\mathcal{M}}$ and $\mathcal{M}_0$ respectively and hence the spaces $\overline{\mathcal{H}}([P];\gamma;\mathcal{X},\mathcal{Y})$ and $\mathcal{H}_0(\gamma;\mathcal{X},\mathcal{Y})$ are representations of $\mathfrak{so}(3)$. Moreover, the commutant $\mathfrak{su}(2)$, in $\mathfrak{so}(4N)$, of the $\mathfrak{sp}(N)$ holonomy (where we denote the real dimension of the moduli space $\overline{\mathcal{M}}$ or $\mathcal{M}_0$ by $4N$) has a lift to the spinor bundle and again acts on the kernel of $\not{D}^{\mathcal{Y}}$. There is therefore an action of $\mathfrak{so}(3)_{\text{rot}} \oplus \mathfrak{su}(2)_R$ on the BPS spaces. The subscripts indicate that the symmetry arises from rotational symmetry and "R-symmetry" in the physical theory.[4]

(2) We have referred to the pair $(\mathcal{X},\mathcal{Y})$ as "defining a vacuum." The precise translation to the physical Coulomb vacuum is described in full detail in [**25**]. In the framed case there are physical parameters (u,ζ), where u is a point on the Coulomb branch and ζ is the phase mentioned above, and the mapping to $(\mathcal{X},\mathcal{Y})$ is defined by

$$\begin{aligned}\mathcal{X} &= \operatorname{Im}(\zeta^{-1}a(u))\\ \mathcal{Y} &= \operatorname{Im}(\zeta^{-1}a_D(u))\end{aligned} \tag{3.19}$$

where on the RHS we have used standard physical notations. The fiber of the mapping does not actually determine a unique vacuum: One can rescale ζ and a by a common phase. In the unframed case the same formulae hold with $-\zeta$ replaced by the phase of the N=2 central charge $Z_\gamma(a)$. The standard collective coordinate supersymmetric quantum mechanics involves a potential energy term $\sim\|\ \mathrm{G}(\mathcal{Y})\ \|^2$. Having a "moduli space with potential" is in fact a contradiction in terms; one is usually confined to working in a small potential energy approximation. In [**25**] it is suggested that (3.19) takes into account all the quantum corrections to the small potential energy approximation.

(3) When hypermultiplet matter in a quaternionic representation R of the gauge group is included the above definitions are extended by modifying the Dirac operator so that it also couples to the bundle associated to the universal bundle by the representation R. This bundle is endowed with a hyperholomorphic connection [**2, 12, 22**].

3.3. Building BPS States From $\mathcal{K}_0$. A natural question to ask, and one which is important to our discussion of Fredholm conditions below, is whether spinors in the space $\mathcal{K}_0$ defined in (3.12) can be used to build BPS states in (3.18). In this section we describe those subspaces of $\mathcal{K}_0$. Accordingly, let us begin with

[3] When $\mathfrak{so}(3)_{\text{rot}} \oplus \mathfrak{su}(2)_R$ are properly defined $\mathcal{K}_{\text{com}}$ is the $(\frac{1}{2};0)\oplus(0;\frac{1}{2})$ representation and is known in physics as the "half-hypermultiplet."

[4] The precise relationship between the $\mathfrak{so}(3)_K$ action induced from the Killing fields, and the canonically defined $\mathfrak{so}(3)_{\text{rot}}$ and $\mathfrak{su}(2)_R$ of the N=2 field theory is that $\mathfrak{so}(3)_K$ is the diagonal subalgebra of $\mathfrak{so}(3)_{\text{rot}} \oplus \mathfrak{su}(2)_R$. This is established through the collective coordinate ansatz; see [**25**].

the Dirac operator $\not{D}_0^{\mathrm{Y}}$ on $\mathcal{M}_0(\gamma_{\mathrm{m}}, \mathcal{X})$ for an element $\mathrm{Y} \in \mathfrak{t}_0$. Only the subspace $\mathfrak{t}_0 \subset \mathfrak{t}$ orthogonal to γ_{m} in the Killing metric will act effectively on $\mathcal{M}_0(\gamma_{\mathrm{m}}, \mathcal{X})$ so our question becomes: When are spinors in the $\mathfrak{t}_0$-isotypical components of $\mathcal{K}_0$ factors of BPS spinors $\Psi \in \mathcal{H}(\gamma; \mathcal{X}, \mathcal{Y})$ for some $\gamma^{\mathrm{e}} \in \Lambda_{\mathrm{rt}}$ and $\mathcal{Y} \in \mathfrak{t}$?

In order to answer this question introduce the rank one projection operator $\mathbf{Q} : \mathfrak{t}^\vee \to \mathfrak{t}^\vee$ dual to $\mathbf{P}$, i.e. $\langle \mathbf{Q}(\gamma^{\mathrm{e}}), h\rangle := \langle \gamma^{\mathrm{e}}, \mathbf{P}(h)\rangle$ for all $h \in \mathfrak{t}$. Then one computes $\mathbf{Q}(\gamma^{\mathrm{e}}) = \frac{\langle \gamma^{\mathrm{e}}, \mathcal{X}\rangle}{\langle \gamma_{\mathrm{m}}, \mathcal{X}\rangle}\gamma_{\mathrm{m}}^\vee$. Since $\mathbf{Q}$ is a projection operator we can use it to give a decomposition

$$\mathfrak{t}^\vee \cong \mathrm{Im}\mathbf{Q} \oplus \mathrm{Im}(1 - \mathbf{Q}) := \mathfrak{t}_{\mathrm{com}}^\vee \oplus \mathfrak{t}_0^\vee \tag{3.20}$$

The isotypical decomposition of the $\mathfrak{t}_0$ action on $\mathcal{K}_0$ is a sum over characters $\gamma_0^{\mathrm{e}} \in \mathfrak{t}_0^\vee$. However, we can only complete a spinor $\Psi_0 \in \mathcal{K}_0$ with a spinor $\Psi_{\mathrm{com}} \in \mathcal{K}_{\mathrm{com}}$ if the electric charge is properly quantized, and the condition for this is

$$\gamma_0^{\mathrm{e}} \in (1 - \mathbf{Q})(\Lambda_{\mathrm{rt}}) \; . \tag{3.21}$$

Let $\mathcal{L}(\gamma_0^{\mathrm{e}}) := \Lambda_{\mathrm{rt}} \cap (1 - \mathbf{Q})^{-1}(\gamma_0^{\mathrm{e}})$. For each choice of $\gamma^{\mathrm{e}} \in \mathcal{L}(\gamma_0^{\mathrm{e}})$ compute q via (3.15), and form $\Psi = \Psi_{\mathrm{com}} \otimes \Psi_0$. Moreover, we define $\mathrm{Y}' \in \mathfrak{t}_{\mathrm{com}}$ to be the solution to (3.16) and set $\mathcal{Y} := \mathrm{Y}' + \mathrm{Y}$. We now have the quartet $(\mathcal{X}, \mathcal{Y}, \gamma_{\mathrm{m}}, \gamma^{\mathrm{e}})$ and $\Psi \in \mathcal{K}^{\gamma^{\mathrm{e}}}$ potentially describes a BPS state. However, we still must require that Ψ descend from $\widetilde{\mathcal{M}}$ to $\mathcal{M}$. Recall that $\phi^*(x^4) = x^4 + 2\pi$ and therefore we must require that

$$\phi^*(\Psi_0) = e^{2\pi i q}\Psi_0. \tag{3.22}$$

Now that we know when we can build BPS states from elements of $\mathcal{K}_0$ we should ask about uniqueness. Suppose that we have found an element of $\mathcal{L}(\gamma_0^{\mathrm{e}})$ so that (3.22) holds. What other charges in $\mathcal{L}(\gamma_0^{\mathrm{e}})$ will also support BPS states? Note that $\mathcal{L}(\gamma_0^{\mathrm{e}})$ is a torsor for $\Lambda_{\mathrm{rt}} \cap \ker(1 - \mathbf{Q}) = \Lambda_{\mathrm{rt}} \cap \mathrm{Im}\mathbf{Q}$ and since $\mathrm{Im}\mathbf{Q}$ is just the real line through $\gamma_{\mathrm{m}}^\vee$ this is simply the set of multiples (necessarily rational) of $\gamma_{\mathrm{m}}^\vee$ that are in the root lattice. Therefore $\mathcal{L}(\gamma_0^{\mathrm{e}}) \subset \Lambda_{\mathrm{rt}}$ is a $\mathbb{Z}$-torsor where the $\mathbb{Z}$-action is generated by $\gamma^{\mathrm{e}} \to \gamma^{\mathrm{e}} + \frac{1}{\ell}\gamma_{\mathrm{m}}^\vee$, that is, any two elements of the torsor are related by shifting

$$\gamma^{\mathrm{e}} \to \gamma^{\mathrm{e}} + \frac{k}{\ell}\gamma_{\mathrm{m}}^\vee \tag{3.23}$$

where k is an integer. By (3.15) this has the effect of shifting: $q \to q + \frac{k}{\ell}$. Therefore, if (3.22) has a solution for some γ^{e} and we shift by integral multiples of $\gamma_{\mathrm{m}}^\vee$, that is, if k is a multiple of ℓ, then the same spinor $\Psi_0 \in \mathcal{K}_0^{\gamma_0^{\mathrm{e}}}$ will satisfy (3.22) and hence we can use the same Ψ_0, but a different Ψ_{com}, to construct a new BPS state. We call the collection of BPS states thus obtained a *Julia-Zee tower*. On the other hand, if we consider a shift (3.23) where k is not a multiple of ℓ then the equivariance condition (3.22) changes. There might, or might not, be a different spinor in $\mathcal{K}_0^{\gamma_0^{\mathrm{e}}}$ transforming according to this new equivariance condition.

4. Mathematical Predictions

In this section we state three mathematical conjectures following from recent results in the physics literature.

4.1. Predictions From The No-Exotics Theorem. As we have remarked, the semiclassical spaces of framed and unframed BPS states:

$$\overline{\underline{\mathcal{H}}}([P];\gamma;\mathcal{X},\mathcal{Y}),$$

and $\mathcal{H}_0(\gamma;\mathcal{X},\mathcal{Y})$, respectively, are representations of $\mathfrak{so}(3)_{\mathrm{rot}} \oplus \mathfrak{su}(2)_R$. The "no exotics conjecture" of [**10**] states that $\mathfrak{su}(2)_R$ should act trivially on these spaces: They are direct sums of $\mathrm{SU}(2)_R$ singlets. This has been proven for gauge groups $\mathrm{SU}(K)$ (using the relation of BPS states to the cohomology of spaces of representations of quivers and the Kontsevich-Soibelman wall-crossing formula) in [**3**] and for all simply laced gauge groups in [**5**]. Recently a very beautiful general argument based on the structure of supersymmetric stress-tensor multiplets has been given [**4**]. To see one simple implication for the Dirac operator we note that under the homomorphism

$$\mathrm{SU}(2)_R \times \mathrm{Sp}(N)_{\mathrm{hol}} \to \mathrm{Spin}(4N) \tag{4.1}$$

the element $(-1,1)$ maps to the Clifford volume form with the natural orientation provided by any of the complex structures on the moduli space. It therefore follows that all the spinors in the kernel of the Dirac operator have positive chirality.

Moreover, choosing any complex structure on the moduli space we can express the no-exotics condition in terms of the Dolbeault operator acting on sections of $\Lambda^{0,*}$:

$$\bar{\partial}^{\mathcal{Y}} := \bar{\partial} + \mathrm{G}(\mathcal{Y})^{0,1}\wedge \tag{4.2}$$

using the isomorphism of the spinor bundle with $\Lambda^{0,*}$ [**15**]. In this formulation a set of generators of $\mathfrak{su}(2)_R$ can be chosen so that

$$\begin{aligned} I^3|_{\Lambda^{0,q}} &= \frac{1}{2}(q-N)\mathbf{1} \\ I^+ = \omega^{0,2}\wedge \qquad & I^- = \iota(\omega^{2,0}) \end{aligned} \tag{4.3}$$

By the no-exotics theorem these operators must act trivially on the cohomology and hence the L^2-cohomology of $H^{0,q}$ with respect to $\bar{\partial}^{\mathcal{Y}}$ is concentrated in the middle degree, $q = N$, and is primitive in the sense that it is annihilated by product with the holomorphic-symplectic form or by contraction with the conjugate of the holomorphic-symplectic form.

In recently completed work [**2**] this discussion is generalized to include gauge theories with hypermultiplet matter. The L^2-cohomology of the corresponding operator $\bar{\partial}^{\mathcal{Y}}$ is still primitive and concentrated in the middle degree. In particular, specializing to $\mathfrak{g} = \mathfrak{su}(2)$ with a hypermultiplet in the adjoint we obtain precisely the renowned prediction of Ashoke Sen [**29**] based on S-duality.

4.2. Predictions From Wall-Crossing Formulae. We now consider the families of Dirac operators as we vary the Higgs fields. For simplicity our discussion will focus on the family parametrized by $\mathcal{Y} \in \mathfrak{t}$ in the framed case, and by $\mathrm{Y} \in \mathfrak{t}_0$ in the unframed case.[5] Generically, $\not{D}^{\mathcal{Y}}$ acting on L^2 spinors over $\overline{\underline{\mathcal{M}}}(P,\gamma_{\mathrm{m}},\mathcal{X})$ (and $\not{D}_0^{\mathrm{Y}}$ acting on L^2 spinors over $\mathcal{M}_0(\gamma_{\mathrm{m}};\mathcal{X})$) is Fredholm, and by the no-exotics property, the character of $\ker \not{D}_+^{\mathcal{Y}}$ is the same as the character-valued index. Therefore, we expect the $\mathfrak{so}(3)$ characters of $\overline{\underline{\mathcal{H}}}([P];\gamma;\mathcal{X},\mathcal{Y})$ and $\mathcal{H}_0(\gamma;\mathcal{X},\mathcal{Y})$ to be piecewise constant as functions of $\mathcal{Y}\in\mathfrak{t}$. However, on real codimension one walls, $\not{D}^{\mathcal{Y}}$, respectively $\not{D}_0^{\mathrm{Y}}$, will fail to be Fredholm.[6] Physically, these walls correspond to walls of marginal stability, and using standard formulae for the $N=2$ central charges we are then led to the following predictions:

First, a translation of the standard criterion for crossing a wall of marginal stability, taking into account simplifications from working in the semiclassical limit, leads to the following criterion: $\not{D}_0^{\mathrm{Y}}$ is not Fredholm only if there are charges $\gamma_1,\gamma_2\in\Gamma$, with $\gamma_{1,\mathrm{m}}+\gamma_{2,\mathrm{m}}=\gamma_{\mathrm{m}}$, "nonlocal" in the sense that $\langle\!\langle\gamma_1,\gamma_2\rangle\!\rangle\neq 0$, and "occupied" in the sense that $\mathcal{H}(\gamma_i;\mathcal{X},\mathcal{Y})\neq 0$, for both $i=1,2$, where $\mathcal{Y}$ is constructed from Y and a choice of $\gamma^{\mathrm{e}}=\gamma_1^{\mathrm{e}}+\gamma_2^{\mathrm{e}}$ as described above. Note that (3.16) implies that

$$(\gamma_{i,\mathrm{m}},\mathcal{Y})+\langle\gamma_i^{\mathrm{e}},\mathcal{X}\rangle=0,\qquad i=1,2, \tag{4.4}$$

but these two equations are not independent of each other because their sum gives the identity (3.16) for the total charge $\gamma=\gamma_1+\gamma_2$. Indeed the codimension wall on which $\not{D}_0^{\mathrm{Y}_0}$ will fail to be Fredholm (due to this pair of charges γ_1,γ_2) is defined by (either) one of the above conditions, which can be written in the form:

$$(\mathrm{Y},\gamma_{\mathrm{m},1})+\langle\gamma_1^{\mathrm{e}}-\gamma_{\mathrm{m},1}\frac{\langle\gamma^{\mathrm{e}},\mathcal{X}\rangle}{(\gamma_{\mathrm{m}},\mathcal{X})},\mathcal{X}\rangle=0. \tag{4.5}$$

Equation (4.5) defines a real codimension one wall in $\mathfrak{t}_0$. If Y crosses this wall the space $\mathcal{H}(\gamma;\mathcal{X},\mathcal{Y})$ will jump in a way dictated by the Kontsevich-Soibelman wall-crossing formula [**20**]. We expect that there is a relatively simple physical argument that shows that the continuous spectrum of $\not{D}_0^{\mathrm{Y}}$ extends down to zero at these real codimension one walls. It entails an analysis of the square of the operator $\not{D}_0^{\mathrm{Y}}$ using the asymptotic form of moduli space metrics discussed in [**14, 34**]. We hope to address this elsewhere.

Second, for singular monopoles, $\not{D}^{\mathcal{Y}}$ is Fredholm except on real codimension one walls defined by

$$W(\gamma_{\mathrm{h}}):=\{\mathcal{Y}:(\gamma_{\mathrm{h,m}},\mathcal{Y})+\langle\gamma_{\mathrm{h,e}},\mathcal{X}\rangle=0\,\&\,\mathcal{H}(\gamma_{\mathrm{h}};\mathcal{X},\mathcal{Y})\neq 0\,\&\,Z_{\gamma_{\mathrm{h}}}(\zeta^{-1}a)<0\} \tag{4.6}$$

(The last condition $Z_{\gamma_{\mathrm{h}}}(\zeta^{-1}a)<0$ is satisfied automatically in the semiclassical region.) In this case a very explicit formula for the jump of $\overline{\underline{\mathcal{H}}}([P],\zeta;\gamma;\mathcal{X},\mathcal{Y})$ follows from [**10**]: We introduce the Heisenberg group extension of $\mathbb{C}[\Gamma]$ defined by $X_{\gamma_1}X_{\gamma_2}=y^{\langle\!\langle\gamma_1,\gamma_2\rangle\!\rangle}X_{\gamma_1+\gamma_2}$ and form the generating function

$$F([P];\mathcal{X},\mathcal{Y}):=\sum_{\gamma\in\Gamma}\overline{\underline{\Omega}}([P],\gamma;\mathcal{X},\mathcal{Y})X_\gamma \tag{4.7}$$

where $\overline{\underline{\Omega}}(P,\gamma;\mathcal{X},\mathcal{Y})$ is the $\mathfrak{so}(3)$ character of $\overline{\underline{\mathcal{H}}}([P];\gamma;\mathcal{X},\mathcal{Y})$, that is, the trace of y^{2J} where J is any generator of $\mathfrak{so}(3)$ normalized to have half-integer eigenvalues.

[5] One can also let $\mathcal{X}$ vary in the fundamental Weyl chamber. Strictly speaking, in order to vary $\mathcal{X}$ we should define a connection on the bundle of hyperkähler manifolds over the chamber. This is not worked out in [**25**], but we trust it is easily taken care of.

[6] Our statements generalize the discussion in [**30**].

In a similar way, define a set of integers $a_m(\gamma_h)$ by the character of $\mathcal{H}(\gamma_h;\mathcal{X},\mathcal{Y})$, $\Omega(\gamma_h;\mathcal{X},\mathcal{Y}) = \sum_m a_m(-y)^m$. Then, crossing a wall $W(\gamma_h)$ we have the transformation law $F \to SFS^{-1}$, where $S = \prod_{n>0} S_{n\gamma_h}$,

$$S_{\gamma_h} = \prod_m \Phi((-y)^m X_{\gamma_h})^{a_m(\gamma_h)} \tag{4.8}$$

and $\Phi(X) = \prod_{k=1}^{\infty}(1 + y^{2k-1}X)^{-1}$ is the quantum dilogarithm.

5. Two Examples

5.1. Example 1: Wall Crossing For Smooth Monopoles. We consider the case when $\mathfrak{g}$ has rank 2, so there are two simple coroots, and we take the magnetic charge to be $\gamma_m = H_1 + H_2$ in this example. What follows is a reinterpretation of the discussion of [**13**]. In this case the strongly centered moduli space $\mathcal{M}_0$ is the Taub-NUT metric. We can represent it as projection $\pi : \mathcal{M}_0 \to \mathbb{R}^3$ where the fiber is a circle, parametrized by $\psi \sim \psi + 2\pi$, except for the fiber over the origin, which is just a point. Splitting $\mathrm{G}(\mathcal{Y})$ to extract the component along $\mathcal{M}_0$ yields

$$\mathrm{G}(\mathcal{Y}_0) = \frac{1}{p}(H_1, \mathcal{Y}_0)\frac{\partial}{\partial\psi} \tag{5.1}$$

where $p = 1, 2, 3$ for $\mathfrak{g} = \mathfrak{su}(3)$, $\mathfrak{so}(5)$, $\mathfrak{g}_2$, respectively and we take α_1 to be the long root. Electric charges of BPS states are of the form

$$\gamma^{\mathrm{e}} = n_{\mathrm{e}}^1\alpha_1 + n_{\mathrm{e}}^2\alpha_2 \tag{5.2}$$

with $n_{\mathrm{e}}^1, n_{\mathrm{e}}^2 \in \mathbb{Z}$. Now $\mathfrak{t}_0$ is one-dimensional and the projection in (3.21) is

$$\gamma_0^{\mathrm{e}} = (pn_{\mathrm{e}}^1 - n_{\mathrm{e}}^2)\cdot\left(\frac{m_2\alpha_1 - m_1 p\alpha_2}{p(m_1 + m_2)}\right) \tag{5.3}$$

where $m_i := (H_i, \mathcal{X})$. Define $N_{\mathrm{e},0} := pn_{\mathrm{e}}^1 - n_{\mathrm{e}}^2$. Then, for fixed γ_0^{e} the preimage of compatible electric charges is the $\mathbb{Z}$ torsor with fixed value of $N_{\mathrm{e},0}$, that is the set of electric charges $\gamma^{\mathrm{e}} = n\alpha_1 + (np - N_{\mathrm{e},0})\alpha_2$ with $n \in \mathbb{Z}$. Elements of $\mathcal{K}_0$ that can be used to build states of definite electric charge are equivariant in the ψ direction: $L_{\frac{\partial}{\partial\psi}}\Psi_0 = -\mathrm{i}N_{\mathrm{e},0}\Psi_0$. The non-Fredholm condition is now based on a pair of electromagnetic charges $\gamma_i = H_i \oplus n_{\mathrm{e}}^i\alpha_i$, $i = 1, 2$ with $\langle\!\langle\gamma_1, \gamma_2\rangle\!\rangle = -N_{\mathrm{e},0} \neq 0$. The spaces $\mathcal{H}_0(\gamma_i;\mathcal{X},\mathcal{Y})$ correspond to the case $\mathcal{M}_0 = \mathrm{pt}$ and are one-dimensional so the non-Fredholm criterion is satisfied. The jump in the L^2 kernel can be verified by direct computation in this case since the determination of the kernel of the Dirac-like operator $\not{D}_0^{\mathrm{Y}}$ on $\mathcal{M}_0$ is a computation going back to [**26**]. The result is illustrated in Figure 1 and is simply the result predicted by the primitive wall-crossing formula of [**7, 8**], which suffices in this case since the γ_i are primitive. Moreover, the boundstate wavefunction, as a function of the (suitably rescaled) radial direction r in $\mathbb{R}^3$, is proportional to $r^{(|N_{\mathrm{e},0}|-1)/2}e^{-|2C+N_{\mathrm{e},0}|r}$ where $\mathrm{G}(\mathcal{Y}_0) = \frac{2\mu}{p^2}C\frac{\partial}{\partial\psi}$, $\mu^{-1} = m_1^{-1} + m_2^{-1}$, and the constant C is proportional to the coordinate y_1 in Figure 1. The wavefunction is peaked at $r_{\mathrm{ext}} = |N_{\mathrm{e},0}|/|2C + N_{\mathrm{e},0}|$. This turns out to be precisely the bound-state radius computed from explicit BPS field configurations by Denef [**6**].

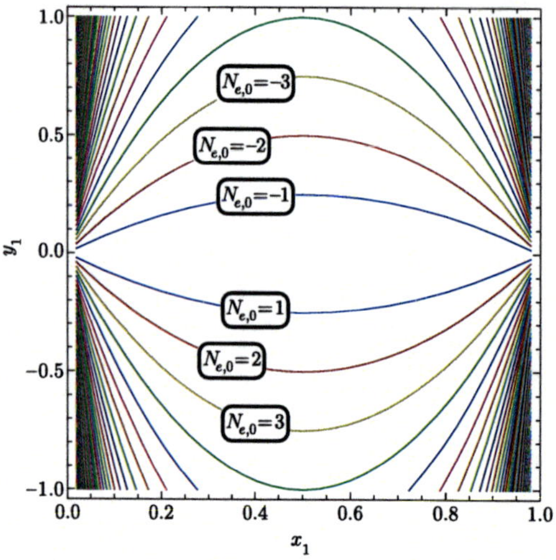

FIGURE 1. Walls of marginal stability for $\mathfrak{g} = \mathfrak{su}(3)$, $\mathfrak{so}(5)$, $\mathfrak{g}_2$, for $p = 1, 2, 3$ respectively. We choose axes to be the scale-invariant quantities $x_1 := (H_1, \mathcal{X})/(\gamma_{\rm m}, \mathcal{X})$ and $y_1 := p(H_1, \mathcal{Y}_0)/(\gamma_{\rm m}, \mathcal{X})$. The chamber containing the line $y_1 = 0$ has no BPS states. Crossing a wall labeled by $N_{\rm e,0}$ in the direction away from the central chamber adds a tower of BPS states with electric charge $\gamma^{\rm e} = n\alpha_1 + (pn - N_{\rm e,0})\alpha_2$ with $n \in \mathbb{Z}$ to the spectrum. The $\mathfrak{so}(3)_{\rm rot}$ representation $\mathcal{H}_0(\gamma; \mathcal{X}, \mathcal{Y})$ gains a spin j representation with $j = \frac{1}{2}(|N_{\rm e,0}| - 1)$. At the walls $x \to 0, 1$ the classical mass of the monopoles goes to zero and the semiclassical analysis is not necessarily valid.

5.2. Example 2: Wall-Crossing For Singular Monopoles. We now consider $\mathfrak{g} = \mathfrak{su}(2)$, but consider a singular monopole with $P = \frac{p}{2} H_\alpha$. We can take $p > 0$. Splitting $\gamma = \gamma_{\rm m} \oplus \gamma^{\rm e}$ with $\gamma_{\rm m} = (\tilde{n}_{\rm m} - \frac{1}{2}p)H_\alpha$, we see that $\overline{\underline{\Omega}}(P, \gamma; \mathcal{X}, \mathcal{Y})$ is the $\mathfrak{so}(3)$ character of the Dirac operator on a moduli space of singular monopoles of dimension $4\tilde{n}_{\rm m}$. The spectrum of the ordinary BPS states is well-known [**28**]: With our choice $p > 0$ the only relevant spaces are $\mathcal{H}_0(\gamma_n; \mathcal{X}, \mathcal{Y}) \cong \mathbb{C}$ with $\gamma_n = H_\alpha + n\alpha$. We can thus divide up the $\mathcal{Y}$-line into chambers separated by walls $W_n = W(\gamma_n)$. In the chamber between walls W_n and W_{n+1} the generating function (4.7) is given by

$$F_n = \left[X_1^{-1/2} X_2^{-n/2} \left(U_n(f_n) - X_2^{-1/2} U_{n-1}(f_n) \right) \right]^p \tag{5.4}$$

where $X_1 = X_{H_\alpha}$ and $X_2 = X_\alpha$ generate the Heisenberg extension of $\mathbb{C}[\Gamma]$,

$$f_n := \frac{1}{2} \left(X_2^{1/2} + X_2^{-1/2} \left(1 + y^{2n+3} X_1 X_2^{n+1} \right) \right), \tag{5.5}$$

and $U_n(\cos\theta) = \sin((n+1)\theta)/\sin\theta$. The result (5.4) is proven in **[25]** by careful application of the wall crossing formula for $F([P]; \mathcal{X}, \mathcal{Y})$ and generalizes the result in **[10]** at $y = 1$. Equation (5.4) makes some fairly nontrivial predictions for L^2-kernels of Dirac operators on singular monopole moduli spaces, some of which are spelled out in detail in **[25]**.

References

[1] Michael Atiyah and Nigel Hitchin, *The geometry and dynamics of magnetic monopoles*, M. B. Porter Lectures, Princeton University Press, Princeton, NJ, 1988. MR934202

[2] T. D. Brennan and G. W. Moore, "A Note On The Semiclassical Formulation Of BPS States In Four-Dimensional N=2 Theories," arXiv:1610.00697 [hep-th].

[3] Wu-yen Chuang, Duiliu-Emanuel Diaconescu, Jan Manschot, Gregory W. Moore, and Yan Soibelman, *Geometric engineering of (framed) BPS states*, Adv. Theor. Math. Phys. **18** (2014), no. 5, 1063–1231. MR3281276

[4] C. Cordova and T. Dumitrescu, to appear.

[5] M. Del Zotto and A. Sen, "About the Absence of Exotics and the Coulomb Branch Formula," arXiv:1409.5442 [hep-th].

[6] Frederik Denef, *Quantum quivers and Hall/hole halos*, J. High Energy Phys. **10** (2002), 023, 42, DOI 10.1088/1126-6708/2002/10/023. MR1952307

[7] Frederik Denef and Gregory W. Moore, *Split states, entropy enigmas, holes and halos*, J. High Energy Phys. **11** (2011), 129, i, 152, DOI 10.1007/JHEP11(2011)129. MR2913216

[8] Duiliu-Emanuel Diaconescu and Gregory W. Moore, *Crossing the wall: branes versus bundles*, Adv. Theor. Math. Phys. **14** (2010), no. 6, 1621–1650. MR2872467

[9] S. K. Donaldson, *Nahm's equations and the classification of monopoles*, Comm. Math. Phys. **96** (1984), no. 3, 387–407. MR769355

[10] Davide Gaiotto, Gregory W. Moore, and Andrew Neitzke, *Framed BPS states*, Adv. Theor. Math. Phys. **17** (2013), no. 2, 241–397. MR3250763

[11] Jerome P. Gauntlett, *Low-energy dynamics of $N = 2$ supersymmetric monopoles*, Nuclear Phys. B **411** (1994), no. 2-3, 443–460, DOI 10.1016/0550-3213(94)90457-X. MR1257834

[12] Jerome P. Gauntlett and Jeffrey A. Harvey, *S-duality and the dyon spectrum in $N = 2$ super Yang-Mills theory*, Nuclear Phys. B **463** (1996), no. 2-3, 287–314, DOI 10.1016/0550-3213(96)00035-1. MR1393642

[13] Jerome P. Gauntlett, Nakwoo Kim, Jaemo Park, and Piljin Yi, *Monopole dynamics and BPS dyons in $N = 2$ super-Yang-Mills theories*, Phys. Rev. D (3) **61** (2000), no. 12, 125012, 9, DOI 10.1103/PhysRevD.61.125012. MR1789909

[14] G. W. Gibbons and N. S. Manton, *The moduli space metric for well-separated BPS monopoles*, Phys. Lett. B **356** (1995), no. 1, 32–38, DOI 10.1016/0370-2693(95)00813-Z. MR1346718

[15] Nigel Hitchin, *Harmonic spinors*, Advances in Math. **14** (1974), 1–55, DOI 10.1016/0001-8708(74)90021-8. MR0358873

[16] N. J. Hitchin, N. S. Manton, and M. K. Murray, *Symmetric monopoles*, Nonlinearity **8** (1995), no. 5, 661–692. MR1355037

[17] Jacques Hurtubise, *The classification of monopoles for the classical groups*, Comm. Math. Phys. **120** (1989), no. 4, 613–641. MR987771

[18] Stuart Jarvis, *Construction of Euclidean monopoles*, Proc. London Math. Soc. (3) **77** (1998), no. 1, 193–214, DOI 10.1112/S0024611598000446. MR1625471

[19] Anton Kapustin and Edward Witten, *Electric-magnetic duality and the geometric Langlands program*, Commun. Number Theory Phys. **1** (2007), no. 1, 1–236, DOI 10.4310/CNTP.2007.v1.n1.a1. MR2306566

[20] Maxim Kontsevich and Yan Soibelman, *Motivic Donaldson-Thomas invariants: summary of results*, Mirror symmetry and tropical geometry, Contemp. Math., vol. 527, Amer. Math. Soc., Providence, RI, 2010, pp. 55–89, DOI 10.1090/conm/527/10400. MR2681792

[21] N. S. Manton, "A Remark on the Scattering of BPS Monopoles," Phys. Lett. B **110**, 54 (1982). doi:10.1016/0370-2693(82)90950-9

[22] N. S. Manton and B. J. Schroers, *Bundles over moduli spaces and the quantisation of BPS monopoles*, Ann. Physics **225** (1993), no. 2, 290–338, DOI 10.1006/aphy.1993.1060. MR1226321

[23] G. W. Moore, A. B. Royston and D. V. d. Bleeken, "Parameter counting for singular monopoles on $\mathbb{R}^3$," JHEP **1410**, 142 (2014) [arXiv:1404.5616 [hep-th]].

[24] Gregory W. Moore, Andrew B. Royston, and Dieter Van den Bleeken, *Brane bending and monopole moduli*, J. High Energy Phys. **10** (2014), 157, front matter+48. MR3325159

[25] Gregory W. Moore, Andrew B. Royston, and Dieter Van den Bleeken, *Semiclassical framed BPS states*, J. High Energy Phys. **7** (2016), 071, front matter+234. MR3547347

[26] C. N. Pope, *Axial-vector anomalies and the index theorem in charged Schwarzschild and Taub-NUT spaces*, Nuclear Phys. B **141** (1978), no. 4, 432–444, DOI 10.1016/0550-3213(78)90038-X. MR508482

[27] Graeme Segal and Alex Selby, *The cohomology of the space of magnetic monopoles*, Comm. Math. Phys. **177** (1996), no. 3, 775–787. MR1385085

[28] N. Seiberg and E. Witten, *Erratum: "Electric-magnetic duality, monopole condensation, and confinement in $N = 2$ supersymmetric Yang-Mills theory"*, Nuclear Phys. B **430** (1994), no. 2, 485–486, DOI 10.1016/0550-3213(94)00449-8. MR1303306

[29] Ashoke Sen, *Dyon-monopole bound states, self-dual harmonic forms on the multi-monopole moduli space, and* $\mathrm{SL}(2,\mathbf{Z})$ *invariance in string theory*, Phys. Lett. B **329** (1994), no. 2-3, 217–221, DOI 10.1016/0370-2693(94)90763-3. MR1281578

[30] Mark Stern and Piljin Yi, *Counting Yang-Mills dyons with index theorems*, Phys. Rev. D (3) **62** (2000), no. 12, 125006, 15, DOI 10.1103/PhysRevD.62.125006. MR1814370

[31] Clifford Henry Taubes, *The existence of multimonopole solutions to the nonabelian, Yang-Mills-Higgs equations for arbitrary simple gauge groups*, Comm. Math. Phys. **80** (1981), no. 3, 343–367. MR626705

[32] E. J. Weinberg, "Parameter Counting for Multi-Monopole Solutions," Phys. Rev. D **20**, 936 (1979). doi:10.1103/PhysRevD.20.936

[33] E. J. Weinberg, "Fundamental Monopoles and Multi-Monopole Solutions for Arbitrary Simple Gauge Groups," Nucl. Phys. B **167**, 500 (1980). doi:10.1016/0550-3213(80)90245-X

[34] Erick J. Weinberg and Piljin Yi, *Magnetic monopole dynamics, supersymmetry, and duality*, Phys. Rep. **438** (2007), no. 2-4, 65–236, DOI 10.1016/j.physrep.2006.11.002. MR2347368

[35] E. Witten and D. I. Olive, "Supersymmetry Algebras That Include Topological Charges," Phys. Lett. B **78**, 97 (1978).

NHETC and Department of Physics and Astronomy, Rutgers University, Piscataway, New Jersey 08855, USA
E-mail address: `gmoore@physics.rutgers.edu`

George P. & Cynthia Woods Mitchell Institute for Fundamental Physics and Astronomy, Texas A&M University, College Station, Texas 77843 USA
E-mail address: `aroyston@physics.tamu.edu`

Physics Department, Boğaziçi University, 34342 Bebek / Istanbul, Turkey
E-mail address: `dieter.van@boun.edu.tr`

$\mathfrak{BPS}/\mathfrak{CFT}$ correspondence: Instantons at crossroads and gauge origami

Nikita Nekrasov

ABSTRACT. Gieseker-Nakajima moduli spaces $\mathcal{M}_k(n)$ parametrize the charge k noncommutative $U(n)$ instantons on $\mathbb{R}^4$ and framed rank n torsion free sheaves $\mathcal{E}$ on $\mathbb{CP}^2$ with $\mathrm{ch}_2(\mathcal{E}) = k$. We define a generalization $\mathfrak{M}_k(\vec{n})$ of $\mathcal{M}_k(n)$, the moduli space of charge k (noncommutative) instantons on origami spacetimes: a union X of (up to six) coordinate complex planes $\mathbb{C}^2$ intersecting in $\mathbb{C}^4$, the instantons of a collection of four dimensional gauge theories sewn along two dimensional defect surfaces and defect points. We also define several quiver versions $\mathfrak{M}^{\gamma}_{\mathbf{k}}(\underline{\vec{\mathbf{n}}})$ of $\mathfrak{M}_k(\vec{n})$, motivated by the considerations of sewn gauge theories on orbifolds $\mathbb{C}^4/\Gamma$.

The geometry of the spaces $\mathfrak{M}^{\gamma}_{\mathbf{k}}(\underline{\vec{\mathbf{n}}})$, more specifically the compactness of the set of torus-fixed points, for various tori, underlies the non-perturbative Dyson-Schwinger identities recently found to be satisfied by the correlation functions of qq-characters viewed as local gauge invariant operators in the $\mathcal{N} = 2$ quiver gauge theories.

The cohomological and K-theoretic operations defined using $\mathfrak{M}_k(\vec{n})$ and their quiver versions as correspondences provide the geometric counterpart of the qq-characters, line and surface defects.

CONTENTS

2010 *Mathematics Subject Classification.* Primary 81T30, 81T13, 81T75, 81T60, 81T70, 57R91.

1. Introduction

Recently [**33**] we introduced a set of observables in quiver $\mathcal{N} = 2$ supersymmetric gauge theories which are useful in organizing the non-perturbative Dyson-Schwinger equations, relating contributions of different instanton sectors to the expectation values of gauge invariant chiral ring observables. In this paper we shall provide the natural geometric setting for these observables. We also explain the gauge and string theory motivations for these considerations.

Notations. In our story we explore certain moduli spaces, which roughly parametrize the sheaves $\mathcal{E}$ supported on a union of coordinate complex two-planes $\approx \mathbb{C}^2$ inside $\mathbb{C}^4$ which is a local model of Calabi-Yau fourfold Z containing a possibly singular surface $X \subset Z$:

$$Z^{loc} = \mathbb{C}^4, \qquad X^{loc} = \bigcup_{A \in \underline{\mathbf{6}}} \mathbb{C}^2_A, \qquad \mathrm{supp}(\mathcal{E}) = \bigcup_{A \in \underline{\mathbf{6}}} n_A \mathbb{C}^2_A \tag{1}$$

We denote by $\underline{\mathbf{4}}$ the set of complex coordinates in $\mathbb{C}^4$:

$$\underline{\mathbf{4}} = \{1,2,3,4\}, \qquad a \in \underline{\mathbf{4}} \leftrightarrow z_a \in \mathbb{C} \tag{2}$$

and by

$$\underline{\mathbf{6}} = \binom{\underline{\mathbf{4}}}{2} = \left\{ \{1,2\},\{1,3\},\{1,4\},\{2,3\},\{2,4\},\{3,4\} \right\} \tag{3}$$

the set of two-element subsets of $\underline{\mathbf{4}}$, i.e. the set of coordinate two-planes in $\mathbb{C}^4$.

We shall sometimes denote the elements of $\underline{\mathbf{6}}$ by the pairs $ab = ba \leftrightarrow \{a, b\} \in \underline{\mathbf{6}}$. We also define, for $A \in \underline{\mathbf{6}}$, $\bar{A} = \underline{\mathbf{4}} \backslash A$, and

$$\epsilon(A) = \varepsilon_{abcd}, \qquad A = \{a,b\},\ \bar{A} = \{c,d\}, \qquad a < b,\ c < d \tag{4}$$

so that, e.g. $\overline{12} = 34$, $\epsilon(23) = \varepsilon_{2314} = 1$, $\epsilon(24) = \varepsilon_{2413} = -1$.

The two-plane $\mathbb{C}^2_A \subset \mathbb{C}^4$ corresponding to $A \in \underline{\mathbf{6}}$ is defined by the equations: $z_{\bar{a}} = 0$, for all $\bar{a} \in \bar{A}$.

We denote by $\underline{\mathbf{3}}$ the quotient $\underline{\mathbf{6}}/\mathbb{Z}_2$ where $\mathbb{Z}_2$ acts by the involution $A \mapsto \bar{A}$. The elements $\mathsf{a} \in \underline{\mathbf{3}}$ are the unordered pairs $(A, \bar{A})$. It will be convenient to visualize the sets $\underline{\mathbf{3}}$, $\underline{\mathbf{4}}$, $\underline{\mathbf{6}}$, using the tetrahedron:

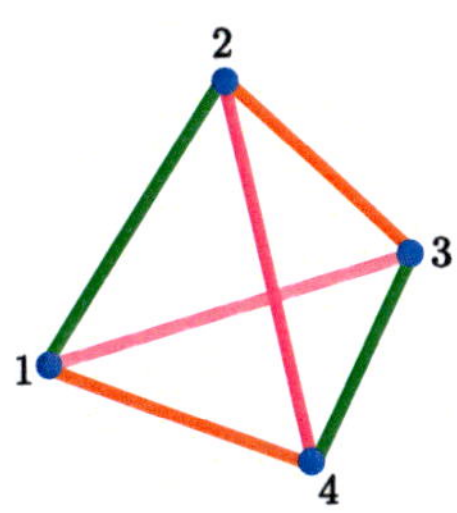

Figure 1: Tetrahedron with the sets $\underline{\mathbf{4}}$ and $\underline{\mathbf{6}}$ of vertices and edges, and the set $\underline{\mathbf{3}}$ = {red, green, orange} of crossed edges

Our story will involve four complex parameters $q_a \in \mathbb{C}^\times$, $a \in \underline{\mathbf{4}}$, obeying

$$\prod_{a \in \underline{\mathbf{4}}} q_a = 1 \tag{5}$$

We shall also use the additive variables $\mathfrak{e}_a \in \mathbb{C}$, $a \in \underline{\mathbf{4}}$, obeying

$$\sum_{a \in \underline{\mathbf{4}}} \mathfrak{e}_a = 0 \tag{6}$$

Define the lattice

$$Z_{\underline{\mathfrak{e}}} = \mathbb{Z}\mathfrak{e}_1 + \mathbb{Z}\mathfrak{e}_2 + \mathbb{Z}\mathfrak{e}_3 \subset \mathbb{C} \tag{7}$$

which is the image of the projection:

$$\mathbb{Z}^4 \to \mathbb{C}\,, \qquad (i,j,k,l) \mapsto \mathfrak{e}_1 i + \mathfrak{e}_2 j + \mathfrak{e}_3 k + \mathfrak{e}_4 l \tag{8}$$

We shall use the following functions on $\underline{\mathbf{6}}$:

$$\begin{aligned} p_A &= \prod_{a \in A} (1 - q_a)\ , \\ q_A &= \prod_{a \in A} q_a\ , \\ \mathfrak{e}_A &= \sum_{a \in A} \mathfrak{e}_a = -\mathfrak{e}_{\bar{A}}\,. \end{aligned} \tag{9}$$

In what follows we denote by $[n]$, for $n \in \mathbb{Z}_{>0}$, the set $\{1, 2, 3, \ldots, n\} \subset \mathbb{Z}_{>0}$.

Let S be a finite set, and $(V_s)_{s \in S}$ a collection of vector spaces. We use the notation

$$\sum_{s \in S} V_s \tag{10}$$

for the vector space which consists of all linear combinations

$$\sum_{s \in S} \psi_s\ , \qquad \psi_s \in V_s\ . \tag{11}$$

1.1. Organization of the paper. We review the gauge and string theory motivation in the section **2**. The moduli space $\mathfrak{M}_k(\vec{n})$ of spiked instantons is introduced in the section **3**. The symmetries of spiked instantons are studied in the section **4**. The moduli space of ordinary $U(n)$ instantons on (noncommutative) $\mathbb{R}^4$ is reviewed in section **5**. The section **6** discusses in more detail two particular cases of spiked instantons, the crossed instantons and the folded instantons. The crossed instantons live on two four-dimensional manifolds transversely intersecting in the eight-dimensional ambient manifold (actually, a Calabi-Yau fourfold), the folded ones live on two four-dimensional manifolds intersecting transversely in the six dimensional ambient manifold. The section **7** constructs the spiked instantons out of the ordinary ones, and studies the toric spiked instantons in some detail. The section **8** is the main result of this paper: the compactness theorem. In section **9** we enter the theory of integration over the spiked and crossed instantons, and relate the analyticity of the partition functions to the compactness theorem. The section **10** discusses the quiver generalizations of crossed instantons. The section **11** is devoted to future directions and open questions.

2. Gauge and string theory motivations

2.1. Generalized gauge theory. Below we introduce the moduli spaces $\mathfrak{M}_{X,G}$ of what might be called supersymmetric gauge fields in the generalized gauge theories, whose space-time X contains several, possibly intersecting, components: see Fig. 2. We call such X the *origami worldvolume*. The gauge groups $G|_{X_A} = G_A$

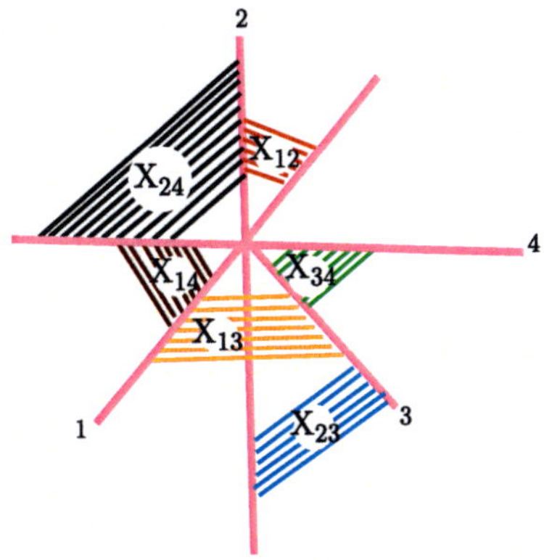

Figure 2: The origami wolrdvolume $X = \bigcup_A X_A$

on different components may be different. The intersections $X_A \cup X_B$ lead to the bi-fundamental matter fields charged under $G_A \times G_B$. The arrangement is motivated by the string theory considerations, where the open string Hilbert space, in the presence of several D-branes, splits into sectors labelled by the boundary conditions. It is well-known [**10, 35**] that some features of the open string theory are captured by the noncommutative gauge theory.

2.2. Gauge origami. We are interested in integrals over the moduli spaces $\mathfrak{M}_{X,G}$. We shall view $\mathfrak{M}_{X,G}$ as the "space, defined by some equations modulo symmetry". More formally, $\mathfrak{M}_{X,G}$ is the quotient of a set of zeroes of some $\mathsf{G_g}$-equivariant section $s : M \to V$ of $\mathsf{G_g}$-equivariant vector bundle $V \to M$ over some smooth space (vector space in our case) with $\mathsf{G_g}$-action, with some Lie group $\mathsf{G_g}$. If M is compact the integral over $\mathfrak{M}_{X,G}$ of a closed differential form can be represented by the $\mathsf{G_g}$-equivariant integral over M of the pull-back of the corresponding form times the Euler class of V. In the non-compact case one uses equivariant cohomology (mass deformation, in the physics language) with respect to both $\mathsf{G_g}$ and some global symmetry group H, and Mathai-Quillen representatives of the Euler class.

The resulting partition functions

$$\mathcal{Z}_{X,G}(\xi) \sim \int_{\mathfrak{M}_{X,G}}^{\mathsf{H}_{\text{equiv}}} 1 \sim \frac{1}{\mathrm{Vol}(\mathsf{G_g})} \int_{\mathrm{Lie}(\mathsf{G_g})} \int_M^{(\mathsf{G_g}\times\mathsf{H})_{\text{equiv}}} \mathrm{Euler}(V) \tag{12}$$

are functions on the Lie algebra of $\mathsf{H_C}$, $\xi \in \mathrm{Lie}(\mathsf{H_C})$. The analytic properties of $\mathcal{Z}_{X,G}(\xi)$ reflect some of the geometric and topological features of $\mathfrak{M}_{X,G}$. They are the main focus of this paper.

The equivariant localization expresses $\mathcal{Z}_{X,G}$ as the sum over the fixed points of H-action, which are typically labelled by multiple partitions, i.e. collections of

Young diagrams [**31**]. The resulting statistical mechanical model is called the gauge origami and is studied in detail in the companion paper [**27**].

2.3. Symmetries, twisting, equivariance. The partition functions $\mathcal{Z}_{X,G}(\xi)$ are analytic functions of $\xi \in \mathrm{Lie}(\mathsf{H}_{\mathbb{C}})$, with possible singularities. Given $\xi \in \mathrm{Lie}(\mathsf{H}_{\mathbb{C}})$, the closure of the subgroup $\exp t\xi$, $t \in \mathbb{C}$ defines a torus T_ξ. The partition function $\mathcal{Z}_{X,G}(\xi)$ can be computed, by Atiyah-Bott fixed point formula, as a sum over the T_ξ-fixed points. Even though the moduli space $\mathfrak{M}_{X,G}$ may be noncompact (it is noncompact for noncompact X), the fixed point set, for suitable ξ, may still be compact, so that the integrals over $\mathfrak{M}_{X,G}$ of the equivariant differential forms converge. The set $\mathfrak{M}^{\mathsf{T}_\xi}_{X,G}$ of T_ξ-fixed points may have several connected components:

$$\mathfrak{M}^{\mathsf{T}_\xi}_{X,G} = \bigcup_f \left(\mathfrak{M}^{\mathsf{T}_\xi}_{X,G}\right)_f \tag{13}$$

The contributions $\mathcal{Z}_f$ of $\left(\mathfrak{M}^{\mathsf{T}_\xi}_{X,G}\right)_f$ are rational functions on $\mathrm{Lie}(\mathsf{H}_{\mathbb{C}})$, they have poles. In the nice situations the component $\left(\mathfrak{M}^{\mathsf{T}_\xi}_{X,G}\right)_f$ has a normal bundle in $\mathfrak{M}_{X,G}$ (or in the ambient smooth variety, as in the case of the obstructed theory), N_f, which inherits an action of T_ξ, and decomposes into the sum of complex line bundles (real rank two bundles) $L_{f,w}$, with w going through the set of T_ξ-weights. The fixed point formula states

$$\mathcal{Z}_f = \int_{\left(\mathfrak{M}^{\mathsf{T}_\xi}_{X,G}\right)_f} \frac{\mathrm{Euler}_\xi(\mathrm{Obs}_f)}{\prod_w \left(w(\xi) + c_1(L_{f,w})\right)} \tag{14}$$

The poles in $\mathcal{Z}_f$ occur when the Lie algebra element ξ crosses the hyperplane $w(\xi) = 0$ for some w occuring in the decomposition of N_f. Geometrically this means that the ξ belongs to a subalgebra of $\mathrm{Lie}\mathsf{T} \subset \mathrm{Lie}(\mathsf{H}_{\mathbb{C}})$ which fixes not only $\left(\mathfrak{M}^{\mathsf{T}_\xi}_{X,G}\right)_f$, but also (at least infinitesimally, at the linearized order) a two-dimensional surface passing through f, in the direction of $L_{f,w}$.

We shall be interested in the analytic properties of $\mathcal{Z}_{X,G}$ and one of the questions we shall be concerned with is whether the poles in $\mathcal{Z}_f$ are cancelled by the poles in the contribution of some other component $\left(\mathfrak{M}^{\mathsf{T}_\xi}_{X,G}\right)_{f'}$ of the fixed point set. More precisely, once $\xi \to \xi_c$ where ξ_c belongs to the hyperplane $w(\xi) = 0$ defined relative to the weight decomposition of N_f, the component of the fixed point set may enhance,

$$\left(\mathfrak{M}^{\mathsf{T}_\xi}_{X,G}\right)_f \subset \left(\mathfrak{M}^{\mathsf{T}_{\xi_c}}_{X,G}\right)_{f''}, \tag{15}$$

reaching out to the other component $\left(\mathfrak{M}^{\mathsf{T}_\xi}_{X,G}\right)_{f'}$

$$\left(\mathfrak{M}^{\mathsf{T}_\xi}_{X,G}\right)_{f'} \cap \left(\mathfrak{M}^{\mathsf{T}_{\xi_c}}_{X,G}\right)_{f''} \neq \emptyset \tag{16}$$

If the enhanced component $\left(\mathfrak{M}^{\mathsf{T}_{\xi_c}}_{X,G}\right)_{f''}$ is compact, then the pole at $\xi = \xi_c$ in $\mathcal{Z}_f$ will be cancelled by the pole in $\mathcal{Z}_{f'}$.

So the issue in question is the compactness of the fixed point set for the torus generated by the non-generic infinitesimal symmetries ξ_c.

In our case we shall choose a class of subgroups $\mathfrak{H} \subset \mathsf{H}$. We shall show that the set of $\mathfrak{H}$-fixed points is compact. It means that for generic choice of $\widehat{\xi} \in \mathrm{Lie}(\mathfrak{H}_{\mathbb{C}})$ the partition function $\mathcal{Z}_{X,G}(\widehat{\xi} + \mathbf{x})$ as a function of $\mathbf{x} \in \mathrm{Lie}(\mathfrak{H}_{\mathbb{C}})^{\perp} \subset \mathrm{Lie}(\mathsf{H})$ has no singularities.

The procedure of restricting the symmetry group of the physical system to a subgroup is well-known to physicists under the name of twisting [**39**]. It is used in the context of topological field theories, which are obtained from the supersymmetric field theories having an R-symmetry group H_R such that the group of rotations $\mathsf{G}_{\mathrm{rot}}$ of flat spacetime can be embedded nontrivially into the direct product

$$\mathsf{G}_{\mathrm{rot}} \longrightarrow \mathsf{G}_{\mathrm{rot}} \times \mathsf{H}_R \ . \tag{17}$$

We shall encounter a lot of instances of the procedure analogous to (17) in what follows.

2.4. Gauge theories on stacks of D-branes. The maximally supersymmetric Yang-Mills theory in $p+1$-dimensions models [**40**] the low energy behavior of a stack of parallel Dp-branes. This description can be made p-blind by turning on a background constant B-field. In the strong B-field the "non-abelian Born-Infeld/Yang-Mills" theory description of the low energy physics of the open strings connecting the Dp-branes crosses over to the noncommutative Yang-Mills description [**35**]. In this paper we shall use the noncommutative Yang-Mills to study the dynamics of intersecting stacks of Dp-branes.

2.4.1. *The Matrix models.* Recall the dimensional reductions of the maximally supersymmetric Yang-Mills theory down to $0+0$, $0+1$ and $1+1$ dimensions [**2**], [**17**], [**9**]. We take the gauge group to be $\mathsf{G}_{\mathbf{g}} = U(\mathbf{N})$ for some large $\mathbf{N}$. Following [**21**] we shall view the model of [**17**]

$$\frac{1}{\mathrm{Vol}\mathsf{G}_{\mathbf{g}}} \int_{\mathbb{R}^{10|16}\otimes \mathrm{Lie}\mathsf{G}_{\mathbf{g}}} DX^m D\theta^\alpha \exp \left(-\frac{1}{4} \sum_{m<n} \mathsf{Tr}\, [X^m, X^n]^2 - \frac{1}{2} \sum_{m,\alpha,\beta} \Gamma^m_{\alpha\beta} \mathsf{Tr}\, \theta^\alpha [X^m, \theta^\beta] \right) \tag{18}$$

with the adjoint bosons X^m and the adjoint fermions θ^α transforming in the representations **10** and **16** of Spin(10), respectively, as the cohomological field theory in 0 dimensions, while [**2**] and [**9**] are obtained by the lift procedure of [**3**].

The approach to the noncommutative gauge theory in which the gauge field A_m is traded for the (infinite) matrix X^m is used in the background-independent (p-uniform) formalism of [**34**].

2.4.2. $(0,1)$-*formalism,* Spin(7)-*instantons.* Let us start in the $1+1$-dimensional case. Let Σ be the worldsheet of our theory, with the local complex coordinates $z, \bar{z}$. The theory has a gauge field $A = A_z dz + A_{\bar{z}} d\bar{z}$, 8 Hermitian adjoint scalars X^m, and 16 adjoint fermions, which split into 8 right ψ^a_+, and 8 left $\chi^{\dot{a}}_-$ ones. Here m, a and $\dot{a}$ are the indices of the $\mathbf{8}_v$, $\mathbf{8}_s$ and $\mathbf{8}_c$ representations of the global symmetry group Spin(8), respectively. The formalism that we shall adopt singles out a particular spinor among $\mathbf{8}_c$. The isotropy subgroup Spin(7) of that spinor has the following significance. The representation $\mathbf{8}_c$, under Spin(7), decomposes as $\mathbf{7} \oplus \mathbf{1}$, the $\mathbf{1}$ being the invariant subspace. Accordingly, we split $\left(\chi^{\dot{a}}_-\right)_{\dot{a}\in\mathbf{8}_c} \longrightarrow \left(\chi^i_-\right)_{i\in\mathbf{7}} \oplus \eta_-$. The representations $\mathbf{8}_s$ and $\mathbf{8}_v$ become the spinor $\mathbf{8}$ of

Spin(7). We shall also need the auxiliary fields h^i, which are the worldsheet scalars, transform in the adjoint of the gauge group, and in the representation **7** of Spin(7). The theory, in this formalism, has one supercharge δ_+, which squares to the chiral translation on the worldsheet

$$\delta_+^2 = D_{++} = \bar{\partial}_{\bar{z}} + A_{\bar{z}} \tag{19}$$

(the conjugate derivative $D_{--} = \partial_z + A_z$) which acts on the fields of the model as follows:

$$\begin{aligned} \delta_+ X^m &= \psi_+^m, & \delta_+ \psi_+^m &= D_{++} X^m = \bar{\partial}_{\bar{z}} X^m + [A_{\bar{z}}, X^m] \\ \delta_+ \chi_-^i &= h^i, & \delta_+ h^i &= D_{++} \chi_-^i = \bar{\partial}_{\bar{z}} \chi_-^i + [A_{\bar{z}}, \chi_-^i] \\ \delta_+ A_z &= \eta_-, & \delta_+ \eta_- &= F_{z\bar{z}} \end{aligned} \tag{20}$$

The Lagrangian

$$L = \delta_+ \int_\Sigma \mathrm{Tr}\, \left(\psi_+^m D_{--} X^m + \chi_-^i \left(\mathrm{i}\wp_{mn}^i [X^m, X^n] - h^i\right) + \eta_- F_{z\bar{z}}\right) \tag{21}$$

becomes that of the standard $\mathcal{N} = 8$ supersymmetric Yang-Mills once the auxiliary fields h^i are eliminated by their equations of motion. Here $\wp_{mn}^i$ is the matrix of the projection $\wp_7 : \Lambda^2 \mathbf{8}_v = \mathrm{LieSpin}(8) \longrightarrow \mathbf{7}$ onto the orthogonal complement to $\mathrm{LieSpin}(7) \subset \mathrm{LieSpin}(8) = \mathrm{LieSpin}(7) \oplus \mathbf{7}$.

Upon the dimensional reduction to 0 dimensions, the gauge field A becomes a complex scalar $\sigma = A_{\bar{z}}$ and its conjugate $\bar{\sigma} = A_z$.

2.4.3. $(0,2)$-*formalism,* SU(4)-*instantons.* In this formalism we have two supercharges $Q_+, \bar{Q}_+$, obeying

$$Q_+^2 = \bar{Q}_+^2 = 0\,, \qquad Q_+ \bar{Q}_+ + \bar{Q}_+ Q_+ = D_{++} \tag{22}$$

so that $\delta_+ = Q_+ + \bar{Q}_+$. We split 8 Hermitian adjoint scalars X^m into 4 complex adjoint scalars Z^a, $a \in \underline{\mathbf{4}}$, and their conjugates $\bar{Z}^{\bar{a}}$, and the same for the fermions $\psi_+^m \longrightarrow \psi_+^a, \bar{\psi}_+^{\bar{a}}$. The **7** χ_-^i's split as $\underline{\mathbf{6}} \oplus \mathbf{1}$: $(\chi_-^i)_{i \in \mathbf{7}} \to \left(\chi_{A,-} = \varepsilon_{A\bar{A}} \bar{\chi}_{\bar{A},-}\right)_{A \in \underline{\mathbf{6}}} \oplus \chi_-$. This splitting breaks the symmetry group $\mathrm{Spin}(8) \times \mathrm{Spin}(2) \subset \mathrm{Spin}(10)$ of (18) down to SU(4).

The Lagrangian (21), in this formalism, reads as follows:

$$\begin{aligned} L = \delta_+ \Psi\,, \qquad \Psi = &\int_\Sigma \mathrm{Tr}\, \left(\psi_+^a D_{--} \bar{Z}^{\bar{a}} + \bar{\psi}_+^{\bar{a}} D_{--} Z^a + \eta_- F_{z\bar{z}}\right) + \\ &+ \mathrm{i} \int_\Sigma \mathrm{Tr}\, \left(\chi_{ab,-} \left([Z^a, Z^b] + \frac{1}{2}\varepsilon^{abcd} [\bar{Z}^{\bar{c}}, \bar{Z}^{\bar{d}}]\right) + \chi_- \mu\right) - \\ &- \int_\Sigma \mathrm{Tr}\, \left(\chi_- h + \chi_{ab,-} h^{ab}\right) \end{aligned} \tag{23}$$

where

$$\mu = \sum_{a \in \underline{\mathbf{4}}} [Z^a, \bar{Z}^{\bar{a}}] \tag{24}$$

The supersymmetric (for flat Σ) solutions of (23) are the covariantly holomorphic matrices, solving the equations

$$D_{\bar{z}} Z^a = 0, \qquad D_{\bar{z}} \bar{Z}^{\bar{a}} = 0, \quad a \in \underline{\mathbf{4}}, \qquad \mu = 0 \tag{25}$$

and

$$[Z^a, Z^b] + \frac{1}{2}\varepsilon^{abcd}[\bar{Z}^{\bar{c}}, \bar{Z}^{\bar{d}}] = 0, \qquad \{a,b\} \in \underline{\mathbf{6}} \tag{26}$$

For finite dimensional $\mathbb{C}^{\mathbf{N}}$ these equations imply that all matrices commute and can be simultaneously diagonalized.

2.4.4. *Noncommutative gauge theory.* We now wish to consider a generalization of the model [**17**] in which the finite dimensional vector space $\mathbb{C}^{\mathbf{N}}$ is replaced by a Hilbert space $\mathcal{H}$. In order to keep the action (21) finite the combination

$$\wp^i_{mn}[X^m, X^n] \tag{27}$$

could be deformed to

$$\wp^i_{mn}[X^m, X^n] - \mathrm{i}\vartheta^i \cdot \mathbf{1}_{\mathcal{H}} \tag{28}$$

for some constants ϑ^i. One possibility to have a finite action configuration (after h^i's are integrated out) is to have the operators X^m obey the Heisenberg algebra:

$$[X^m, X^n] = \mathrm{i}\vartheta^{mn} \cdot \mathbf{1}_{\mathcal{H}} \tag{29}$$

with the c-number valued matrix $\vartheta^{mn} = -\vartheta^{mn}$ obeying

$$\vartheta^i = \sum_{m<n} \wp^i_{mn}\vartheta^{mn} \,. \tag{30}$$

Since there are too many choices of ϑ^{mn} given ϑ^i, the modification (28) is not what we need. A more sensible modification is to define the action (with the auxiliary fields eliminated) to have the bosonic potential:

$$\sum_{m<n} \mathsf{Tr}_{\mathcal{H}} \left([X^m, X^n] - \mathrm{i}\vartheta^{mn} \cdot \mathbf{1}_{\mathcal{H}} \right)^2 \tag{31}$$

whose absolute minimuma are given by the representation of the Heisenberg algebra (29) in $\mathcal{H}$. These are classified, for the non-degenerate ϑ^{mn}, modulo the gauge group $\mathsf{G}_{\mathbf{g}} = U(\mathcal{H})$, by the Stone-von Neumann theorem. Fix a non-negative integer N, and a standard oscillator representation H of the Heisenberg algebra $[\widehat{x}^m, \widehat{x}^n] = \mathrm{i}\vartheta^{mn} \cdot \mathbf{1}_H$. Then $H = L^2(\mathbb{R}^4)$, $\mathcal{H} = \mathbb{C}^N \otimes H$. For example, let us choose a block diagonal basis for ϑ^{mn}, in which

$$\widehat{z}^a = \widehat{x}^{2a-1} + \mathrm{i}\widehat{x}^{2a}, \quad \widehat{z}^{a\dagger} = \widehat{x}^{2a-1} - \mathrm{i}\widehat{x}^{2a}, \qquad a \in \underline{\mathbf{4}} \tag{32}$$

obey

$$\begin{aligned} &[\widehat{z}^a, \widehat{z}^b] = 0, \quad [\widehat{z}^{a\dagger}, \widehat{z}^{b\dagger}] = 0, \qquad \{a,b\} \in \underline{\mathbf{6}} \\ &[\widehat{z}^a, \widehat{z}^{b\dagger}] = -\theta^a\,\delta^{ab} \end{aligned} \tag{33}$$

The supersymmetric solution of the matrix model, the 0-dimensional reduction of (21) is given by the operators

$$\begin{aligned} &X^m = \mathbf{1}_N \otimes \widehat{x}^m, \qquad m = 1, \ldots 8 \\ &\sigma = \mathrm{diag}(\sigma_1, \ldots, \sigma_N) \otimes \mathbf{1}_H \end{aligned} \tag{34}$$

This solution, for $\theta_a \neq 0$ for all $a \in \underline{\mathbf{4}}$ describes a stack of N $D7$ branes whose worldvolume extends in the $1, \ldots, 8$ directions. They are localized in the remaining two dimensions, parametrized by the eigenvalues of the complex matrix σ.

Now let us assume all θ_a equal to $\zeta > 0$. Take $H = L^2(\mathbb{R}^2)$ to be the Fock space representation of the algebra

$$(35)\qquad \begin{aligned}&[c_1, c_2] = [c_1^\dagger, c_2^\dagger] = 0\\ &[c_i, c_j^\dagger] = \zeta\delta_{ij}, \qquad i,j = 1,2\end{aligned}$$

Define:

$$(36)\qquad \begin{aligned}\hat{z}^1 &= \frac{1}{\sqrt{2}}\left(\mathbf{1}_{N_{12}} \otimes c_1^\dagger + \mathbf{1}_{N_{13}} \otimes c_1^\dagger + \mathbf{1}_{N_{14}} \otimes c_1^\dagger\right)\\ \hat{z}^2 &= \frac{1}{\sqrt{2}}\left(\mathbf{1}_{N_{12}} \otimes c_2^\dagger + \mathbf{1}_{N_{23}} \otimes c_1^\dagger + \mathbf{1}_{N_{24}} \otimes c_1^\dagger\right)\\ \hat{z}^3 &= \frac{1}{\sqrt{2}}\left(\mathbf{1}_{N_{13}} \otimes c_2^\dagger + \mathbf{1}_{N_{23}} \otimes c_2^\dagger + \mathbf{1}_{N_{34}} \otimes c_1^\dagger\right)\\ \hat{z}^4 &= \frac{1}{\sqrt{2}}\left(\mathbf{1}_{N_{14}} \otimes c_2^\dagger + \mathbf{1}_{N_{24}} \otimes c_2^\dagger + \mathbf{1}_{N_{34}} \otimes c_2^\dagger\right)\end{aligned}$$

These operators obey:

$$(37)\qquad \begin{aligned}&[\hat{z}^a, \hat{z}^b] = [\hat{z}^{a\dagger}, \hat{z}^{b\dagger}] = 0\\ &\sum_{a\in\underline{4}}[\hat{z}^a, \hat{z}^{a\dagger}] = -\zeta\mathbf{1}_{N\otimes H}\end{aligned}$$

where

$$(38)\qquad N = \bigoplus_{A\in\underline{6}} N_A$$

The solution (37) describes six stacks of $D3$-branes spanning the coordinate two-planes $\mathbb{C}^2 \subset \mathbb{C}^4$, with $n_A = \dim N_A$ branes spanning the two-plane $\mathbb{C}^2_A$. This is a generalization of the "piercing string" and "fluxon" solutions of [**12, 13**].

We can easily produce more general solutions of the BPS equations. Take six solutions $\widehat{C}^1_A, \widehat{C}^2_A$ of non-commutative instanton equations in $\mathbb{R}^4$, viewed as operators in $N_A \otimes H$, obeying:

$$(39)\qquad [\widehat{C}^1_A, \widehat{C}^2_A] = 0\,, \qquad [\widehat{C}^1_A, \widehat{C}^{1\dagger}_A] + [\widehat{C}^2_A, \widehat{C}^{2\dagger}_A] = \zeta$$

Define operators in $\mathcal{H} = \bigoplus_{A\in\underline{6}} N_A \otimes H$:

$$(40)\qquad \widehat{Z}^a = \frac{1}{\sqrt{2}}\bigoplus_{A\ni a} \widehat{C}_A^{h_A(a)\dagger}$$

where $h_{\{a,b\}}(a) = 1$ for $a < b$, and $h_{\{a,b\}}(a) = 2$ for $a > b$. These operators satisfy the higher dimensional analogues of the noncommutative instanton equations (26), (25), [**30, 32**]:

$$(41)\qquad \begin{aligned}&[\widehat{Z}^a, \widehat{Z}^b] + \frac{1}{2}\sum_{c,d}\varepsilon_{abcd}[\widehat{Z}^{d\dagger}, \widehat{Z}^{c\dagger}] = 0, \qquad a,b,c,d \in \underline{4}\\ &\sum_{a\in\underline{4}}[\widehat{Z}^a, \widehat{Z}^{a\dagger}] = -\zeta\mathbf{1}_{N\otimes H}\end{aligned}$$

The aim of the next section is to produce the (almost) finite-dimensional model of the moduli space of finite action solutions to (41). Some of these solutions are of the form (40).

...

Recently the field theory description of two stacks of intersecting $D3$ branes in IIB string theory sharing a common $1+1$-dimensional worldvolume was explored in [**6, 20**]. The theories exibit unusual holographic and renormalization properties.

...

The string theory of two stacks of transversely intersecting $D3$ branes in IIB theory has been recently studied in [**36, 37**], albeit in the $\zeta = 0$ case. None of the beauty (to the trained eye) of the picture presented below seem to survive in this limit.

3. Spiked instantons

We are going to work with the collections of vector spaces and linear maps between them. The vector spaces will be labelled by the coordinate complex two-planes in the four dimensional complex vector space $\mathbb{C}^4$.

3.1. Generalized ADHM equations. We start by fixing seven Hermitian vector spaces: K and N_A, $A \in \underline{\mathbf{6}}$. Let $k = \dim_{\mathbb{C}}(K)$, $n_A = \dim_{\mathbb{C}}(N_A)$. Consider the vector space $\mathcal{A}_k(\vec{n})$ of linear maps $(\mathbf{B}, \mathbf{I}, \mathbf{J})$

$$\begin{aligned} \mathbf{B} &= (B_a)_{a\in\underline{\mathbf{4}}}\,, \quad & B_a &: K \to K\,, \\ \mathbf{I} &= (I_A)_{A\in\underline{\mathbf{6}}}\,, \quad & I_A &: N_A \to K\,, \\ \mathbf{J} &= (J_A)_{A\in\underline{\mathbf{6}}}\,, \quad & J_A &: K \to N_A\,. \end{aligned} \tag{42}$$

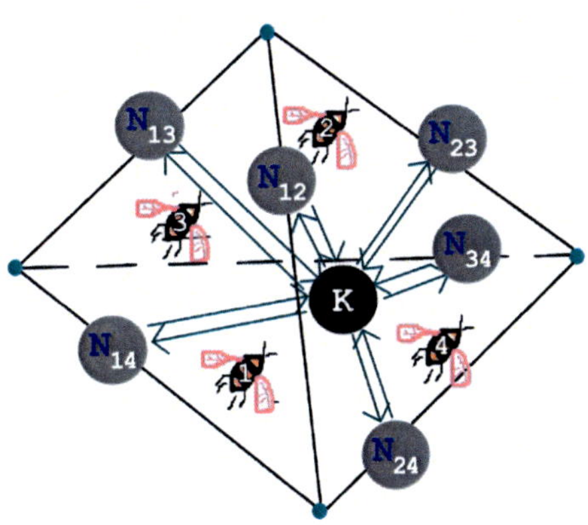

Figure 3: Seven vector spaces and maps between them

The vector spaces and the maps are conveniently summarized by the tetrahedron diagram on Fig. 3. The choice of the matrices can be motivated by the string theory considerations. Namely, consider k $D(-1)$-branes in the vicinity of the six stacks of $D3$-branes (some of these stacks could be $\overline{D3}$-branes) spanning the coordinate two-planes $\mathbb{C}^2_A \subset \mathbb{C}^4$. The number of branes spanning $\mathbb{C}^2_A$ is n_A.

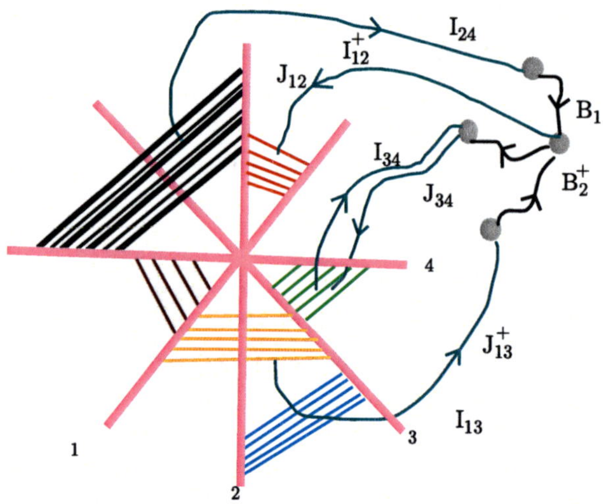

Figure 4: Open string sectors: fields **B**, **I**, **J**

Then the open strings stretched between the $D(-1)$ and $D(-1)$'s produce, upon quantization, the matrices $B_a, B_a^\dagger$, together with their superpartners, and some auxiliary fields, which enter the effective Lagrangian in such a way so as to impose the following

3.1.1. *KK equations.* Define, for $A = \{a, b\}$, $a < b$,

$$\mu_A = [B_a, B_b] + I_A J_A, \tag{43}$$

and

$$s_A = \mu_A + \epsilon(A)\mu_{\bar{A}}^\dagger \ : \ K \to K, \qquad A \in \underline{\mathbf{6}} \tag{44}$$

obeying

$$s_A^\dagger = \epsilon(A) s_{\bar{A}} \tag{45}$$

Define the *real moment map*

$$\mu = \sum_{a \in \underline{\mathbf{4}}} [B_a, B_a^\dagger] + \sum_{A \in \underline{\mathbf{6}}} \left(I_A I_A^\dagger - J_A^\dagger J_A \right) \tag{46}$$

The symmetry (45) allows to view the collection $\vec{s} = (s_A)_{A \in \underline{\mathbf{6}}} \oplus \mu$ as the $U(K)$-equivariant map

$$\vec{s} : \mathcal{A}_k(\vec{n}) \longrightarrow \mathrm{Lie}U(K)^* \otimes \mathbb{R}^7 , \tag{47}$$

as a sort of an octonionic version of the hyperkähler moment map [**14**].

...

Likewise, the open strings stretched between the $D(-1)$ and $D3$'s produce, upon quantization, the matrices $I_A, J_A, I_A^\dagger, J_A^\dagger$, together with their superpartners, and some auxiliary fields,

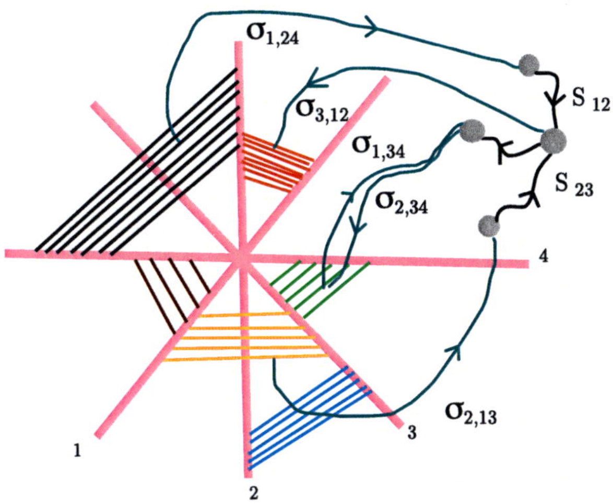

Figure 5: Open string sectors: mutiplets of the equations $\sigma_{\bar{a}A}, s_A$

which enter the effective Lagrangian in such a way so as to impose the following

3.1.2. *KN equations.* For each pair $(\bar{a}, A)$, where $A \in \underline{\mathbf{6}}$, and $\bar{a} \in \bar{A}$, define

$$\sigma_{\bar{a}A} = B_{\bar{a}} I_A + \varepsilon_{\bar{a}\bar{b}A} B_{\bar{b}}^{\dagger} J_A^{\dagger} \; : \; N_A \to K \tag{48}$$

where $\bar{b} \in \bar{A}$, and $\bar{b} \neq \bar{a}$.

3.1.3. *NN equations.* Now, for each $A \in \underline{\mathbf{6}}$ define

$$\Upsilon_A = J_{\bar{A}} I_A - \epsilon(A) I_{\bar{A}}^{\dagger} J_A^{\dagger} \; : \; N_A \to N_{\bar{A}} \tag{49}$$

which obey

$$\Upsilon_A^{\dagger} = -\Upsilon_{\bar{A}} \, . \tag{50}$$

Because of the symmetry (50) the collection of the maps $(\Upsilon_A)_{A \in \underline{\mathbf{6}}}$ takes values in the real vector space of dimension

$$\sum_{A \in \underline{\mathbf{6}}} n_A n_{\bar{A}} \tag{51}$$

The equations (50) result from integrating out the open strings connecting the two stacks of $D3$-branes which intersect only at a point, the origin in $\mathbb{C}^4$.

...

For each pair $A', A'' \in \underline{\mathbf{6}}$, such that $A' \cap A'' = \{a\}$, and $i \geq 1$, define

$$\Upsilon_{A',A'';i} = J_{A'} B_a^{i-1} I_{A''} \, . \tag{52}$$

These equations result (conjecturally) from integrating out the $3-3$ strings connecting the neighbouring stacks $\mathbb{C}^2_{A'}$ and $\mathbb{C}^2_{A''}$, intersecting along a real two-dimensional plane $\mathbb{C}^1_a$.

...

Finally, for each $A = \{a', a''\} \in \underline{\mathbf{6}}$, e.g. $a' < a''$, and $i, j \geq 1$, define

$$\Upsilon_{A;i,j} = J_A B_{a'}^{i-1} B_{a''}^{j-1} I_A \tag{53}$$

3.1.4. *A very useful identity.* Let us compute

$$\sum_{A\in\underline{\mathbf{6}}}\mathrm{Tr}\, s_A s_A^\dagger + \sum_{A\in\underline{\mathbf{6}},\bar{a}\in\bar{A}}\mathrm{Tr}\,\sigma_{\bar{a}A}\sigma_{\bar{a}A}^\dagger + \sum_{A\in\underline{\mathbf{6}}}\mathrm{Tr}\,\Upsilon_A\Upsilon_A^\dagger = \tag{54}$$

$$2\sum_{A\in\underline{\mathbf{6}}}\mathrm{Tr}\,\mu_A\mu_A^\dagger + \sum_{A\in\underline{\mathbf{6}}}\varepsilon(A)\left(\mathrm{Tr}\,\mu_A\mu_{\bar{A}} + \mathrm{Tr}\,\mu_A^\dagger\mu_{\bar{A}}^\dagger\right) +$$

$$\sum_{A\in\underline{\mathbf{6}},a\in\bar{A}}\mathrm{Tr}\,\left(B_a^\dagger B_a\Pi_A^I + B_a B_a^\dagger \Pi_A^J\right) + 2\sum_{A\in\underline{\mathbf{6}}}\mathrm{Tr}\,\Pi_A^J\Pi_{\bar{A}}^I +$$

$$-2\sum_{A\in\underline{\mathbf{6}},\bar{A}=\{\bar{a},\bar{b}\}}\epsilon(A)\mathrm{Tr}\,\left([B_{\bar{a}},B_{\bar{b}}]I_A J_A + c.c.\right) - \sum_{A\in\underline{\mathbf{6}}}\epsilon(A)\mathrm{Tr}\,\left(I_A J_A I_{\bar{A}} J_{\bar{A}} + c.c.\right) =$$

$$= 2\sum_{A\in\underline{\mathbf{6}}}\left(\|\mu_A\|^2 + \|J_{\bar{A}} I_A\|^2\right) + \sum_{A\in\underline{\mathbf{6}},a\in\bar{A}}\|B_a I_A\|^2 + \|J_A B_a\|^2$$

where

$$\Pi_A^J = J_A^\dagger J_A, \qquad \Pi_A^I = I_A I_A^\dagger \tag{55}$$

3.2. Holomorphic equations. Using the identity (54) it is easy to show that the equations

$$\begin{aligned} &s_A = 0, \quad \Upsilon_A = 0, \quad A\in\underline{\mathbf{6}} \\ &\sigma_{\bar{a}A} = 0, \quad \bar{a}\in\bar{A}, \end{aligned} \tag{56}$$

which are not holomorphic in the variables $\mathbf{B},\mathbf{I},\mathbf{J}$, imply stronger holomorphic equations: for each $A\in\underline{\mathbf{6}}$,

$$\begin{aligned} &\mu_A = 0, \quad J_{\bar{A}} I_A = 0 \\ &B_{\bar{a}} I_A = 0\,, \quad J_A B_{\bar{a}} = 0\,, \qquad \bar{a}\in\bar{A}, \end{aligned} \tag{57}$$

3.3. The moduli spaces $\mathfrak{M}_k^*(\vec{n})$. Define $\mathfrak{M}_k^i(\vec{n})$ to be the $U(k)$-quotient of the space of solutions to (56) (which imply, by the above argument, (57)), the additional equations

$$\Upsilon_{A',A'',j} = 0, \qquad 1\leq j\leq i, \tag{58}$$

for all $A',A''\in\underline{\mathbf{6}}$ with $\#A'\cap A''=1$, and the "moment map" equation

$$\mu = \zeta\cdot\mathbf{1}_K \tag{59}$$

The group $U(k)$ acts by:

$$(\mathbf{B},\,\mathbf{I},\,\mathbf{J}) \mapsto (g^{-1}\mathbf{B}g,\, g^{-1}\mathbf{I},\, \mathbf{J}g)\,, \qquad g\in U(k) \tag{60}$$

It is clear that, as a set

$$\mathfrak{M}_k^\infty(\vec{n}) \subset \ldots \subset \mathfrak{M}_k^i(\vec{n}) \subset \mathfrak{M}_k^{i-1}(\vec{n}) \subset \ldots \subset \mathfrak{M}_k^1(\vec{n}) \subset \mathfrak{M}_k^0(\vec{n}) \tag{61}$$

and that the sequence stabilizes at $i\geq k$ (use the fact that a $k\times k$ matrix obeys the degree k polynomial equation).

3.4. Stability. Imposing (59) with $\zeta > 0$ and dividing by $U(k)$ is equivalent to imposing the stability condition and dividing by the action (60) with $g \in GL(k) \equiv GL(k, \mathbb{C})$. Note that we deal with the equations (57) when talking about the $GL(k)$ symmetry. The stability condition reads:

(62)
$$\begin{array}{l}\text{Any subspace } K' \subset K\,, \text{ such that}\\ \qquad I_A(N_A) \subset K'\,, \quad \text{for all } A \in \underline{\mathbf{6}}\\ \qquad\quad \text{and}\\ \qquad B_a(K') \subset K'\,, \quad \text{for all } a \in \underline{\mathbf{4}}\\ \qquad\quad \text{coincides with all of } K\,,\ K' = K\\[1em] \text{in other words}\,, \quad \sum_{A \in \underline{\mathbf{6}}} \mathbb{C}[B_1, B_2, B_3, B_4]\, I_A(N_A) = K\end{array}$$

The proof is standard. In one direction, let us prove (62) holds given that the $GL(k)$-orbit of the tuple $(\mathbf{B}, \mathbf{I}, \mathbf{J})$ of matrices crosses the locus $\mu = \zeta \mathbf{1}_K$. Indeed, assume there is K' which is $\mathbf{B}$-invariant, and contains the image of I_A's. Let K'' be the orthogonal complement $K'' = (K')^{\perp}$. Let P', P'' be the orthogonal projections onto K', K'', respectively:

(63) $$\mathbf{1}_K = P' + P'', \quad P'P'' = P''P' = 0, \quad (P')^2 = (P')^{\dagger} = P', \quad (P'')^2 = (P'')^{\dagger} = P''$$

Since the images of $\mathbf{I}$'s are in K', we have:

(64) $$P'' I_A = 0, \quad A \in \underline{\mathbf{6}}$$

Since $\mathbf{B}$ preserve K', we have:

(65) $$P'' B_a P' = 0, \qquad a \in \underline{\mathbf{4}}$$

Define

(66) $$\begin{aligned} b_a &= P'' B_a P'', \quad b_a^{\dagger} = P'' B_a^{\dagger} P'', \qquad a \in \underline{\mathbf{4}}, \\ j_A &= J_A P'', \quad P'' J_A^{\dagger} = j_A^{\dagger}, \qquad A \in \underline{\mathbf{6}} \end{aligned}$$

Thus:

(67) $$\zeta P'' = P'' \mu P'' = \sum_{a \in \underline{\mathbf{4}}} [b_a, b_a^{\dagger}] - \sum_{a \in \underline{\mathbf{4}}} P'' B_a^{\dagger} P' B_a P'' - \sum_{A \in \underline{\mathbf{6}}} j_A^{\dagger} j_A$$

Now, taking the trace of both sides of (67) we arrive at the conclusion $K'' = 0$:

(68) $$0 \leq \zeta \mathrm{dim} K'' = - \sum_{a \in \underline{\mathbf{4}}} \| P' B_a P'' \|^2 - \sum_{A \in \underline{\mathbf{6}}} \| j_A \|^2 \leq 0 \implies \mathrm{dim} K'' = 0$$

Conversely, assume (62) holds. Let

(69) $$f = \frac{1}{2} \mathrm{Tr}\, (\mu - \zeta \mathbf{1}_K)^2$$

Consider the gradient flow, generated by f with respect to the flat Kähler metric

(70) $$ds^2 = \| d\mathbf{B} \|^2 + \| d\mathbf{I} \|^2 + \| d\mathbf{J} \|^2$$

The function f decreases along the gradient trajectory. Moreover, the trajectory belongs to the $GL(k)$-orbit. Eventually, the trajectory stops at a critical point of f.

Either it is the absolute minimum, i.e. the solution to (59), or the higher critical point, where

$$\langle \xi, \nabla \mu \rangle = 0, \qquad \xi = \mu - \zeta \mathbf{1}_K \neq 0 \tag{71}$$

The one-parametric subgroup $(\exp t\xi)_{t\in\mathbb{C}} \subset GL(K)$, preserves $(\mathbf{B}, \mathbf{I}, \mathbf{J})$,

$$\begin{aligned} &[B_a, \xi] = 0, \qquad a \in \underline{\mathbf{4}} \,, \\ &\xi I_A = 0, \quad J_A \xi = 0, \qquad A \in \underline{\mathbf{6}} \end{aligned} \tag{72}$$

Define $K' = \ker\xi$. The Eq. (72) implies K' is $\mathbf{B}$-invariant, and contains the image of $\mathbf{I}$. Therefore, by (62), $K' = K$, $\xi \equiv 0$, i.e. (59) is satisfied.

Notation.

We denote by $[\mathbf{B}, \mathbf{I}, \mathbf{J}]$ the $GL(k)$-orbit $\left(g^{-1}B_a g, g^{-1}I_A, J_A g\right)_{a\in\underline{\mathbf{4}}, A\in\underline{\mathbf{6}}, g\in GL(k)}$.

4. The symmetries of spiked instantons

The moduli spaces $\mathfrak{M}^*_k(\vec{n})$ are acted on by a group $\mathsf{H} = \mathsf{H}_{\vec{n}}$ of symmetries, defined below. The symmetry of $\mathfrak{M}^*_k(\vec{n})$ will be used in several ways. First, we shall be studying H-equivariant integration theory of the spiked instanton moduli, in cohomology and equivariant K-theory. Second, the shall define new moduli spaces by studying the Γ-fixed loci $\left(\mathfrak{M}^*_k(\vec{n})\right)^{\Gamma}$ in $\mathfrak{M}^*_k(\vec{n})$, for subgroups $\Gamma \subset \mathsf{H}$. These moduli spaces have the commutant $C_\Gamma(\mathsf{H})$ as the symmetry group. Finally, the connected components $\mathfrak{M}^{*,\gamma}_{\mathbf{k}}(\underline{\vec{\mathbf{n}}}) \subset \left(\mathfrak{M}^*_k(\vec{n})\right)^{\Gamma}$ can be defined using only the quiver of Γ, not the group Γ. The definition can be then generalized to define more general quiver spiked instantons. Their symmetry H_γ generalizes the commutant $C_\Gamma(\mathsf{H})$.

4.1. Framing and spatial rotations.

First of all, we can act by a collection $\underline{\mathbf{h}} = (h_A)_{A\in\underline{\mathbf{6}}}$ of unitary matrices $h_A \in U(n_A)$, defined up to an overall $U(1)$ multiple (cf. (9)):

$$\underline{\mathbf{h}} \cdot [B_a\,,\ I_A\,,\ J_A] = \left[B_a\,,\ I_A h_A\,,\ h_A^{-1} J_A\right] \tag{73}$$

We call the symmetry (73) the framing rotation.

Secondly, we can multiply the matrices B_a by the phases $B_a \mapsto q_a B_a$, as long as their product is equal to 1:

$$\prod_{a\in\underline{\mathbf{4}}} q_a = 1 \tag{74}$$

and we supplement this transformation with the transformation $J_A \mapsto q_A J_A$:

$$\underline{\mathfrak{q}} \cdot [B_a\,,\ I_A\,,\ J_A] = [q_a B_a\,,\ I_A\,,\ q_A J_A] \tag{75}$$

We can view $\underline{\mathfrak{q}}$ as the diagonal matrix

$$\underline{\mathfrak{q}} = \mathrm{diag}\,(q_1, q_2, q_3, q_4) \in U(1)^3_{\underline{\epsilon}} \subset SU(4) \tag{76}$$

which belongs to the maximal torus $U(1)^3_{\underline{\epsilon}}$ of the group $SU(4)$ of rotations of $\mathbb{C}^{\underline{\mathbf{4}}}$ preserving some supersymmetry. We call (75) the spatial rotations.

The group

$$\mathsf{H} = \mathrm{P}\left(\bigtimes_{A\in\underline{\mathbf{6}}} U(n_A)\right) \times U(1)^3_{\underline{\epsilon}} \tag{77}$$

is the symmetry of the moduli space of spiked instantons for generic ζ and $\vec{n}$. The complexification $\mathsf{H}_{\mathbb{C}}$ preserves the holomorphic equations (57) and the stability condition (62).

The center Z_{H} of H is the eight dimensional torus

$$Z_{\mathsf{H}} = U(1)^5_{\mathbf{x}} \times U(1)^3_{\underline{\mathfrak{e}}} \tag{78}$$

The maximal torus T_{H} of H is the torus

$$T_{\mathsf{H}} = \left(\left(\times_{A \in \underline{\mathbf{6}}} T_A \right) \Big/ U(1) \right) \times U(1)^3_{\underline{\mathfrak{e}}} \tag{79}$$

where

$$T_A \subset U(n_A) \tag{80}$$

is the group of diagonal $n_A \times n_A$ unitary matrices, the maximal torus of $U(n_A)$, $T_A \approx U(1)^{n_A}$. In the Eq. (79) we divide by the $U(1)$ embedded diagonally into the product of all $\sum_A n_A$ $U(1)$'s.

4.1.1. *Coulomb parameters.* Let $(\underline{\mathfrak{a}}, \underline{\mathfrak{e}}) \in \mathrm{Lie}(T_{\mathsf{H}}) \otimes \mathbb{C}$,

$$\begin{aligned} &\underline{\mathfrak{e}} = (\mathfrak{e}_1, \mathfrak{e}_2, \mathfrak{e}_3, \mathfrak{e}_4)\,, \qquad \mathfrak{e}_a \in \mathbb{C}, \ \sum_{a \in \underline{\mathbf{4}}} \mathfrak{e}_a = 0\,, \\ &\underline{\mathfrak{a}} = (\mathfrak{a}_A)_{A \in \underline{\mathbf{6}}}\,, \qquad \mathfrak{a}_A = \mathrm{diag}\,(\mathfrak{a}_{A,1}, \ldots, \mathfrak{a}_{A,n_A}) \in \mathrm{Lie}(T_A) \otimes \mathbb{C} \end{aligned} \tag{81}$$

The eigenvalues $\mathfrak{a}_{A,\alpha} \in \mathbb{C}$ are defined modulo the overall shift $\mathfrak{a}_{A,\alpha} \mapsto \mathfrak{a}_{A,\alpha} + \mathbf{x}$, $\mathbf{x} \in \mathbb{C}$.

The integrals (12) which we define below are meromorphic functions of $(\underline{\mathfrak{e}}, \underline{\mathfrak{a}})$.

4.1.2. *Symmetry enhancements.* Sometimes the symmetry of the spiked ADHM equations enhances. First of all, if all $\mathbf{I} = \mathbf{J} = 0$ (for $N_A = 0$, for all A), then the $\underline{\mathfrak{q}}$-transformations can be generalized to the action of the full $\mathrm{SU}(4) = \mathrm{Spin}(6)$:

$$B_a \mapsto \sum_{c \in \underline{\mathbf{4}}} g_{a\bar{c}} B_c\,, \qquad g g^{\dagger} = 1, \ \det(g) = 1 \tag{82}$$

In the case of less punitive restrictions on N_A's, e.g. in the crossed instanton case, the symmetry enhances to $\mathrm{SU}(2) \times U(1) \times \mathrm{SU}(2)$, and, if $\zeta = 0$, to $\mathrm{SU}(2)^3$. Let us assume, for definiteness, that only N_{12} and N_{34} are non-zero. Then the transformations:

$$\begin{gathered} (B_1, B_2, B_3, B_4, I_{12}, J_{12}, I_{34}, J_{34}) \mapsto \\ \left(uaB_1 + ubB_2, -u\bar{b}B_1 + u\bar{a}B_2, \bar{u}cB_3 + \bar{u}dB_4, -\bar{u}\bar{d}B_3 + \bar{u}\bar{c}B_4, uI_{12}, uJ_{12}, \bar{u}I_{34}, \bar{u}J_{34} \right), \\ \begin{pmatrix} a & b \\ -\bar{b} & \bar{a} \end{pmatrix} \in \mathrm{SU}(2)_{12}\,, \quad \begin{pmatrix} c & d \\ -\bar{d} & \bar{c} \end{pmatrix} \in \mathrm{SU}(2)_{34}\,, \quad \begin{pmatrix} u & 0 \\ 0 & \bar{u} \end{pmatrix} \in U(1)_{\Delta} \subset \mathrm{SU}(2)_{\Delta} \\ a\bar{a} + b\bar{b} = c\bar{c} + d\bar{d} = u\bar{u} = 1 \end{gathered} \tag{83}$$

preserve the crossed instanton equations (56). When $\zeta = 0$ the $U(1)_{\Delta}$ symmetry enhances to the full $\mathrm{SU}(2)_{\Delta}$, acting by:

$$\begin{aligned}
&(B_1, B_2, B_3, B_4),\ (I_{12}, J_{12}, I_{34}, J_{34}) \mapsto \\
&\qquad \left(uB_1 - vB_2^{\dagger},\ vB_1^{\dagger} + uB_2,\ \bar{u}B_3 - \bar{v}B_4^{\dagger},\ \bar{v}B_3^{\dagger} + \bar{u}B_4\right), \\
&\qquad\qquad \left(uI_{12} - vJ_{12}^{\dagger},\ uJ_{12} + vI_{12}^{\dagger},\ \bar{u}I_{34} - \bar{v}J_{34}^{\dagger},\ \bar{u}J_{34} + \bar{v}I_{34}^{\dagger}\right), \\
&\qquad \begin{pmatrix} u & \bar{v} \\ -v & \bar{u} \end{pmatrix} \in \mathrm{SU}(2)_{\Delta}, \qquad u\bar{u} + v\bar{v} = 1
\end{aligned} \tag{84}$$

The equation $\Upsilon_{12} \equiv J_{34}I_{12} - I_{34}^{\dagger}J_{12}^{\dagger}$, the equations $s_{13} = -s_{24}^{\dagger}, s_{14} = s_{23}^{\dagger}$, the equations $\sigma_{3,12}, \sigma_{4,12}$ as well as the equations $\sigma_{1,34}, \sigma_{2,34}$ are $\mathrm{SU}(2)_{\Delta}$-invariant, while the equations $s_{12}, \mu, s_{34} = s_{12}^{\dagger}$ form a doublet.

4.2. Subtori. In what follows we shall encounter the arrangement of hyperplanes $\mathfrak{H}_l$ in $\mathrm{Lie}(T_{\mathsf{H}}) \otimes \mathbb{C}$ defined by the system of linear equations:

$$L_i(\underline{\mathfrak{a}}, \underline{\mathfrak{e}}) = \sum_{A \in \underline{6}} \sum_{\alpha \in [n_A]} \varpi_{i;A,\alpha} \mathfrak{a}_{A,\alpha} + \sum_{a \in \underline{4}} n_{i;a} \mathfrak{e}_a = 0 \tag{85}$$

with $\varpi_{i;A,\alpha} \in \{-1, 0, +1\}$, $n_{i;a} \in \mathbb{Z}$ and the matrix $\varpi_{i;A,\alpha}$ of maximal rank. Such equations (85) can be interpreted as defining a subtorus $\mathfrak{H} = T_L \subset T_{\mathsf{H}}$: simply solve (85) for the subset of $\mathfrak{a}_{A,\alpha}$'s for which the matrix $\varpi_{i;A,\alpha}$ is invertible. We shall not worry about the integrality of the inverse matrix in this paper, by using the covering tori, if necessary.

One of the reasons we need to look at the subtori T_L is the following construction.

4.3. Orbifolds, quivers, defects. In this section the global symmetry group H is equal to

$$\mathsf{H} = P\left(\bigtimes_{A \in \underline{6}} U(n_A) \right) \times \mathsf{G}_{\mathrm{rot}}$$

where

(1)
$$\mathsf{G}_{\mathrm{rot}} = U(1)^3_{\underline{\mathfrak{e}}}$$
if there are at least two $A' \neq A'' \in \underline{6}$ with non-empty intersection with $n_{A'}n_{A''} \neq 0$, and to

(2)
$$\mathsf{G}_{\mathrm{rot}} = \mathrm{SU}(2)_A \times U(1)_{\Delta} \times \mathrm{SU}(2)_{\bar{A}}$$
otherwise, i.e. there is at most one pair $A, \bar{A}$ with $n_A n_{\bar{A}} \neq 0$.

In all cases

$$\mathsf{G}_{\mathrm{rot}} \subset \mathrm{SU}(4)\,,$$

so that to every $\gamma \in \Gamma$ one associates a unitary 4×4 matrix $\underline{q} = \| q_a^b(\gamma) \|_{a,b \in \underline{4}}$ with unit determinant. In the first case this matrix is diagonal, in the second case it is a 2×2 block-diagonal matrix with unitary 2×2 blocks of inverse determinants.

The symmetry of $\mathfrak{M}_k^*(\vec{n})$ can be used to define new moduli spaces. Suppose $\Gamma \subset \mathsf{H}$ is a discrete subgroup. Let $\mathsf{H}^{\Gamma} \subset \mathsf{H}$ be the maximal subgroup commuting with Γ, the centralizer of Γ. Let $\Gamma^{\vee}$ be the set of irreducible unitary representations

$(\mathcal{R}_\omega)_{\omega\in\Gamma^\vee}$ of Γ, and $\vec{k}\in\mathbb{Z}_{\geq 0}^{\Gamma^\vee}$. The representations N_A, $A\in\underline{\mathbf{6}}$ of H decompose as representations of Γ

(86) $$N_A = \bigoplus_{\omega\in\Gamma^\vee} N_{A,\omega}\otimes\mathcal{R}_\omega$$

Let $\underline{\vec{\mathbf{n}}}$ now denote the collection $(n_{A,\omega})_{A\in\underline{\mathbf{6}},\omega\in\Gamma^\vee}$ of dimensions

(87) $$n_{A,\omega} = \mathrm{dim}N_{A,\omega}$$

of multiplicity spaces. The vector $\mathbf{k} = (k_\omega)_{\omega\in\Gamma^\vee}$ defines a representation of Γ:

(88) $$\gamma\in\Gamma\mapsto g_\gamma\in U(K)\,, \qquad K = \bigoplus_{\omega\in\Gamma^\vee} K_\omega\otimes\mathcal{R}_\omega\,, \qquad k_\omega = \mathrm{dim}K_\omega$$

We call the components k_ω of the vector $\mathbf{k}$ fractional instanton charges. The moduli space of Γ-folded spiked instantons of charge $\vec{k}$ is the component $\mathfrak{M}^{*,\Gamma}_{\mathbf{k}}(\underline{\vec{\mathbf{n}}})$ set of Γ-fixed points $\left(\mathfrak{M}^*_{\vec{k}}(\vec{n})\right)^\Gamma$. The representation (88) enters the realization of the Γ-fixed locus in the space of matrices $(\mathbf{B},\mathbf{I},\mathbf{J})$:

(89) $$\gamma\cdot(\mathbf{B},\mathbf{I},\mathbf{J}) = \left(g_\gamma\mathbf{B}g_\gamma^{-1}, g_\gamma\mathbf{I}, \mathbf{J}g_\gamma^{-1}\right)\,, \qquad g_\gamma\in U(K)$$

where

(90) $$\gamma\cdot(\mathbf{B},\mathbf{I},\mathbf{J}) \equiv \left(q_a^b(\gamma)B_b, I_A h_A(\gamma)^{-1}, q_A(\gamma)h_A(\gamma)J_A\right)\,,$$
$$\gamma\in\Gamma\mapsto (h_A(\gamma))_{A\in\underline{\mathbf{6}}}\times\|q_a^b(\gamma)\|_{a,b\in\underline{\mathbf{4}}}\in\mathsf{H}$$

is the defining representation of Γ, with $q_A(\gamma)$ given by (9) in the case (1), and by the projection to $U(1)_\Delta$ in the case (2). The equations (89) are invariant under the subgroup

(91) $$\bigtimes_{\omega\in\Gamma^\vee} U(K_\omega)\ \subset\ U(K)$$

of unitary transformations of K commuting with Γ. The holomorphic equations (57) restricted onto the locus of Γ-equivariant i.e. obeying (89) matrices $\mathbf{B},\mathbf{I},\mathbf{J}$ become the holomorphic equations defining $\mathfrak{M}^\Gamma_{\mathbf{k}}(\underline{\vec{\mathbf{n}}})$. The stability condition (62) can be further refined, analogously to the refinement of the real moment map equation $\mu = \sum_{\omega\in\Gamma^\vee}\zeta_\omega\mathbf{1}_{K_\omega}\otimes\mathbf{1}_{\mathcal{R}_\omega}$. We shall work in the chamber where all $\zeta_\omega>0$.

The moduli spaces $\mathfrak{M}^\Gamma_{\mathbf{k}}(\underline{\vec{\mathbf{n}}})$ in the case (1) parametrize the spiked instantons in the presence of $U(1)^3_{\underline{\varepsilon}}$-invariant surface operators, while in the case (2) they parametrize the instantons in supersymmetric quiver gauge theories on the ALE spaces, with additional defect.

The commutant H^Γ acts on $\mathfrak{M}^\Gamma_{\mathbf{k}}(\underline{\vec{\mathbf{n}}})$, so that the partition functions we study are meromorphic functions on Lie $\left(\mathsf{H}^\Gamma\right)\otimes\mathbb{C}$.

Note that if Γ has trivial projection to $\mathsf{G}_{\mathrm{rot}}$ then the moduli space of Γ-folded instantons is simply the product of the moduli spaces of spiked instantons for N_ω's. In what follows we assume the projection to $\mathsf{G}_{\mathrm{rot}}$ to be non-trivial.

4.3.1. *Subtori for Γ-folds.* Let us now describe the maximal torus T_{H^Γ} of the Γ-commutant as T_L. In other words, the choice of a discrete subgroup $\Gamma\subset\mathsf{H}$ defines the hyperplanes $L_i(\mathfrak{a},\underline{\mathfrak{e}}) = 0$ in Lie(T_{H}).

In the case (1) the $\mathsf{G}_{\mathrm{rot}}$-part of Γ is abelian, i.e. it is a product of cyclic groups (if Γ is finite) or it is a torus itself. In either case there is no restriction on the $\underline{\mathfrak{e}}$-parameters. The framing part of Γ reduces $P\left(\times_A U(n_A)\right)$ to $P\left(\times_{A,\omega}U(n_{A,\omega})\right)$ which means that some of the eigenvalues $\mathfrak{a}_{A,\alpha}$, viewed as the generators of Lie$U(n_A)$, must coincide, more precisely to be of multiplicity $\dim\mathcal{R}_\omega$. The minimal case, when Γ is abelian, imposes no restrictions on $(\mathfrak{a}, \underline{\mathfrak{e}})$, so that $T_{\mathsf{H}^\Gamma} = T_{\mathsf{H}}$.

In the case (2) the $\mathsf{G}_{\mathrm{rot}}$-part of Γ need not be abelian. Let us assume, for definiteness, that $A = 12$, $\bar{A} = 34$. If the image of Γ in $SU(2)_{12}$ is non-abelian, then $\mathfrak{e}_1 = \mathfrak{e}_2$. Likewise if the image of Γ in $SU(2)_{34}$ is non-abelian then $\mathfrak{e}_3 = \mathfrak{e}_4$. The non-abelian discrete subgroups of $SU(2)$ have irreducible representations of dimensions 2 and higher, up to 6. Thus the corresponding $\mathfrak{a}_{A,\alpha}$ eigenvalues will have the multiplicity up to 6.

4.3.2. *Subtori for sewing.* Let us specify the integral data for the subtori, i.e. the explicit solutions to the constraints (85). Let $\underline{e} = (e_a)_{a\in\underline{4}}$, $e_a \in \mathbb{Z}_{\neq 0}$, be a 4-tuple of non-zero integers, with no common divisors except for ± 1, which sum up to zero:

$$\sum_{a\in\underline{4}} e_a = 0 \ . \tag{92}$$

Such a collection $\underline{e}$ defines a split $\underline{6} = \underline{6}^+ \amalg \underline{6}^-$, where $\underline{6}^\pm$ being the set of $A = \{a, b\}$ such that $\pm e_a e_b > 0$.

For $A = \{a, b\} \in \underline{6}^-$, i.e. $e_a e_b < 0$, let $p_A = \gcd(|e_a|, |e_b|) > 0$. Let us also fix for such $A \in \underline{6}^-$ a partition $\nu_A = (\nu_{A,\iota})$ of size n_A, whose parts do not exceed e_A: $1 \leq \nu_{A,\iota} \leq p_A$. Let $\ell_A = \ell(\nu_A)$ be its length.

Given ν_A we partition the set $[n_A]$ as the union of nonintersecting subsets

$$[n_A] = \bigcup_{\iota} [n_A]_\iota \ , \qquad \#[n_A]_\iota = \nu_{A,\iota} \ , \tag{93}$$

$[n_A]_{\iota'} \cap [n_A]_{\iota''} = \emptyset$ for $\iota' \neq \iota''$. Fix a map $c_{A,\iota} : [n_A]_\iota \to \mathbb{Z}$ obeying, for any $a', a'' \in [n_A]_\iota$, $a' \neq a''$:

$$c_{A,\iota}(a') - c_{A,\iota}(a'') \neq 0 \, (\mathrm{mod}\, p_A) \tag{94}$$

When $p_A = 1$ the condition (94) is empty.

For $A = \{a, b\} \in \underline{6}^+$, let us fix a map $c_A : [n_A] \to \mathbb{Z}$, obeying, for any $a', a'' \in [n_A]$,

$$c_A(a') - c_A(a'') \notin \mathbb{Z}_{>0} e_a + \mathbb{Z}_{>0} e_b \tag{95}$$

Note that (95) does not forbid the situation where $c_A(a') = c_A(a'')$ for some $a', a'' \in [n_A]$. To make the notation uniform we assign to such A, $\nu_A = (n_A)$, $\ell_A = 1$.

The final piece of data is the choice $\iota_A \in [\ell_A]$ for each $A \in \underline{6}^-$. Define the set $\lambda_A = [\ell_A] \backslash \{\iota_A\}$, of cardinality $\ell_A - 1$.

Now, we associate to the data

$$\mathfrak{L} = (\nu, \lambda, c, \underline{e}) \ , \tag{96}$$

the torus

$$\mathfrak{H} = T_{\mathfrak{L}} = \left(\bigtimes_{A\in\underline{6}} U(1)^{\ell_A - 1} \right) \times U(1) \tag{97}$$

Note that only $A \in \underline{6}^-$ contribute to the product in (97). This torus is embedded into T_{H} as follows: the element

$$(e^{\mathrm{i}\xi}, e^{\mathrm{i}t}) \equiv \bigtimes_{A\in\underline{6}^-} \left(e^{\mathrm{i}\xi_{A,i}}\right)_{i\in\lambda_A} \times e^{\mathrm{i}t} \in T_{\mathfrak{L}} \tag{98}$$

is mapped to

$$\left[\bigtimes_{A\in\underline{\mathbf{6}}^-} \mathrm{diag}_A^- \times \bigtimes_{A\in\underline{\mathbf{6}}^+} \mathrm{diag}_A^+ \right] \times \mathrm{diag}\left(e^{\mathrm{i}e_a t}\right)_{a\in\underline{\mathbf{4}}} \in \mathrm{P}\left(\bigtimes_{A\in\underline{\mathbf{6}}} U(n_A) \right) \times \mathrm{SU}(4) \tag{99}$$

where $\mathrm{diag}_A^{\pm} \in U(n_A)$ are the diagonal matrices with the eigenvalues
(100)

$$\mathrm{Eigen}\left(\mathrm{diag}_A^-\right) = \left\{ e^{\mathrm{i}c_{A,\iota_A}(\alpha)t} \mid \alpha \in [n_A]_{\iota_A} \right\} \cup \left\{ e^{\mathrm{i}(\xi_{A,\iota}+c_{A,\iota}(\alpha)t)} \mid \iota \in \lambda_A, \alpha \in [n_A]_\iota \right\}$$

$$\mathrm{Eigen}\left(\mathrm{diag}_A^+\right) = \left\{ e^{\mathrm{i}c_A(\alpha)t} \mid \alpha \in [n_A] \right\} \tag{101}$$

Thus, the torus $T_{\mathfrak{L}}$ corresponds to the solution of the Eqs. (85) with

$$\begin{aligned} \mathfrak{e}_a &= e_a u, \ a \in \underline{\mathbf{4}}, \\ \mathfrak{a}_{A,\alpha} &= c_A(\alpha) u, \ A \in \underline{\mathbf{6}}^+, \alpha \in [n_A] \\ \mathfrak{a}_{A,\alpha} &= c_{A,\iota_A}(\alpha) u, \ A \in \underline{\mathbf{6}}^-, \alpha \in [n_A]_{\iota_A} \\ \mathfrak{a}_{A,\alpha} &= \xi_{A,\iota} + c_{A,\iota}(\alpha) u, \ A \in \underline{\mathbf{6}}^-, \ \alpha \in [n_A]_\iota, \ \iota \in \lambda_A \end{aligned} \tag{102}$$

In other words, the Ω-background parameters are maximally rationally dependent (the worst way to insult the rotational parameters), the framing of the spaces N_A, $A \in \underline{\mathbf{6}}^+$ is completely locked with space rotations (spin-color locking), while the framing of the spaces N_A, $A \in \underline{\mathbf{6}}^-$ is locked partially.

4.4. Our goal: compactness theorem. Our goal is to establish the compactness of the fixed point sets $\mathfrak{M}_k(\vec{n})^{T_{\mathfrak{L}}}$ and $\left(\mathfrak{M}^{\Gamma}_{\mathbf{k}}(\underline{\vec{\mathbf{n}}})\right)^{T_{\mathrm{H}^{\Gamma}}}$. Before we attack this problem we shall discuss a little bit the ordinary instantons, then look at a few examples of the particular types of spiked instantons: the crossed and the folded instantons, and then proceed with the analysis of the general case. The reader interested only in the compactness theorem can skip the next two sections at the first reading.

5. Ordinary instantons

In this section we discuss the relations between the ordinary four dimensional $U(n)$ instantons and the spiked instantons.

5.1. ADHM construction and its fine print. In the simplest case only one of six vector spaces is non-zero, e.g.

$$N_A = 0, \qquad A \neq \{1,2\}. \tag{103}$$

Let $n = n_{12}$. We shall now show that, set theoretically, the moduli space of spiked instantons in this case is $\mathcal{M}_k(n)$, the ADHM moduli space (more precisely, its Gieseker-Nakajima generalization).

Recall the ADHM construction of the $U(n)$ framed instantons of charge k on (noncommutative) $\mathbb{R}^4$ [**1, 22, 29**]. It starts by fixing Hermitian vector spaces N and K of dimensions n and k, respectively. Consider the space of quadruples (B_1, B_2, I, J),

$$I : N \to K, \ J : K \to N, \ B_\alpha : K \to K, \ \alpha = 1, 2 \tag{104}$$

obeying

$$\vec{\mu}_{12} \equiv \left(2\mu_{12}^{\mathbb{R}},\ \mu_{12}^{\mathbb{C}} + \mu_{12}^{\mathbb{C}\dagger},\ \mathrm{i}\left(\mu_{12}^{\mathbb{C}} - \mu_{12}^{\mathbb{C}\dagger}\right) \right) = (\zeta, 0, 0) \cdot \mathbf{1}_K \ , \tag{105}$$

where

$$\mu_{12}^{\mathbb{C}} = [B_1, B_2] + IJ, \qquad 2\mu_{12}^{\mathbb{R}} = [B_1, B_1^{\dagger}] + [B_2, B_2^{\dagger}] + II^{\dagger} - J^{\dagger}J \tag{106}$$

Note that the number of equations (105) plus the number of symmetries is less then the number of variables. The moduli space $\mathcal{M}_k(n)$ of solutions to (105) modulo the $U(k)$ action

$$(B_1, B_2, I, J) \mapsto (g^{-1}B_1 g, g^{-1}B_2 g, g^{-1}I, Jg), \quad g \in U(k) \tag{107}$$

has the positive dimension

$$\mathrm{dim}_{\mathbb{R}}\mathcal{M}_k(n) = 4k(n+k) - 3k^2 - k^2 = 4nk \tag{108}$$

Again, the $\mu_{12}^{\mathbb{R}}$-equation, with $\zeta > 0$, can be replaced by the stability condition, and the $GL(k)$-symmetry:

(109)

Any subspace $K' \subset K$, such that $I(N) \subset K'$,
and $B_\alpha(K') \subset K'$, for all $\alpha = 1, 2$
coincides with all of K, $K' = K$
in other words, $\mathbb{C}[B_1, B_2]\, I(N) = K$

Notation.
We denote by $[B_1, B_2, I, J]$ the $GL(k)$-orbit $\left(g^{-1}B_1 g, g^{-1}B_2 g, g^{-1}I, Jg\right)_{g \in GL(k)}$.

5.2. Ordinary instantons from spiked instantons. Now, to show that the spiked instantons reduce to the ordinary instantons when (103) is obeyed, we need to show that $B_3 = B_4 = 0$ on the solutions of our equations (57). This is easy:

$$B_3 f(B_1, B_2) I = f(B_1, B_2) B_3 I = 0 \tag{110}$$

where we used $[B_1, B_3] = \mu_{13} = 0$, $[B_2, B_3] = \mu_{23} = 0$, and $B_3 I = 0$. Therefore B_3 acts by zero on all of K. The same argument proves the vanishing of B_4.

5.3. Uhlenbeck spaces. Then $\zeta = 0$, the equations $\vec{\mu}_{12} = 0$ on the quadruple of matrices (B_1, B_2, I, J) with the moment maps given by (106) define the so-called Uhlenbeck (partial) compactification $\overline{\mathcal{M}}_k(n)$ of the moduli space $\mathcal{M}_k(n)$ of true $U(n)$ instantons on $\mathbb{R}^4$. This space is stratified:

$$\overline{\mathcal{M}}_k(n) = \mathcal{M}_k(n) \cup \mathcal{M}_{k-1}(n) \times \mathbb{R}^4 \cup \mathcal{M}_{k-2}(n) \times Sym^2(\mathbb{R}^4) \cup \ldots \cup Sym^k(\mathbb{R}^4) \tag{111}$$

where the stratum $\mathcal{M}_{k-l}(n) \times Sym^l(\mathbb{R}^4)$ parameterizes the quadruples $(B_1.B_2, I, J)$ where

$$\mathrm{dim}(K_{12}) = k - l, \qquad K_{12} = \mathbb{C}[B_1, B_2]\, I(N)\,, \tag{112}$$

and B_1, B_2 are commuting normal matrices on $K_{12}^{\perp}$. This is proven in the same way we proved the stability of the spiked instantons: define K_{12} as in (112), the orthogonal projectors $P_{K_{12}}$, $P_{K_{12}^{\perp}}$, so that $P_{K_{12}^{\perp}} B_\alpha P_{K_{12}} = 0$, $\alpha = 1, 2$ and the operators:

$$b_\alpha = P_{K_{12}^{\perp}} B_\alpha P_{K_{12}^{\perp}}\,, \quad \tilde{b}_\alpha = P_{K_{12}} B_\alpha P_{K_{12}^{\perp}}\,, \quad j = J P_{K_{12}^{\perp}}\,. \tag{113}$$

Now compute $P_{K_{12}^{\perp}}\mu_{12}^{\mathbb{R}}P_{K_{12}^{\perp}}$:

$$P_{K_{12}^{\perp}}\mu_{12}^{\mathbb{R}}P_{K_{12}^{\perp}} = -j^{\dagger}j + \sum_{\alpha=1}^{2}\left([b_{\alpha}, b_{\alpha}^{\dagger}] - \tilde{b}_{\alpha}^{\dagger}\tilde{b}_{\alpha}\right) \tag{114}$$

Imposing $\mu_{12}^{\mathbb{R}} = 0$ implies, by taking the trace:

$$0 = \|j\|^2 + \|\tilde{b}_1\|^2 + \|\tilde{b}_2\|^2 \tag{115}$$

therefore $J|_{K_{12}^{\perp}} = 0$, $B_{\alpha}\left(K_{12}^{\perp}\right) \subset K_{12}^{\perp}$, and

$$[b_1, b_1^{\dagger}] + [b_2, b_2^{\dagger}] = 0\,, \quad [b_1, b_2] = 0\,, \tag{116}$$

which, in turn, implies $[b_{\alpha}, b_{\beta}^{\dagger}] = 0$, $\alpha, \beta = 1, 2$.

5.4. One-instanton example. Let $k = 1$. We can solve the equations (106) explicitly. The matrices B_1, B_2 are just complex numbers, e.g. $b_1, b_2 \in \mathbb{C}$. The pair I, J obeys $IJ = 0$, $\|I\|^2 - \|J\|^2 = \zeta$. Assuming $\zeta > 0$ define the vectors $w_1 = J(K) \in N$, $w_2 = \frac{1}{\sqrt{\zeta + \|J\|^2}} I^{\dagger}(K) \in N$. They obey $\langle w_2, w_1 \rangle = 0$, $\langle w_2, w_2 \rangle = 1$. Dividing by the $U(1) = U(K)$ symmetry we arrive at the conclusion:

$$\mathcal{M}_1(n) = \mathbb{C}^2 \times T^*\mathbb{CP}^{n-1} \tag{117}$$

The first factor parametrizes (b_1, b_2), the base $\mathbb{CP}^{n-1}$ of the second factor is the space of w_2's obeying $\|w_2\|^2 = 1$ modulo $U(1)$ symmetry.

5.5. The canonical complex $\mathcal{S}$. Let us go back to the $\zeta > 0$ case now. Recall that the ADHM data (B_1, B_2, I, J) defines the complex $\mathcal{S}$ of holomorphic vector bundles over $\mathbb{CP}^2 = \mathbb{C}^2 \cup \mathbb{CP}^1_{\infty}$:

$$\mathcal{S} = [0 \to K \otimes \mathcal{O}(-1) \xrightarrow{\tau_z} \quad (K \oplus K \oplus N) \xrightarrow{\sigma_z} \quad K \otimes \mathcal{O}(1) \to 0] \tag{118}$$

where for

$$(u : z_1 : z_2) \in \mathbb{CP}^2 \tag{119}$$

the maps

$$\tau_z(\psi) = (uB_1 - z_1)\psi \oplus (uB_2 - z_2)\psi \oplus uJ\psi\,,$$

$$\sigma_z(\psi_+ \oplus \psi_- \oplus \xi) = -(uB_2 - z_2)\psi_+ + (uB_1 - z_1)\psi_- + uI\xi\,,$$

obey $\tau_z \circ \sigma_z = 0$. The "physical spacetime" $\mathbb{C}^2$ is the locus $u = 1$. We assign degree -1 to the K term on the left in (118), degree 0 to $K \oplus K \oplus N$, degree $+1$ to the K term on the right.

Over the line $\mathbb{CP}^1_{\infty}$ at infinity, defined by $u = 0$, the complex (118) becomes the direct sum of the trivial vector bundle $N \otimes \mathcal{O}_{\mathbb{CP}^1_{\infty}}$ and the acyclic complex

$$K \otimes [\mathcal{O}_{\mathbb{CP}^1_{\infty}}(-1) \to \mathcal{O}_{\mathbb{CP}^1_{\infty}} \otimes \mathbb{C}^2 \to \mathcal{O}_{\mathbb{CP}^1_{\infty}}(1)]\,.$$

The complex (118), for all z, has no cohomology in degree $+1$, i.e. $\mathrm{im}(\sigma_z) = K$. Indeed, the subspace $K' = \mathrm{im}(\sigma_z)$ contains the image of I (take $\psi_+ = \psi_- = 0$), and is B_1, B_2-invariant (if $\psi \in \mathrm{im}(\sigma_z)$ then $B_1\psi = z_1\psi + \sigma_z(0 \oplus \psi \oplus 0) \in \mathrm{im}(\sigma_z)$). Therefore, by (109) $\mathrm{im}(\sigma_z) = K$.

The cohomology of $H^{\bullet}\mathcal{S}$ in degrees 0 and 1 will be denoted by $\mathcal{S}^-$ and $\mathcal{S}^+$, respectively.

5.6. P-spaces. We shall need the fibers of $\mathcal{S}^{\pm}$ at $z_1 = z_2 = 0$:

$$P_{12}^{-} = \ker(\tau_0), \quad P_{12}^{+} = \ker(\sigma_0)/\mathrm{im}(\tau_0) \tag{120}$$

In more detail:

$$\psi \in P_{12}^{-} \Leftrightarrow B_1\psi = B_2\psi = 0, \; J\psi = 0 \tag{121}$$

and

$$[\psi_+ \oplus \psi_- \oplus \xi] = [(\psi_+ + B_1\chi) \oplus (\psi_- + B_2\chi) \oplus (\xi + J\chi)] \in P_{12}^{+} \Leftrightarrow \\ B_1\psi_- - B_2\psi_+ + I\xi = 0\,, \; \chi \in K \tag{122}$$

The spaces P_{12}^{-}, P_{12}^{+} play important rôle in the construction of *Hecke correspondences* in **[22–26]**.

5.7. Stratification and correspondences. We can stratify the moduli space $\mathcal{M}_k(n)$:

$$\mathcal{M}_k(n) = \bigcup_{s=0}^{k} \mathcal{M}_{k;s}(n)\,, \\ \mathcal{M}_{k;s}(n) = \left\{ [B_1, B_2, I, J] \in \mathcal{M}_k(n)\,,\, \dim P_{12} = s \right\} \tag{123}$$

We shall also need the space of quaduples

$$\mathcal{N}_{k;s,v}^{-}(n) = \left\{ B_1, B_2, I, J, V \;\middle|\; [B_1, B_2, I, J] \in \mathcal{M}_k(n)\,,\, \dim P_{12} = s\,,\, V \subset P_{12}\,,\, \dim V = v\,, \right\} \tag{124}$$

which is a total space of the fiber bundle $\mathcal{N}_{k;s,v}(n) \to \mathcal{M}_{k;s}(n)$ with the fibers $\mathrm{Gr}_v(s)$ the Grassmanian of v-dimensional subspaces in the s-dimensional vector space, and the space of quadruples:

$$\mathcal{N}_{k;s,v}^{!}(n) = \left\{ B_1, B_2, I, J, V \;\middle|\; [B_1, B_2, I, J] \in \mathcal{M}_k(n)\,,\, \dim P_{12} = s\,,\, V \subset P_{12}'\,,\, \dim V = v\,, \right\} \tag{125}$$

which is a total space of the fiber bundle $\mathcal{N}_{k;s,v}^{+}(n) \to \mathcal{M}_{k;s}(n)$ with the fibers $\mathrm{Gr}_v(s+n)$ the Grassmanian of v-dimensional subspaces in the $s+n$-dimensional vector space (recall that $\mathrm{dim}P_{12}' = \mathrm{dim}P_{12} + n$).

The correspondences of **[22–26]** are built on the maps: $\phi_s^{\pm} : \mathcal{N}_{k;s,v}^{\pm}(n) \longrightarrow \mathcal{M}_{k\pm v}(n)$, given by:

$$\phi_s^{\pm}[B_1, B_2, I, J, V] = [B_1^{\pm}, B_2^{\pm}, I^{\pm}, J^{\pm}] \tag{126}$$

with the operators $B_1^{\pm}, B_2^{\pm} \in \mathrm{End}(K^{\pm})$, $I^{\pm} : N \to K^{\pm}$, $J^{\pm} : K^{\pm} \to N$, and the spaces $K^{\pm}$ defined by:

$$\begin{aligned} &K^- = K/V\,, && V \subset P_{12} = \ker(\tau_0) \subset K \\ &K^+ \subset \ker(\sigma_0)\,, && \text{s.t. } K^+/\mathrm{im}(\tau_0) = V \subset P'_{12} = \ker(\sigma_0)/\mathrm{im}(\tau_0) \end{aligned} \tag{127}$$

and by

$$\begin{aligned} &B_\alpha^-(\chi) = B_\alpha \chi \,\mathrm{mod} V\,,\ \alpha = 1,2\,, \\ &I^- = I \,\mathrm{mod} V\,,\ J^-(\chi) = J\chi \end{aligned} \tag{128}$$

and (recall that $(\psi_1 \oplus \psi_2 \oplus \xi) \in \ker(\sigma_0)$ iff $B_1\psi_2 - B_2\psi_1 + I\xi = 0$)

$$\begin{aligned} &B_\alpha^+\,(\psi_1 \oplus \psi_2 \oplus \xi) = (B_1\psi_\alpha \oplus B_2\psi_\alpha \oplus J\psi_\alpha) = \tau_0(\psi_\alpha) \subset \ker(\sigma_0) \\ &J^+\,(\psi_1 \oplus \psi_2 \oplus \xi) = \xi \in N \ \text{ and, for } \eta \in N \\ &I^+\eta = (B_1 I\eta \oplus B_2 I\eta \oplus JI\eta) = \tau_0\,(I\eta) \subset \ker(\sigma_0) \end{aligned} \tag{129}$$

which are well-defined thanks to (121) and obey

$$[B_1^{\pm}, B_2^{\pm}] + I^{\pm}J^{\pm} = 0$$

as they should.

Now collect the spaces with different s:

$$\mathcal{N}_{k;v}^{\pm}(n) = \bigcup_s \mathcal{N}_{k;s,v}^{\pm}(n) \tag{130}$$

and define the maps

$$\varphi^{\pm} : \mathcal{N}_{k;v}^{\pm}(n) \longrightarrow \mathcal{M}_{k\pm v}(n) \tag{131}$$

to be equal to $\phi_s^{\pm}$ on the corresponding strata $\mathcal{N}_{k;s,v}^{\pm}(n)$. Thus we have defined the typical Hecke correspondence diagram:

$$\begin{array}{ccc} & \mathcal{N}_{k;v}^{\pm}(n) & \\ {\scriptstyle f^{\pm}}\swarrow & & \searrow{\scriptstyle \varphi^{\pm}} \\ \mathcal{M}_k(n) & & \mathcal{M}_{k\pm v}(n) \end{array} \tag{132}$$

which can be used to define Nakajima's operations $(\varphi^{\pm})_*(f^{\pm})^*$ in cohomology or K-theory of $\mathcal{M}_k(n)$'s.

5.8. *L*-spaces. We shall also need the spaces $L_{1,2} \subset K$ and $L_{2,1} \subset K$,

$$L_{a,b} = \ker\left(B_b|_K\right) \cap \bigcap_{j=1}^{\infty} \ker\left(J\,\left(B_a|_K\right)^{j-1}\right) \subset K, \tag{133}$$

so that

$$P_{12} = L_{1,2} \cap L_{2,1}\ . \tag{134}$$

The space $L_{a,b}$ carries an action of the linear operator B_a:

$$B_a\,(L_{a,b}) \subset L_{a,b} \tag{135}$$

Indeed, if $\psi \in L_{a,b}$, then, by definition,

$$B_b\psi = 0, \qquad JB_a^{j-1}\psi = 0, \qquad j = 1,2,3,\ldots \tag{136}$$

Then

$$B_b(B_a\psi) = [B_b, B_a]\psi = \pm IJ\psi = 0 \tag{137}$$

We can decompose $L_{a,b}$ into the root subspaces of B_a (the Jordan block decomposition):

$$L_{a,b} = \bigoplus_{\lambda,j} L_{a,b}^{\lambda,j} \tag{138}$$

where $L_{a,b}^{\lambda,j}$ is the irreducible B_a-invariant subspace, a Jordan block of length d_j:

$$L_{a,b}^{\lambda,j} = \mathbb{C}\widehat{\chi}_{\lambda,j} \oplus \mathbb{C}(B_a - \lambda)\widehat{\chi}_{\lambda,j} \oplus \mathbb{C}(B_a - \lambda)^2\widehat{\chi}_{\lambda,j} \ldots \oplus \mathbb{C}(B_a - \lambda)^{d_j - 1}\widehat{\chi}_{\lambda,j} \tag{139}$$

for some vector $\widehat{\chi}_{\lambda,j} \in \ker(B_a - \lambda)^{d_j}$.

5.9. The symmetries of the ADHM space. The group $\mathsf{H}_{12} = PU(n) \times SU(2)_{12} \times U(1)'_{12}$ acts on $\mathcal{M}_k(n)$:

$$(B_1, B_2, I, J) \mapsto \left(u(\alpha B_1 + \beta B_2), u(-\bar{\beta} B_1 + \bar{\alpha} B_2), uI\tilde{h}, u\tilde{h}^{-1}J\right) \tag{140}$$

with

$$\begin{aligned} &\tilde{h} \in U(n), \qquad \tilde{h} \sim \tilde{h}\tilde{u}, \qquad \tilde{u} \in U(1), \ |\tilde{u}|^2 = 1, \\ &\begin{pmatrix} \alpha & \beta \\ -\bar{\beta} & \bar{\alpha} \end{pmatrix} \in SU(2)_{12}, \ |\alpha|^2 + |\beta|^2 = 1, \\ &u \in U(1)'_{12}, \ |u|^2 = 1 \end{aligned} \tag{141}$$

This symmetry is enhanced to $PU(n) \times SU(2)_{12} \times SU(2)'_{12}$ when $\zeta \to 0$.

5.10. ***U*** **versus** ***PU*.** In describing the action of H_{12} in (141) we use an element $\tilde{h}$ of the group $U(n)$ yet we claim it is the group $PU(n) = U(n)/U(1) = SU(n)/\mathbb{Z}_n$ which acts faithfully on $\mathcal{M}_k(n)$. Indeed, multiplying $\tilde{h}$ by a scalar matrix

$$\tilde{h} \to \tilde{h}\tilde{u}, \qquad \tilde{u} \in U(1) \tag{142}$$

does not change the transformation (140) since it can be undone by the $U(k)$-transformation (107) with $g = \tilde{u}^{-1}\mathbf{1}_K \in U(k)$.

5.11. Tangent space. Let $m \in \mathcal{M}_k(n)$. Let (B_1, B_2, I, J) be the representative of $m = [B_1, B_2, I, J]$. Consider the nearby quadruple

$$(B_1 + \delta B_1, B_2 + \delta B_2, I + \delta I, J + \delta J) \tag{143}$$

Assuming it solves the ADHM equations to the linear order, the variations $\delta B_1, \delta B_2, \delta I, \delta J$ are subject to the linearized $\mu_{12}^{\mathbb{C}}$ equation:

$$d_2\left(\delta B_1, \delta B_2, \delta I, \delta J\right) := [B_1, \delta B_2] + [\delta B_1, B_1] + (\delta I)J + I(\delta J) = 0 \tag{144}$$

and we identify the variations which differ by an infinitesimal $GL(K)$-transformation of (B_1, B_2, I, J):

$$\begin{aligned} &(\delta B_1,\, \delta B_2,\, \delta I,\, \delta J) \sim (\delta B_1,\, \delta B_2,\, \delta I,\, \delta J) + d_1\left(\delta\sigma\right), \\ &\qquad d_1\left(\delta\sigma\right) := \left(\, [B_1, \delta\sigma]\,,\, [B_2, \delta\sigma],\, -\delta\sigma \cdot I\,,\, J \cdot \delta\sigma\, \right) \end{aligned} \tag{145}$$

Since $d_2 \circ d_1 = 0$, the tangent space is the degree 1 cohomology, $T_m \mathcal{M}_k(n) = \ker d_2/\mathrm{im} d_1 = H^1 \mathcal{T}_m \mathcal{M}_k(n)$, of the complex

(146) $$\mathcal{T}_m \mathcal{M}_k(n) = [0 \to \mathrm{End}(K) \xrightarrow{d_1} \mathrm{End}(K) \otimes \mathbb{C}^2_{12} \oplus \mathrm{Hom}(N,K) \oplus \mathrm{Hom}(K,N) \xrightarrow{d_2} \mathrm{End}(K) \otimes \wedge^2 \mathbb{C}^2_{12} \to 0]$$

5.12. Fixed locus. In applications we will be interested in the fixed point set $\mathcal{M}_k(n)^{\mathrm{T}_{12}}$ with $\mathrm{T}_{12} \subset \mathsf{H}_{12}$ a commutative subgroup. The maximal torus $\mathrm{T}_{12} \subset \mathsf{H}_{12}$ is the product of the maximal torus $\mathbf{T}_n \subset PU(n)$ and the two dimensional torus $U(1)_{12} \times U(1)'_{12} \subset SU(2)_{12} \times U(1)'_{12}$. Let

(147) $$\mathfrak{a} = \mathrm{i}\,\mathrm{diag}(\mathfrak{a}_1, \ldots, \mathfrak{a}_n)\,, \qquad \mathfrak{a}_\alpha \in \mathbb{R}$$

be the generic element of Lie$\,(\mathbf{T}_n)$. It means that the numbers $\mathfrak{a}_\alpha$ are defined up to the simultaneous shift

(148) $$\mathfrak{a}_\alpha \sim \mathfrak{a}_\alpha + a, \qquad a \in \mathbb{R}$$

and we assume $\mathfrak{a}_\alpha \neq \mathfrak{a}_\beta$, for $\alpha \neq \beta$. Let

(149) $$\underline{\mathfrak{e}}_{12} = \frac{1}{2}\,(\,\mathfrak{e}_1 - \mathfrak{e}_2,\ \mathfrak{e}_1 + \mathfrak{e}_2\,)$$

be the generic element of Lie$\,(U(1)_{12} \times U(1)'_{12})$. The pair $(\mathfrak{a}, \underline{\mathfrak{e}}_{12})$ generates an infinitesimal transformation (140) of the quadruple (B_1, B_2, I, J):

(150) $$\delta_{\mathfrak{a},\underline{\mathfrak{e}}_{12}}(B_1, B_2, I, J) = (\,\mathfrak{e}_1 B_1, \mathfrak{e}_2 B_2, I\mathfrak{a}, (\mathfrak{e}_1 + \mathfrak{e}_2 - \mathfrak{a})\,J\,)$$

For the $U(k)$-equivalence class $f = [B_1, B_2, I, J]$ to be fixed under the infinitesimal transformation generated by the generic pair $(\mathfrak{a}, \underline{\mathfrak{e}})$ there must exist an infinitesimal $U(k)$-transformation (107)

(151) $$\delta_\sigma(B_1, B_2, I, J) = (\,[B_1, \sigma],\ [B_2, \sigma],\ -\sigma I,\ J\sigma\,)$$

undoing it: $\delta_\sigma(B_1, B_2, I, J) + \delta_{\mathfrak{a},\underline{\mathfrak{e}}_{12}}(B_1, B_2, I, J) = 0$. In other words, there must exist the operators $\widehat{d}_\alpha, \widehat{\mathrm{d}}_1, \widehat{\mathrm{d}}_2 \in \mathrm{End}(K)$, such that:

(152) $$\sigma = \sum_{\alpha \in [n]} \mathfrak{a}_\alpha \widehat{d}_\alpha + \mathfrak{e}_1 \widehat{\mathrm{d}}_1 + \mathfrak{e}_2 \widehat{\mathrm{d}}_2\,, \qquad \sum_{\alpha \in [n]} \widehat{d}_\alpha = \mathbf{1}_K$$

obeys:

(153) $$\mathfrak{e}_a B_a = [\sigma, B_a]\,,\ a \in \{1,2\} \qquad I\mathfrak{a} = \sigma I\,, \quad -(\mathfrak{e}_1 + \mathfrak{e}_2)\,J + \mathfrak{a}J = J\sigma\,,$$

or, in the group form:

(154) $$q_a B_a = g_t B_a g_t^{-1}\,, \quad q_1 q_2 h_t^{-1} J = J g_t^{-1}\,, \quad I h_t = g_t I$$

where $q_a = e^{t\mathfrak{e}_a}$, $g_t = e^{t\sigma}$, $h_t = e^{t\mathfrak{a}}$. Here t is an arbitrary complex number, the map $t \mapsto g_t$ defines the representation $\mathbf{T}_{12} \to GL(k)$. The Eqs. (154) imply:

(155) $$g_t \left(B_1^{i-1} B_2^{j-1} I\right) = q_1^{i-1} q_2^{j-1} \left(B_1^{i-1} B_2^{j-1} I\right) h_t$$

The Eqs. (153) for generic $(\mathfrak{a}, \underline{\mathfrak{e}})$ imply:

(156)
$$K = \bigoplus_{\alpha\in[n]} K_\alpha\,, \qquad \widehat{d}_\alpha|_{K_\beta} = \delta_{\alpha,\beta}$$

The eigenspace K_α is generated by $I_\alpha = I(N_\alpha)$, where $N_\alpha \subset N$ is the eigenline of $\mathfrak{a}$ with the eigenvalue $\mathfrak{a}_\alpha$:

(157)
$$K_\alpha = \sum_{i,j\geq 1} B_1^{i-1} B_2^{j-1} I_\alpha\,.$$

The subspace $I_\alpha \subset K$ (it is one-dimensional for generic $\mathfrak{a}$) obeys:

(158)
$$\widehat{d}_\beta I_\alpha = \delta_{\alpha,\beta}\,, \ \widehat{\mathrm{d}}_1 I_\alpha = \widehat{\mathrm{d}}_2 I_\alpha = 0$$

On K_α the operators $\widehat{\mathrm{d}}_1, \widehat{\mathrm{d}}_2$ have non-negative spectrum:

(159)
$$\sigma\left(B_1^{i-1}B_2^{j-1}I_\alpha\right) = \left(\mathfrak{a}_\alpha + \mathfrak{e}_1(i-1) + \mathfrak{e}_2(j-1)\right)\left(B_1^{i-1}B_2^{j-1}I_\alpha\right)$$

Therefore, as long as $\mathfrak{a}_\beta - \mathfrak{a}_\alpha \notin \mathfrak{e}_1\mathbb{Z}_{>0} + \mathfrak{e}_2\mathbb{Z}_{>0}$,

(160)
$$J\left(B_1^{i-1}B_2^{j-1}I_\alpha\right) = 0\,,$$

as follows from the last Eq. in (153).

Thus, we have shown that

(161)
$$J = 0, \qquad [B_1, B_2] = 0$$

Define an ideal $\mathcal{I}^{(\alpha)} \subset \mathbb{C}[x,y]$ in the ring of polynomials in two variables by:

(162)
$$P(x,y) \in \mathcal{I}^{(\alpha)} \Leftrightarrow P(B_1,B_2)I_\alpha = 0$$

Define the partition $\lambda^{(\alpha)} = \left(\lambda_1^{(\alpha)} \geq \lambda_2^{(\alpha)} \geq \ldots \geq \lambda_{\ell_{\lambda^{(\alpha)}}}^{(\alpha)}\right)$ by

(163)
$$\lambda_i^{(\alpha)} = \min\{\, j \mid B_1^{i-1}B_2^{j}I_\alpha = 0\,\}$$

Thus, $K_\alpha = \mathbb{C}[z_1,z_2]/\mathcal{I}_{\lambda^{(\alpha)}}$. Here we denote by $\mathcal{I}_\lambda \subset \mathbb{C}[x,y]$ the ideal generated by the monomials $x^{i-1}y^{\lambda_i}$, $i = 1,2,\ldots,\ell_\lambda$.

Conversely, given the monomial ideal $\mathcal{I}_{\lambda^{(\alpha)}}$ define the vector $I_\alpha \subset K_\alpha$ to be the image of the polynomial 1 in the quotient $\mathbb{C}[x,y]/\mathcal{I}_{\lambda^{(\alpha)}}$. The operators B_1, B_2 act by multiplication by the coordinates x, y, respectively. Furthermore,

(164)
$$K_\alpha = \bigoplus_{i=1}^{\ell_\lambda}\bigoplus_{j=1}^{\lambda_i} K_{\alpha;i,j}$$

where

(165)
$$\widehat{\mathrm{d}}_1|_{K_{\alpha;i,j}} = i-1, \qquad \widehat{\mathrm{d}}_2|_{K_{\alpha;i,j}} = j-1$$

The map $(\mathfrak{a}, \underline{\mathfrak{e}}_{12}) \mapsto \sigma \in \mathrm{End}(K)$ makes the space K a $\mathbf{T}_{12}$-representation. Its character is easy to compute:

(166)
$$K_\chi := \mathsf{Tr}_K\, g_t = \sum_{\alpha\in[n]} e^{t\mathfrak{a}_\alpha} \sum_{i=1}^{\ell_{\lambda^{(\alpha)}}} q_1^{i-1} \sum_{j=1}^{\lambda_i^{(\alpha)}} q_2^{j-1}$$

The set of eigenvalues of σ is a union of n collections of centers of boxes of Young diagrams $\lambda^{(1)},\ldots,\lambda^{(n)}$

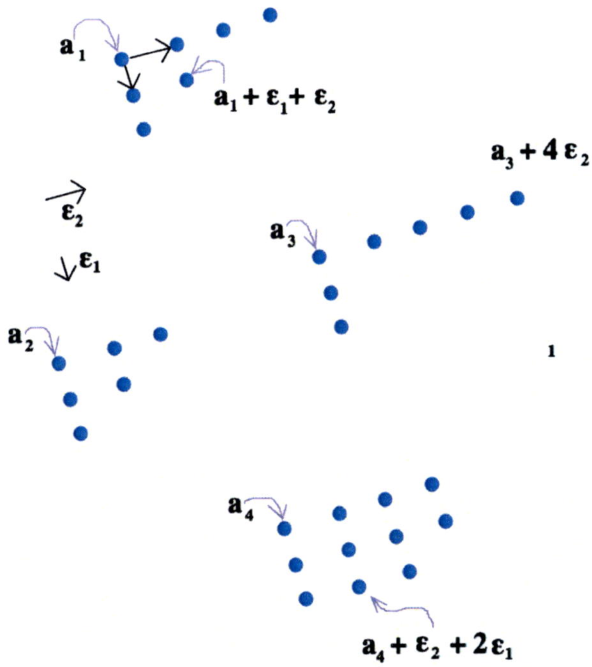

Figure 6: Eigenvalues of σ

The space N is a $\mathbf{T}_{12}$-representation by definition:

(167) $$N_{\chi} := \mathrm{Tr}_{N}\, h_t = \sum_{\alpha\in[n]} e^{ta_{\alpha}}$$

We also define:

(168) $$K^{*}_{\chi} = \mathrm{Tr}_{K}\, g_t^{-1}\ ,\ N^{*}_{\chi} = \mathrm{Tr}_{N}\, h_t^{-1}$$

5.13. Tangent space at the fixed point. Finally, the tangent space $T_f\mathcal{M}_k(n)$ to the moduli space at the fixed point f is also a $\mathbf{T}_{12}$-representation. Let us compute its character. Let $f = [B_1, B_2, I, J]$. The quadruple (B_1, B_2, I, J) is fixed by the composition of the $\mathbf{T}_{12}$ transformation $(e^{t\mathbf{a}}, e^{t\underline{\varepsilon}_{12}}) \in \mathbf{T}^{\mathbb{C}}_{12}$ and the $GL(k)$-transformation $e^{t\sigma} \in GL(k)$, for any complex number t, cf. (153). Now take the nearby quadruple

$$\left(\tilde{B}_1 = B_1 + \delta B_1, \tilde{B}_2 = B_2 + \delta B_2, \tilde{I} = I + \delta I, \tilde{J} = J + \delta J\right)$$

and act on it by the combination of the $\mathbf{T}^{\mathbb{C}}_{12}$ transformation $(e^{t\mathbf{a}}, e^{t\underline{\varepsilon}_{12}})$:

$$\left(\tilde{B}_1, \tilde{B}_2, \tilde{I}, \tilde{J}\right) \mapsto \left(q_1\tilde{B}_1,\, q_2\tilde{B}_2,\, \tilde{I},\, q_1q_2\tilde{J}\right)$$

and the $GL(k)$-transformation g_t:

$$\left(\tilde{B}_1,\, \tilde{B}_2,\, \tilde{I},\, \tilde{J}\right) \mapsto \left(g_t^{-1}\tilde{B}_1 g_t,\, g_t^{-1}\tilde{B}_2 g_t,\, g_t^{-1}\tilde{I},\, \tilde{J}g_t\right)\ ,$$

defining the $\mathbf{T}_{12}$-action on $(\delta B_1, \delta B_2, \delta I, \delta J)$:

(169) $$e^{t}\cdot[\delta B_1, \delta B_2, \delta I, \delta J] = \left[q_1 g_t^{-1}\delta B_1 g_t\, ,\, q_2 g_t^{-1}\delta B_2 g_t\, ,\, g_t^{-1}\delta I h_t, q_1q_2 h_t^{-1}\delta J g_t\right]\ .$$

So the space $\mathcal{T}_f^1\mathcal{M}_k(n)$ of variations $(\delta B_1, \delta B_2, \delta I, \delta J)$ is a $\mathbf{T}_{12}$ representation, with the character:

$$\mathrm{Tr}_{\mathcal{T}_f^1\mathcal{M}_k(n)}(h,q) = (q_1+q_2)K_\chi K_\chi^* + N_\chi K_\chi^* + N_\chi^* K_\chi q_1 q_2 \tag{170}$$

The first two terms on the right hand side account for $\delta B_1, \delta B_2$, the third term corresponds to the δI variations, and the last term accounts for the δJ variations.

Now, the tangent space $T_f\mathcal{M}_k(n)$ is the degree 1 cohomology $H^1\mathcal{T}_f\mathcal{M}_k(n)$ of the complex (146), which has no H^0 or H^2 cohomology (for $\zeta > 0$). The character of $T_f\mathcal{M}_k(n)$ can be therefore computed by taking the alternating sum of the characters of $\mathcal{T}_f^0\mathcal{M}_k(n)$, $\mathcal{T}_f^1\mathcal{M}_k(n)$ and $\mathcal{T}_f^2\mathcal{M}_k(n)$, giving:

$$\mathrm{Tr}_{T_f\mathcal{M}_k(n)}(h,q) = N_\chi K_\chi^* + q_{12}N_\chi^* K_\chi - p_{12}K_\chi K_\chi^* \tag{171}$$

Thus,

$$\mathrm{Tr}_{T_f\mathcal{M}_k(n)}(h,q) = \sum_{\alpha,\beta\in[n]} e^{t(\mathfrak{a}_\alpha - \mathfrak{a}_\beta)}\, T_\chi(\lambda^{(\alpha)}, \lambda^{(\beta)}) \tag{172}$$

where

$$T_\chi(\mu,\lambda) = \sum_{(i,j)\in\lambda} q_1^{i-\lambda_j^t} q_2^{\mu_i+1-j} + \sum_{(i,j)\in\mu} q_1^{\mu_j^t+1-i} q_2^{j-\lambda_i} \tag{173}$$

We see that, as long as there is no rational relation between $\mathfrak{e}_1$ and $\mathfrak{e}_2$, and $\mathfrak{a}_\alpha - \mathfrak{a}_\beta \notin \mathfrak{e}_1\mathbb{Z} + \mathfrak{e}_2\mathbb{Z}$ the weights which appear in the character of the tangent space are non-zero. In other words, the tangent space $T_f\mathcal{M}_k(n)$ does not contain trivial representations of $\mathbf{T}_{12}$, i.e. f is an isolated fixed point.

5.14. Canonical complex at the fixed point. Let $f = \left(\lambda^{(\alpha)}\right)_{\alpha\in[n]} \in \mathcal{M}_k(n)$ be one of the $\mathbf{T}_{12}$-fixed points. The complex $\mathcal{S}_z$ is $\mathbf{T}_{12}$-equivariant, while its fibers over $z=0$ and the cohomology are $\mathbf{T}_{12}$-invariant. Accordingly, $P_{12}^{\pm}$ are $\mathbf{T}_{12}$-graded. Let us discuss this structure in more detail. First of all, K is the $\mathbf{T}_{12}$-representation, $e^t \cdot \psi = g_t\psi$, where $g_t = e^{t\sigma}$. The space P_{12}^-, being a subspace of K, is endowed with the $\mathbf{T}_{12}$-action. Indeed, if $\psi \in P_{12}^-$, i.e. $B_a\psi = 0$, then $B_a(e^t\cdot\psi) = q_a^{-1}g_t B_a\psi = 0$, for $a = 1,2$.

Now, let us endow P_{12}^+ with the action of $\mathbf{T}_{12}$. Let $\psi_1 \oplus \psi_2 \oplus \xi \in K \oplus K \oplus N$. We define

$$e^t \cdot [\psi_1 \oplus \psi_2 \oplus \xi] = [q_2 g_t\psi_1 \oplus q_1 g_t \psi_2 \oplus h\xi] \tag{174}$$

so that we have

(175)

$$\sigma_0\left(e^t\cdot(\psi_1\oplus\psi_2\oplus\xi)\right) = q_1B_1g_t\psi_2 - q_2B_2g_t\psi_1 + Ih\xi = g_t\sigma_0\left(\psi_1\oplus\psi_2\oplus\xi\right),$$
$$\tau_0\left(e^t\cdot\psi\right) = \left(B_1g_t\psi \oplus B_2g_t\psi \oplus Jg_t\psi\right) = q_1^{-1}q_2^{-1}\left(e^t\cdot\tau_0(\psi)\right)$$

We can now compute the $\mathbf{T}_{12}$-characters of $P_{12}^{\pm}$:

(176) $$\mathrm{Tr}_{P_{12}^+}\left(e^t\right) - q_1q_2\mathrm{Tr}_{P_{12}^-}\left(e^t\right) = \mathrm{Tr}_{S^+}(g_t) - \mathrm{Tr}_{S^-}(g_t) =$$

$$\sum_{\alpha\in[n]} e^{t\mathfrak{a}_\alpha}\left(\sum_{(i,j)\in\partial_+\lambda^{(\alpha)}} q_1^{i-1}q_2^{j-1} - \sum_{(i,j)\in\partial_-\lambda^{(\alpha)}} q_1^{i-1}q_2^{j-1}\right)$$

where the sets $\partial_{\pm}\lambda^{(\alpha)}$ of squares can be found near the Young diagram of $\lambda^{(\alpha)}$,

$$S^{+} = \bigoplus_{\alpha\in[n]} \bigoplus_{\square\in\partial_{+}\lambda^{(\alpha)}} S^{+}_{\alpha;\square} \qquad S^{-} = \bigoplus_{\alpha\in[n]} \bigoplus_{\square\in\partial_{-}\lambda^{(\alpha)}} S^{-}_{\alpha;\square}$$

(177)

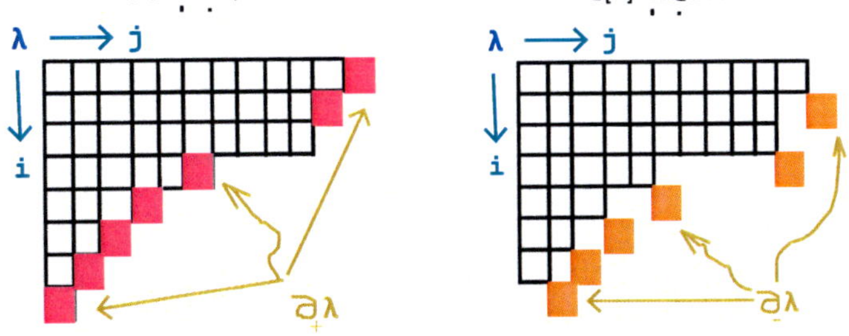

Fig. 5.14

see the picture 5.14 above. For a partition λ the set $\partial_{+}\lambda$ consists of squares which can be added to the Young diagram of λ making it another Young diagram. The squares in $\partial_{-}\lambda$ shifted one step North and one step West can be removed from the Young diagram of λ making it a smaller Young diagram.

The operation which adds to $\lambda^{(\alpha)}$ of adding a few squares from $\partial_{+}\lambda^{(\alpha)}$ or removes a few squares which are obtained by the $(-1,-1)$ shift from $\partial_{-}\lambda^{(\alpha)}$ is the fixed point version of the Hecke correspondences (132).

5.15. Smaller tori. Let $(\Gamma^{\vee}, d)$ be a pair consisting of a finite or a countable set $\Gamma^{\vee}$ (the meaning of the notation will become clear later), and a function $d : \Gamma^{\vee} \to \mathbb{N}$, which we shall call the dimension. We assign to each $\omega \in \Gamma^{\vee}$ a vector space

(178) $$\mathcal{R}_{\omega} = \mathbb{C}^{d(\omega)}$$

of the corresponding dimension.

Let $\underline{\mathbf{n}}$ be a d-partition of n,

(179) $$n = \sum_{\omega\in\Gamma^{\vee}} n_{\omega} d(\omega)\,, \qquad n_{\omega} \geq 0$$

with only a finite number of $n_{\omega} > 0$. Let

(180) $$\ell_{\underline{\mathbf{n}}} = \#\{\, \omega \,|\, n_{\omega} > 0\,\}$$

We associate to $\underline{\mathbf{n}}$ a decomposition of N into the direct sum of tensor products:

(181) $$N = \bigoplus_{\omega\in\Gamma^{\vee}} N_{\omega} \otimes \mathcal{R}_{\omega}$$

with n_{ω}-dimensional complex vector spaces N_{ω}.

Define, for the d-partition $\underline{\mathbf{n}}$ and a pair (e_1, e_2) of non-zero integers, the subtorus

(182) $$\mathrm{T}_{\underline{\mathbf{n}};\underline{\mathbf{e}}} \approx \mathbf{T}_{\underline{\mathbf{n}}} \times U(1)_{\underline{\mathbf{e}}} \subset PU(N) \times \mathrm{Spin}(4)_{12}\,.$$

Here $U(1)_{\underline{\mathbf{e}}}$ is embedded into $U(1)_{12} \times U(1)'_{12} \subset U(2)_{12} \subset \mathrm{Spin}(4)_{12}$ by

(183) $$U(1)_{\underline{\mathbf{e}}}\,:\; e^{\mathrm{i}\vartheta} \mapsto \left(e^{\frac{\mathrm{i}}{2}(e_1+e_2)\vartheta},\; e^{\frac{\mathrm{i}}{2}(e_1-e_2)\vartheta} \right)\,,$$

in other words, it acts on $\mathbb{C}^2_{12}$ by:

(184) $$(z_1, z_2) \mapsto \left(e^{\mathrm{i}e_1\vartheta} z_1, e^{\mathrm{i}e_2\vartheta} z_2 \right)\,.$$

The torus $\mathbf{T}_{\underline{\mathbf{n}}} \subset \mathbf{T}_n$ is defined to be a quotient of the product of the maximal tori of $U(n_\omega)$ by the overall center $U(1)$:

$$(185)\quad h = \bigoplus_{\omega\in\Gamma^\vee} h_\omega \otimes \mathbf{1}_{\mathcal{R}_\omega} \in U(N)\,, \quad \left(h_1 : \ldots : h_{\ell_{\underline{\mathbf{n}}}}\right) = \left(h_1 u : \ldots : h_{\ell_{\underline{\mathbf{n}}}} u\right) \in \mathbf{T}_{\underline{\mathbf{n}}}\,,$$

where $h_\omega h_\omega^\dagger = 1$, $\omega \in \Gamma^\vee$, $h_\omega = \mathrm{diag}\,(u_{\omega,1}, \ldots, u_{\omega,n_\omega})$, and $|u|^2 = 1$.

5.16. Fixed points of smaller tori. Let us start with $n = 1$, so that $\mathbf{T}_{12} = \mathrm{T}_{\underline{\mathbf{e}}}$. The $\mathrm{T}_{\underline{\mathbf{e}}}$-fixed points on $\mathcal{M}_k(1) = \mathrm{Hilb}^{[k]}(\mathbb{C}^2)$ are isolated for $e_1e_2 < 0$ and non-isolated for $e_1e_2 > 0$, as we see from the $T_\chi(\lambda,\lambda)$ character (173). Indeed, as soon as the partition λ has a box $\square = (i,j)$ whose arm plus one-to-leg, or leg plus one-to-arm ratio is equal to $e_1 : e_2$,

$$(186)\quad e_1(i - \lambda_j^t) + e_2(\lambda_i + 1 - j) = 0\,, \qquad \text{or} \qquad e_1(\lambda_j^t + 1 - i) + e_2(j - \lambda_i) = 0$$

then $T_\lambda \mathcal{M}_k(1)$ contains trivial $\mathrm{T}_{\underline{\mathbf{e}}}$-representations, i.e. λ is not an isolated fixed point. Geometrically, the fixed points of the $\mathrm{T}_{\underline{\mathbf{e}}}$-action for $e_1e_2 > 0$ are the (e_1, e_2)-graded ideals $\mathcal{I}$ in $\mathbb{C}[x,y]$, i.e. the ideals which are invariant under the $\mathbb{C}^\times$-action:

$$(187)\qquad (x,y) \to (t^{e_1}x, t^{e_2}y)$$

For such an ideal $\mathcal{I}$ the quotient $K = \mathbb{C}[x,y]/\mathcal{I}$ is also a graded vector space:

$$(188)\qquad K = \bigoplus_{s=0}^{d_K} K_s$$

For general $\Gamma^\vee$ and the general partition $\underline{\mathbf{n}}$ the $\mathbf{T}_{\underline{\mathbf{n}};\underline{\mathbf{e}}}$-fixed point set is a finite union of finite product

$$(189)\qquad \mathcal{M}_k(n)^{\mathbf{T}_{\underline{\mathbf{n}};\underline{\mathbf{e}}}} = \bigcup_{\sum_{\omega\in\Gamma^\vee,\alpha\in[n_\omega]} k_{\alpha,\omega}=k} \ \bigtimes_{\omega\in\Gamma^\vee} \bigtimes_{\alpha\in[n_\omega]} \mathcal{M}_{k_{\alpha,\omega}}(d(\omega))^{\mathbf{T}_{\underline{\mathbf{e}}}}$$

of the $\mathbf{T}_{\underline{\mathbf{e}}}$-fixed point sets on the moduli spaces $\mathcal{M}_{k'}(n')$. This is easy to show using the same methods as we employed so far. It suffices then to analyze the structure of $\mathcal{M}_k(n)^{\mathbf{T}_{\underline{\mathbf{e}}}}$ where the torus $\mathbf{T}_{\underline{\mathbf{e}}} \approx U(1)$ acts on the matrices (B_1, B_2, I, J) via:

$$(190)\qquad e^{\mathrm{i}t} \cdot [B_1, B_2, I, J] = \left[e^{\mathrm{i}e_1 t}B_1, e^{\mathrm{i}e_2 t}B_2, I, e^{\mathrm{i}(e_1+e_2)t}J\right]$$

As usual, the $GL(K)$-equivalence class of the quadruple $[B_1, B_2, I, J]$ is $\mathbf{T}_{\underline{\mathbf{e}}}$-invariant if for every $e^t \in \mathbb{C}^\times$ there is an operator $g_t \in GL(K)$ which undoes (190), i.e.

$$(191)\qquad \begin{aligned} &e^{\mathrm{i}e_1 t} g_t^{-1} B_1 g_t = B_1\,, \quad e^{\mathrm{i}e_2 t} g_t^{-1} B_2 g_t = B_2 \\ &g_t^{-1} I = I\,, \quad e^{\mathrm{i}(e_1+e_2)t} J g_t = J \end{aligned}$$

The correspondence $e^t \mapsto g_t$ splits K as the sum of irreducible representations of $\mathbf{T}_{\underline{\mathbf{e}}}$

$$(192)\qquad K = \bigoplus_{s\geq 0} K_s \otimes \mathcal{R}_s$$

with K_s being the multiplicity space of the charge s representation $\mathcal{R}_s$: $e^t \mapsto e^{ts}$. Let

$$(193)\qquad k_s = \mathrm{dim} K_s\ .$$

We have:

$$\sum_{\mathfrak{s}} k_{\mathfrak{s}} = k \tag{194}$$

The grade 0 component is 1-dimensional:

$$K_0 = I(N)\,. \tag{195}$$

The operators B_1, B_2 raise the grading by e_1 and e_2, respectively:

$$B_a : K_{\mathfrak{s}} \to K_{\mathfrak{s}+e_a}\,,\ a = 1,2 \tag{196}$$

The complex $\mathcal{S}_0$ and its cohomology $P_{12}^{\pm}$ are also graded:

$$P_{12}^{\pm} = \bigoplus_{\mathfrak{s}} P_{12,\mathfrak{s}}^{\pm}\,. \tag{197}$$

The dimensions $k_{\mathfrak{s}}$ are constant throughout the connected component of the set of (e_1, e_2)-homogeneous ideals. In fact, for $e_1 = e_2$ the component is a smooth projective variety, [**16**]. See also [**15**], [**18**].

5.17. Compactness of the fixed point set. The topology of the fixed point set $\mathcal{M}_k(n)^{\mathbf{T}_{12}}$ depends on the choice of the torus $\mathbf{T}_{12}$. In other words, it depends on how non-generic the choice of $(\underline{\mathfrak{a}}, \underline{\mathfrak{e}})$ is.

If there is no rational relation between $\underline{\mathfrak{a}}$ and $\mathfrak{e}_1, \mathfrak{e}_2$, more precisely, if for any $\alpha, \beta \in [n]$ and $p, q \in \mathbb{Z}$

$$\mathfrak{a}_\alpha - \mathfrak{a}_\beta + \mathfrak{e}_1 p + \mathfrak{e}_2 q = 0 \qquad \Longrightarrow \qquad \alpha = \beta\,, \quad p = q = 0 \tag{198}$$

then the fixed points f are isolated, $f \leftrightarrow \left(\lambda^{(\alpha)}\right)_{\alpha \in [n]}$. Their total number, for fixed k is finite, therefore the set of fixed points is compact.

What if there is a rational relation between $\mathfrak{a}_\alpha - \mathfrak{a}_\beta$ and $\mathfrak{e}_1, \mathfrak{e}_2$? That is for some non-trivial $\alpha, \beta \in [n]$ and $p, q \in \mathbb{Z}$,

$$\mathfrak{a}_\alpha - \mathfrak{a}_\beta + \mathfrak{e}_1 p + \mathfrak{e}_2 q = 0\,. \tag{199}$$

We shall assume all the rest of the parameters $\mathfrak{a}_\gamma, \mathfrak{e}_1, \mathfrak{e}_2$ generic. In particular we assume both $\mathfrak{e}_1, \mathfrak{e}_2$ non-zero. There are three cases to consider:

(1) $\alpha \neq \beta$ and $pq > 0$;
(2) $\alpha = \beta$ and $pq > 0$;
(3) $pq < 0$ and no restriction on α, β ;

In the case (1) the fixed locus is non-compact. It is parametrized by the value of the invariant

$$J_\beta B_1^{p-1} B_2^{q-1} I_\alpha \tag{200}$$

We therefore must make sure, in what follows, that the eigenvalues $(\mathfrak{a}_\alpha)_{\alpha \in [n]}$ of the infinitesimal framing rotations and the parameters $(\mathfrak{e}_1, \mathfrak{e}_2)$ of the spatial rotation do not land on the hyperplanes:

$$\mathfrak{a}_\alpha - \mathfrak{a}_\beta + \mathfrak{e}_1 p + \mathfrak{e}_2 q \neq 0, \tag{201}$$

for all $\alpha \neq \beta$, and integer $p, q \geq 1$.

In the case (2) the fixed points corresponding to the monomial ideals are isolated, since the weights in (173) have the form $\mathfrak{e}_1 p' + \mathfrak{e}_2 q'$ with $p'q' \leq 0$. We shall show below that

> the $U(1)$-fixed points in the case $pq > 0$ correspond to the monomial ideals. In other words they are $U(1) \times U(1)$-invariant

For fixed k the sizes of the Young diagrams $\lambda^{(\alpha)}$ are bounded above, since

$$\sum_{\alpha=1}^{n} |\lambda^{(\alpha)}| = k \tag{202}$$

Since the number of collections of Young diagrams which obey (202) is finite, the set of points fixed by the action of the maximal torus $\mathbf{T}$ is compact. This set, as we just showed, is in one-to-one correspondence with the collections

$$f \leftrightarrow \left(\lambda^{(\alpha)}\right)_{\alpha\in[n]} \tag{203}$$

obeying (202).

In the case (3) the fixed points are not isolated, but the fixed point set is compact. Let us show the $\mathbf{T}_{\underline{\mathbf{n}};\underline{\mathbf{e}}}$-fixed point set is compact. There are two cases:

(1) $e_1 e_2 > 0$. In this case the minimal torus corresponds to $\Gamma^{\vee} = \{0\}$, $d(0) = n$, $n_0 = 1$, i.e. for $\mathbf{T}_{\underline{\mathbf{n}}} = 1$, $\mathbf{T}_{\underline{\mathbf{n}};\underline{\mathbf{e}}} = \mathbf{T}_{\underline{\mathbf{e}}}$. The corresponding Coulomb parameter vanishes, $\mathfrak{a} = 0$.

We are going to demonstrate that for all $\mathbf{T}_{\underline{\mathbf{n}};\underline{\mathbf{e}}}$-fixed points on $\mathcal{M}_k(n)$ the L^2-norm of (B_1, B_2, I, J) is bounded above by a constant which depends only on n, k, and ζ. We use the real moment map equation (46):

$$\begin{aligned} k\zeta &= \mathrm{Tr}_K(\mu) = \|I\|^2 - \|J\|^2 \\ \zeta \mathrm{Tr}_K \sigma &= \mathrm{Tr}_K(\sigma\mu) = e_1\|B_1\|^2 + e_2\|B_2\|^2 + (e_1+e_2)\|J\|^2 \end{aligned} \tag{204}$$

where we used the Eqs. (153) with the specialization $\mathfrak{e}_1 = e_1$, $\mathfrak{e}_2 = e_2$:

$$\begin{aligned} & e_a B_a = [\sigma, B_a], \quad e_a B_a^{\dagger} = [B_a^{\dagger}, \sigma], \quad a = 1, 2, \\ & 0 = \sigma I, \quad 0 = I^{\dagger}\sigma, \quad (e_1+e_2)J = -J\sigma, \quad J^{\dagger}(e_1+e_2) = -\sigma J^{\dagger}, \end{aligned} \tag{205}$$

The Eqs. (205) imply, by the same arguments as before, that the spectrum of σ has the form:

$$\mathfrak{s} = e_1(i-1) + e_2(j-1), \qquad (i,j) \in \Sigma \tag{206}$$

for a finite set Σ of pairs (i,j) of positive integers, and that J maps the eigenvectors of σ in K to zero, unless the eigenvalue is equal to $-e_1 - e_2$. Now, the eigenvalues of σ are of the form (206), which are never equal to $-e_1 - e_2$. Thus, $J|_K = 0$, therefore B_1 and B_2 commute on K.

Now, $\sigma|_{I(N)} = 0$, i.e. $m_0 = \dim(\mathrm{im} I) \leq n$. Now, the vector spaces

$$K_{i,j} = \mathbb{C} \cdot B_1^{i-1} B_2^{j-1} I(N) \tag{207}$$

if non-zero, contribute $\mathrm{dim}K_{i,j} \leq n$ to $k_{\mathfrak{s}}$ with $\mathfrak{s} = e_1(i-1) + e_2(j-1)$. It is clear that

$$k_{\mathfrak{s}} = \dim \sum_{e_1(i-1)+e_2(j-1)=\mathfrak{s}} \mathbb{C}\left(B_1^{i-1}B_2^{j-1}I(N)\right) \leq \tag{208}$$

$$\sum_{e_1(i-1)+e_2(j-1)=\mathfrak{s}} \dim \mathbb{C}\left(B_1^{i-1}B_2^{j-1}I(N)\right) \leq n\ \mathrm{Coeff}_{t^{\mathfrak{s}}} \frac{(1-t^{ke_1})(1-t^{ke_2})}{(1-t^{e_1})(1-t^{e_2})},$$

since both i and j cannot be greater then k. The trace $\mathrm{Tr}_K\sigma$ can be estimated by

$$\mathrm{Tr}_K\,\sigma = \sum_{\mathfrak{s}\in\Sigma} \mathfrak{s}\, k_{\mathfrak{s}} \leq \tag{209}$$

$$n\sum_{\mathfrak{s}=0}^{\infty} \mathfrak{s}\ \mathrm{Coeff}_{t^{\mathfrak{s}}} \frac{(1-t^{ke_1})(1-t^{ke_2})}{(1-t^{e_1})(1-t^{e_2})} = n\,t\frac{d}{dt}\bigg|_{t=1} \frac{(1-t^{ke_1})(1-t^{ke_2})}{(1-t^{e_1})(1-t^{e_2})} = \frac{1}{2}k^2(k-1)n(e_1+e_2)$$

Thus, $J = 0$, the norms $\|B_{1,2}\|^2$ of the operators $B_{1,2}$ are bounded above, while the norm of the operator I is fixed:

$$\|I\|^2 = \zeta k\ , \quad \frac{e_1}{e_1+e_2}\|B_1\|^2 + \frac{e_2}{e_1+e_2}\|B_2\|^2 \leq \frac{\zeta}{2}k^2(k-1)n \tag{210}$$

(2) $e_1e_2 < 0$. In this case we take $\Gamma^{\vee} = [n]$, $d(\omega) = 1$ for all $\omega \in [n]$. The Coulomb parameters are the generic n complex numbers $\mathfrak{a}_\alpha \in \mathbb{C}$, $\alpha \in [n]$, defined up to an overall shift. Below we further restrict the parameters $\mathfrak{a}_\alpha$ to be real, so that they belong to the Lie algebra of the compact torus $\mathbf{T}_{\underline{\mathbf{n}};\underline{\mathbf{e}}}$. The equations (205) generalize to:

(211)
$$e_aB_a = [\sigma, B_a]\,, \quad e_aB_a^{\dagger} = [B_a^{\dagger}, \sigma]\,, \quad a = 1, 2\,,$$
$$I\mathfrak{a} = \sigma I\,, \quad \mathfrak{a}I^{\dagger} = I^{\dagger}\sigma\,, \quad (e_1+e_2-\mathfrak{a})J = -J\sigma\,, \quad J^{\dagger}(e_1+e_2-\mathfrak{a}) = -\sigma J^{\dagger}\,,$$

The fixed point set $\mathcal{M}_k(n)^{\mathbf{T}_{\underline{\mathbf{n}};\underline{\mathbf{e}}}}$ splits:

$$\mathcal{M}_k(n)^{\mathbf{T}_{\underline{\mathbf{n}};\underline{\mathbf{e}}}} = \bigcup_{k_1+\ldots+k_n=k} \bigtimes_{\alpha\in[n]} \mathcal{M}_{k_\alpha}(1)^{\mathbf{T}_{\underline{\mathbf{e}}}} \tag{212}$$

The fixed points are isolated, these are our friends $(\lambda^{(\alpha)})_{\alpha\in[n]}$, the n-tuples of partitions with the total size equal to k. Since it is a finite set, it is compact.

Note that we couldn't restrict the torus $\mathbf{T}_{\underline{\mathbf{n}};\underline{\mathbf{e}}}$ any further in this case. Indeed, the crucial ingredient in arriving at (212) is vanishing of the J matrix for the $\mathbf{T}_{\underline{\mathbf{n}};\underline{\mathbf{e}}}$-invariant solutions of the ADHM equations. The argument below the Eq. (206) we used before would not work for $e_1e_2 < 0$, since $\mathfrak{a}_\alpha - (e_1+e_2)$ may be equal to $\mathfrak{a}_\beta + e_1(i-1) + e_2(j-1)$ for some $\alpha, \beta \in [n]$, $i, j \geq 1$. In this case J may have a non-trivial matrix element, giving rise to a non-compact fixed locus. Now, insisting on the $\mathbf{T}_{\underline{\mathbf{n}};\underline{\mathbf{e}}}$-invariance with $\mathbf{T}_{\underline{\mathbf{n}}} = U(1)^{n-1}$ means $\mathfrak{a}_\alpha$'s in (211) are completely generic, in particular, for $\alpha \neq \beta$, $\mathfrak{a}_\alpha - \mathfrak{a}_\beta \notin \mathbb{Z}$. This still leaves the case $\alpha = \beta$ as a potential source of noncompactness. But this is the case of the $\mathbf{T}_{\underline{\mathbf{e}}}$-action on $\mathcal{M}_k(1) = \mathrm{Hilb}^{[k]}(\mathbb{C}^2)$. In this case J vanishes not because of the toric symmetry, but rather because of the stability condition [**26**]:

$$
\begin{aligned}
(213)\quad & JI = \mathrm{Tr}\, IJ = \mathrm{Tr}\,[B_2, B_1] = 0, \\
& J(xB_1 + yB_2)^l I = \mathrm{Tr}\,(xB_1 + yB_2)^l [B_2, B_1] = \\
& \qquad \mathrm{Tr}\,(xB_1 + yB_2)^l [xB_1 + yB_2, x'B_1 + y'B_2] = 0, \\
& \qquad \text{for any } x, y, x', y', \text{ s.t. } x'y - xy' = 1, \\
& Jf(B_1, B_2)B_1B_2g(B_1, B_2)I = Jf(B_1, B_2)B_2B_1g(B_1, B_2)I + \\
& \qquad (Jf(B_1, B_2)I)\,(Jg(B_1, B_2)I) = 0, \qquad \text{by induction} \\
& \qquad \Longrightarrow J\mathbb{C}[B_1, B_2]I = 0 \Longrightarrow J = 0, \quad \text{by stability}
\end{aligned}
$$

The compactness of $\mathcal{M}_k(n)^{\mathbf{T}_{\underline{n};\epsilon}}$ is thus established.

5.18. Ordinary instantons as the fixed set. Let us now consider the particular $\mathbf{T}_{\mathbf{x}} = U(1)^5$ symmetry of the spiked instanton equations,

$$(214)\qquad (I_A\,,\, J_A) \longrightarrow \left(e^{\mathrm{i}\vartheta_A} I_A\,,\, e^{-\mathrm{i}\vartheta_A} J_A\right)$$

where $\vec{\vartheta} = (\vartheta_1, \vartheta_2, \vartheta_3, \vartheta_4, \vartheta_5, \vartheta_6) \sim (\vartheta_1 + \vartheta, \vartheta_2 + \vartheta, \vartheta_3 + \vartheta, \vartheta_4 + \vartheta, \vartheta_5 + \vartheta, \vartheta_6 + \vartheta)$ for any ϑ. The $\mathbf{T}_{\mathbf{x}}$-invariant configuration $[\mathbf{B}, \mathbf{I}, \mathbf{J}]$ defines a homomorphism of the covering torus $\tilde{\mathbf{T}}_{\mathbf{x}} \approx U(1)^6 \longrightarrow U(k)$, via the compensating $U(k)$-transformation $g(\vec{\vartheta})$ obeying:

$$(215)\qquad e^{\mathrm{i}\vartheta_A} I_A = g(\vec{\vartheta}) I_A\,,\; e^{-\mathrm{i}\vartheta_A} J_A = J_A g(\vec{\vartheta})^{-1}\,,\; g(\vec{\vartheta}) B_a g(\vec{\vartheta})^{-1} = B_a$$

The space K splits as the orthogonal direct sum
(216)

$$K = \bigoplus_{A \in \underline{\mathbf{6}}} K_A\,, \qquad g(\vec{\vartheta})|_{K_A} = e^{\mathrm{i}\vartheta_A}\,, \qquad K_A = \mathbb{C}[B_a, B_b] I_A(N_A)\,, \qquad \text{for } A = \{a, b\}$$

This decomposition is preserved by the matrices $\mathbf{B}, \mathbf{I}, \mathbf{J}$. Thus the solution is the direct sum of the solutions of ADHM equations:

$$(217)\qquad \mathfrak{M}^*_k(\vec{n})^{\mathbf{T}_{\mathbf{x}}} = \bigcup_{\sum_{A \in \underline{\mathbf{6}}} k_A = k} \; \bigtimes_{A \in \underline{\mathbf{6}}} \mathcal{M}_{k_A}(n_A)$$

6. Crossed and folded instantons

Distorted shadows fell
Upon the lighted ceiling:
Shadows of crossed arms, of crossed legs-
Of crossed destiny.†

The next special case is where only two e.g. $N_{A'}$ and $N_{A''}$ out of six vector spaces N_A are non-zero. There are two basic cases.

† *Winter night, from "Dr.Zhivago", B.Pasternak, English translation by Bernard G.Guerney*

6.1. Crossed instantons. Suppose $A' \cup A'' = \emptyset$, e.g. $A' = \{1,2\}$ and $A'' = \{3,4\}$. In this case we define

$$\mathfrak{M}^{+}_{k}(n,w) = \mathfrak{M}^{0}_{k}(n_{12} = n, 0, 0, 0, 0, n_{34} = w) \tag{218}$$

We call the space $\mathfrak{M}^{+}_{k}(n,w)$ the space of (n,w)-crossed instantons.

The virtual dimension of the space $\mathfrak{M}^{+}_{k}(n,w)$ is independent of k, it is equal to $-2nw$. As a set, $\mathfrak{M}^{+}_{k}(n,w)$ is stratified

$$\mathfrak{M}^{+}_{k}(n,w) = \bigcup_{k'+k''\geq k} \mathfrak{M}^{+}_{k',k'';k}(n,w)\ . \tag{219}$$

The stratum

$$\mathfrak{M}^{+}_{k',k'';k}(n,w) = \left\{ \ [\mathbf{B},\mathbf{I},\mathbf{J}] \mid \mathrm{dim}K_{12} = k',\ \mathrm{dim}K_{34} = k'' \ \right\} \tag{220}$$

parametrizes the crossed instantons, whose ordinary instanton components have the charges k' and k'', respectively: the crossed instanton $[\mathbf{B},\mathbf{I},\mathbf{J}]$ defines two ordinary instantons, $[B_1, B_2, I_{12}, J_{12}]$ on $\mathbb{C}^2_{12}$ and $[B_3, B_4, I_{34}, J_{34}]$ on $\mathbb{C}^2_{34}$, of the charges

$$k' = \mathrm{dim}K_{12}\,, \quad k'' = \mathrm{dim}K_{34} \tag{221}$$

We have two maps:

$$\begin{array}{ccc} & \mathfrak{M}^{+}_{k',k'';k}(n,w) & \\ {}^{e_1}\swarrow & & \searrow^{e_2} \\ \mathcal{N}^{-}_{k';k'+k''-k}(n) & & \mathcal{N}^{-}_{k'';k'+k''-k}(w) \end{array} \tag{222}$$

defined in an obvious way:

$$\begin{aligned} e_1[\mathbf{B},\mathbf{I},\mathbf{J}] &= [B_1, B_2, I_{12}, J_{12}, K_{12}\cap K_{34}] \\ e_2[\mathbf{B},\mathbf{I},\mathbf{J}] &= [B_3, B_4, I_{34}, J_{34}, K_{12}\cap K_{34}] \end{aligned} \tag{223}$$

The intersection $K_{12,34} := K_{12}\cap K_{34}$ has dimension $k' + k'' - k$. It is annihilated by the operators B_a for all $a \in \underline{4}$, and by the operators J_{12}, J_{34}. Therefore

$$K_{12,34} \subset P_{12} \cap P_{34} \tag{224}$$

Conversely, starting with the GL(k')-equivalence class $[B_1, B_2, I_{12}, J_{12}] \in \mathcal{M}_{k'}(n)$, $B_1, B_2 \in \mathrm{End}(K')$ etc. and the GL(k'')-equivalence class $[B_3, B_4, I_{34}, J_{34}] \in \mathcal{M}_{k''}(w)$, $B_3, B_4 \in \mathrm{End}(K'')$ etc., with $\mathrm{dim}K' = k'$, $\mathrm{dim}K'' = k''$ and a choice of dimension $\mathrm{dim}P = k' + k'' - k$ vector space P, together with two non-degenerate (injective) linear maps,

$$i_{12} : P \to P_{12} \subset K'\,, \qquad i_{34} : P \to P_{34} \subset K'' \tag{225}$$

defined up to the automorphisms of P:

$$(i_{12}, i_{34}) \sim (i_{12}\circ h^{-1}, i_{34}\circ h^{-1})\,, \qquad h \in GL(k' + k'' - k) \tag{226}$$

we define:

$$K = (K' \oplus K'')\ /\ P \tag{227}$$

where $\chi \in P$ acts on $K' \oplus K''$ via:

$$\chi \cdot (\psi_{12}, \psi_{34}) = (\psi_{12} + i_{12}(\chi)\,,\, \psi_{34} - i_{34}(\chi))\,, \qquad \psi_{12} \in K',\, \psi_{34} \in K''. \tag{228}$$

The maximal strata are $\mathfrak{M}^{+}_{k',k'';k}(n,w)$ with $k'+k'' = k$. The closure $\overline{\mathfrak{M}^{+}_{k',k'';k}(n,w)}$, as a set, is the product $\mathcal{M}_{k'}(n) \times \mathcal{M}_{k''}(w)$. The obstruction bundle, restricted on such a stratum $\mathfrak{M}^{+}_{k',k'';k}(n,w)$ splits as the sum

$$\mathrm{Obs}|_{\mathfrak{M}^{+}_{k',k'';k}(n,w)} = p_1^* T^*_{\mathcal{M}_{k'}(n)} \oplus p_2^* T^*_{\mathcal{M}_{k''}(w)} \oplus \mathrm{Hom}\left(p_1^* S^+, p_2^* S^+\right) \tag{229}$$

where p_1, p_2 are the projections onto the first and the second factors:

$$\begin{array}{ccc} & \mathfrak{M}^{+}_{k',k'';k}(n,w) & \\ {}^{p_1}\swarrow & & \searrow^{p_2} \\ \mathcal{M}_{k'}(n) & & \mathcal{M}_{k''}(w) \end{array} \tag{230}$$

6.2. One-instanton crossed example. When $k = 1$ the matrices B_a are just complex numbers $b_a \in \mathbb{C}$, $a \in \underline{4}$. The equations $b_1 I_{34} = b_2 I_{34} = 0$ and $b_3 I_{12} = b_4 I_{12} = 0$ imply that if $(b_1, b_2) \neq 0$, then $I_{34} = 0$, $I_{12} \neq 0$, $(b_3, b_4) = (0,0)$, $K = K_{12}$ and the rest of the matrices define the ordinary charge 1 $U(n)$ instanton, parametrized by the space (117). Likewise, if $(b_3, b_4) \neq 0$, then $I_{12} = 0$, $I_{34} \neq 0$, $(b_1, b_2) = (0,0)$, $K = K_{34}$ and the rest of the matrices define the ordinary charge 1 $U(w)$ instanton. Finally, if $(b_1, b_2, b_3, b_4) = 0$, then both I_{12}, I_{34} need not vanish. If, indeed, both I_{12}, I_{34} do not vanish, then J_{12} and J_{34} vanish, by the υ-equaitons, while I_{12}, I_{34} obey

$$\|I_{12}\|^2 + \|I_{34}\|^2 = \zeta\,, \tag{231}$$

which, modulo $U(1) = U(K)$ symmetry, define a subset in $\mathbb{CP}^{n+w-1}$, the complement to the pair of "linked" projective spaces, $\mathbb{CP}^{n-1}$ and $\mathbb{CP}^{w-1}$, corresponding to the vanishing of I_{34} and I_{12}, respectively. The result is, then

$$\mathfrak{M}^{+}_{1}(n,w) = \mathbb{C}^2 \times T^*\mathbb{CP}^{n-1} \cup \mathbb{CP}^{n+w-1} \cup \mathbb{C}^2 \times T^*\mathbb{CP}^{w-1}\,, \tag{232}$$

the first and second components intersect along $(0,0) \times \mathbb{CP}^{w-1}$ the second and the third components intersect along $(0,0) \times \mathbb{CP}^{n-1}$, where $\mathbb{CP}^{n-1} \cup \mathbb{CP}^{w-1} \subset \mathbb{CP}^{n+w-1}$ are non-intersecting $\mathbb{CP}^{n-1} \cap \mathbb{CP}^{w-1} = \emptyset$ projective subspaces.

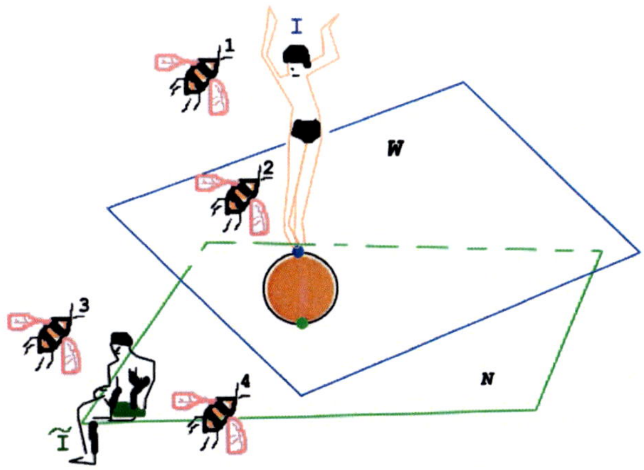

Figure 7: Charge one crossed instanton moduli space: the planes represent the complex plane factors $\mathbb{C}^2_{12}$ and $\mathbb{C}^2_{34}$, the girl represents the $T^*\mathbb{CP}^{w-1}$ factor, the man represents the $T^*\mathbb{CP}^{n-1}$ factor, the orange ball is the $\mathbb{CP}^{n+w-1}$ component, the blue and green dots are the $\mathbb{CP}^{n-1}$ and $\mathbb{CP}^{w-1}$ loci of the intersections of components

6.3. Folded instantons. In this case $A' \cap A'' = \{a\}$, e.g. $A' = \{1,2\}$, $A'' = \{1,3\}$, $a = 1$.

We define:

$$\mathfrak{M}^{\perp}_k(n,w) = \mathfrak{M}^k_k(n_{12} = n, n_{13} = w, 0, 0, 0, 0) \tag{233}$$

We call the space $\mathfrak{M}^{\perp}_k(n,w)$ the space of (n,w)-folded instantons.

There exists an analogue of the stratification (219) for the folded instantons. Again, the folded instanton data $[B_1, B_2, B_3, B_4, I_{12}, I_{13}, J_{12}, J_{13}]$ defines two ordinary noncommutative instantons on $\mathbb{R}^4$, one on $\mathbb{C}^2_{12}$, $[B_1, B_2, I_{12}, J_{12}]$, another on $\mathbb{C}^2_{13}$, $[B_1, B_3, I_{13}, J_{13}]$. The stability implies that B_4 vanishes. The spaces $K_{12} = \mathbb{C}[B_1, B_2]I_{12}(N_{12})$ and $K_{13} = \mathbb{C}[B_1, B_3]I_{13}(N_{13})$ generate all of K,

$$K = K_{12} + K_{13} \tag{234}$$

Their intersection $K_{12,13} = K_{12} \cap K_{13}$ must belong to the smaller subspace $L_{1,2} \cap L_{1,3} \subset K_{12} \cap K_{13}$. The subspace $L_{1,2} \subset K_{12}$ is the kernel of B_2 and $J_{12}\mathbb{C}[B_1]$ while $L_{1,3} \subset K_{13}$ is the kernel of B_3 and $J_{13}\mathbb{C}[B_1]$. As a consequence the operator B_1 preserves the subspaces $L_{1,2}$, $L_{1,3}$ and $K_{12,13}$.

Let us denote, for the vector space W and the linear operator $B \in \mathrm{End}(W)$ by $\mathrm{Gr}(W,B)$ the variety parameterizing B-invariant vector subspaces $V \subset W$, $B(V) \subset V$. This is a particular case of the A-Grassmanian of [**4**, **5**]. To describe the intersection $K_{12,13}$ we need a version $\mathrm{Gr}_2(W,B)$ of the variety $\mathrm{Gr}(W,B)$. The Gr_2-version parametrizes pairs (W_1, W_2) of B-invariant subspaces $W_1, W_2 \subset W$, such that $W = W_1 + W_2$. The strata of $\mathfrak{M}^{\perp}_k(n,w)$ are Gr_2-fibrations over the products $\mathcal{L}^-_{k';k'+k''-k}(n)_{12} \times \mathcal{L}^-_{k'';k'+k''-k}(w)_{13}$ of spaces $\mathcal{L}^-_{k;s}(n)_{ab}$ which parameterize the ordinary instantons $[B_a, B_b, I_{ab}, J_{ab}]$ and subspaces $V \subset L_{a,b}$ of dimension s.

For the purposes of this paper this description of the moduli space of folded instantons would suffice.

6.4. One-instanton folded example. When $k = 1$, as before, the matrices B_a are the complex numbers b_a, $a \in \underline{4}$, except that b_4 vanishes. Now, the equation $b_2 I_{13} = 0$ implies that if $b_2 \neq 0$ then $I_{13} = 0$, and we have the charge one ordinary $U(n)$ instanton on $\mathbb{C}^2_{12}$. Likewise the equation $b_3 I_{12} = 0$ implies that if $b_3 \neq 0$ then $I_{12} = 0$ and we have the charge one ordinary $U(w)$ instaton on $\mathbb{C}^2_{13}$. Finally, when both $b_2 = b_3 = 0$, the remaining equations $J_{12}I_{13} = J_{13}I_{12} = J_{12}I_{12} = J_{13}I_{13} = 0$, and $\|I_{12}\|^2 + \|I_{13}\|^2 - \|J_{12}\|^2 - \|J_{13}\|^2 = \zeta$, define the variety which is a product of a copy of $\mathbb{C}^1$ (parametrized by b_1) and our friend the union of three pieces: $\mathbb{CP}^{n+w-1}$ (this is the locus where $J_{12} = J_{13} = 0$), $T^*\mathbb{CP}^{n-1}$ (the locus where $I_{13} = J_{13} = 0$) and $T^*\mathbb{CP}^{w-1}$ (the locus where $I_{12} = J_{12} = 0$):

$$\mathfrak{M}^{\perp}_1(n, w) = \mathbb{C}^2 \times T^*\mathbb{CP}^{w-1} \cup \mathbb{C}^1 \times \mathbb{CP}^{n+w-1} \cup \mathbb{C}^2 \times T^*\mathbb{CP}^{n-1}, \tag{235}$$

the first and second components intersect along $\mathbb{C}^1 \times \mathbb{CP}^{w-1}$ the second and the third components intersect along $\mathbb{C}^1 \times \mathbb{CP}^{n-1}$, where $\mathbb{CP}^{n-1} \cup \mathbb{CP}^{w-1} \subset \mathbb{CP}^{n+w-1}$ are non-intersecting $\mathbb{CP}^{n-1} \cap \mathbb{CP}^{w-1} = \emptyset$ projective subspaces.

6.5. Fixed point sets: butterflies and zippers. Let us now discuss the fixed point sets of toric symmetries of the crossed and folded instantons. The torus $\mathbf{T}_{n,w} \times U(1)^3_{\mathfrak{e}}$ acts on $\mathfrak{M}^{+}_k(n, w)$ and $\mathfrak{M}^{\perp}_k(n, w)$:

$$(B_a, I_A, J_A) \mapsto \left(e^{it\mathfrak{e}_a} B_a, I_A e^{-it\mathfrak{a}_A}, e^{it\mathfrak{e}_A} e^{it\mathfrak{a}_A} J_A\right) \tag{236}$$

Here $\mathfrak{a}_A = \mathrm{diag}\,(\mathfrak{a}_{A,1}, \ldots, \mathfrak{a}_{A,n_A})$ where the complex numbers $\mathfrak{a}_{A,\alpha}$, $\alpha \in [n_A]$ are defined up to the overall shift

$$\mathfrak{a}_{A,\alpha} \sim \mathfrak{a}_{A,\alpha} + y\,, \tag{237}$$

with $y \in \mathbb{C}$. Let $\underline{\mathfrak{e}} = (\mathfrak{e}_1, \mathfrak{e}_2, \mathfrak{e}_3, \mathfrak{e}_4)$,

$$\sum_{a \in \underline{4}} \mathfrak{e}_a = 0 \tag{238}$$

We assume $\mathfrak{e}_a \neq 0$ for each $a \in \underline{4}$.

6.5.1. *Toric crossed instantons.* The fixed point set $\mathfrak{M}^{+}_k(n, w)^{\mathbf{T}_{n,w} \times U(1)^3_{\mathfrak{e}}}$ is easy to describe. The infinitesimal transformation generated by $(\mathfrak{a}, \underline{\mathfrak{e}})$ is compensated by the infinitesimal $GL(k)$ transformation, generated by $\sigma \in \mathrm{End}(K)$. As in the previous section this makes K a representation of $\mathbf{T}_{n,w} \times U(1)^3_{\mathfrak{e}}$. The space K contains two subspaces, K_{12} and K_{34}, whose intersection $K_{12,34} = K_{12} \cap K_{34}$ belongs to both P_{12} and P_{34}:

$$K_{12,34} \subset P_{12} \cap P_{34} \tag{239}$$

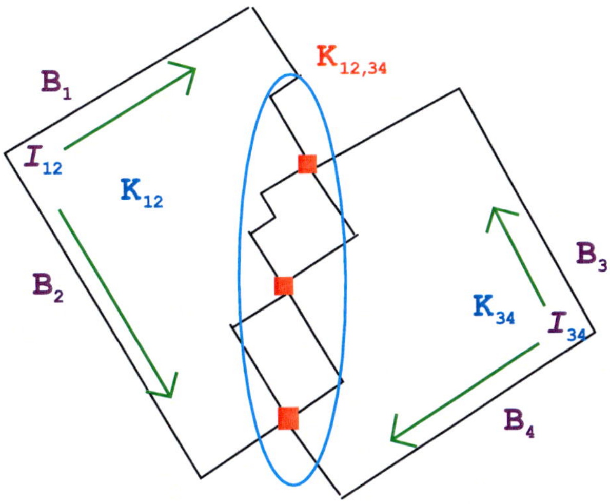

Figure 8: The butterfly

The eigenvalues of σ on K_{12} have the form:

$$\text{(240)}\quad \mathrm{Eigen}\,(\sigma|_{K_{12}}) = \left\{ \mathfrak{a}_{12,\alpha} + \mathfrak{e}_1(i-1) + \mathfrak{e}_2(j-1) \;\middle|\; \alpha \in [n]\,,\, (i,j) \in \lambda^{(12,\alpha)} \right\}$$

The eigenvalues of σ on K_{34} have the form:

$$\text{(241)}\quad \mathrm{Eigen}\,(\sigma|_{K_{34}}) = \left\{ \mathfrak{a}_{34,\beta} + \mathfrak{e}_3(i-1) + \mathfrak{e}_4(j-1) \;\middle|\; \beta \in [w]\,,\, (i,j) \in \lambda^{(34,\beta)} \right\}$$

These two sets do not overlap when all the parameters $\mathfrak{a}_{A,\alpha}, \mathfrak{e}_a$ are generic. Therefore $K_{12} \perp K_{34}$ and the $\mathbf{T}_{n,w} \times U(1)^3_{\mathfrak{e}}$-fixed points are isolated. These fixed points are, therefore, in one-to-one correspondence with the pairs

$$\text{(242)}\qquad (\lambda^{12}; \lambda^{34})$$

consisting of n- and w-tuples

$$\lambda^{12} = \left(\lambda^{(12,1)}, \ldots, \lambda^{(12,n)}\right) ; \qquad \lambda^{34} = \left(\lambda^{(34,1)}, \ldots, \lambda^{(34,w)}\right) ,$$

of partitions, obeying

$$\text{(243)}\qquad \sum_{\alpha\in[n]} \left|\lambda^{(12,\alpha)}\right| + \sum_{\beta\in[w]} \left|\lambda^{(34,\alpha)}\right| = k$$

Their number is finite, therefore the set $\mathfrak{M}^+_k(n)^{\mathbf{T}_{n,w}\times U(1)^3_{\mathfrak{e}}}$ of fixed points is compact.

Now let us try to choose a sub-torus $\mathbf{T}' \subset \mathbf{T}_{n,w} \times U(1)^3_{\mathfrak{e}}$, restricted only by the condition that $J_{12} = J_{34} = 0$ for the $\mathbf{T}'$-invariant solutions of (57). We wish to prove that the set of $\mathbf{T}'$-fixed points is compact in this case as well. In the next sections we shall describe such tori in more detail.

We start by the observation that if $K_{12,34} \neq 0$ then the two sets (240) and (241) of σ-eigenvalues must overlap. Therefore, for some $(\alpha, \beta) \in [n] \times [w]$, and for some $(i', j') \in \lambda^{(12,\alpha)}$, $(i'', j'') \in \lambda^{(34,\beta)}$

$$\text{(244)}\qquad \mathfrak{a}_{12,\alpha} + \mathfrak{e}_1(i'-1) + \mathfrak{e}_2(j'-1) = \mathfrak{a}_{34,\beta} + \mathfrak{e}_3(i''-1) + \mathfrak{e}_4(j''-1)$$

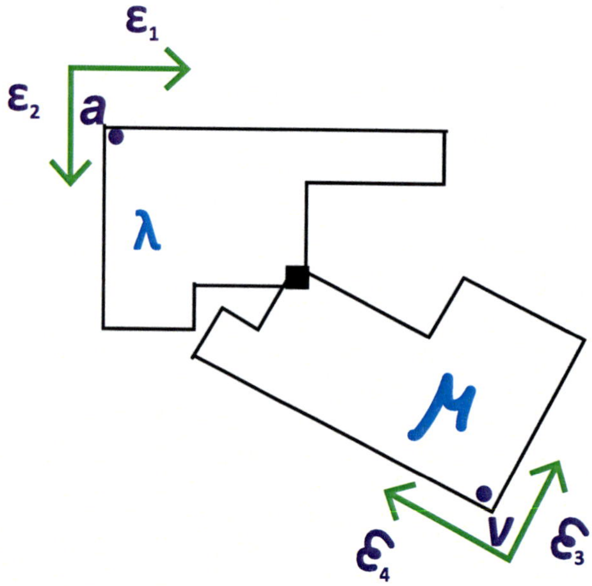

Figure 9: ■ $= K_{12,34}$, $\mathfrak{a} = \mathfrak{a}_{12,\alpha}$, $\nu = \mathfrak{a}_{34,\beta}$, $\lambda = \lambda^{(12,\alpha)}$, $\mu = \lambda^{(34,\beta)}$

Note that (244) is invariant under the shifts (237). Moreover, if (cf. (7))

$$
\begin{aligned}
&\mathfrak{a}_{12,\alpha'} - \mathfrak{a}_{12,\alpha''} \notin Z_{\underline{\varepsilon}}\,, \qquad \alpha' \neq \alpha'' \\
&\mathfrak{a}_{34,\beta'} - \mathfrak{a}_{34,\beta''} \notin Z_{\underline{\varepsilon}}\,, \qquad \beta' \neq \beta''
\end{aligned}
\tag{245}
$$

and $0 \notin Z_{\underline{\varepsilon}}$, then the condition (244) determines (α, β) and $i'j'$ and $i''j''$ uniquely, up to the shifts $(i', j', i'', j'') \mapsto (i' + k, j' + k, i'' - k, j'' - k)$, $k \in \mathbb{Z}$. The relation (244) defines the codimension 1 subtorus $\mathbf{T}' \subset \mathbf{T}_{n,w} \times U(1)^3_{\underline{\varepsilon}}$. Let us describe its fixed locus.

If the condition (244) on $\mathfrak{a}, \underline{\varepsilon}$ is obeyed, it does not imply that $K_{12,34} \neq 0$. However, if in addition to (244) also the condition (245) is obeyed, then the intersection $K_{12,34}$ is not more then one-dimensional. Let $H_{12} = P_{12;\alpha;i'j'} \subset P_{12}$, $H_{34} = P_{34;\beta;i''j''} \subset P_{34}$ be the one-dimensional eigenspaces of σ corresponding to the eigenvalue (244). If an eigenbasis of N_{12} for $\mathfrak{a}_{12}$ and the eigenbasis of N_{34} for $\mathfrak{a}_{34}$ are chosen, then H_{12} and H_{34} are endowed with the basis vectors as well (act on the eigenvector of $\mathfrak{a}_{12}$ corresponding to $\mathfrak{a}_{12,\alpha}$ by $B_1^{i'-1}B_2^{j'-1}I_{12}$ to get the basis vector of H_{12}).

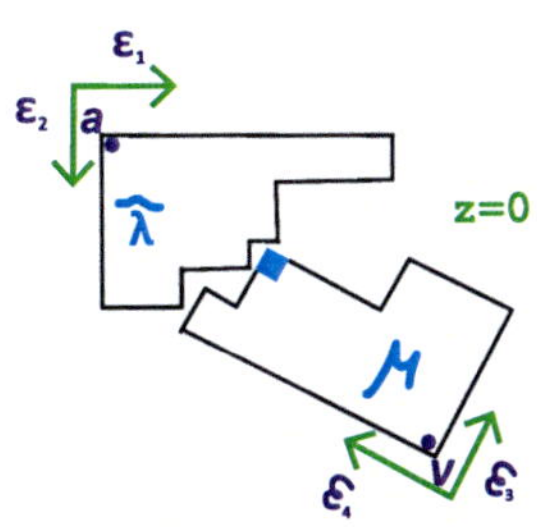

The component of the fixed point set corresponding to (244) is a copy of the complex projective line: $\mathbb{P}(H_{12} \oplus H_{34})$. It parametrizes rank one linear relations between H_{12} and H_{34}

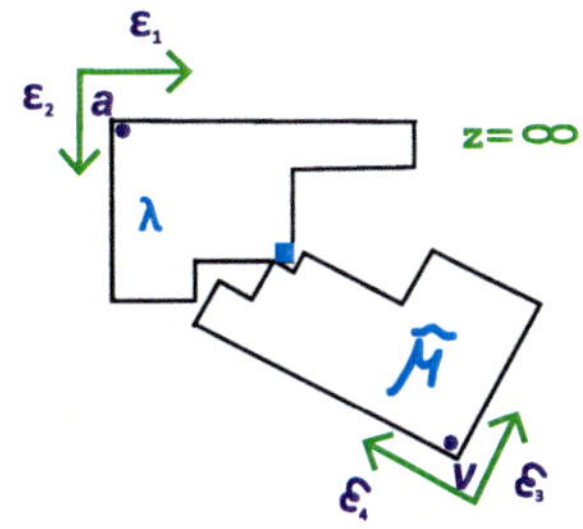

Figure 10: The component $\mathbb{P}(H_{12} \oplus H_{34})$

Let z be the coordinate on $\mathbb{P}(H_{12} \oplus H_{34})$ such that $z = \infty$ corresponds to the line H_{12} while $z = 0$ corresponds to the line H_{34}. For $z \neq 0, \infty$ the linear spaces H_{12} and H_{34} coincide, z being the isomorphism. When $z \to 0$ the image (cf. (230)) $(p_1(z=0), p_2(z=0))$ is the pair $(\tilde{\lambda}^{12}, \lambda^{34})$, the image $(p_1(z=\infty), p_2(z=\infty))$ is the pair $(\lambda^{12}, \tilde{\lambda}^{34})$. Here $\tilde{\lambda}^{12}$ is the n-tuple of partitions which differs from λ^{12} in that the Young diagram of $\tilde{\lambda}^{(12,\alpha)}$ ($\tilde{\lambda}$ on the Fig. 10) is obtained by removing the (i', j') square from the Young diagram of $\lambda^{(12,\alpha)}$. Similarly, the w-tuple $\tilde{\lambda}^{34}$ ($\tilde{\mu}$ on the Fig. 10) is obtained by modifying $\lambda^{(34,\beta)}$ by removing the box (i'', j'').

In the next chapters we shall relax the condition (245). In other words, we shall consider a subtorus in $\mathbf{T}_{n,w} \times U(1)^3_{\underline{\mathfrak{e}}}$.

6.5.2. *Toric folded instantons.* Now let us explore the folded instantons invariant under the action of the maximal torus $\mathbf{T}_{n,w} \times U(1)^3_{\mathfrak{e}}$. It is easy to see that these are again the pairs $(\lambda^{12}, \lambda^{13})$, with $\lambda^{12} = (\lambda^{(12,1)}, \ldots, \lambda^{(12,n)})$, $\lambda^{13} = (\lambda^{(13,1)}, \ldots, \lambda^{(13,w)})$. The spaces K_{12} and K_{13} do not intersect, $K = K_{12} \oplus K_{13}$. In other words, the only $\mathbf{T}_{n,w} \times U(1)^3_{\mathfrak{e}}$-invariant folded instantons are the superpositions of the ordinary instantons on $\mathbb{C}^2_{12}$ and $\mathbb{C}^2_{13}$, of the charges k_{12} and k_{13}, respectively, with $k = k_{12} + k_{13}$.

Now let us consider the non-generic case, such that $K_{12,13} \neq \emptyset$. We call the corresponding fixed point "the zipper", see the Fig. 11. The codimension one subtorus for which this is possible corresponds to the relation $\mathfrak{a}_{12,\alpha} - \mathfrak{a}_{13,\beta} \notin Z_{\underline{\mathfrak{e}}}$ between the parameters of the infinitesimal torus transformation.

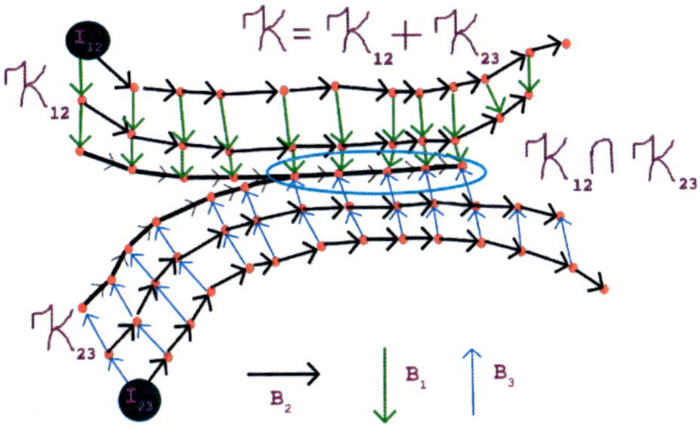

Figure 11: The zipper

The non-empty overlap $K_{12} \cap K_{13}$ implies the sets of eigenvalues of σ on K_{12} and K_{13} overlap, leading to

$$\mathfrak{a}_{12,\alpha} + \mathfrak{e}_1(i'-1) + \mathfrak{e}_2(j'-1) = \mathfrak{a}_{13,\beta} + \mathfrak{e}_1(i''-1) + \mathfrak{e}_3(j''-1) \tag{246}$$

for some $\alpha \in [n], \beta \in [w], i', j', i'', j''$. Unlike the Eq. (244) the Eq. (246) the integers i', i'' are not uniquely determined. Since the left hand side of (246) is the eigenvalue of $\sigma|_{L_{1,2}}$, while the right hand side is the eigenvalue of $\sigma|_{L_{1,3}}$, we conclude:

$$(i', j') \in \lambda^{(12,\alpha)}, (i'', j'') \in \lambda^{(13,\beta)}, (i', j'+1) \notin \lambda^{(12,\alpha)}, (i'', j''+1) \notin \lambda^{(13,\beta)} \tag{247}$$

i.e. $j' = \lambda_{i'}^{(12,\alpha)t}$, $j'' = \lambda_{i''}^{(13,\beta)t}$. The change $(i', i'') \mapsto (i' \pm 1, i'' \pm 1)$ maps the solution of (246) to another solution of (246). Let $l \geq 0$ be the maximal integer such that $j' = \lambda_{i'-l}^{(12,\alpha)}$, $j'' = \lambda_{i''-l}^{(13,\beta)}$, and let $l \geq m \geq 0$ be the maximal integer such that $j' = \lambda_{i'-m}^{(12,\alpha)}$, $j'' = \lambda_{i''-m}^{(13,\beta)}$, and $e_{12} = B_1^{i'-m-1} B_2^{j'-1} I_{12,\alpha} \in K_{12,13}$, $e_{13} = B_1^{i''-m-1} B_3^{j''-1} I_{13,\beta} \in K_{12,13}$. In other words the vectors e_{12} and e_{13} are linearly dependent, $e_{12} = z e_{13}$. Consequenly, the arm-lengths $a_{i'-m,j'} = \lambda_{j'}^{(12,\alpha)t} - i' + m$, $a_{i'-m,j'} = \lambda_{j''}^{(13,\beta)t} - i'' + m$ must be equal:

$$\lambda_{j'}^{(12,\alpha)t} - i' + m = a_{i'-m,j'} = \lambda_{j''}^{(13,\beta)t} - i'' + m \tag{248}$$

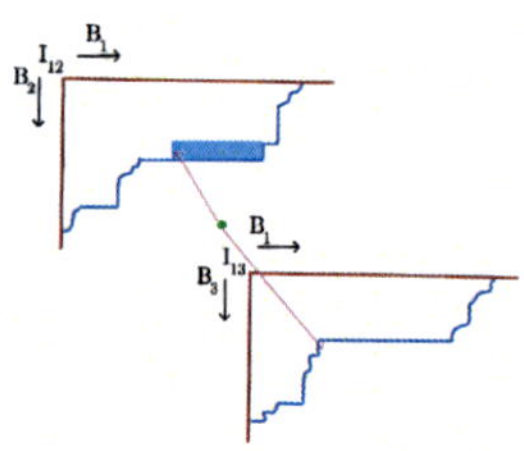

The component of the fixed point set corresponding to (246) is a copy of the complex projective line: $\mathbb{P}(\mathbb{C}e_{12} \oplus \mathbb{C}e_{13})$. It parametrizes rank one linear relations between e_{12} and e_{13}

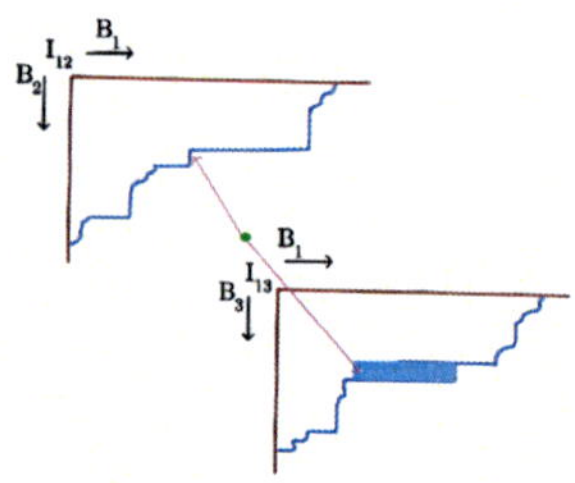

Let z be the coordinate on $\mathbb{P}(\mathbb{C}e_{12} \oplus \mathbb{C}e_{13})$ such that $z = \infty$ corresponds to the line $\mathbb{C}e_{12}$ while $z = 0$ corresponds to the line $\mathbb{C}e_{13}$. Then the image $(p_1(z = 0), p_2(z = 0))$ is the pair $(\tilde{\lambda}^{(12,\alpha)}, \lambda^{(13,\beta)})$, the image $(p_1(z = \infty), p_2(z = \infty))$ is the pair $(\lambda^{(12,\alpha)}, \tilde{\lambda}^{(13,\beta)})$. Here we defined $\tilde{\lambda}^{(12,\alpha)}$ to be the partition whose Young diagram is obtained by removing the block of squares $(i' - m, j') \ldots (\lambda_{j'}^{(12,\alpha)t}, j')$ from the Young diagram of $\lambda^{(12,\alpha)}$. Similarly, the Young diagram of $\tilde{\lambda}^{(13,\beta)}$ is obtained by removing the block of squares $(i'' - m, j'') \ldots (\lambda_{j''}^{(13,\beta)t}, j'')$ from the Young diagram of $\lambda^{(13,\beta)}$.

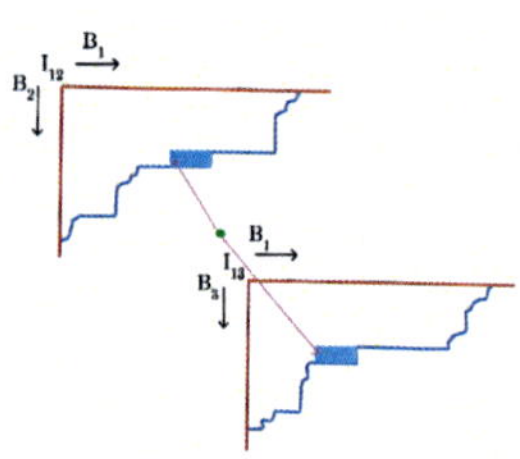

Note that the pair of Young diagrams $\lambda^{(12,\alpha)}$, $\lambda^{(34,\beta)}$ gives rise to several components of the fixed point set, isomorphic to $\mathbb{CP}^1$, e.g. the ones corresponding to the blocks of horizontal boxes of different length,

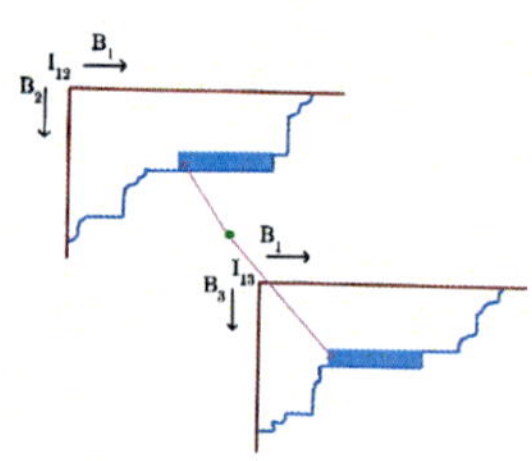

see the pictures above on the left and on the right. But they actually belong to moduli spaces of folded instantons of different charges (in computing the charge k we subtract the length of the block from the sum of the sizes of Young diagrams). So despite the similarity in graphic design, these are pieces of different architectures.

7. Reconstructing spiked instantons

In this section we describe the sewing procedure, which produces a spiked instanton out of six ordinary noncommutative instantons. We then use the stitching to describe the spiked instantons invariant under the toric symmetry, i.e. the $T_{\mathfrak{L}}$-fixed locus.

7.1. The local K-spaces. For $A = \{a, b\} \in \underline{\mathbf{6}}$ we define:

$$K_A = \mathbb{C}[B_a, B_b] \,\mathrm{im}(I_A) \subset K \tag{249}$$

By definition, this is the minimal B_a, B_b-invariant subspace of K, containing the image $\mathrm{im}(I_A) = I_A(N_A)$ of N_A.

The equations (57), (58) for $i \geq k$ imply that

$$J_B(K_A) = 0, \qquad B \neq A \tag{250}$$

and

$$B_b(K_A) = 0, \qquad b \notin A \tag{251}$$

7.1.1. *The intersections of local spaces.* Let $\underline{\mathbf{A}} = (A_1, \ldots, A_i) \in \underline{\mathbf{6}}$ be a collection of distinct subsets, $A_i \neq A_j$ for $i \neq j$. Define

$$K_{\underline{\mathbf{A}}} = K_{A_1} \cap \ldots \cap K_{A_i} \tag{252}$$

We can picture the collection $\underline{\mathbf{A}}$ as an oriented graph with $\underline{\mathbf{4}}$ being the set of vertices, and the edges A_1, A_2, ..., A_i. The graph $\underline{\mathbf{A}}$ is connected whenever $i \neq 2$. For $i = 2$ there are precisely three disconnected graphs: $(12, 34), (13, 24), (14, 23)$. Here are the example of $i = 3$ graphs (in red) for the collections $(14, 24, 34)$, $(12, 24, 14)$, $(14, 24, 23)$, respectively:

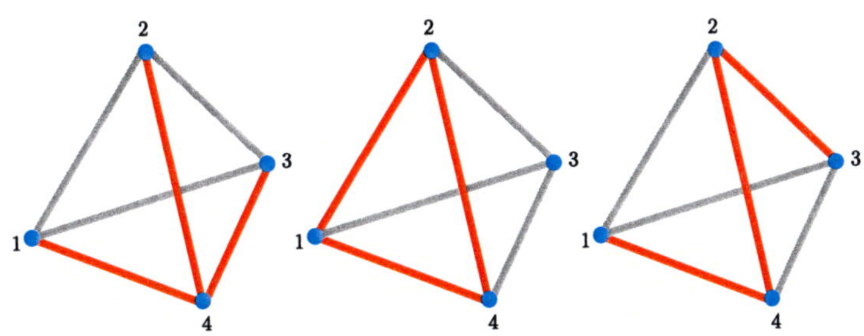

Figure 12: Examples of triple intersections

Let us now enumerate all possible intersections:

(1) Pair – wise intersections:

For $A', A'' \in \underline{\mathbf{6}}$, $\#A' \cap A'' = 1$, we define:

$$L_{A,a} = K_{(A',A'')}, \qquad \text{where } A' \cap A'' = \{a\}\,,\; A' \cup A'' = A \cup \{a\} \tag{253}$$

The only operator which acts non-trivially on $L_{A,a}$ is B_a. It follows from the Eqs. (57),(58) (with $i \geq k$) that

$$J_B f(B_a) L_{A,a} = 0 \tag{254}$$

for any polynomial $f(x)$ and any $B \in \underline{\mathbf{6}}$. For $\alpha = (A, \bar{A}) \in \underline{\mathbf{3}}$ we define

$$P_\alpha = K_{(A,\bar{A})} \tag{255}$$

For all $a \in \underline{\mathbf{4}}$, $B \in \underline{\mathbf{6}}$, $\alpha \in \underline{\mathbf{3}}$ the operators B_a and J_B vanish on P_α.

Twelve spaces $L_{A,a}$, $a \notin A$, $A \in \underline{\mathbf{6}}$, and three spaces P_α, $\alpha \in \underline{\mathbf{3}}$ describe all fifteen pair-wise intersections $K_A \cap K_B$.

(2) Triple intersections:

There are at most twenty triple intersections, $K_A \cap K_B \cap K_C$:
For $a, \bar{a} \in \underline{\mathbf{4}}$, we define

$$\bar{\mathcal{V}}_a = \bigcap_{A \in \underline{\mathbf{6}},\, a \in A} K_A\,, \qquad \mathcal{V}_{\bar{a}} = \bigcap_{A \in \underline{\mathbf{6}},\, \bar{a} \notin A} K_A \tag{256}$$

For $A = a'a'' \in \underline{\mathbf{4}}$, $a \in \bar{A}$, we define $\widehat{\mathcal{V}}_{a,A}$ to be:

$$\widehat{\mathcal{V}}_{a,A} = K_{aa'} \cap K_{a'a''} \cap K_{a''b}\,, \quad b \in \underline{\mathbf{4}} \backslash \{a, a', a''\} \tag{257}$$

In the Fig. 12 we give the examples of the spaces $\bar{\mathcal{V}}_4, \mathcal{V}_3, \widehat{\mathcal{V}}_{3,24}$.

(3) $4, 5, 6$ – tuple intersections:

There are fifteen quadruple intersections. They are, of course, in one-to-one correspondence with the pair-wise intersections:

$$\begin{aligned} & L^\vee_{A,a} = K_{\underline{\mathbf{6}} \backslash \{A',A''\}}, \qquad \text{where } A' \cap A'' = \{a\}\,,\; A' \cup A'' = A \cup \{a\} \\ & P^\vee_\alpha = K_{\underline{\mathbf{6}} \backslash \{A,\bar{A}\}}\,, \;\; \text{for } \alpha = (A, \bar{A}) \end{aligned} \tag{258}$$

There are four 5-fold intersections

$$K^\vee_A = K_{\underline{\mathbf{6}} \backslash \{A\}} \tag{259}$$

There is one 6-fold intersection:

$$K^\vee = K_{\underline{\mathbf{6}}} \tag{260}$$

On all 4- and higher-fold intersections the operators B_a, $a \in 4$ and J_B, $B \in \underline{\mathbf{6}}$, vanish identically.

7.2. Toric spiked instantons. Now let us describe the spiked instantons, invariant under the torus action. The tori in question are the subgroups of H, the global symmetry group. We consider first the maximal torus T_{H} (cf. (79)) and then its subtori $T_{\mathfrak{L}}$ for various choices of the $\mathfrak{L}$ data.

7.2.1. *Maximal torus.* First of all, let us consider the T_{H}-fixed points. Let $\underline{\mathfrak{a}} = (\mathfrak{a}_A)_{A\in\underline{6}}$ be the collection of diagonal matrices $\mathfrak{a}_A = \mathrm{diag}(\mathfrak{a}_{A,1}, \ldots, \mathfrak{a}_{A,n_A}) \in \mathrm{Lie}U(n_A)$. The spiked instanton $[\mathbf{B}, \mathbf{I}, \mathbf{J}]$ is T_{H}-invariant iff for any $\underline{\mathfrak{a}}$ and $\underline{\mathfrak{e}}$ there exists an operator $\sigma \in \mathrm{End}(K)$, such that:

$$
\begin{aligned}
&\mathfrak{e}_a B_a = [\sigma, B_a]\,, \qquad a \in \underline{4} \\
&(\mathfrak{e}_A - \mathfrak{a}_A) J_A = -J_A \sigma\,, \\
&I_A \mathfrak{a}_A = \sigma I_A\,, \qquad A \in \underline{6}
\end{aligned}
\tag{261}
$$

Let $N_{A,\alpha}$, $\alpha \in [n_A]$, be the eigenspace of $\mathfrak{a}_A$ with the eigenvalue $\mathfrak{a}_{A,\alpha}$. Let $I_{A,\alpha} = I_A(N_{A,\alpha})$. We have (for $A = \{a, b\}$, $a < b$):

$$
K_A = \sum_{\alpha\in[n_A],\, i,j\geq 1} K^{i,j}_{A,\alpha}\,, \qquad K^{i,j}_{A,\alpha} = B_a^{i-1} B_b^{j-1} I_{A,\alpha}
\tag{262}
$$

The eigenvalue of σ on $K^{i,j}_{A,\alpha}$ is equal to

$$
\sigma|_{K^{i,j}_{A,\alpha}} = \mathfrak{a}_{A,\alpha} + \mathfrak{e}_a(i-1) + \mathfrak{e}_b(j-1)
\tag{263}
$$

On the other hand, Eq. (261) implies that the vector

$$
\psi = J_A(K^{i,j}_{A,\alpha}) \in N_A
\tag{264}
$$

is an eigen-vector of $\mathfrak{a}_A$ with the eigen-value:

$$
\mathfrak{a}_A \psi = (\mathfrak{a}_{A,\alpha} + \mathfrak{e}_a i + \mathfrak{e}_b j)\psi
\tag{265}
$$

The T_{H}-invariance means we are free to choose the parameters $\mathfrak{a}_{A,\alpha}$, $\mathfrak{e}_a$ in an arbitrary fashion. It means, that $\mathfrak{a}_{A,\alpha} + \mathfrak{e}_a i + \mathfrak{e}_b j \neq \mathfrak{a}_{A,\beta}$ for $i, j \geq 1$, $\alpha, \beta \in [n_A]$. Therefore J_A vanishes on all $K^{i,j}_{A,\alpha}$ subspaces, and therefore on all of K_A. Therefore, all B_a's commute with each other. Also, the eigenvalues (263) are different for different $(A, \alpha; i, j)$. Therefore, the spaces $K^{i,j}_{A,\alpha}$ are orthogonal to each other.

Define, for $A \in \underline{6}, \alpha \in [n_A]$ the partition $\lambda^{(A,\alpha)}$, by:

$$
\lambda^{(A,\alpha)}_i = \sup \left\{ j \,|\, j \geq 1, K^{i,j}_{A,\alpha} \neq 0 \right\}
\tag{266}
$$

We have:

$$
k_A = \sum_{\alpha\in[n_A]} |\lambda^{(A,\alpha)}|
\tag{267}
$$

The T_{H}-fixed points are, therefore, in one-to-one correspondence with the collections

$$
\mathbf{\Lambda} = \left(\lambda^{(A,\alpha)}\right)_{A\in\underline{6},\alpha\in[n_A]}
\tag{268}
$$

of

$$
\mathbf{n} = \sum_{A\in\underline{6}} n_A
\tag{269}
$$

Young diagrams. In the companion paper **[27]** we shall be studying the statistical mechanical model, where the random variables are the collections $\mathbf{\Lambda}$, while the complexified Boltzman weights are the contributions of $\mathbf{\Lambda}$ to the gauge partition function, defined below.

7.2.2. *Subtori.* Now fix the data $\mathfrak{L}$ and consider the $T_{\mathfrak{L}}$-invariant spiked instantons $[\mathbf{B}, \mathbf{I}, \mathbf{J}]$. As usual, these come with the homomorphism $T_{\mathfrak{L}} \to U(K)$ which associates the compensating $U(K)$-transformation $g_{t,\xi}$ for every $(e^{i\xi}, e^{it}) \in T_{\mathfrak{L}}$. Since K decomposes into the direct sum of weight subspaces

$$K = \bigoplus_{w,n} K^{w,n} \tag{270}$$

where

$$g_{t,\xi}|_{K^{w,n}} = e^{i\langle w,\xi\rangle + int} \tag{271}$$

where $n \in \mathbb{Z}$, while w belongs to the weight lattice of $\bigtimes_{A\in\underline{\mathbf{6}}^-} U(1)^{\ell_A - 1}$.

The relations:

$$\begin{aligned} &e^{ie_a t} B_a = g_{t,\xi} B_a g_{t,\xi}^{-1} , \\ &I_A(N_{A,\alpha}) e^{ic_A(\alpha)t} = g_{t,\xi} I_A(N_{A,\alpha}) , \ A \in \underline{\mathbf{6}}^+ , \alpha \in [n_A] \\ &I_A(N_{A,\iota,\alpha}) e^{ic_{A,\iota}(\alpha)t} = g_{t,\xi} I_A(N_{A,\iota,\alpha}) , \ A \in \underline{\mathbf{6}}^- , \alpha \in [n_A]_\iota \end{aligned} \tag{272}$$

imply:

$$\begin{aligned} I_A(N_{A,\alpha}) \in K^{0,c_A(\alpha)} , \ \alpha \in [n_A], \ A \in \underline{\mathbf{6}}^+ , \\ I_A(N_{A,\iota_A,\alpha}) \in K^{0,c_{A,\iota_A}(\alpha)} , \ \alpha \in [n_A]_{\iota_A}, \ A \in \underline{\mathbf{6}}^- , \\ I_A(N_{A,\iota,\alpha}) \in K^{\varpi_{A,\iota}, c_{A,\iota}(\alpha)} , \ \alpha \in [n_A]_\iota, \ \iota \in \lambda_A, \ A \in \underline{\mathbf{6}}^- \\ B_a(K^{w,n}) \subset K^{w,n+e_a} , \ a \in \underline{\mathbf{4}} \end{aligned} \tag{273}$$

where $\varpi_{A,\iota}$ is the fundamental weight, $\langle \varpi_{A,\iota}, \xi \rangle = \xi_{A,\iota}$.

Finally, the $T_{\mathfrak{L}}$-invariance translates to

$$\begin{aligned} J_A(K^{w,n}) \in N_{A,\alpha} , \ A \in \underline{\mathbf{6}}^+, \alpha \in [n_A] \ \Leftrightarrow n = c_A(\alpha) - e_A \ , w = 0 , \\ J_A(K^{w,n}) \in N_{A,\iota_A,\alpha} , \ A \in \underline{\mathbf{6}}^-, \alpha \in [n_A]_{\iota_A} \ \Leftrightarrow n = c_{A,\iota_A}(\alpha) - e_A \ , w = 0 , \\ J_A(K^{w,n}) \in N_{A,\iota,\alpha} , \ A \in \underline{\mathbf{6}}^-, \alpha \in [n_A]_\iota \ \Leftrightarrow n = c_{A,\iota}(\alpha) - e_A \ , w = \varpi_{A,\iota} . \end{aligned} \tag{274}$$

which imply, with our choice of $T_{\mathfrak{L}}$, that $J_A = 0$. This is shown using the same arguments as we used around the Eq. (265).

7.2.3. *K-spaces for toric instantons.* Let $A \in \underline{\mathbf{6}}^+$. The local space K_A is $g_{t,\xi}$-invariant, and decomposes as

$$K_A = \bigoplus_n K_A^n \tag{275}$$

with integer n, via

$$g_{t,\xi}|_{K_A^n} = e^{int} \tag{276}$$

where $n \geq c_A^- = \inf_{\alpha\in[n_A]} c_A(\alpha)$, when $e_a, e_b > 0$ and $n \leq c_A^+ = \sup_{\alpha\in[n_A]} c_A(\alpha)$ when $e_a, e_b < 0$. For $e_a, e_b > 0$ both operators B_a, $a \in A$ raise the grading. For $e_a, e_b < 0$ both operators B_a, $a \in A$ lower the grading. Let $k_{A,n} = \dim K_A^n$. Since K_A is finite dimensional, $k_{A,n}$ vanish for $|n| > C_A$ for some some constant $C_A \leq k$.

Let $A \in \underline{\mathbf{6}}^-$. The local space K_A is $g_{t,\xi}$-invariant, and decomposes as

$$K_A = \bigoplus_n K_A^n \oplus \bigoplus_{\iota\in\lambda_A, n} K_{A,\iota}^n \tag{277}$$

with

$$g_{t,\xi}|_{K^n_{A,\iota}} = e^{int+i\xi_{A,\iota}} \tag{278}$$

for $i \in \lambda_A$, and

$$g_{t,\xi}|_{K^n_A} = e^{int} \tag{279}$$

Since the eigenvalues of $g_{t,\xi}$ on $K^n_{A,\iota}$ for $\iota \in \lambda_A$ differ from each other and from those on $K^{n'}_B$ for all $B \in \underline{\mathbf{6}}$, $n' \in \mathbb{Z}$, the spaces $K^n_{A,\iota}$ are orthogonal to $K^{n'}_B$ and to each other:

$$\begin{aligned} &K^{n'}_{A,\iota'} \perp K^{n''}_{A,\iota''}, \qquad \iota' \neq \iota'' \\ &K^{n'}_{A,\iota} \perp K^{n''}_A \end{aligned} \tag{280}$$

The action of $\mathbf{B}, \mathbf{I}$-operators respects the orthogonal decomposition (280).

We now prove that the spaces K^n_A and $K^n_{A,i}$ have an additional $U(1)$-action. Indeed, let f_a, f_b be the two positive mutually prime integers, such that

$$e_a f_a + e_b f_b = 0\,, \tag{281}$$

so that $e_a = p_{ab} f_b, \quad e_b = -p_{ab} f_a$ (assuming $e_a > 0 > e_b$). Then the operator

$$\mathcal{E} = B_a^{f_a} B_b^{f_b} \tag{282}$$

commutes with $g_{t,\xi}$, thanks to (272). Since all the eigenvalues of B_a and B_b vanish (again, thanks to (272)), the operators B_a, B_b, and $\mathcal{E}$ are nilpotent. By Jacobson-Morozov theorem, $\mathcal{E}$ can be included into the sl_2-triple, i.e. for each K^n_A, $K^n_{A,i}$ there are operators $\mathcal{H}$, $\mathcal{E}^\vee$, such that

$$[\mathcal{E}, \mathcal{E}^\vee] = \mathcal{H}, \ [\mathcal{H}, \mathcal{E}] = 2\mathcal{E}, \ [\mathcal{H}, \mathcal{E}^\vee] = -2\mathcal{E}^\vee \tag{283}$$

so that

$$K^n_A = \bigoplus_h K^{n,h}_A, \qquad K^n_{A,\iota} = \bigoplus_h K^{n,h}_{A,\iota} \tag{284}$$

with h standing for the eigenvalue of $\mathcal{H}$. Now, it is not difficult to prove that the (n,h)-grading is equivalent to the (i,j)-grading, with $i, j \geq 1$:

$$\begin{aligned} &i = i' + (h - h'_A(\alpha))f_a\,, \\ &j = j' + (h - h'_A(\alpha))f_b\,, \\ &n = c_A(\alpha) + p_A(f_b(i'-1) - f_a(j'-1))\,, \\ &i', j' \geq 1\,,\ i' \leq f_a \text{ and/or } j' \leq f_b \\ &h'_A(\alpha) = \inf \operatorname{Spec}(\mathcal{H}|_{K^n_A}) \end{aligned} \tag{285}$$

and α is uniquely determined by $n \bmod p_A = c_A(\alpha) \bmod p_A$. Thus,

$$K_A = \bigoplus_{i,j\geq 1} K^{i,j}_A\,, \qquad K_{A,\iota} = \bigoplus_{i,j\geq 1} K^{i,j}_{A,\iota} \tag{286}$$

with

$$B_a(K^{i,j}_A) \subset K^{i+1,j}_A\,, \qquad B_b(K^{i,j}_A) \subset K^{i,j+1}_A \tag{287}$$

Now we are ready for the final push:

8. The compactness theorem

We now prove the compactness theorem which establishes the analyticity of the partition function defined in the next chapter. To this end we estimate the norm of $(\mathbf{B}, \mathbf{I}, \mathbf{J})$ whose $U(k)$-orbit is invariant with respect to the action of any minimal torus $T_{\mathfrak{L}}$.

Since J_A's vanish, the real moment map equation reads as follows:

$$\sum_{a\in\underline{4}}[B_a, B_a^{\dagger}] + \sum_{A\in\underline{6}} I_A I_A^{\dagger} = \zeta \mathbf{1}_K \tag{288}$$

The trace of this equation gives the norm of I_A's:

$$\sum_{A\in\underline{6}} \|I_A\|^2 = \zeta k \tag{289}$$

But we need to estimate the norms $\|B_a\|^2$ which drop out of trace. However, it is not too difficult to chase them down. We have:

$$\begin{gathered}\sum_{a\in\underline{4}}\|B_a\|^2 + \sum_{A\in\underline{6}}\|I_A\|^2 = \sum_{a\in\underline{4}} \mathrm{Tr}_K B_a B_a^{\dagger} + \sum_{A\in\underline{6}} \mathrm{Tr}_K I_A I_A^{\dagger} \leq \\ \sum_{A\in\underline{6}} \mathrm{Tr}_{K_A}\left(\sum_{a\in\underline{4}} B_a B_a^{\dagger} + \sum_{C\in\underline{6}} I_C I_C^{\dagger}\right) = \sum_{A\in\underline{6}}\left(\zeta k_A + \mathrm{Tr}_{K_A}\sum_{a\in\underline{4}} B_a^{\dagger} B_a\right) = \\ \zeta\sum_{A\in\underline{6}} k_A + \sum_{A\in\underline{6}, a\in A} \mathrm{Tr}_{K_A} B_a^{\dagger} B_a\end{gathered} \tag{290}$$

where we used the moment map equation (59), projected onto K_A, and the Eq. (251). Define:

$$\begin{gathered}\delta_{A,n} = \frac{1}{\zeta}\mathrm{Tr}_{K_A^n}\left(\sum_{a\in\underline{4}} B_a B_a^{\dagger} + \sum_{A'\in\underline{6}} I_{A'} I_{A'}^{\dagger}\right) \\ \delta_A = \frac{1}{\zeta}\mathrm{Tr}_{K_A}\left(\sum_{a\in\underline{4}} B_a B_a^{\dagger} + \sum_{A'\in\underline{6}} I_{A'} I_{A'}^{\dagger}\right) = \sum_n \delta_{A,n}\end{gathered} \tag{291}$$

Now for $A \in \underline{6}^{+}$ use the decomposition (275), and (251) to show, that for $e_a \neq e_b$:

$$\delta_{A,n} = k_{A,n} + \frac{1}{\zeta}\sum_{a\in A}\mathrm{Tr}_{K_A^{n+e_a}}\left(B_a B_a^{\dagger}\right) \leq k_{A,n} + \sum_{a\in A}\delta_{A,n+e_a} \tag{292}$$

where we very conservatively estimated:

$$\mathrm{Tr}_{K_A^n}\left(B_a B_a^{\dagger}\right) \leq \delta_{A,n}\,, \tag{293}$$

for any $a \in \underline{4}$. This very conservative inequality can be used to show the boundeness of $\delta_{A,n}$.

From now on let us assume $e_a > e_b > 0$. The case of negative e_a, e_b is treated analogously. First of all, let introduce the sequence of generalized Fibonacci numbers $F_n^{p,q}$, $n \in \mathbb{Z}$, for positive integeres $p > q > 0$, by:

$$
\begin{aligned}
&F_n^{p,q} = 0\,, \qquad\qquad 1-p \leq n \leq 0 \\
&F_1^{p,q} = 1\,, \\
&F_n^{p,q} = F_{n-p}^{p,q} + F_{n-q}^{p,q}\,,\ n > 0
\end{aligned}
\tag{294}
$$

It is easy to write the formula for $F_n^{p,q}$ in terms of the roots λ_ω, $\omega = 0, \ldots, p-1$, of the characteristic equation

$$
\lambda_\omega^p = \lambda_\omega^{p-q} + 1\,,
\tag{295}
$$

$$
F_n^{p,q} = \sum_\omega f_\omega \lambda_\omega^n
\tag{296}
$$

where the coefficients f_ω are to be found from the linear equations $F_n^{p,q} = \delta_{n,1}$, $1 - p \leq n \leq 1$.

Now, we can estimate $\delta_{A,n}$ by induction in n:

$$
\delta_{A,n} \leq \sum_{n' \geq n} k_{A,n'} F_{n'+1-n}^{e_a,e_b}
\tag{297}
$$

This leads to the following, also very conservative, bound on δ_A:

$$
\delta_A < k_A F_{k_A}^{e_a,e_b}
\tag{298}
$$

When $e_a = e_b = e_A/2 \geq 1$ we can make a better estimate:

$$
\delta_{A,n} = k_{A,n} + \frac{1}{\zeta} \sum_{a \in A} \mathsf{Tr}_{K_A^{n+e_a}} \left(B_a B_a^\dagger\right) \leq k_{A,n} + \delta_{A,n+e_A/2}
\tag{299}
$$

which, by iteration, implies:

$$
\delta_{A,n} \leq k_{A,n} + k_{A,n+e_A/2} + k_{A,n+e_A} + \ldots
\tag{300}
$$

which in turns implies the upper bound on

$$
\delta_A \leq \frac{1}{e_A} k_A (k_A + 1)
\tag{301}
$$

It remains to estimate δ_A for $A \in \underline{\mathbf{6}}^-$. This is easy to do using the (i,j)-grading (286). Define:

$$
\delta_{A,n} = \frac{1}{\zeta} \sum_{i+j=n+2} \mathsf{Tr}_{K_A^{i,j}} \left(\sum_{a \in \underline{\mathbf{4}}} B_a B_a^\dagger + \sum_{A' \in \underline{\mathbf{6}}} I_{A'} I_{A'}^\dagger \right)
\tag{302}
$$

Then, using (59), projected onto $K_A^{i,j}$, and (287), we derive the estimate:

$$
\delta_{A,n} \leq \delta_{A,n+1} + \sum_{i+j=n+2} \mathrm{dim} K_A^{i,j}
\tag{303}
$$

from which we get the estimate:

$$
\delta_A \leq \frac{1}{2} k_A (k_A + 1)
\tag{304}
$$

9. Integration over the spiked instantons

The moduli spaces $\mathfrak{M}^{i}_{k}(\vec{n})$ are not your favorite smooth varieties. They can be stratified by smooth varieties of various dimensions. Over these smooth components the obstruction bundles keep track of the non-genericity of the equations we used to define the spaces $\mathfrak{M}^{i}_{k}(\vec{n})$.

In applications we need to compute the integrals over the spaces $\mathfrak{M}^{\infty}_{k}(\vec{n})$, as well as to define and compute the equivariant indices of various twisted Dirac operators (for five dimensional theories compactified on a circle).

Mathematically one can take the so-called virtual approach **[11]**, where the fundamental cycle $[\mathfrak{M}^{\infty}_{k}(\vec{n})]$ is replaced by the virtual fundamental cycle $[\mathfrak{M}^{\infty}_{k}(\vec{n})]^{\mathrm{vir}}$, which is defined as the Euler class of the bundle of equations over the smooth variety of the original variables $(\mathbf{B}, \mathbf{I}, \mathbf{J})$. There are two difficulties with this definition: *i*) the space of $(\mathbf{B}, \mathbf{I}, \mathbf{J})$, being a vector space, is non-compact; *ii*) the bundle of equations is infinite dimensional, unless we are in the situation with only the crossed or ordinary instantons.

The problem *i*) is solved by passing to the equivariant cohomology. The problem *ii*) is cured by working with $\mathfrak{M}^{i}_{k}(\vec{n})$ for large but finite i, and then regularize the limit $i \to \infty$ by using the Γ-functions.

Physically, the problem is solved by the considerations of the matrix integral (matrix quantum mechanics) of the system of k $D(-1)$-branes (k $D0$-branes whose worldlines wrap $\mathbb{S}^1$) in the vicinity of six stacks of $D3$ branes ($D4$ branes) wrapping coordinate two-planes $\mathbb{C}^2_A$ (times a circle $\mathbb{S}^1$) in the IIB (IIA) background $\mathbb{R}^2 \times \mathbb{C}^4$ ($\mathbb{R}^1 \times \mathbb{S}^1 \times \mathbb{C}^4$).

One can also define the elliptic genus by the study of the two-dimensional gauge theory corresponding to the stack of k $D1$-strings wrapping a $\mathbf{T}^2$ shared by six stacks of $D5$ branes in IIB string theory, wrapping $\mathbf{T}^2 \times \mathbb{C}^2_A$.

9.1. Cohomological field theory. Let us briefly recall the physical approach. For every variable, i.e. for every matrix element of the matrices B_a, I_A, J_A, we introduce the fermionic variables $\Psi^{\mathbf{B}}_{a}, \Psi^{\mathbf{I}}_{A}, \Psi^{\mathbf{J}}_{A}$ with the same $\mathsf{H} \times U(K)$ transformation properties.

For every equation $\mathfrak{s}_A, \mu, \sigma_{\bar{a}A}, \Upsilon_A, \Upsilon_{A,B;i}$ we introduce a pair of fermion-boson variables valued in the dual space: $(\chi_A, h_A), (\chi, \mu), (\chi_{\bar{a}A}, h_{\bar{a}A}), (\xi_A, r_A), (\varpi_{A',A'';i}, y_{A',A'';i})$.

Finally, we need a triplet of variables $(\sigma, \bar{\sigma}, \eta)$ (two bosons and a fermion), valued in the Lie algebra of $U(K)$.

Our model has the fermionic symmetry acting by:

$$
\begin{aligned}
\delta B_a &= \Psi^{\mathbf{B}}_{a}, \quad \delta\Psi^{\mathbf{B}}_{a} = -[\sigma, B_a] + \mathfrak{e}_a B_a \\
\delta I_A &= \Psi^{\mathbf{I}}_{A}, \quad \delta\Psi^{\mathbf{I}}_{A} = -\sigma I_A + I_A \mathfrak{a}_A \\
\delta J_A &= \Psi^{\mathbf{J}}_{A}, \quad \delta\Psi^{\mathbf{J}}_{A} = -\mathfrak{a}_A J_A + J_A \sigma + \mathfrak{e}_A J_A
\end{aligned}
\tag{305}
$$

(cf. (261)) and

$$
\begin{aligned}
&\delta\chi_A = h_A\,, \quad \delta h_A = -[\sigma,\chi_A] + \mathfrak{e}_A\chi_A\,,\\
&\qquad\qquad \delta\chi = h\,, \quad \delta h = -[\sigma,\chi]\,,\\
&\delta\chi_{\bar{a}A} = h_{\bar{a}A}\,, \quad \delta h_{\bar{a}A} = -\mathfrak{a}_A\chi_{\bar{a}A} + \chi_{\bar{a}A}\sigma - \mathfrak{e}_{\bar{a}}\chi_{\bar{a}A}\,,\\
&\delta\xi_A = r_A \quad \delta r_A = \mathfrak{a}_{\bar{A}}\xi_A - \xi_A\mathfrak{a}_A + \mathfrak{e}_A\xi_A\,,\\
&\qquad \delta\varpi_{A',A'';i} = y_{A',A'';i}\,,\\
&\qquad \delta y_{A',A'';i} = -\mathfrak{a}_{A'}\varpi_{A',A'';i} + \varpi_{A',A'';i}\mathfrak{a}_{A''} + (i-1)\mathfrak{e}_a\varpi_{A',A'';i}\,,\\
&\qquad\qquad \text{whenever} \qquad A'\cap A'' = \{a\}
\end{aligned}
\tag{306}
$$

and

$$
\delta\bar{\sigma} = \eta, \quad \delta\eta = [\sigma,\bar{\sigma}], \qquad \delta\sigma = 0 \tag{307}
$$

Now we can define the partition function

$$
\mathcal{Z}^i_k(\vec{n};\underline{\mathfrak{e}},\underline{\mathfrak{a}}) = \int e^{-\mathbf{S}_i}\frac{D\sigma}{\mathrm{Vol}(U(k))} D\mathbf{B}D\mathbf{B}^{\dagger}D\Psi^{\mathbf{B}}D\Psi^{\mathbf{B}\dagger}\ldots D\chi Dh\ldots D\bar{\sigma}D\eta \prod_{j=1}^{i-1} D\chi_{A,B;j}Dh_{A,B;j}D\bar{\chi}_{A,B;j}D\bar{h}_{A,B;j} \tag{308}
$$

where

$$
\mathbf{S}_i = \delta\Xi_i\,, \Xi_i = \Xi^s + \Xi^f + \Xi^K + \sum_A \Xi^N_A + \sum_{A',A'';\#A'\cap A''=1}\sum_{j=1}^{i-1}\Xi^N_{A'A'',j}\,, \tag{309}
$$

$$
\Xi^s = \mathrm{Tr}_K\,\eta[\sigma,\bar{\sigma}]\,, \quad \Xi^f = \mathrm{Tr}_K\sum_a\left(\Psi^{\mathbf{B}}_a\left(-[\bar{\sigma},B^{\dagger}_a] + \bar{\varepsilon}_a B^{\dagger}_a\right) + c.c.\right) +
$$

$$
\mathrm{Tr}_K\sum_A\Big((-\bar{\sigma}I_A + I_A\bar{\mathfrak{a}}_A)\Psi^{\mathbf{I}\dagger}_A + c.c.\Big) + \mathrm{Tr}_K\sum_A\Big(\Psi^{\mathbf{J}\dagger}_A(J_A\bar{\sigma} - \bar{\mathfrak{a}}_A J_A + \bar{\varepsilon}_A J_A) + c.c.\Big)\,,
$$

$$
\Xi^K = \mathrm{Tr}_K\,\chi\left(-\mathrm{i}\left(\mu - \zeta\mathbf{1}_K\right) + h\right) + \sum_A\left(\chi_{\bar{A}}\left(-\mathrm{i}s_A + h_A\right) + \sum_{\bar{a}\in\bar{A}}\chi^{\dagger}_{\bar{a}A}\left(-\mathrm{i}\sigma_{\bar{a}A} + h_{\bar{a}A}\right)\right).
$$

$$
\Xi^N_A = \mathrm{Tr}_{N_A}\left(\xi^{\dagger}_A\left(-\mathrm{i}\Upsilon_A + r_A\right) + c.c.\right)\,,
$$

$$
\Xi^N_{A',A'';j} = \mathrm{Tr}_{N_A}\left(\varpi^{\dagger}_{A',A'';j}\left(-\mathrm{i}\Upsilon_{A',A'';j} + y_{A',A'';j}\right) + c.c.\right)
$$

Here $\bar{\varepsilon}_a$, $\bar{\mathfrak{a}}_A$ are auxiliary elements of the Lie algebra of H, which can be chosen arbitrary, as long as the integral (308) converges.

9.2. Localization and analyticity. The usual manipulations with the integral (308), for generic $(\underline{\mathfrak{e}},\underline{\mathfrak{a}})$, express it as a sum over the fixed points, which we enumerated in the Eq. (268). Each fixed point contributes a homogeneous (degree zero) rational function of $\mathfrak{a}_{A,\alpha}$'s, $1 \leq \alpha \leq n_A$ and $\mathfrak{e}_a$, times the product

$$\prod_{A\in\underline{\mathbf{6}},\, 4\in\bar{A}} \prod_{\alpha=1}^{n_A}\prod_{\beta=1}^{n_{\bar{A}}} \left(\mathfrak{a}_{A,\alpha} - \mathfrak{a}_{\bar{A},\beta} + \mathfrak{e}_{\bar{A}}\right) \quad \times \tag{310}$$

$$\prod_{A,B\in\underline{\mathbf{6}},\, A\cap B=\{c\}} \prod_{\alpha=1}^{n_A}\prod_{\beta=1}^{n_B}\prod_{j=1}^{i-1} \left(\mathfrak{a}_{A,\alpha} - \mathfrak{a}_{B,\beta} + \mathfrak{e}_c(j-1)\right)$$

The compactness theorem of the previous chapter implies, among other things, that the partition functions $\mathcal{Z}^i_k(\vec{n}; \underline{\mathfrak{e}}, \underline{\mathfrak{a}})$, for $i > k$, have no singularities in

$$\mathfrak{x}_A = \frac{1}{n_A}\sum_{\alpha=1}^{n_A} \mathfrak{a}_{A,\alpha} \tag{311}$$

with fixed $\mathfrak{e}_a$'s and $\tilde{\mathfrak{a}}_{A,\alpha} = \mathfrak{a}_{A,\alpha} - \mathfrak{x}_A$. In other words, they are polynomials of $\mathfrak{x}_A$.

10. Quiver crossed instantons

10.1. Crossed quivers. Let γ be an oriented graph, with the set V_γ of vertices, and the set E_γ of edges. Let $s, t : \mathsf{E}_\gamma \to \mathsf{V}_\gamma$ be the source and the target maps. The crossed quiver is the data $(\mathsf{V}_\gamma, {\mathsf{E}_\gamma}^+, {\mathsf{E}_\gamma}^-, s, t; p)$, where: $\mathsf{E}_\gamma = {\mathsf{E}_\gamma}^+ \amalg {\mathsf{E}_\gamma}^-$ is the partition of the set of edges into two (possibly empty) sets, $p \in \mathbb{Z}_{\geq 0}$ is a non-negative integer. Let $\Xi_p = \mathbb{Z}/p\mathbb{Z}$. Fix four dimension vectors $\underline{\mathbf{k}}, \underline{\mathbf{n}}^+, \underline{\mathbf{n}}^-, \underline{\mathbf{m}} : \mathsf{V}_\gamma \times \Xi_p \to \mathbb{Z}_{\geq 0}$. Let $(K_{\omega,n}, N^\pm_{\omega,n}, M_{\omega,n})_{\omega\in\mathsf{V}_\gamma,\, n\in\mathbb{Z}/p\mathbb{Z}}$ be a collection of complex vector spaces whose dimensions are given by the components of the corresponding dimension vectors, e.g. $\dim K_{\omega,n} = k_{\omega,n} \equiv \underline{\mathbf{k}}(\omega, n)$. We view the spaces $N^\pm_{\omega,n}, M_{\omega,n}$ as fixed, e.g. with some fixed basis, while the spaces $K_{\omega,n}$ are varying, i.e. defined up to the automorphisms. We also fix a decomposition $M_{\omega,n} = M'_{\omega,n} \oplus M''_{\omega,n}$ as an additional refinement of our structure.

Consider the vector superspace $\mathcal{A}^\gamma_{\underline{\mathbf{k}}}(\underline{\mathbf{n}}^\pm, \underline{\mathbf{m}})$ of linear maps $(\mathbf{B}, \mathbf{I}, \mathbf{J}, \mathbf{\Theta})$

(312)

$$\begin{aligned}
&\mathbf{B} = (B_{e,n}, \tilde{B}_{e,n})_{e\in\mathsf{E}_\gamma, n\in\Xi_p}\,, \\
&\quad B_{e,n} : K_{s(e),n} \to K_{t(e),n+1}\,, \quad \tilde{B}_{e,n} : K_{t(e),n} \to K_{s(e),n\pm 1}\,, \qquad \text{when } e \in {\mathsf{E}_\gamma}^\pm \\
&\mathbf{I} = (I^+_{\omega,n}, I^-_{\omega,n})_{\omega\in\mathsf{V}_\gamma, n\in\Xi_p}\,, \qquad I^\pm_{\omega,n} : N^\pm_{\omega,n} \to K_{\omega,n\pm 1}\,, \\
&\mathbf{J} = (J^+_{\omega,n}, J^-_{\omega,n})_{\omega\in\mathsf{V}_\gamma, n\in\Xi_p}\,, \qquad J^\pm_{\omega,n} : K_{\omega,n} \to N^\pm_{\omega,n\pm 1}\,, \\
&\qquad\qquad \mathbf{\Theta} = (\Theta'_{\omega,n}, \Theta''_{\omega,n})_{\omega\in\mathsf{V}_\gamma, n\in\Xi_p}\,, \\
&\qquad\qquad \Theta'_{\omega,n} \in \Pi\mathrm{Hom}(K_{\omega,n}, M'_{\omega,n})\,, \\
&\Theta''_{\omega,n} \in \Pi\mathrm{Hom}(M''_{\omega,n-1}, K_{\omega,n})
\end{aligned}$$

Let $\mathcal{G}^\gamma_{\underline{\mathbf{k}}}, \mathcal{G}^{\mathbb{C}\gamma}_{\underline{\mathbf{k}}}$ be the groups:

$$\mathcal{G}^\gamma_{\underline{\mathbf{k}}} = \bigtimes_{\omega\in\mathsf{V}_\gamma, n\in\Xi_p} U(k_{\omega,n})\,, \qquad \mathcal{G}^{\mathbb{C}\gamma}_{\underline{\mathbf{k}}} = \bigtimes_{\omega\in\mathsf{V}_\gamma, n\in\Xi_p} GL(k_{\omega,n}, \mathbb{C})\,, \tag{313}$$

which act on $\mathcal{A}^{\gamma}_{\underline{\mathbf{k}}}(\underline{\mathbf{n}}, \underline{\mathbf{w}}, \underline{\mathbf{m}})$ via:

$$\times_{\omega\in\mathsf{V}_\gamma, n\in\Xi_p}\, g_{\omega,n}\cdot(\mathbf{B},\mathbf{I},\mathbf{J},\mathbf{\Theta}) = \Big(g_{t(e),n\pm1}B_{e,n}g^{-1}_{s(e),n},\, g_{s(e),n\pm1}\tilde{B}_{e,n}g^{-1}_{t(e),n},\, {}_{e\in\mathsf{E}_\gamma{}^{\pm}};\; g_{\omega,n\pm1}I^{\pm}_{\omega,n},\, J^{\pm}_{\omega,n}g^{-1}_{\omega,n}\,;\Theta'_{\omega,n}g^{-1}_{\omega,n}, g_{\omega,n}\Theta''_{\omega,n}\Big) \tag{314}$$

We want to impose the following analogues of the Eqs. (56):

$$\begin{aligned} &\mu_{\omega,n} = \zeta_{\omega,n}\mathbf{1}_{K_{\omega,n}}\,, && \omega\in\mathsf{V}_\gamma,\ n\in\Xi_p, \qquad \zeta_{\omega,n}>0\\ &s^{\mathsf{a}}_{\omega',\omega'';n} = 0\,, && \mathsf{a}\in\underline{\mathbf{3}},\ \omega',\omega''\in\mathsf{V}_\gamma,\ n\in\Xi_p\\ &\Sigma_{e,n}=0\,,\ \tilde{\Sigma}_{e,n}=0\,, && e\in\mathsf{E}_\gamma\,,\ n\in\Xi_p\,,\\ &\Upsilon_{\omega,n} = 0, && \omega\in\mathsf{V}_\gamma,\ n\in\Xi_p \end{aligned} \tag{315}$$

where

$$\begin{aligned} &\Sigma_{e,n} = \sigma_{e,n} + \widehat{\sigma}^{\dagger}_{e,n}\,, \qquad \tilde{\Sigma}_{e,n} = \tilde{\sigma}_{e,n} - \widehat{\tilde{\sigma}}^{\dagger}_{e,n}\,,\\ &s^{A,\bar{A}}_{\omega',\omega'';n} = \mu^{A}_{\omega',\omega'';n} + \epsilon(A)\left(\mu^{\bar{A}}_{\omega'',\omega';n}\right)^{\dagger}\,,\\ &\Upsilon_{\omega,n} = \upsilon^{+}_{\omega,n-1} - \left(\upsilon^{-}_{\omega,n+1}\right)^{\dagger} \end{aligned} \tag{316}$$

with the linear maps

$$\begin{aligned} &\sigma_{e,n} : N^{\pm}_{s(e),n}\to K_{t(e),n}\,, && \text{for}\quad e\in\mathsf{E}_\gamma{}^{\mp}\\ &\tilde{\sigma}_{e,n} : N^{\pm}_{t(e),n}\to K_{s(e),n}\,, && \text{for}\quad e\in\mathsf{E}_\gamma{}^{\mp}\\ &\upsilon^{\pm}_{\omega,n\mp1} : N^{\mp}_{\omega,n}\to N^{\pm}_{\omega,n}\,, \end{aligned} \tag{317}$$

and

$$\begin{aligned} &\mu_{\omega,n} : K_{\omega,n}\to K_{\omega,n}\,,\\ &\mu^{12}_{\omega',\omega'';n} : K_{\omega'',n}\to K_{\omega',n+2}\,, \qquad \mu^{34}_{\omega',\omega'';n} : K_{\omega'',n}\to K_{\omega',n-2}\,,\\ &\mu^{13}_{\omega',\omega'';n} : K_{\omega'',n}\to K_{\omega',n}\,, \qquad \mu^{24}_{\omega',\omega'';n} : K_{\omega'',n}\to K_{\omega',n}\,,\\ &\mu^{14}_{\omega',\omega'';n} : K_{\omega'',n}\to K_{\omega',n}\,, \qquad \mu^{23}_{\omega',\omega'';n} : K_{\omega'',n}\to K_{\omega',n}\,. \end{aligned} \tag{318}$$

The maps (317) are given by, for $e\in\mathsf{E}_\gamma{}^{\mp}$:

$$\begin{aligned} &\sigma_{e,n} = B_{e,n\pm1}I^{\pm}_{s(e),n}\,, \qquad \widehat{\sigma}_{e,n} = J^{\pm}_{s(e),n\mp1}\tilde{B}_{e,n}\,,\\ &\tilde{\sigma}_{e,n} = \tilde{B}_{e,n\pm1}I^{\pm}_{t(e),n}\,, \qquad \widehat{\tilde{\sigma}_{e,n}} = J^{\pm}_{t(e),n\mp1}B_{e,n}\,, \end{aligned} \tag{319}$$

and

$$\upsilon^{\pm}_{\omega,n} = J^{\pm}_{\omega,n\mp1}I^{\mp}_{\omega,n} \tag{320}$$

The analogues of real and complex moment maps (318) are given by:

(321)

$$\mu_{\omega,n} = I^+_{\omega,n-1}\left(I^+_{\omega,n-1}\right)^\dagger + I^-_{\omega,n+1}\left(I^-_{\omega,n+1}\right)^\dagger - \left(J^+_{\omega,n}\right)^\dagger J^+_{\omega,n} - \left(J^-_{\omega,n}\right)^\dagger J^-_{\omega,n} +$$
$$\sum_{e\in s^{-1}(\omega)} B^\dagger_{e,n}B_{e,n} + \sum_{e\in t^{-1}(\omega)} \tilde{B}^\dagger_{e,n}\tilde{B}_{e,n} -$$
$$\sum_{e\in s^{-1}(\omega)\cap \mathsf{E}_\gamma{}^+} \tilde{B}_{e,n-1}\tilde{B}^\dagger_{e,n-1} - \sum_{e\in s^{-1}(\omega)\cap \mathsf{E}_\gamma{}^-} \tilde{B}_{e,n+1}\tilde{B}^\dagger_{e,n+1} -$$
$$\sum_{e\in t^{-1}(\omega)\cap \mathsf{E}_\gamma{}^+} B_{e,n-1}B^\dagger_{e,n-1} - \sum_{e\in t^{-1}(\omega)\cap \mathsf{E}_\gamma{}^-} B_{e,n+1}B^\dagger_{e,n+1},$$

and

(322)

$\mu^{12}_{\omega',\omega'';n} =$  $-$ 

$$I^+_{\omega',n+1}J^+_{\omega',n}\,\delta_{\omega',\omega''} + \\ + \sum_{\substack{e',e''\in \mathsf{E}_\gamma{}^+,\\ t(e')=\omega'',\ t(e'')=\omega' \\ s(e')=s(e'')}} B_{e'',n+1}\tilde{B}_{e',n} - \sum_{\substack{e',e''\in \mathsf{E}_\gamma{}^+,\\ s(e')=\omega'',\ s(e'')=\omega' \\ t(e')=t(e'')}} \tilde{B}_{e'',n+1}B_{e',n}$$

(323)

$\mu^{34}_{\omega',\omega'';n} =$ 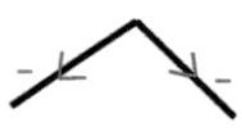 $-$ 

$$I^-_{\omega',n-1}J^-_{\omega',n}\,\delta_{\omega',\omega''} + \\ + \sum_{\substack{e',e''\in \mathsf{E}_\gamma{}^-,\\ t(e')=\omega'',\ t(e'')=\omega' \\ s(e')=s(e'')}} B_{e'',n-1}\tilde{B}_{e',n} - \sum_{\substack{e',e''\in \mathsf{E}_\gamma{}^-,\\ s(e')=\omega'',\ s(e'')=\omega' \\ t(e')=t(e'')}} \tilde{B}_{e'',n-1}B_{e',n}$$

(324)

$\mu^{13}_{\omega',\omega'';n} =$ 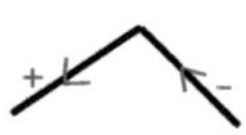 $-$ 

$$\sum_{\substack{e'\in \mathsf{E}_\gamma{}^-,\ e''\in \mathsf{E}_\gamma{}^+,\\ s(e')=\omega'',\ t(e'')=\omega' \\ t(e')=s(e'')}} B_{e'',n-1}B_{e',n} - \sum_{\substack{e'\in \mathsf{E}_\gamma{}^+,\ e''\in \mathsf{E}_\gamma{}^-,\\ s(e')=\omega'',\ t(e'')=\omega' \\ t(e')=s(e'')}} B_{e'',n+1}B_{e',n}$$

(325)

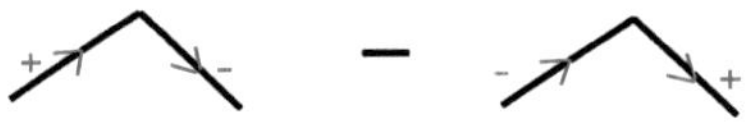

$$\mu^{24}_{\omega',\omega'';n} = \sum_{\substack{e'\in E_\gamma^-,\ e''\in E_\gamma^+,\\ t(e')=\omega'',\ s(e'')=\omega'\\ s(e')=t(e'')}} \tilde{B}_{e'',n-1}\tilde{B}_{e',n} - \sum_{\substack{e'\in E_\gamma^+,\ e''\in E_\gamma^-,\\ t(e')=\omega'',\ s(e'')=\omega'\\ s(e')=t(e'')}} \tilde{B}_{e'',n+1}\tilde{B}_{e',n}$$

(326)

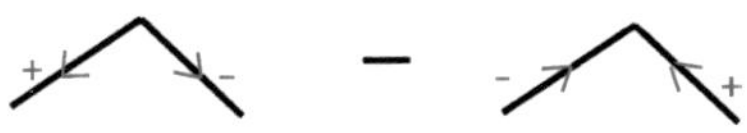

$$\mu^{14}_{\omega',\omega'';n} = \sum_{\substack{e'\in E_\gamma^-,\ e''\in E_\gamma^+,\\ t(e')=\omega'',\ t(e'')=\omega'\\ s(e')=s(e'')}} B_{e'',n-1}\tilde{B}_{e',n} - \sum_{\substack{e'\in E_\gamma^+,\ e''\in E_\gamma^-,\\ s(e')=\omega'',\ s(e'')=\omega'\\ t(e')=t(e'')}} \tilde{B}_{e'',n+1}B_{e',n}$$

(327)

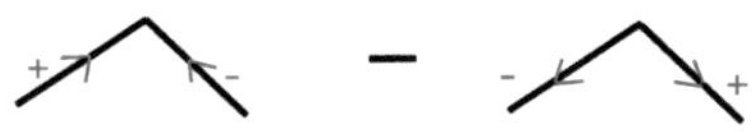

$$\mu^{23}_{\omega',\omega'';n} = \sum_{\substack{e'\in E_\gamma^-,\ e''\in E_\gamma^+,\\ s(e')=\omega'',\ s(e'')=\omega'\\ t(e')=t(e'')}} \tilde{B}_{e'',n-1}B_{e',n} - \sum_{\substack{e'\in E_\gamma^+,\ e''\in E_\gamma^-,\\ t(e')=\omega'',\ t(e'')=\omega'\\ s(e')=s(e'')}} B_{e'',n+1}\tilde{B}_{e',n}$$

The moduli space of quiver crossed instantons $\mathfrak{M}^{\gamma}_{\underline{\mathbf{k}}}(\underline{\mathbf{n}},\underline{\mathbf{w}},\underline{\mathbf{m}})$ is the space of solutions to (315) modulo the action (313) of $\mathcal{G}^{\gamma}_{\underline{\mathbf{k}}}$.

Again, the identity

$$(328)\quad \sum_{\omega',\omega''\in V_\gamma}\sum_{A\in\underline{\mathbf{6}}} \mathrm{Tr}_{K_{\omega',n}}\left(\mu^{A}_{\omega',\omega'';n+n_A}\mu^{A}_{\omega'',\omega';n}\right) + \sum_{\substack{e',e''\in E_\gamma^{\pm},\\ s(e')=s(e''),\ t(e')=t(e'')}} \mathrm{Tr}_{K_{t(e'),n}}\left(\sigma_{e',n}\widehat{\sigma}_{e'',n}\right) - \mathrm{Tr}_{K_{s(e'),n}}\left(\tilde{\sigma}_{e',n}\widehat{\tilde{\sigma}}_{e'',n}\right) - \sum_{\omega\in V_\gamma}\mathrm{Tr}_{N^-_{\omega,n}}\left(v^+_{\omega,n-1}v^-_{\omega,n+1}\right) = 0$$

where $n_{12} = -n_{34} = 1$, $n_{13} = n_{24} = n_{14} = n_{23} = 0$, demonstrates by the argument identical to (54), that the equations (315) imply the holomorphic equations

$$(329)\quad \sigma_{e,n} = \tilde{\sigma}_{e,n} = \widehat{\sigma}_{e,n} = \widehat{\tilde{\sigma}}_{e,n} = 0$$

$$\mu^{A}_{\omega',\omega'';n} = 0$$

$$v^{\pm}_{\omega,n} = 0$$

Thus, $\mathfrak{M}^{\gamma}_{\underline{\mathbf{k}}}(\underline{\mathbf{n}}, \underline{\mathbf{w}}, \underline{\mathbf{m}})$ is the space of stable solutions of (329) modulo the action (313) of $\mathcal{G}^{\mathbb{C}\gamma}_{\underline{\mathbf{k}}}$. Here, the stability condition can be formulated as follows:

(330)
$$\boxed{\begin{array}{l}\text{Any collection of subspaces } K'_{\omega,n} \subset K_{\omega,n}\,, \text{ such that} \\ \qquad I^{\pm}_{\omega,n}(N^{\pm}_{\omega,n}) \subset K'_{\omega,n\pm 1}\,, \\ \qquad\quad \text{and} \\ \qquad B_{e,n}(K'_{s(e),n}) \subset K'_{t(e),n\pm 1}\,, \quad \text{for all } e \in \mathsf{E}_{\gamma}{}^{\pm} \\ \qquad \tilde{B}_{e,n}(K'_{t(e),n}) \subset K'_{s(e),n\pm 1}\,, \quad \text{for all } e \in \mathsf{E}_{\gamma}{}^{\pm} \\ \qquad\quad \text{coincides with all of } K_{\omega,n}\,,\ K'_{\omega,n} = K_{\omega,n} \\ \\ \text{in other words}\,, \quad K^{+}_{\omega,n} + K^{-}_{\omega,n} = K_{\omega,n}\end{array}}$$

where $K^{\pm}_{\omega,n}$ is the subspace, generated by acting with arbitrary (noncommutative) polynomials in $B_{e,n'}$, $\tilde{B}_{e,n'}$'s with $e \in \mathsf{E}_{\gamma}{}^{\pm}$ on the image $I^{\pm}_{\omega',n''}(N_{\omega',n''})$ for some $\omega' \in \mathsf{V}_{\gamma}$, $n', n'' \in \Xi_p$. The space $\mathfrak{M}^{\gamma}_{\underline{\mathbf{k}}}(\underline{\mathbf{n}}, \underline{\mathbf{w}}, \underline{\mathbf{m}})$ is acted upon by the group

(331)
$$\mathsf{H}^{\gamma} = \left(\mathsf{H}^{\gamma}_{\mathsf{frame}} \times \mathsf{H}^{\gamma}_{\mathsf{edg}}\right) / \mathsf{H}^{\gamma}_{\mathsf{ver}} \times U(1)\,,$$

where

(332)
$$\mathsf{H}^{\gamma}_{\mathsf{frame}} = \bigtimes_{\omega \in \mathsf{V}_{\gamma}, n \in \Xi_p} U(N^{+}_{\omega,n}) \times U(N^{-}_{\omega,n})\,,$$
$$\mathsf{H}^{\gamma}_{\mathsf{edg}} = \bigtimes_{e \in \mathsf{E}_{\gamma}, n \in \Xi_p} U(1)\,, \quad \mathsf{H}^{\gamma}_{\mathsf{ver}} = \bigtimes_{\omega \in \mathsf{V}_{\gamma}, n \in \Xi_p} U(1)$$

10.2. Orbifolds and defects: ADE $\times$ $U(1)$ $\times$ ADE. The construction above can be motivated by the following examples.

Recall that the moduli space $\mathcal{M}^{+}_{k}(n, w)$ of crossed instantons has an $\mathrm{SU}(2) \times \mathrm{U}(1) \times \mathrm{SU}(2)$ symmetry.

10.2.1. *Space action.* Let Γ be a discrete subgroup of $\mathrm{SU}(2)_{12} \times \mathrm{U}(1)_{\Delta} \times \mathrm{SU}(2)_{34}$,

(333)
$$\iota : \Gamma \longrightarrow \mathsf{G}_{\mathrm{rot}} = \mathrm{SU}(2)_{12} \times \mathrm{U}(1)_{\Delta} \times \mathrm{SU}(2)_{34}\,,$$

10.2.2. *Framing action.* Now let us endow the spaces N_{12} and N_{34} with the structure of Γ module:

(334)
$$N_{12} = \bigoplus_{\omega \in \Gamma^{\vee}} N_{\omega} \otimes \mathcal{R}_{\omega}\,,$$
$$N_{34} = \bigoplus_{\omega \in \Gamma^{\vee}} W_{\omega} \otimes \mathcal{R}_{\omega}\,,$$

in other words let us fix the homomorphisms

(335)
$$\rho_A : \Gamma \longrightarrow U(n_A)$$

Let us denote by $\underline{\mathbf{n}}, \underline{\mathbf{w}}$ the vectors of dimensions $(\mathrm{dim}N_{12,\omega})_{\omega \in \Gamma^{\vee}}$, $(\mathrm{dim}N_{34,\omega})_{\omega \in \Gamma^{\vee}}$, respectively.

The data (334), (333) defines the embedding of the group Γ into H, the symmetry group of $\mathcal{M}^{+}_{k}(n, w)$.

10.2.3. *New moduli spaces.* The set of Γ-fixed points in $\mathcal{M}_k^+(n,w)$ splits into components

$$\mathcal{M}_k^+(n,w)^\Gamma = \bigcup_{\underline{\mathbf{k}}} \mathcal{M}_{\underline{\mathbf{k}}}^{+,\gamma_\Gamma}(\underline{\mathbf{n}},\underline{\mathbf{w}}) \tag{336}$$

This is a particular case of the space $\mathfrak{M}_{\underline{\mathbf{k}}}^{\gamma}(\underline{\mathbf{n}},\underline{\mathbf{w}},0)$.

The requirement that Γ preserves the U(k)-orbit of $(\mathbf{B},\mathbf{I},\mathbf{J})$ translates to the fact that Γ is unitary represented in K, so that

$$\gamma\cdot(\mathbf{B},\mathbf{I},\mathbf{J}) = (g_\gamma \mathbf{B} g_\gamma^{-1}, g_\gamma \mathbf{I}, \mathbf{J} g_\gamma^{-1}), \qquad \gamma\mapsto g_\gamma \in \mathrm{U}(K) \tag{337}$$

Thus, we can decompose K into the irreps of Γ:

$$K = \bigoplus_{\omega\in\Gamma^\vee} K_\omega \otimes \mathcal{R}_\omega \tag{338}$$

The operators $\mathbf{B},\mathbf{I},\mathbf{J}$ then become linear maps between the spaces $K_{\omega'}, N_{A,\omega''}$, which can be easily classified by unraveling the equivariance conditions (337).

The components $\mathcal{M}_{\underline{\mathbf{k}}}^{+,\gamma_\Gamma}(\underline{\mathbf{n}},\underline{\mathbf{w}})$ can then be deformed by modifying the real moment map equation to

$$\mu = \sum_{\omega\in\Gamma^\vee} \zeta_\omega \mathbf{1}_{K_\omega} \tag{339}$$

In the particular case $\Gamma\subset \mathrm{SU}(2)_{34}$ the orbifold produces the moduli spaces of supersymmetric gauge configurations in the

$$\mathsf{G}_{\mathbf{g}} = \bigtimes_{\omega\in\Gamma^\vee} \mathrm{U}(n_\omega) \tag{340}$$

gauge theory in the presence of a point-like defect, the *qq*-character

$$\mathcal{X}_{(w_\omega)_{\omega\in\Gamma^\vee}}(x,\nu_{\omega,\beta}) \tag{341}$$

The gauge theory in question is the affine ADE quiver gauge theory.

10.2.4. *Odd dimensions and finite quivers.* We can also obtain the moduli space of supersymmetric gauge field configurations in the quiver gauge theories built on finite quivers. The natural way to do that is to start with an affine quiver and send some of the gauge couplings to zero, i.e. by making some of dimensions k_ω vanishing.

Remarkably, this procedure produces the superspace, the odd variables originating in the multiplet of the cohomological field theory. Let us explain this in more detail. Let us consider, for simplicity, the group $\Gamma\subset SU(2)_{34}$, so that $\mathsf{E}_\gamma{}^+$ has one element.

The linear algebra data

$$\left(B_{1,\omega}\,,\, B_{2,\omega}\,,\, I_{12,\omega}\,,\, J_{12,\omega}\,,\, I_{34,\omega}\,,\, J_{34,\omega}\,,\, B_e\,,\, \tilde{B}_e\right) \in \mathcal{A}_{\underline{\mathbf{k}}}^{\gamma}(\underline{\mathbf{n}},\underline{\mathbf{w}},0)$$

is parametrized by the

$$2\sum_{\omega\in\Gamma^\vee} k_\omega\left(k_\omega + n_\omega + w_\omega\right) + 2\sum_{e\in E} k_{t(e)}k_{s(e)} \tag{342}$$

complex dimensional space. The Eqs. (315) plus the $\mathrm{GL}_\gamma(\underline{\mathbf{k}})$-invariance remove

$$\sum_{\omega\in\Gamma^\vee}\left(n_\omega w_\omega + 2k_\omega(k_\omega+w_\omega)\right) + \sum_{e\in E} 2k_{s(e)}k_{t(e)} + k_{s(e)}n_{t(e)} + k_{t(e)}n_{s(e)} \tag{343}$$

dimensions (this is half the number of equations (329)). The result is $\underline{\mathbf{k}}$-linear,

$$\text{virtual } \dim \mathcal{M}^{+,\gamma}_{\underline{\mathbf{k}}}(\underline{\mathbf{n}}, \underline{\mathbf{w}}) = \sum_{\omega \in \Gamma^{\vee}} \left(k_{\omega} m_{\omega} - n_{\omega} w_{\omega} \right) \tag{344}$$

where

$$m_{\omega} = 2n_{\omega} - \sum_{e \in s^{-1}(\omega)} n_{t(e)} - \sum_{e \in t^{-1}(\omega)} n_{s(e)} \tag{345}$$

Now, if for all $\omega \in \Gamma^{\vee}$ the deficits m_{ω} are non-negative, and at least for one vertex the deficit is positive then the quiver is, in fact, a finite ADE Dynkin diagram. In this case we can add the odd variables taking values in the spaces $\mathrm{Hom}(K_{\omega}, M_{\omega})$ with the complex vector space M_{ω} of dimension m_{ω}, and define the moduli space to be the supermanifold which is the total space of the odd vector bundle $\Pi\mathrm{Hom}(K_{\omega}, M_{\omega})$ over the previously defined bosonic moduli space. In practice this means that the integration over the "true" moduli space is the integral over the coarse moduli space of the equivariant Euler class of the vector bundle $\mathrm{Hom}(K_{\omega}, M_{\omega})$. This is what the cohomological field theory applied to the affine quiver case with the subsequent setting $k_{\omega'} = 0$ for some $\omega' \in \Gamma^{\vee}$ would amount to. With the "compensator" vector bundle in place the virtual dimension of the moduli space becomes $\underline{\mathbf{k}}$-independent. This is the topological counterpart of the asymptotic conformal invariance of the gauge theory.

11. Spiked instantons on orbifolds and defects

Now let us go back to the general case of spiked instantons. Choose a discrete subgroup Γ of $U(1)^3_{\underline{\epsilon}}$, e.g. $\Gamma \approx \Xi_{p_1} \times \Xi_{p_2} \times \Xi_{p_3}$, $\Gamma^{\vee} \approx \Gamma$. Fix the framing homomorphisms: $\rho_{A,\Gamma} \to U(n_A)$:

$$N_A = \bigoplus_{\vec{\omega} \in \Gamma^{\vee}} N_{\vec{\omega}} \otimes \mathcal{R}_{\vec{\omega}} \tag{346}$$

The set of Γ-fixed points in $\mathfrak{M}_k(\vec{n})$ splits into components

$$\mathfrak{M}_k(\vec{n})^{\Gamma} = \bigcup_{\underline{\mathbf{k}}} \mathfrak{M}^{\gamma_{\Gamma}}_{\underline{\mathbf{k}}}(\underline{\vec{\mathbf{n}}}) \tag{347}$$

It describes the moduli spaces of spiked instantons in the presence of additional surface and point-like conical defects. The compactness theorem holds in this case. Its proof is a simple extension of the proof of section **8**: $K_A \to K_{A,\vec{\omega}}$, where $\vec{\omega} = (\omega_1, \omega_2, \omega_3)$, $\omega_i \in \Xi_{p_i}$.

12. Conclusions and future directions

In this paper we introduced several moduli spaces: $\mathfrak{M}^{+}, \mathfrak{M}^{\natural}, \mathfrak{M}^{*}$ of matrices solving quadratic equations modulo symmetries. These moduli spaces generalize the Gieseker-Nakajima partial compactification $\mathcal{M}$ of the ADHM moduli space of $U(n)$ instantons on $\mathbb{R}^4$. We gave some motivations for these constructions and proved the compactness theorem which we shall use in the next papers to establish useful identities on the correlation functions of supersymmetric gauge theories in four dimensions.

In this concluding section we would like to make a few remarks.

First of all, one can motivate the crossed instanton construction by starting the with the ordinary ADHM construction and adding the co-fields [**7**, **8**] which mirror the embedding of the $\mathcal{N} = 2$ super-Yang-Mills vector multiplet into the $\mathcal{N} = 4$ super-Yang-Mills vector multiplet [**38**].

Secondly, we would like to find the crossed instanton analogue of the stable envelopes of [**19**].

Third, it would be nice to generalize the spiked instanton construction to allow more general orbifold groups $\Gamma \subset SU(4)$, and more general (Lagrangian?) subvarieties in $\mathbb{C}^4/\Gamma$.

The purpose of the ADHM construction, after all, is the construction of the solutions to the instanton equations

$$F_A^+ = 0$$

We didn't find the analogue of the ADHM construction for the spiked instantons. Conjecturally, the matrices $[\mathbf{B}, \mathbf{I}, \mathbf{J}]$ solving the Eqs. (56) are in one-to-one correspondence with the finite action solutions to Eqs. (41). A step in this direction is the string theory realization of crossed and spiked instantons [**28**].

Finally, we have proposed a definition of quiver crossed instantons, which are defined for quivers more general then the products of ADE Dynkin diagrams. It would be interesting to find the precise restrictions on these quivers compatible with the compactness theorem.

In the forthcoming papers the compactness theorem will be used to derive the main statements of the theory of qq-characters [**33**].

Acknowledgements

Research was supported in part by the NSF grant PHY 1404446. The author thanks the organizers of String-Math-2015 for hospitality in Sanya and to the participants of String-Math-2015 for interesting questions and suggestions.

References

[1] M. F. Atiyah, N. J. Hitchin, V. G. Drinfeld, and Yu. I. Manin, *Construction of instantons*, Phys. Lett. A **65** (1978), no. 3, 185–187, DOI 10.1016/0375-9601(78)90141-X. MR598562

[2] T. Banks, W. Fischler, S. H. Shenker, and L. Susskind, *M theory as a matrix model: a conjecture*, Phys. Rev. D (3) **55** (1997), no. 8, 5112–5128, DOI 10.1103/PhysRevD.55.5112. MR1449617

[3] Laurent Baulieu, Andrei Losev, and Nikita Nekrasov, *Chern-Simons and twisted supersymmetry in various dimensions*, Nuclear Phys. B **522** (1998), no. 1-2, 82–104, DOI 10.1016/S0550-3213(98)00096-0. MR1638896

[4] Ionuţ Ciocan-Fontanine and Mikhail Kapranov, *Derived Quot schemes* (English, with English and French summaries), Ann. Sci. École Norm. Sup. (4) **34** (2001), no. 3, 403–440, DOI 10.1016/S0012-9593(01)01064-3. MR1839580

[5] Ionuţ Ciocan-Fontanine and Mikhail M. Kapranov, *Derived Hilbert schemes*, J. Amer. Math. Soc. **15** (2002), no. 4, 787–815, DOI 10.1090/S0894-0347-02-00399-5. MR1915819

[6] Neil R. Constable, Johanna Erdmenger, Zachary Guralnik, and Ingo Kirsch, *Intersecting D3-brane and holography*, Phys. Rev. D (3) **68** (2003), no. 10, 106007, 24, DOI 10.1103/PhysRevD.68.106007. MR2071072

[7] Stefan Cordes, Gregory Moore, and Sanjaye Ramgoolam, *Lectures on 2D Yang-Mills theory, equivariant cohomology and topological field theories*, Nuclear Phys. B Proc. Suppl. **41** (1995), 184–244, DOI 10.1016/0920-5632(95)00434-B. String theory, gauge theory and quantum gravity (Trieste, 1994). MR1352757

[8] Stefan Cordes, Gregory Moore, and Sanjaye Ramgoolam, *Large N 2D Yang-Mills theory and topological string theory*, Comm. Math. Phys. **185** (1997), no. 3, 543–619, DOI 10.1007/s002200050102. MR1463054

[9] Robbert Dijkgraaf, Erik Verlinde, and Herman Verlinde, *Matrix string theory*, Nuclear Phys. B **500** (1997), no. 1-3, 43–61, DOI 10.1016/S0550-3213(97)00326-X. MR1471648

[10] Michael R. Douglas and Nikita A. Nekrasov, *Noncommutative field theory*, Rev. Modern Phys. **73** (2001), no. 4, 977–1029, DOI 10.1103/RevModPhys.73.977. MR1878801

[11] T. Graber and R. Pandharipande, *Localization of virtual classes*, Invent. Math. **135** (1999), no. 2, 487–518, DOI 10.1007/s002220050293. MR1666787

[12] David J. Gross and Nikita A. Nekrasov, *Dynamics of strings in noncommutative gauge theory*, J. High Energy Phys. **10** (2000), Paper 21, 28, DOI 10.1088/1126-6708/2000/10/021. MR1811454

[13] Gross, D. J., and Nekrasov, N. A. Solitons in noncommutative gauge theory. JHEP 03 (2001), 044.

[14] N. J. Hitchin, A. Karlhede, U. Lindström, and M. Roček, *Hyper-Kähler metrics and supersymmetry*, Comm. Math. Phys. **108** (1987), no. 4, 535–589. MR877637

[15] Iarrobino, A., and Yameogo, J. The family G_T of graded quotients of k[x,y] of given Hilbert function.

[16] Anthony A. Iarrobino, *Punctual Hilbert schemes*, Mem. Amer. Math. Soc. **10** (1977), no. 188, viii+112, DOI 10.1090/memo/0188. MR0485867

[17] Nobuyuki Ishibashi, Hikaru Kawai, Yoshihisa Kitazawa, and Asato Tsuchiya, *A large-N reduced model as superstring*, Nuclear Phys. B **498** (1997), no. 1-2, 467–491, DOI 10.1016/S0550-3213(97)00290-3. MR1459082

[18] Konstantin Loginov, *Hilbert-Samuel sequences of homogeneous finite type*, J. Pure Appl. Algebra **221** (2017), no. 4, 821–831, DOI 10.1016/j.jpaa.2016.08.005. MR3574209

[19] Maulik, D., and Okounkov, A. Quantum Groups and Quantum Cohomology. ArXiv/1211.1287 (Nov. 2012).

[20] Mintun, E., Polchinski, J., and Sun, S. The Field Theory of Intersecting D3-branes. hep-th/1402.6327 (2014).

[21] Gregory Moore, Nikita Nekrasov, and Samson Shatashvili, *D-particle bound states and generalized instantons*, Comm. Math. Phys. **209** (2000), no. 1, 77–95, DOI 10.1007/s002200050016. MR1736943

[22] Hiraku Nakajima, *Gauge theory on resolutions of simple singularities and simple Lie algebras*, Internat. Math. Res. Notices **2** (1994), 61–74, DOI 10.1155/S1073792894000085. MR1264929

[23] Hiraku Nakajima, *Instantons on ALE spaces, quiver varieties, and Kac-Moody algebras*, Duke Math. J. **76** (1994), no. 2, 365–416, DOI 10.1215/S0012-7094-94-07613-8. MR1302318

[24] Hiraku Nakajima, *Heisenberg algebra and Hilbert schemes of points on projective surfaces*, Ann. of Math. (2) **145** (1997), no. 2, 379–388, DOI 10.2307/2951818. MR1441880

[25] Hiraku Nakajima, *Quiver varieties and Kac-Moody algebras*, Duke Math. J. **91** (1998), no. 3, 515–560, DOI 10.1215/S0012-7094-98-09120-7. MR1604167

[26] Hiraku Nakajima, *Lectures on Hilbert schemes of points on surfaces*, University Lecture Series, vol. 18, American Mathematical Society, Providence, RI, 1999. MR1711344

[27] Nekrasov, N. BPS/CFT Correspondence III: Gauge Origami partition function and qq-characters.

[28] Nekrasov, N., and Prabhakar, N. S. Spiked Instantons from Intersecting D-branes. Nucl. Phys. B914 (2017), 257–300.

[29] Nikita Nekrasov and Albert Schwarz, *Instantons on noncommutative* $\mathbf{R}^4$, *and* $(2,0)$ *superconformal six-dimensional theory*, Comm. Math. Phys. **198** (1998), no. 3, 689–703, DOI 10.1007/s002200050490. MR1670037

[30] N. A. Nekrasov, *Lectures on open strings, and noncommutative gauge theories*, Unity from duality: gravity, gauge theory and strings (Les Houches, 2001), NATO Adv. Study Inst., EDP Sci., Les Ulis, 2003, pp. 477–495, DOI 10.1007/3-540-36245-2_8. MR2010977

[31] Nikita A. Nekrasov, *Seiberg-Witten prepotential from instanton counting*, Adv. Theor. Math. Phys. **7** (2003), no. 5, 831–864. MR2045303

[32] Nekrasov, Nikita. Trieste lectures on solitons in noncommutative gauge theories. hep-th/0011095 (2000).

[33] Nekrasov, Nikita. BPS/CFT correspondence: Non-perturbative Dyson-Schwinger equations and *qq*-characters. arXiv:1512.05388 (2015).

[34] Nathan Seiberg, *A note on background independence in noncommutative gauge theories, Matrix model, and tachyon condensation*, J. High Energy Phys. **9** (2000), Paper 3, 16, DOI 10.1088/1126-6708/2000/09/003. MR1789107

[35] Nathan Seiberg and Edward Witten, *String theory and noncommutative geometry*, J. High Energy Phys. **9** (1999), Paper 32, 93, DOI 10.1088/1126-6708/1999/09/032. MR1720697
[36] TONG, D. The holographic dual of $AdS_3 \times S^3 \times S^3 \times S^1$. JHEP 1404 (2014), 193.
[37] TONG, D., AND WONG, K. ADHM Revisited: Instantons and Wilson Lines. arXiv:1410.8523 (2014).
[38] Cumrun Vafa and Edward Witten, *A strong coupling test of S-duality*, Nuclear Phys. B **431** (1994), no. 1-2, 3–77, DOI 10.1016/0550-3213(94)90097-3. MR1305096
[39] Edward Witten, *Topological sigma models*, Comm. Math. Phys. **118** (1988), no. 3, 411–449. MR958805
[40] Edward Witten, *Bound states of strings and p-branes*, Nuclear Phys. B **460** (1996), no. 2, 335–350, DOI 10.1016/0550-3213(95)00610-9. MR1377168

SIMONS CENTER FOR GEOMETRY AND PHYSICS, STONY BROOK UNIVERSITY, STONY BROOK NEW YORK 11794-3636
E-mail address: nikitastring@gmail.com

ON LEAVE OF ABSENCE FROM: IHES, BURES-SUR-YVETTE, FRANCE, ITEP AND IITP, MOSCOW, RUSSIA

Balanced embedding of degenerating Abelian varieties

Xiaowei Wang and Yuguang Zhang

ABSTRACT. For a certain maximal unipotent family of Abelian varieties over the punctured disc, we show that after a base change, one can complete the family over a disc such that the whole degeneration can be simultaneously balanced embedded into a projective space by the theta functions. Then we study the relationship between the balanced filling-in and the Gromov-Hausdorff limit of flat Kähler metrics on the nearby fibers.

1. Introduction

Let $X \subset \mathbb{CP}^\nu$ be a n-dimensional projective variety. The embedding $X \hookrightarrow \mathbb{CP}^\nu$ is called balanced if

$$\int_X \left(\frac{z_i \bar{z}_j}{|z|^2} - \frac{\delta_{ij}}{\nu+1}\right)\omega_{FS}^n = 0,$$

where $z_0, \cdots, z_\nu$ are the homogenous coordinates of $\mathbb{CP}^\nu$, $|z|^2 = \sum |z_i|^2$, and ω_{FS} is the Fubini-Study metric (cf. [**D**, **BLY**]). In [**Z**], S. Zhang proved that X, as an algebraic cycle in $\mathbb{CP}^\nu$, is Chow polystable if and only if the embedding $X \subset \mathbb{CP}^\nu$ can be translated to a balanced one via an element $u \in \mathrm{SL}(\nu+1)$.

A theorem due to Donaldson [**D**] shows the connection of the balanced embedding and the existence of Kähler metric with constant scalar curvature. More precisely, let (X, L) be a polarized manifold of dimension n such that the automorphism group $\mathrm{Aut}(X, L)$ is finite. If there is a Kähler metric ω with constant scalar curvature representing $c_1(L)$, then Donaldson's theorem asserts that for $k \gg 1$, L^k induces a balanced embedding $\Phi_k : X \hookrightarrow \mathbb{CP}^{\nu_k}$ with $\Phi_k^* \mathcal{O}_{\mathbb{CP}^{\nu_k}}(1) = L^k$. Furthermore,

$$\|\omega - k^{-1}\Phi_k^*\omega_{FS}\|_{C^r(X)} \to 0,$$

when $k \to \infty$, in the C^r-sense for any $r > 0$. The Kähler metric $k^{-1}\Phi_k^*\omega_{FS}$ is called a *balanced metric.*

If (X, L) is a polarized Calabi-Yau manifold, Yau's theorem on the Calabi conjecture says that there exists a unique Ricci-flat Kähler-Einstein metric ω with $\omega \in c_1(L)$, i.e. the Ricci curvature $\mathrm{Ric}(\omega) \equiv 0$ (cf. [**Y**]). By Donaldson's theorem,

---

2010 *Mathematics Subject Classification.* Primary 14K99; Secondary 53C55.

Key words and phrases. Abelian variety, balanced embedding, theta function.

The first named author was partially supported by a Collaboration Grants for Mathematicians from Simons Foundation and NSF:DMS-1609335, and the second named author is supported in part by grant NSFC-11271015.

L^k induces a balanced embedding $X \hookrightarrow \mathbb{CP}^{\nu_k}$ for $k \gg 1$, and the Ricci-flat Kähler-Einstein metric ω can be approximated by the balanced metrics. If (X, L) is a principally polarized Abelian variety, it is proven first in [**N**] (and also independently [**WY**]) that the standard embedding induced by the classical theta functions of level k is balanced. Moreover, in this case the convergence of balanced metrics to the flat metric can be verified via a complete elementary way without quoting [**D**].

In [**SYZ**], Strominger, Yau and Zaslow propose a geometric way of constructing mirror Calabi-Yau manifolds via dual special lagrangian fibration, which is the celebrated SYZ conjecture. Later, a new version of the SYZ conjecture is proposed by Gross, Wilson, Kontsevich, and Soibelman, (cf. [**GW, KS1, KS2**]) by using the collapsing of Ricci-flat Kähler-Einstein metrics. Let $(\mathcal{X} \to \Delta, \mathcal{L})$ be a maximal unipotent degeneration of polarized Calabi-Yau n-manifolds, i.e. the relative canonical bundle $\mathcal{K}_{\mathcal{X}/\Delta}$ is trivial, such that $0 \in \Delta$ is a large complex limit point, and ω_t be the Ricci-flat Kähler-Einstein metric satisfying $\omega_t \in c_1(\mathcal{L}|_{X_t})$ for $t \in \Delta^\circ$. The collapsing version of SYZ conjecture asserts that

$$(X_t, \operatorname{diam}^{-2}_{\omega_t}(X_t)\omega_t) \to (B, d_B)$$

in the Gromov-Hausdorff sense, when $t \to 0$, where (B, d_B) is a compact metric space. Furthermore, there is an open dense subset B_0 of B, which is smooth, and is of real dimension n, and admits a real affine structure. The metric d_B is induced by a Monge-Ampère metric g_B on B_0, i.e. under affine coordinates $y_1, \cdots, y_n$, there is a potential function φ such that

$$g_B = \sum_{ij} \frac{\partial^2 \varphi}{\partial y_i \partial y_j} dy_i dy_j, \quad \text{and} \quad \det\left(\frac{\partial^2 \varphi}{\partial y_i \partial y_j}\right) = \text{const.}.$$

Clearly it is true for Abelian varieties. This conjecture is verified by Gross and Wilson for fibred K3 surfaces with only type I_1 singular fibers in [**GW**], and is studied for higher dimensional HyperKähler manifolds in [**GTZ1, GTZ2**].

Bernd Siebert raises a question to relate the balanced embeddings of X_t to the metric limit of rescaled Ricci-flat Kähler-Einstein metrics $\varepsilon_t \omega_t$ for a certain family of constants ε_t. In the Gross-Siebert program (cf. [**G, GS1**]), theta functions are constructed on certain degenerations of polarized Calabi-Yau manifolds $(\mathcal{X} \to \operatorname{Spec}\mathbb{C}[[t]], \mathcal{L})$ as the canonical basis of the space of sections for $\mathcal{L}$ (cf. [**GHKS1, GS2, GHKS2**]), which is predicted by the Homological Mirror Symmetry conjecture. In particular, these theta functions recover the classical theta functions in the case of principally polarized Abelian varieties. If $(\mathcal{X}, \mathcal{L})$ is an analytic family (cf. [**RS**]), Siebert asks whether the theta functions give the balanced embeddings of polarized Calabi-Yau manifolds, and furthermore whether there is a family version of the Donaldson's theorem for the degeneration of Calabi-Yau manifolds near large complex limits. In this note, we study this question in the case of principally polarized Abelian varieties, and establish the connection between the limit metric g_B and the balanced embeddings.

In Section 2, let us recall the basic setup for a maximal unipotent family of principally polarized Abelian varieties over the punctured disc, and then state our main result. Theorem 2.1 says that after a certain base change, one can find a filling-in to complete the family of Abelian varieties to a degeneration, such that the whole degeneration can be simultaneously balanced embedded in a projective space over a disc by the *canonical* theta functions constructed via Gross-Siebert program. Theorem 2.2 studies the relationship between the balanced filling-in and the Gromov-Hausdorff limit of flat Kähler metrics on the nearby fibers. In Section 3, we review the construction of theta functions on degenerations of Abelian varieties in the Gross-Siebert program. Finally, Theorem 2.1 and Theorem 2.2 are proved in Section 4.

2. Setup and Main Theorems

In this paper, we always denote $M \cong \mathbb{Z}^n$, $M_{\mathbb{R}} = M \otimes_{\mathbb{Z}} \mathbb{R}$, $N = \mathrm{Hom}_{\mathbb{Z}}(M, \mathbb{Z})$, $N_{\mathbb{R}} = N \otimes_{\mathbb{Z}} \mathbb{R}$, $T_{\mathbb{C}} = N \otimes_{\mathbb{Z}} \mathbb{C}^*$, and $\langle \cdot, \cdot \rangle$ the pairing between $N_{\mathbb{R}}$ and $M_{\mathbb{R}}$.

2.1. A family of Abelian varieties. This subsection gives the basic setup of this paper, which is a family of Abelian varieties over the punctured disc approaching to a large complex limit.

Let $Z(\cdot, \cdot) : M_{\mathbb{R}} \times M_{\mathbb{R}} \to \mathbb{R}$ be a positive definite bilinear form satisfying $Z(M, M) \subset \mathbb{Z}$. If we define the quadratic function

$$\overline{\varphi}(y) = \frac{1}{2} Z(y, y) \tag{2.1}$$

on $M_{\mathbb{R}}$, and the affine linear function

$$\alpha_\gamma(\cdot) = Z(\gamma, \cdot) + \frac{1}{2} Z(\gamma, \gamma), \tag{2.2}$$

for any $\gamma \in M$, then

$$\overline{\varphi}(y + \gamma) = \overline{\varphi}(y) + \alpha_\gamma(y). \tag{2.3}$$

The couple $\{M, Z\}$ determines a family of principally polarized Abelian varieties over the punctured disc $(\pi : \mathcal{X}_\eta \to \Delta^\circ, \mathcal{L}_\eta)$ as the following.

We define an M-action on $M \times \mathbb{Z}$ via $(m, r) \mapsto (m, r + Z(\gamma, m))$ for any $\gamma \in M$, which induces an M-action on $T_{\mathbb{C}}$ by

$$\mathcal{Z}^m \mapsto \mathcal{Z}^m s^{Z(\gamma, m)}, \quad \gamma \in M, \tag{2.4}$$

for any $s \in \Delta^\circ$. More explicitly, let

$$e_1, \cdots, e_n \in M$$

be a basis of M, we have coordinates $z_1 = \mathcal{Z}^{e_1}, \cdots, z_n = \mathcal{Z}^{e_n}$ on $T_{\mathbb{C}} \cong (\mathbb{C}^\times)^n$, and the M-action is that $(z_1, \cdots, z_n) \mapsto (z_1 s^{Z(\gamma, e_1)}, \cdots, z_n s^{Z(\gamma, e_n)})$ for any $\gamma \in M$. We claim that the quotient $X_s = T_{\mathbb{C}}/M$ is a principally polarized Abelian variety with period matrix

$$\left[I, \frac{\log s}{2\pi\sqrt{-1}} Z_{ij} \right]$$

where $Z_{ij} = Z(e_i, e_j) \in \mathbb{Z}$.

Denote $e_1^*, \cdots, e_n^*$ the dual basis of N, and $N_{\mathbb{C}} = N \times_{\mathbb{Z}} \mathbb{C} \cong \mathbb{C}^n$. We have a natural embedding $N \hookrightarrow N_{\mathbb{C}}$ as the real part, and by abusing notions, we regard $e_1^*, \cdots, e_n^*$ as a $\mathbb{C}$-basis of $N_{\mathbb{C}}$. The universal covering $N_{\mathbb{C}} \to T_{\mathbb{C}}$ is given by $w_i \mapsto z_i = \exp 2\pi\sqrt{-1} w_i$, $i = 1, \cdots, n$, where $w_1, \cdots, w_n$ are coordinates of $N_{\mathbb{C}}$ respecting to $e_1^*, \cdots, e_n^*$. Then $T_{\mathbb{C}} \cong N_{\mathbb{C}}/N$ where N acts on $N_{\mathbb{C}}$ given by $w_i \mapsto w_i + \langle \mu, e_i \rangle = w_i + \mu_i$ for any $\mu = \sum_i \mu_i e_i^* \in N$. We have an $N \times M$-action on $\mathbb{C}^n$ by

$$w_i \mapsto w_i + \langle \mu, e_i \rangle + \frac{\log s}{2\pi\sqrt{-1}} Z(\gamma, e_i),$$

for any $(\mu,\gamma)\in N\times M$, and we obtain $X_s = T_{\mathbb{C}}/M = \mathbb{C}^n/\Lambda_s$, where the lattice $\Lambda_s = \mathrm{span}_{\mathbb{Z}}\{e_1^*,\cdots,e_n^*,\frac{\log s}{2\pi\sqrt{-1}}Z(e_1,\cdot),\cdots,\frac{\log s}{2\pi\sqrt{-1}}Z(e_n,\cdot)\}$. Furthermore, we construct a family of Abelian varieties $\mathcal{X}_\eta = (T_{\mathbb{C}}\times\Delta^\circ)/M \to \Delta^\circ$ over the punctured disc Δ° with fiber X_s.

We extend the $N\times M$-action on $\mathbb{C}^n$ to $\mathbb{C}^n\times\mathbb{C}$ by

$$(w,\lambda)\mapsto (w+\mu+\frac{\log s}{2\pi\sqrt{-1}}Z(\gamma,\cdot),\lambda\exp\pi\sqrt{-1}(-\frac{\log s}{2\pi\sqrt{-1}}Z(\gamma,\gamma)-2\langle w,\gamma\rangle)),$$

for any $(\mu,\gamma)\in N\times M$, where $w=(w_1,\cdots,w_n)\in\mathbb{C}^n$. The quotient $(\mathbb{C}^n\times\mathbb{C})/(N\times M)$ is the relative ample bundle $\mathcal{L}_\eta$. The classical Riemann theta function (cf. [**BL**])

$$\vartheta = \sum_{\gamma\in M}\exp\pi\sqrt{-1}\left(\frac{\log s}{2\pi\sqrt{-1}}Z(\gamma,\gamma)+2\langle w,\gamma\rangle\right) \tag{2.5}$$

is the distinguished section of $\mathcal{L}_\eta$. On any X_s, the first Chern class $c_1(\mathcal{L}_\eta|_{X_s})$ is represented by the flat Kähler metric

$$\omega_s = \frac{-\pi\sqrt{-1}}{\log|s|}\sum_{ij}Z^{ij}dw_i\wedge d\bar{w}_j, \tag{2.6}$$

where Z^{ij} denotes the inverse of the matrix $Z_{ij}=Z(e_i,e_j)$.

Let $x_1,\cdots,x_n,y_1,\cdots,y_n$ denote the coordinates on $N_{\mathbb{C}}$ with respect to the basis e_i^*, $1\leqslant i\leqslant n$, $(\frac{\log|s|}{2\pi\sqrt{-1}}+\arg(s))Z(e_j,\cdot)$, $1\leqslant j\leqslant n$, where $0\leqslant\arg(s)<2\pi$. We also regard $y_1,\cdots,y_n$ (resp. $x_1,\cdots,x_n$) as coordinates on $M_{\mathbb{R}}$ (resp. $N_{\mathbb{R}}$) respecting to $e_1,\cdots,e_n$. Then

$$w_i = x_i + (\arg(s)-\sqrt{-1}\frac{\log|s|}{2\pi})\sum_{j=1}^n Z_{ij}y_j,\ i=1,\cdots,n,$$

and as a symplectic form,

$$\omega_s = \sum_i dx_i\wedge dy_i.$$

The corresponding Riemannian metric is

$$g_s = \frac{-2\pi}{\log|s|}\sum_{i,j}Z^{ij}dx_idx_j - \frac{\log|s|}{2\pi}\sum_{i,j}Z_{ij}dy_idy_j,$$

and, when $s\to 0$,

$$(X_s,\frac{-2\pi}{\log|s|}g_s)\longrightarrow (B,g_B=\sum_{ij}Z_{ij}dy_idy_j)$$

in the Gromov-Hausdorff topology, where $B=M_{\mathbb{R}}/M$.

We can regard $\overline{\varphi}$ as a multivalued function on $B=M_{\mathbb{R}}/M$ by (2.3), and on any small open subset on B, the difference of any two sheets of $\overline{\varphi}$ is a linear function defined in (2.2). Note that the Hessian matrix of $\overline{\varphi}$ is well-defined, and $\frac{\partial^2\overline{\varphi}}{\partial y_i\partial y_j}=Z_{ij}$. The Riemannian metric g_B is the Monge-Ampère metric with potential $\overline{\varphi}$ respecting to the affine coordinates $y_1,\cdots,y_n$, i.e.

$$g_B = \sum_{ij}\frac{\partial^2\overline{\varphi}}{\partial y_i\partial y_j}dy_idy_j,\quad\text{and}\quad \det(\frac{\partial^2\overline{\varphi}}{\partial y_i\partial y_j})\equiv\det(Z_{ij}). \tag{2.7}$$

2.2. Main results. For any $k \gg 1$, let $M_k = kM$, and φ be an M_k-periodic convex piecewise linear function such that the slopes of φ are in N, and

$$\varphi(y+\gamma) = \varphi(y) + \alpha_\gamma(y) \tag{2.8}$$

for any $\gamma \in kM$, which induces an M_k-invariant rational polyhedral decomposition $\mathcal{P}$ of $M_{\mathbb{R}}$, i.e. σ is a cell of $\mathcal{P}$ if and only if φ is linear on σ.

The Mumford's construction (cf. [**M1**]) gives a degeneration of principally polarized Abelian varieties $(\pi : \mathcal{X} \to \Delta, \mathcal{L})$ from the data $(M_k, \mathcal{P}, \varphi)$ such that $\pi : \mathcal{X}_{\Delta^\circ} \to \Delta^\circ$ is the base change of $\mathcal{X}_\eta$ via $t \mapsto t^k = s$, where $\mathcal{X}_{\Delta^\circ} = \pi^{-1}(\Delta^\circ)$, and $\mathcal{L}|_{\mathcal{X}_{\Delta^\circ}}$ is the pull-back of $\mathcal{L}_\eta^k$ (See Section 3 for details). The central fiber $X_0 = \pi^{-1}(0)$ is reduced and reducible with only toric singularities. The intersection complex of X_0 is $(B_k, \tilde{\mathcal{P}})$ where $B_k = M_{\mathbb{R}}/M_k$ and $\tilde{\mathcal{P}}$ is the quotient rational polyhedron decomposition of $\mathcal{P}$. The irreducible components are one to one corresponding to n-cells in $\tilde{\mathcal{P}}$, and for any n-cells $\sigma \in \tilde{\mathcal{P}}$, the respective irreducible component is the toric variety X_σ defined by σ, where we regard σ as a polytope in $M_{\mathbb{R}}$. Furthermore, the restriction of $\mathcal{L}$ on X_σ is the toric ample line bundle defined by σ.

For any $m \in B_k(\mathbb{Z}) = M/M_k$, a section ϑ_m of $\mathcal{L}$ is constructed in Section 6 of [**GHKS1**] such that the restriction of $t^{\varphi(m)-\bar{\varphi}(m)}\vartheta_m$ on any $X_t = \pi^{-1}(t)$, $t \neq 0$, is a classical Riemann theta function (See also [**GS2**]). And the restriction of ϑ_m on any component X_σ of X_0 is a monomial section of $\mathcal{L}|_{X_\sigma}$.

Note that the choice of $(\mathcal{P}, \varphi)$ is not unique, and different choices give different filling-ins X_0. However there is a canonical one studied in [**AN**], which satisfies $\varphi(m) = \bar{\varphi}(m)$ for any $m \in M$. More precisely, let $\varphi : M_{\mathbb{R}} \to \mathbb{R}$ be the convex piecewise linear function such that the graph of φ is the lower bound of the convex hull of $\{(m, \overline{\varphi}(m)) | m \in M\} \subset M_{\mathbb{R}} \times \mathbb{R}$, and let $\mathcal{P}$ be the rational polyhedral decomposition of $M_{\mathbb{R}}$ induced by φ, i.e. a cell $\sigma \in \mathcal{P}$ if and only if φ is linear on σ. It is clear that (2.8) is satisfied, and

$$\varphi(m) = \overline{\varphi}(m), \quad \text{for any } m \in M. \tag{2.9}$$

Let $\overline{\mathcal{P}}$ be the boundary of the convex hull of the lattice points on the graph of $\overline{\varphi}$, then the polyhedral decomposition $\mathcal{P}$ is obtained in such a way that each cell of $\mathcal{P}$ is precisely the projection of a face of $\overline{\mathcal{P}}$ onto $M_{\mathbb{R}}$. The decomposition $\mathcal{P}$ is the *mostly divided* polyhedral decomposition that one can have. Any cell σ of $\mathcal{P}$ intersects with the lattice M only at its vertices, i.e. there is no integral point in the interior of σ.

We further assume that for any $\sigma \in \mathcal{P}$, the slope of $\varphi|_\sigma$ is integral, i.e.

$$d\varphi|_\sigma \in N \tag{2.10}$$

for simplicity. It is not a further restriction, and the reason is as the following. Note that $d\varphi|_\sigma \in N \otimes_{\mathbb{Z}} \mathbb{Q}$. For any $m \in M$, there is a $\nu \in \mathbb{N}$ such that $\nu d\varphi|_\sigma \in N$ for any cell $\sigma \in \mathcal{P}$ with $m \in \sigma$, i.e. σ belongs to the star of m. By the M-action and (2.8), $\nu d\varphi|_\sigma + \nu d\alpha_\gamma \in N$ for any $\gamma \in M$, and thus $\nu\varphi$ satisfies (2.10). If we replace Z by νZ, then the new family of Abelian varieties constructed from $\nu\overline{\varphi}$ is the base change of the original family by $s \mapsto s^\nu$. Hence we assume (2.10).

Before we present the main theorems, we look at some lower dimensional cases. If $\dim_{\mathbb{R}} M_{\mathbb{R}} = 1$, B_k is a cycle, 1-cells in $\mathcal{P}$ are intervals, and X_0 is the Kodaira type I_k fiber, i.e. a cycle of k rational curves. It is well known to experts that X_0 can be balanced embedded. When $\dim_{\mathbb{R}} M_{\mathbb{R}} = 2$, there are only two possible choices of

$\mathcal{P}$ by [**AN**]. One is that any 2-cell in $\mathcal{P}$ is a standard simplex, and the other one is that $\mathcal{P}$ consists of squares. Hence X_0 consists either finite many $\mathbb{CP}^2$ or finite many $\mathbb{CP}^1 \times \mathbb{CP}^1$.

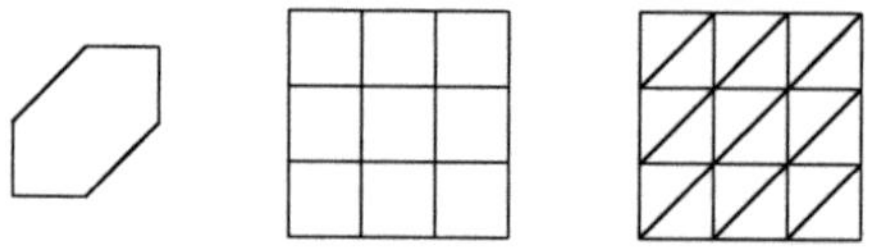

The first result of this paper shows that the embedding of the canonical degeneration $\mathcal{X}$ of [**AN**] by theta functions constructed in [**GHKS1**] is balanced.

THEOREM 2.1. *For any $k \gg 1$, $B_k = M_{\mathbb{R}}/M_k$, $B_k(\mathbb{Z}) = M/M_k$, and let $(\pi : \mathcal{X} \to \Delta, \mathcal{L})$ be the degeneration of principally polarized Abelian varieties from the triple $(M_k, \mathcal{P}, \varphi)$. The theta functions $\{\vartheta_m | m \in B_k(\mathbb{Z})\}$ define a relative balanced embedding*

$$\Phi_k = [\vartheta_{m_1}, \cdots, \vartheta_{m_{k^n}}] : \mathcal{X} \to \mathbb{CP}^{k^n-1} \times \Delta,$$

i.e. for any $t \in \Delta$, $\Phi_k|_{X_t} : X_t \hookrightarrow \mathbb{CP}^{k^n-1}$ is a balanced embedding.

Again it was proven in [**WY**] and [**N**] that any individual Abelian variety X_t, $t \neq 0$, can be balanced embedded in certain $\mathbb{CP}^N$ via theta functions. If one set $t \to 0$, the limit variety X_0 in the projective space is also balanced. However Theorem 2.1 uses a group of different theta functions that guarantee the balanced embedding varying holomorphicly when t approaches to 0. Furthermore, Theorem 2.1 identifies the balanced limit X_0 to be the canonical filling-in of $(\mathcal{X}_{\Delta^\circ}, \mathcal{L}|_{\mathcal{X}_{\Delta^\circ}})$ constructed in [**AN**].

Now we study the connection between the metric limit (B, g_B) and the balanced filling-in when $k \to \infty$. For a fixed $k \gg 1$, B_k has a natural affine structure induced by $M_{\mathbb{R}}$. If f is a convex function on an open subset U of B_k with respect to the affine structure, where we also regard U as a subset of $M_{\mathbb{R}}$ by the quotient, then for any $y_0 \in U$, we let

$$\partial f(y_0) = \{v \in N_{\mathbb{R}} | f(y) \geqslant \langle v, y - y_0 \rangle + f(y_0) \ \text{ for all } \ y \in M_{\mathbb{R}}\},$$

and we define the Monge-Ampère measure

$$\mathrm{MA}(f)(E) = \mathrm{Vol}(\bigcup_{y \in E} \partial f(y)),$$

for any Borel subset $E \subset U$, where Vol denotes the standard Euclidean measure on $N_{\mathbb{R}}$ (cf. [**BPS**] and [**TW**]). It is well known that $\mathrm{MA}(f + \alpha) = \mathrm{MA}(f)$ for any linear function α on $M_{\mathbb{R}}$, and if f is smooth,

$$\mathrm{MA}(f) = \det\left(\frac{\partial^2 f}{\partial y_i \partial y_j}\right) dy_1 \wedge \cdots \wedge dy_n.$$

Note that we can regard the function φ as a multivalued function on B_k, and by (2.8), the difference between any two sheets of φ is a linear function. Thus we have a well-defined Monge-Ampère measure $\mathrm{MA}(\varphi)$. The next theorem shows that after some rescaling, this Monge-Ampère measure converges to the Monge-Ampère measure of the potential function $\bar{\varphi}$ of g_B when $k \to \infty$, and furthermore, the rescaled potential function φ also converges to $\bar{\varphi}$ in the C^0-sense, which shows the link between the balanced embeddings and the Gromov-Hausdorff limit.

THEOREM 2.2. *For any $k \gg 1$, the Monge-Ampère measure of φ is*

$$\mathrm{MA}(\varphi) = \det(Z_{ij}) \sum_{m \in B_k(\mathbb{Z})} \delta_m,$$

where δ_m is the Dirac measure at $m \in B_k(\mathbb{Z})$. If $\chi_k : B \to B_k$ is induced by the dilation $y_i \mapsto ky_i$, $i = 1, \cdots, n$, of $M_\mathbb{R}$, then $\frac{1}{k^2}\chi_k^*\varphi - \overline{\varphi}$ is a well-defined function on B, and

$$\sup_B \left| \frac{1}{k^2}\chi_k^*\varphi - \overline{\varphi} \right| \to 0,$$

when $k \to \infty$. Furthermore,

$$\frac{1}{k^n}\chi_k^*\mathrm{MA}(\varphi) \rightharpoonup \mathrm{MA}(\bar{\varphi}) = \det(Z_{ij}) dy_1 \wedge \cdots \wedge dy_n$$

in the weak sense.

In [**L**], the non-archimedean Monge-Ampère equation is solved for degenerations of Abelian varieties, where the approximation of continuous potential functions by piecewise linear functions are also used. See [**BFJ**] for the non-archimedean Monge-Ampère equations for more general cases. Here in Theorem 2.2, we are working on the intersection complexes instead of the dual intersection complexes as in [**BFJ**,**L**], and our piecewise linear functions are from the balanced embedded degenerations of Abelian varieties.

We end this section by giving a remark of Calabi-Yau manifolds balanced embedded by theta functions. In Section 3 of [**DKLR**], it is shown that the Calabi-Yau hypersurfaces

$$X_t = \{[z_0, \cdots, z_4] \in \mathbb{CP}^n | t(z_0^{n+1} + \cdots + z_n^{n+1}) + z_0 \cdots z_n = 0\}$$

are balanced embedded for any $t \in \mathbb{C}$. The proof involves two finite group actions on $(X_t, \mathcal{O}_{\mathbb{CP}^n}(1)|_{X_t})$. The first one is $Ab_{n+1} = \{(a_0, \cdots, a_n) | a_i \in \mathbb{Z}_{n+1}, a_0 + \cdots + a_n = 0\}/\mathbb{Z}_{n+1}$, which acts on X_t by

$$(z_0, \cdots, z_n) \mapsto (\zeta^{a_0} z_0, \cdots, \zeta^{a_n} z_n)$$

where $\zeta = \exp \frac{2\pi\sqrt{-1}}{n+1}$. The second one is the symmetric group $\mathbb{S}_{n+1}$ on $n+1$ elements, which acts on X_t by translating $z_0, \cdots, z_n$. The same argument as in the proof of Theorem 2.1 also shows the result of X_t being balanced. On the other hand, $z_0, \cdots, z_n$ as sections of $\mathcal{O}_{\mathbb{CP}^n}(1)|_{X_t}$ are theta functions constructed in the Gross-Siebert program for $|t| \ll 1$ (at least for $n = 3$) by Example 6.3 in [**GHKS1**], and thus X_t are balanced embedded by theta functions.

3. Construction of Theta Functions

We recall the construction of theta functions on degenerations of Abelian varieties, and we follow the arguments in Example 6.1 of [**GHKS1**] and Section 2 of [**GS2**] closely. See also [**ABW**] for the elliptic curve case.

Let $(\mathcal{P}, \varphi)$ be the same as in the above section, i.e. the graph of φ is the lower boundary of the upper convex hull

$$\mathrm{conv}\{(m, \overline{\varphi}(m)) | m \in M\} \subset M_\mathbb{R} \times \mathbb{R},$$

and $\mathcal{P}$ be the rational polyhedral decomposition of $M_\mathbb{R}$ induced by φ. If we define

$$\Delta_\varphi = \{(m, r) \in M_\mathbb{R} \times \mathbb{R} | r \geqslant \varphi(m)\},$$

the standard construction for toric degenerations gives a toric variety X_{Δ_φ} with a line bundle L_{Δ_φ}.

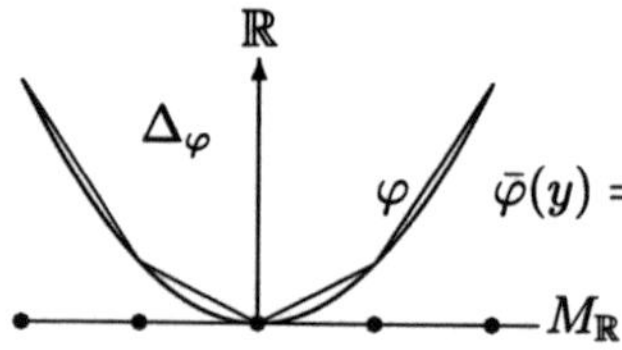

For any $l \in \mathbb{N}$, $H^0(X_{\Delta_\varphi}, L^l_{\Delta_\varphi})$ is generated by *monomial* sections

$$\{\mathcal{Z}^{(m,r,l)} \mid \quad \forall (m,r,l) \in C(\Delta_\varphi) \cap (M \times \mathbb{Z} \times \{l\})\} \tag{3.1}$$

where

$$C(\Delta_\varphi) = \overline{\{(lm', lr', l) | (m', r') \in \Delta_\varphi, l \in \mathbb{R}_{\geqslant 0}\}} \subset M_\mathbb{R} \times \mathbb{R} \times \mathbb{R}.$$

Note that we have a *canonical* regular function

$$\bar{\pi} = \mathcal{Z}^{(0,1)} = \mathcal{Z}^{(0,1,0)} : X_{\Delta_\varphi} \to \mathbb{C}.$$

The toric boundary is $\bar{X}_0 = \{\mathcal{Z}^{(0,1)} = 0\}$, a toric variety with infinite many irreducible components, and $\bar{X}_t = \{\mathcal{Z}^{(0,1)} = t\} \cong T_\mathbb{C}$, for any $t \neq 0$. We have a family of toric varieties $\bar{X}_t$ degenerating to a singular toric varieties $\bar{X}_0$.

The degeneration of principally polarized Abelian varieties $(\pi : \mathcal{X} \to \Delta, \mathcal{L})$ is constructed as the quotient of an M_k-action on $(X_{\Delta_\varphi}, L_{\Delta_\varphi})$ as the following.

LEMMA 3.1. *There is an M-action on $(X_{\Delta_\varphi}, L_{\Delta_\varphi})$ such that the projection $\bar{\pi}$ is M-invariant, i.e. $\bar{\pi}(m\cdot) = \bar{\pi}(\cdot)$ for any $m \in M$; the induced M-action on monomial rational functions is given by*

$$\mathcal{Z}^{(m,r)} \mapsto \mathcal{Z}^{(m,r+Z(\gamma,m))}, \quad \gamma \in M, \quad (m,r) \in M \times \mathbb{Z};$$

and the induced M-action on $H^0(X_{\Delta_\varphi}, L_{\Delta_\varphi})$ is given by

$$\mathcal{Z}^{(m,r,1)} \mapsto \mathcal{Z}^{(m+\gamma, r+\alpha_\gamma(m),1)}, \quad \gamma \in M, \quad (m,r) \in \Delta_\varphi$$

for monomial sections.

PROOF. If $\Sigma \subset N_\mathbb{R} \times \mathbb{R}$ denotes the normal fan of Δ_φ, then one-dimensional rays of Σ are one-to-one correspondence to the maximal dimensional cells in $\mathcal{P}$, and the primitive generator of a ray has the form $(-d\varphi|_\sigma, 1)$ for an n-dimensional cell σ of $\mathcal{P}$. Then M acts on $N \times \mathbb{Z}$ by $\check{T}^0_\gamma : (\mu, l) \mapsto (\mu - l d\alpha_\gamma, l)$ for any $\gamma \in M$, which preserves Σ, and thus induces an M-action on X_{Δ_φ}. The dual M-action on $M \times \mathbb{Z}$ is the transpose, i.e. $T^0_\gamma : (m,r) \mapsto (m, r + d\alpha_\gamma(m))$ for any $\gamma \in M$. Thus the M-action preserves the regular function $\bar{\pi} = \mathcal{Z}^{(0,1)}$, and the induced M-action on monomial rational functions is given by

$$\mathcal{Z}^{(m,r)} \mapsto \mathcal{Z}^{T^0_\gamma(m,r)} = \mathcal{Z}^{(m,r+Z(\gamma,m))},$$

for any $\gamma \in M$ and $(m,r) \in M \times \mathbb{Z}$.

For constructing the M-action on L_{Δ_φ}, we consider the M action on $C(\Delta_\varphi)$ defined by

$$T_\gamma : (m,r,l) \mapsto (m + l\gamma, r + d\alpha_\gamma(m) + lc_\gamma, l) \qquad \gamma \in M, \tag{3.2}$$

where $d\alpha_\gamma(m) = Z(\gamma, m)$ by (2.2) and $c_\gamma = \frac{1}{2}Z(\gamma,\gamma)$. We have $T_\gamma(m, \varphi(m), 1) = (m+\gamma, \varphi(m+\gamma), 1)$ by (2.8). This action lifts the M-action on X_{Δ_φ} to an M-action on $L^l_{\Delta_\varphi}$. More precisely the M-action on $L^l_{\Delta_\varphi}$ is given by

$$(3.3) \qquad \mathcal{Z}^{(m,r,l)} \mapsto \mathcal{Z}^{T_\gamma(m,r,l)} = \mathcal{Z}^{(m+l\gamma, r+d\alpha_\gamma(m)+lc_\gamma, l)}$$

for monomial sections. □

For any $k \in \mathbb{N}$, $M_k = kM$ is a subgroup, and acts on $(X_{\Delta_\varphi}, L_{\Delta_\varphi})$ induced by the M-action in the above lemma. Note that the map $\bar{\pi}$ is M_k-invariant, and M_k acts properly and discontinuously on $\bar{\pi}^{-1}(\Delta)$ for the unit disc $\Delta \subset \mathbb{C}$. The quotient is the degeneration of principally polarized Abelian varieties $\pi : \mathcal{X} = \bar{\pi}^{-1}(\Delta)/M_k \to \Delta$, and $\pi^{-1}(t) = X_t = \bar{X}_t/M_k$. The central fiber X_0 of $\mathcal{X}$ is a union of finite irreducible toric varieties, and the corresponding intersection complex is $B_k = M_\mathbb{R}/M_k$ with rational polyhedron decomposition $\tilde{\mathcal{P}}$ induced by $\mathcal{P}$. There is a one to one corresponding between the n-dimensional cells of $\tilde{\mathcal{P}}$ and the irreducible components of X_0. More precisely, for any n-dimensional cell σ of $\tilde{\mathcal{P}}$, we regard it as a rational polytope in $M_\mathbb{R}$, and it defines a polarized toric variety (X_σ, L_σ). The irreducible component of X_0 corresponding to σ is isomorphic to X_σ. The quotient of L_{Δ_φ} by the M_k-action is the relative ample line bundle $\mathcal{L}$ on $\mathcal{X}$. The restriction of $\mathcal{L}$ on any irreducible component X_σ of X_0 is L_σ. The M_k-invariant sections descend to sections of $\mathcal{L}$.

We claim that $\mathcal{X}_{\Delta^\circ} = \pi^{-1}(\Delta^\circ)$ is a base change of $\mathcal{X}_\eta$ (c.f. Section 2.1) via $t \mapsto t^k = s$. Note that for any $t \neq 0$, M_k acts on $\bar{X}_t = T_\mathbb{C}$ by

$$(3.4) \qquad \mathcal{Z}^{(m,r)} := \mathcal{Z}^m t^r \mapsto \mathcal{Z}^{T^0_\gamma(m,r)} := \mathcal{Z}^{(m,r)} t^{Z(\gamma,m)}.$$

Since $t^{Z(\gamma,m)} = t^{kZ(k^{-1}\gamma,m)} = s^{Z(k^{-1}\gamma,m)}$, $k^{-1}\gamma \in M$, we obtain that $\bar{X}_t/M_k = X_s$ by (2.4) where $t^k = s$.

For any $m \in B_k(\mathbb{Z}) = M/M_k$, we define the theta function

$$(3.5) \qquad \vartheta_m = \sum_{\gamma \in M_k} \mathcal{Z}^{T_\gamma(m,\varphi(m),1)} = \sum_{\gamma \in M_k} \mathcal{Z}^{(m+\gamma, \varphi(m+\gamma), 1)},$$

which is a section of $\mathcal{L}$ (cf. Example 6.1 of **[GHKS1]** and Section 2 of **[GS2]**). By abusing of notation, we will use m to denote *both* a point in M and its image under the quotient map

$$M \longrightarrow B_k(\mathbb{Z}) = M/kM \subset B_k = M_\mathbb{R}/kM$$

without any confusing. We obtain a basis $\{\vartheta_m | m \in B_k(\mathbb{Z})\}$ of $H^0(\mathcal{X}, \mathcal{L})$. For any irreducible component $X_\sigma \subset X_0$, $\vartheta_m|_{X_\sigma}$ is a monomial section of L_σ, and it is not a zero section if and only if $m \in \sigma$. For any $t \neq 0$,

$$(3.6) \qquad \begin{aligned} \vartheta_m(w) &= \sum_{\gamma \in M_k} \mathcal{Z}^{m+\gamma} t^{\varphi(m+\gamma)} \\ &= \sum_{\gamma \in M_k} \exp \pi\sqrt{-1}(2\langle w, m+\gamma\rangle + \frac{\log t}{2\pi\sqrt{-1}} Z(m+\gamma, m+\gamma)), \end{aligned}$$

by $\varphi(m+\gamma) = \bar{\varphi}(m+\gamma)$, where $w = w_1 e_1^* + \cdots + w_n e_n^*$. Thus it is the classical theta function

$$\vartheta_m(w) = \vartheta \begin{bmatrix} m \\ 0 \end{bmatrix} (w, \frac{\log t}{2\pi\sqrt{-1}} Z_{ij})$$

on X_t. If we regard $\vartheta_m|_{X_t}$ as on X_s, $s = t^k$, then

$$\vartheta_m(w) = \sum_{\gamma' \in M} \exp \pi\sqrt{-1}(2k\langle w, \frac{m}{k} + \gamma'\rangle + k\frac{\log s}{2\pi\sqrt{-1}} Z(\frac{m}{k} + \gamma', \frac{m}{k} + \gamma')),$$

and a direct calculation shows

$$\vartheta_m(w + \mu + \frac{\log s}{2\pi\sqrt{-1}} Z(p, \cdot)) = \vartheta_m(w) \exp k\pi\sqrt{-1}(-2\langle w, p\rangle - \frac{\log s}{2\pi\sqrt{-1}} Z(p,p)),$$

for any $\mu \in N$ and $p \in M$. Thus $\mathcal{L}|_{X_t} \cong \mathcal{L}_\eta^k|_{X_s}$, i.e. $\mathcal{L}|_{\mathcal{X}_{\Delta^*}}$ is the pull-back of $\mathcal{L}_\eta^k$.

REMARK 3.2. Note that there are many line bundles that are the k-power of a principal polarization. Here we choose specifically both $\mathcal{L}$ and $\mathcal{L}_\eta$ such that the theta functions (3.6) and (2.5) are their sections respectively.

REMARK 3.3. Notice that, in particular, the monodromy action $\log t \to \log t + 2\pi\sqrt{-1}$ acts trivially on ϑ_m via

$$\begin{aligned} & \vartheta \begin{bmatrix} m \\ 0 \end{bmatrix} (w, \left(\frac{\log t}{2\pi\sqrt{-1}} + 1\right) Z_{ij}) \\ = & \sum_{\gamma \in M_k} \exp \pi\sqrt{-1}(2\langle w, m+\gamma\rangle + (\frac{\log t}{2\pi\sqrt{-1}} + 1) Z(m+\gamma, m+\gamma)) \\ = & \vartheta \begin{bmatrix} m \\ 0 \end{bmatrix} (w, \frac{\log t}{2\pi\sqrt{-1}} Z_{ij}) \end{aligned}$$

since $Z(m+\gamma, m+\gamma) = 2\bar{\varphi}(m+\gamma) \in 2\mathbb{Z}$ for $m \in B_k(\mathbb{Z})$.

EXAMPLE 3.4. We illustrate the explicit formula of the theta function ϑ_m in (3.5) in local coordinates for a special 1-dimensional family. Let $M \cong \mathbb{Z}$, $k = 3$, and $\bar{\varphi}(y) = y^2$ on $M_{\mathbb{R}}$. Note that $(0,0)$ is a vertex of Δ_φ, and the $\mathbb{C}$-algebra $\mathbb{C}[T_{(0,0)}\Delta_\varphi \cap M]$ is generated by $z_1 = \mathcal{Z}^{(1,1)}$, $z_2 = \mathcal{Z}^{(-1,1)}$ and $t = \mathcal{Z}^{(0,1)}$, where $T_{(0,0)}\Delta_\varphi$ is the tangent cone of Δ_φ at $(0,0)$. The toric variety $Y_0 = \mathrm{Spec}(\mathbb{C}[T_{(0,0)}\Delta_\varphi \cap M])$ is defined in $\mathbb{C}^3$ by equation $z_1 z_2 = t^2$, and an open subset of Y_0 is biholomorphic to a neighborhood U_0 of the zero strata of X_0 in $\mathcal{X}$ corresponding to the vertex $(0,0)$. Let us fix a trivialization of $\mathcal{L}|_{U_0} \cong \mathcal{O}_{Y_0}|_{U_0}$ via the identification $\mathcal{Z}^{(m,r,1)} \mapsto \mathcal{Z}^{(m,r)}$ for any $(m,r) \in \Delta_\varphi \cap M$. Note that for $\gamma = 3\nu > 0$, $\mathcal{Z}^{(\gamma,\varphi(\gamma))} = z_1^\gamma t^{\gamma^2-\gamma} = z_1^{3\nu} t^{9\nu^2-3\nu}$, and the similar direct calculations show the formulas of $\mathcal{Z}^{(m+\gamma,\varphi(m+\gamma))}$ when $m = 0, 1, 2$, and $\gamma \in 3\mathbb{Z}$. Then we have

$$\begin{aligned} \vartheta_0 &= 1 + \sum_{\nu \in \mathbb{Z}, \nu > 0} (z_1^{3\nu} + z_2^{3\nu}) t^{9\nu^2 - 3\nu}, \\ \vartheta_1 &= z_1 + \sum_{\nu \in \mathbb{Z}, \nu > 0} (z_1^{1+3\nu} t^{9\nu^2+3\nu} + z_2^{3\nu-1} t^{9\nu^2-9\nu+2}), \\ \vartheta_2 &= z_2 + \sum_{\nu \in \mathbb{Z}, \nu > 0} (z_1^{3\nu-1} t^{9\nu^2-9\nu+2} + z_2^{3\nu+1} t^{9\nu^2+3\nu}), \end{aligned}$$

by (3.5). In particular, ϑ_i extends to the central fiber.

Notice that the central extention of the product group $B_k(\mathbb{Z}) \times T_k$ with $T_k := N/N_k$ is precisely the *finite Heisenberg group* $\mathbb{H}_k = \mu_k \times B_k(\mathbb{Z}) \times T_k$ (cf. [**M2**,

Section 3]) with the multiplication rule:

$$(\mu, a, b) \cdot (\mu', a', b') = (\mu\mu' \exp \frac{2\pi\sqrt{-1}\langle b, a\rangle}{k}, a + a', b + b') \tag{3.7}$$

for any $(\mu, a, b), (\mu', a', b') \in \mu_k \times B_k(\mathbb{Z}) \times T_k$, where μ_k is the cyclotomic group of order k.

LEMMA 3.5. *The group $B_k(\mathbb{Z}) \times T_k$ acts on $(\mathcal{X}, \mathcal{L})$ and induces a representation of $\mathbb{H}_k$ on the $H^0(\mathcal{X}, \mathcal{L}) = \mathrm{Span}_{\mathcal{H}}\{\vartheta_m\}_{m \in B_k(\mathbb{Z})}$ via*

i) *For any $a \in B_k(\mathbb{Z})$*

$$\mathfrak{T}_a \vartheta_m = \vartheta_{a+m}$$

ii) *For any $b \in T_k$*

$$\mathfrak{S}_b \vartheta_m = \vartheta_m \exp \frac{2\pi\sqrt{-1}\langle b, m\rangle}{k},$$

for all $m \in B_k(\mathbb{Z})$, where $\mathcal{H}$ denotes the ring of holomorphic functions on Δ. In particular, the representation of $\mathbb{H}_k$ on $H^0(\mathcal{X}, \mathcal{L})$ is irreducible.

PROOF. The M-action on $(X_{\Delta_\varphi}, L_{\Delta_\varphi})$ in Lemma 3.1 induces the $B_k(\mathbb{Z})$-action on $(\mathcal{X}, \mathcal{L})$, which acts on $H^0(\mathcal{X}, \mathcal{L})$ given by

$$\begin{aligned} \mathfrak{T}_a \vartheta_m &= \sum_{\gamma \in M_k} \mathcal{Z}^{T_a(m+\gamma, \varphi(m+\gamma), 1)} \\ &= \sum_{\gamma \in M_k} \mathcal{Z}^{(m+\gamma+a, \varphi(m+\gamma)+\alpha_a(m+\gamma), 1)} \\ &= \sum_{\gamma \in M_k} \mathcal{Z}^{(m+\gamma+a, \varphi(m+\gamma+a), 1)} \\ &= \vartheta_{a+m} \end{aligned}$$

for any $m \in B_k(\mathbb{Z})$ and $a \in B_k(\mathbb{Z})$ by (3.5). We obtain i), and next we prove ii).

Note that there is a natural injective homeomorphism $\iota_k : T_k \hookrightarrow T_{\mathbb{C}}$ such that

$$\mathcal{Z}^m(\iota_k(b)) = \exp \frac{2\pi\sqrt{-1}\langle b, m\rangle}{k}$$

for any $b \in T_k$ and $m \in M$. The standard $T_{\mathbb{C}}$-action on $(X_{\Delta_\varphi}, L^l_{\Delta_\varphi})$ induces a T_k-action, which preserves the regular function $\bar{\pi} = \mathcal{Z}^{(0,1)}$, acts on monomial rational functions by

$$\mathcal{Z}^{(m,r)} \mapsto \mathcal{Z}^{(m,r)} \exp \frac{2\pi\sqrt{-1}\langle b, m\rangle}{k}$$

for any $(m, r) \in M \times \mathbb{Z}$, and acts on monomial sections of $L^l_{\Delta_\varphi}$ via

$$\mathcal{Z}^{(m,r,l)} \mapsto \mathcal{Z}^{(m,r,l)} \exp \frac{2\pi\sqrt{-1}\langle b, m\rangle}{k}$$

for any $(m, r, l) \in C(\Delta_\varphi)$. The induced T_k-action on $\bigoplus_{l=0}^{\infty} H^0(X_{\Delta_\varphi}, L^l_{\Delta_\varphi})$ commutes with the M_k-action by

$$\mathcal{Z}^{T_\gamma(m,r,l)} \exp \frac{2\pi\sqrt{-1}\langle b, m\rangle}{k} = \mathcal{Z}^{(m+l\gamma, r+d\alpha_\gamma(m)+lc_\gamma, l)} \exp \frac{2\pi\sqrt{-1}\langle b, m+l\gamma\rangle}{k},$$

for any $\gamma \in M_k$ by (3.3). Thus the T_k-action commutes with the M_k-action on $(X_{\Delta_\varphi}, L^l_{\Delta_\varphi})$, which induces a T_k-action on $(\pi : \mathcal{X} \to \Delta, \mathcal{L})$. We denote $\mathfrak{S}$ the induced T_k-action on $H^0(\mathcal{X}, \mathcal{L})$, which satisfies

$$\mathfrak{S}_b \vartheta_m = \sum_{\gamma \in M_k} z^{(m+\gamma, \varphi(m+\gamma), 1)} \exp \frac{2\pi\sqrt{-1}\langle b, m+\gamma\rangle}{k} = \vartheta_m \exp \frac{2\pi\sqrt{-1}\langle b, m\rangle}{k},$$

for any $b \in T_k$ and any $m \in B_k(\mathbb{Z})$, by (3.5). $\square$

REMARK 3.6. The direct calculations show that on any X_t, $t \neq 0$, the action of $B_k(\mathbb{Z}) \times T_k$ on $(X_t, \mathcal{L}|_{X_t})$ is given by $w \mapsto w + \frac{b}{k} + \frac{\log t}{2\pi\sqrt{-1}} Z(\cdot, a)$,

$$\begin{aligned}
(\mathfrak{T}_a \vartheta_m)(w) &= \exp \pi\sqrt{-1}(2\langle w, a\rangle + \frac{\log t}{2\pi\sqrt{-1}} Z(a,a))\vartheta_m(w + \frac{\log t}{2\pi\sqrt{-1}} Z(\cdot, a)) \\
&= \vartheta_{a+m}(w), \\
(\mathfrak{S}_b \vartheta_m)(w) &= \vartheta_m(w + \frac{b}{k}) = \exp \frac{2\pi\sqrt{-1}\langle b, m\rangle}{k} \vartheta_m(w),
\end{aligned}$$

for any $m \in B_k(\mathbb{Z})$ and $(a, b) \in B_k(\mathbb{Z}) \times T_k$. In particular, there the finite torus $B_k(\mathbb{Z}) \times T_k$ acts on the image of the projective embedding $\Phi_k(X_t)$.

4. Proofs of Main Theorems

Before we start the proof, let us recall the Hermitian metric on $\mathcal{L}|_{X_t} \to X_t$ with $s = t^k$ is given by

$$h(w) := \exp\left(\frac{2\pi}{\log|t|} Z^{ij} y_i y_j\right) = \exp k \left(\frac{2\pi}{\log|s|} Z^{ij} y_i y_j\right) \tag{4.1}$$

with

$$w_i = x_i + (\arg(s) - \sqrt{-1}\frac{\log|s|}{2\pi}) \sum_{j=1}^{n} Z_{ij} y_j, \ i = 1, \cdots, n\ .$$

On $X_t \cong X_s$, we have

$$\omega_t = -\sqrt{-1}\partial\bar{\partial} \log h = k\omega_s,$$

by (2.6).

LEMMA 4.1. *$\forall a \in M, b \in N$ and function $f : N_{\mathbb{C}} \to \mathbb{C}$ we have*
(4.2)

$$h(w)|\mathfrak{S}_b f|^2 = (h|f|^2)(w + \frac{b}{k});\ h(w)|\mathfrak{T}_a f|^2 = (h|f|^2)\left(w + \frac{\log s}{2\pi\sqrt{-1}} Z(\cdot, \frac{a}{k})\right).$$

PROOF. By Remark 3.6, the actions $\mathfrak{T}_a$ and $\mathfrak{S}_b$ on theta functions of smooth fibers are exactly the same with those in [**WY**]. Thus the proof follows from [**WY**, Proposition 3]. $\square$

As a consequence, if we let $\|\cdot\|_{\mathrm{FS}}$ denote the L^2-metric on $H^0(X_t, \mathcal{L}|_{X_t})$ induced from the restriction of the Fubini-Study metric on $\mathcal{O}_{\mathbb{CP}^N}(1)|_{\Phi_k(X_t)}$ then for $a \in$

$B_k(\mathbb{Z})$, $b \in T_k$ and $f \in H^0(X_t, \mathcal{L}|_{X_t})$ we have

$$\begin{aligned}
\|(\mathfrak{S}_b f)(w)\|^2_{\mathrm{FS}} &= \int \frac{|(\mathfrak{S}_b f)(w)|^2}{\sum_m |\vartheta_m(w)|^2} \Phi_k^* \omega_{\mathrm{FS}}^n \\
&= \int \frac{h|(\mathfrak{S}_b f)(w)|^2}{\sum_m h|\mathfrak{S}_b \vartheta_m(w)|^2} \Phi_k^* \omega_{\mathrm{FS}}^n \\
&= \int \frac{(h|f|^2)\left(w + \frac{b}{k}\right)}{\left(\sum_m h|\vartheta_m|^2\right)\left(w + \frac{b}{k}\right)} (\mathfrak{S}_b \circ \Phi_k)^* \omega_{\mathrm{FS}}^n \\
&= \int \left(\frac{|f|^2}{\sum_m |\vartheta_m|^2} \Phi_k^* \omega_{\mathrm{FS}}^n\right)\left(w + \frac{b}{k}\right) \\
&= \|f(w)\|^2_{\mathrm{FS}}
\end{aligned}$$

and

$$\begin{aligned}
\|(\mathfrak{T}_a f)(w)\|^2_{\mathrm{FS}} &= \int \frac{|(\mathfrak{T}_a f)(w)|^2}{\sum_m |\vartheta_m(w)|^2} \Phi_k^* \omega_{\mathrm{FS}}^n \\
&= \int \frac{h|(\mathfrak{T}_a f)(w + \frac{\log s}{2\pi\sqrt{-1}} Z(\cdot, \frac{a}{k}))|^2}{\sum_m h|\mathfrak{T}_a \vartheta_m(w + \frac{\log s}{2\pi\sqrt{-1}} Z(\cdot, \frac{a}{k}))|^2} (\mathfrak{T}_a \circ \Phi_k)^* \omega_{\mathrm{FS}}^n \\
&= \int \left(\frac{|f|^2}{\sum_m |\vartheta_m|^2} \Phi_k^* \omega_{\mathrm{FS}}^n\right)(w + \frac{\log s}{2\pi\sqrt{-1}} Z(\cdot, \frac{a}{k})) \\
&= \|f(w)\|^2_{\mathrm{FS}} \cdot
\end{aligned}$$

Hence the finite group generated by image of

$$\{\mathfrak{T}_a, \mathfrak{S}_b \mid (a,b) \in B_k(\mathbb{Z}) \times T_k\} \subset \mathrm{GL}(H^0(X_s, \mathcal{L}|_{X_s}))$$

actually lies in $\mathrm{U}(k^n)$ with respect to the Fubini-Study metric induced via the embedding

$$\Phi_k = [\vartheta_{m_1}, \cdots, \vartheta_{m_{k^n}}] : \mathcal{X} \longrightarrow \mathbb{CP}^{k^n - 1} .$$

Proof of Theorem 2.1. It follows from above that the action generated by $B_k(\mathbb{Z}) \times T_k$ via $\mathfrak{T}$ and $\mathfrak{S}$ lies in $\mathrm{U}(k^n)$. And Lemma 3.5 implies that sections $\{\vartheta_m\}_{m \in B_k(\mathbb{Z})}$ forms an orthonormal basis with respect to the pull back of Fubini-Study metric via the map Φ_k, that is, the embedding Φ_k is balanced for each $t \in \Delta^\circ$.

On the other hand, $\Phi_k(X_t)$ being balanced for each $t \neq 0$ implies that the Chow point for $\Phi_k(X_t)$ lies on the 0-level set of the moment map

$$\begin{array}{cccc}
\mu_{\mathrm{SU}} : \mathrm{Chow}_{\mathbb{CP}^{k^n-1}}(d,n) & \longrightarrow & \mathfrak{su}(k^n) \\
X_t & \longmapsto & \sqrt{-1} \cdot \displaystyle\int_{X_t} \left(\frac{\vartheta_m \bar{\vartheta}_{m'}}{\sum_m |\vartheta_m|^2} - \frac{\delta_{mm'}}{k^n}\right) \frac{\Phi_k^* \omega_{\mathrm{FS}}^n}{n!}
\end{array}$$

of the $\mathrm{SU}(k^n)$-action on the Chow variety of n-dimensional degree d cycles in $\mathbb{CP}^{k^n-1}$ (c.f. [**W**, Proposition 17] and [**Z**, Theorem 1.4]), which is proper via standard Kirwan-Kempf-Ness theory in [**K**] (c.f. also in [**T**]). Notice that

$$\{\vartheta_m(\cdot, \frac{\log t}{2\pi\sqrt{-1}} Z)\}_m$$

vary holomorphically with respect to t and can be extended to X_0 by the construction of theta functions in Section 3, these imply that the image of $\Phi_k(X_t)$

has continuous extension in $\mathbb{CP}^{k^n-1}$ as $t \to 0$, by Riemann extension Theorem, the *unique* continuous extension $\Phi_k(X_0)$ must lies in $\mu_{\mathrm{SU}}^{-1}(0)$, that is, the embedding $\Phi_k(X_0)$ is balanced as well. □

REMARK 4.2. In the appendix, we state a general criterion for an embedding $\Phi : X \hookrightarrow \mathbb{CP}^N$ being balanced when it possesses a large symmetry.

PROOF OF THEOREM 2.2. For any $m \in M$, we denote $\check{m} \subset N_{\mathbb{R}}$ the dual polytope of m with respect to φ. More precisely, if $\check{m}^o \subset N_{\mathbb{R}} \times \mathbb{R}$ denotes the dual cone of the tangent cone $T_{(m,\varphi(m))}\Delta_\varphi \subset M_{\mathbb{R}} \times \mathbb{R}$ of Δ_φ at the vertex $(m, \varphi(m))$, then $\check{m} = \check{m}^o \cap (N_{\mathbb{R}} \times \{1\})$. The Monge-Ampère measure of φ is

$$\mathrm{MA}(\varphi) = \sum_{m \in B_k(\mathbb{Z})} \mathrm{Vol}(\check{m})\delta_m,$$

where $\mathrm{Vol}(\check{m})$ is the Euclidean volume of $\check{m}$ (cf. Proposition 2.7.4 in [**BPS**]). Since the M-action $\check{T}^0$ on $N_{\mathbb{R}} \times \mathbb{R}$ preserves the fan Σ of Δ_φ, we obtain $\check{T}^0_{m-m'}(\check{m}') = \check{m}$ for any two m and $m' \in M$, and thus $\mathrm{Vol}(\check{m}) = V$ is independent of m.

Let D be the fundamental domain of the M_k-action $\check{T}^0$ on $N_{\mathbb{R}} \times \{1\}$. Since $\check{T}^0_\gamma(0,1) = (-d\alpha_\gamma, 1) = (-Z(\gamma,\cdot),1)$ for any $\gamma \in M$, we let D be the convex hull of $(0,1), (-Z(ke_1,\cdot),1), \cdots, (-Z(ke_n,\cdot),1)$ in $N_{\mathbb{R}} \times \{1\}$, where $e_1, \cdots, e_n$ is a basis of M. The Euclidean volume of D is $\mathrm{Vol}(D) = k^n \det(Z_{ij})$, where $Z_{ij} = Z(e_i, e_j)$, and we have

$$\mathrm{Vol}(D) = \sum_{m \in B_k(\mathbb{Z})} \mathrm{Vol}(\check{m}) = k^n V.$$

We obtain $V = \det(Z_{ij})$, and the conclusion

$$\mathrm{MA}(\varphi) = \det(Z_{ij}) \sum_{m \in B_k(\mathbb{Z})} \delta_m.$$

Let $\chi_k : B \to B_k$ be the diffeomorphism induced by the dilation $y_i \mapsto ky_i$, $i = 1, \cdots, n$, of $M_{\mathbb{R}}$. For any smooth function f on B,

$$\frac{1}{k^n}\int_B f\chi_k^*\Big(\sum_{m \in B_k(\mathbb{Z})} \delta_m\Big) = \frac{1}{k^n}\sum_{m' \in (\frac{1}{k}M)/M} f(m') \to \int_B f dy_1 \wedge \cdots \wedge dy_n$$

when $k \to \infty$, which implies that

$$\frac{1}{k^n}\chi_k^*\mathrm{MA}(\varphi) \to \det(Z_{ij})dy_1 \wedge \cdots \wedge dy_n = \mathrm{MA}(\bar{\varphi})$$

in the weak sense.

By (2.3) and (2.8), $(\varphi - \bar{\varphi})(y + \gamma) = (\varphi - \bar{\varphi})(y)$ for any $y \in M_{\mathbb{R}}$ and any $\gamma \in M$, and $\varphi - \bar{\varphi}$ is a well-defined function on B_k. Since

$$\sup_{M_{\mathbb{R}}} |\varphi - \bar{\varphi}| = \sup_{y \in M_{\mathbb{R}}/M} |\varphi - \bar{\varphi}|(y),$$

we obtain

$$\sup_B \left|\frac{1}{k^2}\chi_k^*\varphi - \frac{1}{k^2}\chi_k^*\bar{\varphi}\right| = \frac{1}{k^2}\sup_{M_{\mathbb{R}}} |\varphi - \bar{\varphi}| \to 0,$$

when $k \to \infty$, and the conclusion by $\frac{1}{k^2}\bar{\varphi}(ky) = \bar{\varphi}(y)$. □

5. Appendix

In this section, we state the following Proposition which give a unified interpretation of balanced embedding of both projective spaces and principally polarized Abelian varieties.

PROPOSITION 5.1. *Let $X \subset \mathbb{CP}^N$ be a subvariety and $G < \mathrm{SU}(N+1)$ be a compact (possibly finite) subgroup which leaves the embedding $\Phi : X \hookrightarrow \mathbb{CP}^N$ invariant. Suppose the centralizer $c_G < \mathrm{SU}(N+1)$ of G inside $\mathrm{SU}(N+1)$ is trivial. Then the embedding $X \subset \mathbb{P}^N$ is balanced, i.e.*

$$\mu_{\mathrm{SU}}(\Phi) = \int_X \Phi^* \left(\mu_{\mathbb{CP}^N} \frac{\omega_{\mathrm{FS}}^n}{n!} \right) = \int_{\Phi(X)} \mu_{\mathbb{CP}^N} \frac{\omega_{\mathrm{FS}}^n}{n!} = 0. \tag{5.1}$$

where $\mu_{\mathbb{CP}^N}(z) = \sqrt{-1} \left(\frac{zz^*}{|z|^2} - \frac{I_{N+1}}{N+1} \right) \in \mathfrak{su}(N+1)$ for $z \in \mathbb{C}^{N+1}$ is the moment map of $\mathrm{SU}(N+1)$-action on $\mathbb{CP}^N$.

PROOF. Notice that the moment map of $\mathrm{SU}(N+1)$-action on the Chow variety $\mathrm{Chow}_{\mathbb{CP}^N}(n,d)$ of dimension n and degree d cycle in $\mathbb{CP}^N$ is precisely given by (5.1). In particular, it is $\mathrm{SU}(N+1)$, and hence G-equivariant. This implies that for any $g \in G$

$$\begin{aligned} \int_{\Phi(X)} \mu_{\mathbb{CP}^N} \frac{\omega_{\mathrm{FS}}^n}{n!} &= \int_{\Phi(g\cdot X)} \mu_{\mathbb{CP}^N} \frac{\omega_{\mathrm{FS}}^n}{n!} = \int_{\Phi(X)} \mu_{\mathbb{CP}^N} \circ g \frac{\omega_{\mathrm{FS}}^n}{n!} \\ &= \mathrm{Ad}_g \left(\int_{\Phi(X)} \mu_{\mathbb{CP}^N} \frac{\omega_{\mathrm{FS}}^n}{n!} \right) \end{aligned}$$

where the last identity follows from the Adjoint-equivariance of the moment map $\mu_{\mathbb{CP}^N} : \mathbb{CP}^N \to \mathfrak{su}(N+1)$. By our assumption, we have $\int_X \mu_{\mathbb{P}^N} \frac{\omega_{\mathrm{FS}}^n}{n!} \in \mathfrak{c}_G = 0$, where $\mathfrak{c}_G = \mathrm{Lie}(c_G)$ is the Lie algebra of the centralizer c_G. And our proof is completed. □

REMARK 5.2. For the degree d Veronese embedding of $\mathbb{CP}^n \hookrightarrow \mathbb{CP}^{\binom{n+d}{d}}$, let $G = \mathrm{SU}(n+1)$; and for principally polarized Abelian variety $\Phi_k : X \hookrightarrow \mathbb{CP}^{k^n-1}$, let G be the image of $\mathbb{H}_k \to \mathrm{PGL}(k^n)$. Then we have $\mathfrak{c}_G = 0$ in both cases, and hence the embeddings are balanced.

Acknowledgements

The authors would like to thank Helge Ruddat, Mark Gross and Bernd Siebert for many helpful discussions. The work was done when both authors are in residence at the Mathematical Sciences Research Institute in Berkeley, California, during the 2016 Spring semester.

References

[ABW] Paul S. Aspinwall, Tom Bridgeland, Alastair Craw, Michael R. Douglas, Mark Gross, Anton Kapustin, Gregory W. Moore, Graeme Segal, Balázs Szendrői, and P. M. H. Wilson, *Dirichlet branes and mirror symmetry*, Clay Mathematics Monographs, vol. 4, American Mathematical Society, Providence, RI; Clay Mathematics Institute, Cambridge, MA, 2009. MR2567952

[AN] Valery Alexeev and Iku Nakamura, *On Mumford's construction of degenerating abelian varieties*, Tohoku Math. J. (2) **51** (1999), no. 3, 399–420, DOI 10.2748/tmj/1178224770. MR1707764

[BL] Christina Birkenhake and Herbert Lange, *Complex abelian varieties*, 2nd ed., Grundlehren der Mathematischen Wissenschaften [Fundamental Principles of Mathematical Sciences], vol. 302, Springer-Verlag, Berlin, 2004. MR2062673

[BFJ] Sébastien Boucksom, Charles Favre, and Mattias Jonsson, *Solution to a non-Archimedean Monge-Ampère equation*, J. Amer. Math. Soc. **28** (2015), no. 3, 617–667, DOI 10.1090/S0894-0347-2014-00806-7. MR3327532

[BLY] Jean-Pierre Bourguignon, Peter Li, and Shing-Tung Yau, *Upper bound for the first eigenvalue of algebraic submanifolds*, Comment. Math. Helv. **69** (1994), no. 2, 199–207, DOI 10.1007/BF02564482. MR1282367

[BPS] José Ignacio Burgos Gil, Patrice Philippon, and Martín Sombra, *Arithmetic geometry of toric varieties. Metrics, measures and heights* (English, with English and French summaries), Astérisque **360** (2014), vi+222. MR3222615

[D] S. K. Donaldson, *Scalar curvature and projective embeddings. I*, J. Differential Geom. **59** (2001), no. 3, 479–522. MR1916953

[DKLR] Michael R. Douglas, Robert L. Karp, Sergio Lukic, and René Reinbacher, *Numerical Calabi-Yau metrics*, J. Math. Phys. **49** (2008), no. 3, 032302, 19, DOI 10.1063/1.2888403. MR2406784

[G] Mark Gross, *Mirror symmetry and the Strominger-Yau-Zaslow conjecture*, Current developments in mathematics 2012, Int. Press, Somerville, MA, 2013, pp. 133–191. MR3204345

[GHKS1] M. Gross, P. Hacking, S. Keel, B. Siebert, *Theta functions on varieties with effective anti-canonical class*, arXiv:1601.07081.

[GHKS2] M. Gross, P. Hacking, S. Keel, B. Siebert, *Theta functions for K3 surface*, preprint.

[GS1] Mark Gross and Bernd Siebert, *From real affine geometry to complex geometry*, Ann. of Math. (2) **174** (2011), no. 3, 1301–1428, DOI 10.4007/annals.2011.174.3.1. MR2846484

[GS2] Mark Gross and Bernd Siebert, *Theta functions and mirror symmetry*, Surveys in differential geometry 2016. Advances in geometry and mathematical physics, Surv. Differ. Geom., vol. 21, Int. Press, Somerville, MA, 2016, pp. 95–138. MR3525095

[GW] Mark Gross and P. M. H. Wilson, *Large complex structure limits of $K3$ surfaces*, J. Differential Geom. **55** (2000), no. 3, 475–546. MR1863732

[GTZ1] Mark Gross, Valentino Tosatti, and Yuguang Zhang, *Collapsing of abelian fibered Calabi-Yau manifolds*, Duke Math. J. **162** (2013), no. 3, 517–551, DOI 10.1215/00127094-2019703. MR3024092

[GTZ2] Mark Gross, Valentino Tosatti, and Yuguang Zhang, *Gromov-Hausdorff collapsing of Calabi-Yau manifolds*, Comm. Anal. Geom. **24** (2016), no. 1, 93–113, DOI 10.4310/CAG.2016.v24.n1.a4. MR3514555

[K] Frances Clare Kirwan, *Cohomology of quotients in symplectic and algebraic geometry*, Mathematical Notes, vol. 31, Princeton University Press, Princeton, NJ, 1984. MR766741

[KS1] Maxim Kontsevich and Yan Soibelman, *Homological mirror symmetry and torus fibrations*, Symplectic geometry and mirror symmetry (Seoul, 2000), World Sci. Publ., River Edge, NJ, 2001, pp. 203–263, DOI 10.1142/9789812799821_0007. MR1882331

[KS2] Maxim Kontsevich and Yan Soibelman, *Affine structures and non-Archimedean analytic spaces*, The unity of mathematics, Progr. Math., vol. 244, Birkhäuser Boston, Boston, MA, 2006, pp. 321–385, DOI 10.1007/0-8176-4467-9_9. MR2181810

[L] Yifeng Liu, *A non-Archimedean analogue of the Calabi-Yau theorem for totally degenerate abelian varieties*, J. Differential Geom. **89** (2011), no. 1, 87–110. MR2863913

[M1] David Mumford, *An analytic construction of degenerating abelian varieties over complete rings*, Compositio Math. **24** (1972), 239–272. MR0352106

[M2] David Mumford, *Tata lectures on theta. I*, Progress in Mathematics, vol. 28, Birkhäuser Boston, Inc., Boston, MA, 1983. With the assistance of C. Musili, M. Nori, E. Previato and M. Stillman. MR688651

[N] Yuichi Nohara, *Projective embeddings and Lagrangian fibrations of abelian varieties*, Math. Ann. **333** (2005), no. 4, 741–757, DOI 10.1007/s00208-005-0685-8. MR2195141

[TW] Neil S. Trudinger and Xu-Jia Wang, *The Monge-Ampère equation and its geometric applications*, Handbook of geometric analysis. No. 1, Adv. Lect. Math. (ALM), vol. 7, Int. Press, Somerville, MA, 2008, pp. 467–524. MR2483373

[RS] H. Ruddat, B. Siebert, Canonical Coordinates in Toric Degenerations, arXiv:1409.4750.

[T] R. P. Thomas, *Notes on GIT and symplectic reduction for bundles and varieties*, Surveys in differential geometry. Vol. X, Surv. Differ. Geom., vol. 10, Int. Press, Somerville, MA, 2006, pp. 221–273, DOI 10.4310/SDG.2005.v10.n1.a7. MR2408226

[SYZ] Andrew Strominger, Shing-Tung Yau, and Eric Zaslow, *Mirror symmetry is T-duality*, Nuclear Phys. B **479** (1996), no. 1-2, 243–259, DOI 10.1016/0550-3213(96)00434-8. MR1429831

[W] Xiaowei Wang, *Moment map, Futaki invariant and stability of projective manifolds*, Comm. Anal. Geom. **12** (2004), no. 5, 1009–1037. MR2103309

[WY] Xiaowei Wang and Hok Pun Yu, *Theta function and Bergman metric on abelian varieties*, New York J. Math. **15** (2009), 19–35. MR2481306

[Y] Shing Tung Yau, *On the Ricci curvature of a compact Kähler manifold and the complex Monge-Ampère equation. I*, Comm. Pure Appl. Math. **31** (1978), no. 3, 339–411, DOI 10.1002/cpa.3160310304. MR480350

[Z] Shouwu Zhang, *Heights and reductions of semi-stable varieties*, Compositio Math. **104** (1996), no. 1, 77–105. MR1420712

Department of Mathematics and Computer Sciences, Rutgers University, Newark New Jersey 07102-1222

E-mail address: `xiaowwan@rutgers.edu`

Yau Mathematical Sciences Center, Tsinghua University, Beijing 100084, People's Republic of China.

E-mail address: `yuguangzhang76@yahoo.com`

The modularity/automorphy of Calabi–Yau varieties of CM type

Noriko Yui

ABSTRACT. We consider Calabi–Yau varieties of dimension $d \leq 3$ defined over $\mathbb{Q}$, and address the modularity/automorphy of such Calabi–Yau varieties. When the dimension of the associated Galois representations are large, e.g., > 2, the problem poses a serious challenge and is out of reach in the general situations.

In this paper, I will concentrate on Calabi–Yau varieties of CM type, and establish their (motivic) modularity/automorphy. The expositions are focused on two examples: K3 surfaces with non-symplectic automorphisms, and Calabi-Yau threefolds of Borcea–Voisin type. We will briefly discuss arithmetic mirror symmetry for quite specific examples of K3 surfaces and Calabi–Yau threefolds of Borcea–Voisin type.

1. Introduction

This is a longer version of my talk at the String Math 2015 at TSIMF, Sanya, China. My hope is to give string theorists some idea on the modularity/automorphy questions about Calabi–Yau varieties defined over $\mathbb{Q}$ or number fields (finite extensions of $\mathbb{Q}$). The modularity/automorphy à la Langlands **[24]** is a web of conjectures connecting Galois representations to algebraic/arithmetic geometry, complex geometry and Lie groups. There is no way we can cover them all. There is a nice review article on automorphic forms and Galois representations by R. Taylor **[42]** for a more exhaustive exposition.

In this paper, we will be concentrating on the special class of algebraic varieties, namely, *Calabi–Yau varieties*, defined over $\mathbb{Q}$ (or number fields), and try to give some evidence to the modularity/automorphy conjectures for some quite specific examples.

Let X be a Calabi–Yau variety defined over $\mathbb{Q}$ of dimension $d \leq 3$. Our goal is to establish the modularity/automorphy of X, or that of some motives associated to X. We will discuss the two cases of Calabi–Yau varieties defined over $\mathbb{Q}$, where our goal may be achieved:

2010 *Mathematics Subject Classification.* Primary 14J32, 14J28, 14J33, 14J20, 14K22, 11G40, 11G42, 11F80, 11F70.

Key words and phrases. K3 surfaces, non-symplectic group actions, Calabi–Yau threefolds of Borcea–Voisin type, finite Fermat quotients, Calabi–Yau varieties of CM type, Galois representations, modularity, automorphy, arithmetic mirror symmetry.

The author was supported in part by Discovery Grant of the Natural Sciences and Engineering Research Council (NSERC) of Canada.

(1) X is an algebraic K3 surface with non-symplectic automorphisms. This is based on the article by Livné–Schütt–Yui [**26**].
(2) X is a Calabi–Yau threefolds of Borcea–Voisin type. This is based on the article by Goto–Livné–Yui [**17**].

The reader is referred to the aforementioned two articles for details.

2. The (cohomological) L-series

The modularity/automorphy of Calabi–Yau varieties defined over $\mathbb{Q}$ are formulated in terms of their (cohomological) L-series. This might be the right place to introduce the L-series of smooth projective varieties defined over $\mathbb{Q}$, with special reference to Calabi–Yau varieties.

2.1. The zeta functions. We will consider a Calabi–Yau variety of dimension d defined over $\mathbb{Q}$, say, by vanishing of a finite number of polynomials with coefficients in $\mathbb{Q}$. We say that $X/\mathbb{Q}$ is a Calabi–Yau variety if $X\otimes_{\mathbb{Q}}\mathbb{C}$ is Calabi–Yau variety. Let $X/\mathbb{Q}$ be a Calabi–Yau variety with a defining equation with coefficients in $\mathbb{Z}[1/m]$ for some $m\in\mathbb{N}$. Let p be a prime $(p,m)=1$, and let $X_p:=X \bmod p$ be the reduction of X modulo p. We say that p is *good* if X_p is smooth over $\overline{\mathbb{F}}_p$, otherwise *bad*.

Let $\#X(\mathbb{F}_{p^k})$ be the number of rational points on X_p over $\mathbb{F}_{p^k}$. The *local (congruent) zeta function* of X_p is defined by taking the formal sum

$$Z_p(X,T):=\exp\left(\sum_{k=1}^{\infty}\frac{\#X(\mathbb{F}_{p^k})}{k}T^k\right)\in\mathbb{Q}[[T]]$$

where T is an indeterminate variable.

Let ℓ be a prime $\neq p$. There is a Weil cohomology theory, the ℓ-adic étale chomology theory, that assigns to $\overline{X}_p:=X_p\otimes_{\mathbb{F}_p}\overline{\mathbb{F}}_p$, or to $\overline{X}:=X\otimes_{\mathbb{Q}}\overline{\mathbb{Q}}$, certain (finite dimensional) $\mathbb{Q}_\ell$-vector spaces $H^i_{et}(\overline{X},\mathbb{Q}_\ell)$, $0\le i\le 2d$. The Frobenius morphism Fr_p $(x\mapsto x^p)$ on X_p induces an endomorphism Fr_p^* on the étale cohomology groups $H^i_{et}(\overline{X}_p,\mathbb{Q}_\ell)$ for each i, $0\le i\le 2d$. The Grothendieck specialization theorem gives an isomorphism

$$\psi: H^i_{et}(\overline{X},\mathbb{Q}_\ell)\longrightarrow H^i_{et}(\overline{X}_p,\mathbb{Q}_\ell).$$

By the comparison theorem,

$$H^i_{et}(\overline{X},\mathbb{Q}_\ell)\otimes_{\mathbb{Q}_\ell}\mathbb{C}\simeq H^i(X\otimes_{\mathbb{Q}}\mathbb{C},\mathbb{C})$$

so that $\dim_{\mathbb{Q}_\ell}H^i_{et}(\overline{X},\mathbb{Q}_\ell)=B_i(X)$ (the i-th Betti number).

There is the Poincaré duality:

$$H^i_{et}(\overline{X},\mathbb{Q}_\ell)\times H^{2d-i}_{et}(\overline{X},\mathbb{Q}_\ell)\to H^{2d}_{et}(X,\mathbb{Q}_\ell)\cong\mathbb{Q}_\ell$$

is a perfect pairing for every i, $0\le i\le d$. Thus, $B_i(X)=B_{2d-i}(X)$ for every i, $0\le i\le d$. Let

$$E:=\sum_{i=0}^{2d}(-1)^iB_i(X)$$

be the Euler characteristic of X.

Now the Lefschetz fixed point formula holds, so that the number of $\mathbb{F}_p$-rational points on X can be computed by the following trace formula

$$\#X(\mathbb{F}_p) = \sum_{k=0}^{2d}(-1)^k \mathrm{trace}(\mathrm{Fr}_p^* \mid H^k_{et}(\overline{X}, \mathbb{Q}_\ell)).$$

Let

$$P^i_p(X,T) := \det(1 - \mathrm{Fr}_p^* T \mid H^i_{et}(\overline{X}, \mathbb{Q}_\ell))$$

be the characteristic polynomial of Fr_p^*.

THEOREM 2.1 (Weil's Conjecture (Theorem)).

(1) $P^i_p(X,T) \in 1 + T\mathbb{Z}[T]$.
(2) $P^i_p(X,T)$ *does not depend on the choice of* ℓ *different from* p.
(3) $deg\, P^i_p(X,T) = B_i(X)$ *for every* $i, 0 \le i \le 2d$.
(4) *If we write* $P^i_p(X,T) = \prod_{k=1}^{B_i}(1 - \alpha_k T) \in \overline{\mathbb{Q}}[T]$, *then* α_k *is an algebraic integer with* $|\alpha_k| = p^{i/2}$. (This is the Riemann Hypothesis which was proved by Deligne [**10**],[**11**].)
(5) $Z_p(X,T)$ *satisfies the functional equation:*
$$Z_p(X, p^{-d}T^{-1}) = \pm p^{dE/2} T^E Z_p(X,T).$$
(6) $Z_p(X,T)$ *is a rational function:*
$$Z_p(X,T) = \frac{\prod_{i=1}^d P_p^{2i-1}(X,T)}{\prod_{i=0}^d P_p^{2i}(X,T)}.$$

REMARK 2.2. If X is defined over a number field F, then the definition of the local zeta-function and the assertions in Theorem 2.1 are valid for the reduction of X modulo a "good" prime ideal $\mathfrak{p}$ of the ring $\mathcal{O}_F$ of integers of F over a rational prime p. The reduction $X_\mathfrak{p} := X \pmod{\mathfrak{p}}$ is defined over the finite field $\mathbb{F}_q$ with $q = p^m$ for some $m \in \mathbb{N}$. Replace p by $q = p^m$ in the definition and Theorem 2.1.

2.2. The L-series. For a good prime p, the Galois group $\mathrm{Gal}(\overline{\mathbb{F}}_p/\mathbb{F}_p)$ acts on $\overline{X}_p$ thus making $H^i_{et}(\overline{X}_p, \mathbb{Q}_\ell)$ a $\mathrm{Gal}(\overline{\mathbb{F}}_p/\mathbb{F}_p)$-module.

Let $G_\mathbb{Q} = \mathrm{Gal}(\overline{\mathbb{Q}}/\mathbb{Q})$ be the absolute Galois group. Then $G_\mathbb{Q}$ acts on $\overline{X}$ giving $H^i_{et}(\overline{X}, \mathbb{Q}_\ell)$ a $G_\mathbb{Q}$-module structure. This makes the Poincaré duality a Galois equivariant pairing.

For every embedding $\overline{\mathbb{Q}} \to \overline{\mathbb{Q}}_p$ there is an embedding $\mathrm{Gal}(\overline{\mathbb{Q}}_p/\mathbb{Q}_p) \hookrightarrow \mathrm{Gal}(\overline{\mathbb{Q}}/\mathbb{Q})$. Let Fr_p^* be a lift of Fr_p. Then the action of $\psi \circ \mathrm{Fr}_p^*$ on $H^i_{et}(\overline{X}_p, \mathbb{Q}_\ell)$ is conjugated to the action of Fr_p, where ψ is the ($\mathbb{Q}_\ell$-vector space) isomorphism defined above.

Let

$$\rho^i_{X,\ell} := \rho^i : G_\mathbb{Q} \to GL(H^i_{et}(\overline{X}, \mathbb{Q}_\ell)).$$

be the natural compatible system of ℓ-adic Galois representations. Then we can consider lifts of Fr_p and their trace and determinant at good primes in connection with $G_\mathbb{Q}$-representations.

DEFINITION 2.3. The i-th (cohomological) L-series (or L-function) of $X/\mathbb{Q}$ is defined by

$$\begin{aligned} L_i(X,s) &:= L(H^i_{et}(\overline{X}, \mathbb{Q}_\ell), s) \\ &:= (*) \prod_{p \neq \ell:\mathrm{good}} P^i_p(p^{-s})^{-1} \times (\text{the factor corresponding to } \ell = p) \end{aligned}$$

where the product is taken over all good primes different from ℓ and $(*)$ corresponds to the Euler factors of bad primes. For $\ell = p$, take different ℓ and use Theorem 2.1, or we may use some p-adic cohomology (e.g., crystalline cohomology).

The most interesting cohomology of a d-dimensional variety X is the middle cohomology. For this reason, we define the L-series of X to be the d-th L-series of X and we simply write

$$L(X,s) := L_d(H^d_{et}(\bar{X}, \mathbb{Q}_\ell), s).$$

The Galois representation ρ^d is simply denoted by ρ.

The modularity/automorphy question for a Calabi-Yau variety X defined over $\mathbb{Q}$ is formulated in terms of the L-series $L(X,s)$.

QUESTION 2.4. (a) Suppose that the Galois representation ρ is irreducible.

Are there global functions that determine the L-series $L(X,s)$? That is, are there modular (automorphic) functions that determine the L-series $L(X,s)$?

(b) Suppose that ρ is not irreducible, then we will consider irreducible subrepresentations. Let $\mathcal{M}$ be a motive of X (associated to some $\mathbb{Q}_\ell$-vector subspace of $H^d_{et}(\overline{X}, \mathbb{Q}_\ell)$).

Are there modular (automorphic) functions that determine the L-series $L(\mathcal{M}, s)$?

3. The modularity results

We should mention here the modularity results obtained in the last two decades (plus a couple of years) for Calabi–Yau varieties of dimension at most 3. Calabi–Yau varieties of dimension 1 and 2 are elliptic curves and K3 surfaces, respectively. Calabi–Yau varieties of dimension 3 are Calabi–Yau threefolds.

DEFINITION 3.1 (Modular forms). First we give a quick definition of modular (cusp) forms. Let $\mathfrak{H}$ denote the upper-half complex plane. The modular group $SL(2,\mathbb{Z})$ acts on $\mathfrak{H}$ by the linear fractional transformations:

$$SL(2,\mathbb{Z}) \times \mathfrak{H} \to \mathfrak{H} : \begin{pmatrix} a & b \\ c & d \end{pmatrix}(z) = \frac{az+b}{cz+d}.$$

Let $\mathfrak{H}^*$ denote the union of $\mathfrak{H}$ and the fixed points of this action (called cusps). So $\mathfrak{H}^* = \mathfrak{H} \cup \mathbb{P}^1(\mathbb{Q})$. For a fixed integer $N \in \mathbb{N}$, define the subgroup

$$\Gamma_0(N) := \{\begin{pmatrix} a & b \\ c & d \end{pmatrix} \in SL(2,\mathbb{Z}) \,|\, c \equiv 0 \pmod{N}\} \subset SL(2,\mathbb{Z}).$$

Let $f : \mathfrak{H} \to \mathbb{C}$ be a holomorphic function. For $k \in \mathbb{N}$ and $\begin{pmatrix} a & b \\ c & d \end{pmatrix} \in \Gamma_0(N)$, if f transforms as

$$f\left(\frac{az+b}{cz+d}\right) = (cz+d)^k f(z)$$

and f is holomorphic at cusps, f is said to be a *modular form of weight k and level N*. Since the matrix $\begin{pmatrix} 1 & 1 \\ 0 & 1 \end{pmatrix} \in \Gamma_0(N)$, f satisfies $f(z+1) = f(z)$. Writing $q = e^{2\pi i z}$ with $z \in \mathfrak{H}$, we obtain the Fourier expansion $f(z) = f(q) = \sum_n b_n q^n$. The cusp $i\infty$ corresponds to $q = 0$ and f is holomorphic at the cusp $i\infty$ if $b_n = 0$

for $n < 0$. Further f vanishes at $i\infty$ if $b_0 = 0$. A *cusp* form is a modular form which vanishes at all cusps.

Modular (cusp) forms with a character χ satisfy the transformation formula

$$f\left(\frac{az+b}{cz+d}\right) = \chi(d)(cz+d)^k f(z).$$

Modular (cusp) forms of weight k for $\Gamma_0(N)$, and similarly, modular (cusp) forms of weight k for $\Gamma_0(N)$ with character χ form finite-dimensional $\mathbb{C}$-vector spaces.

3.1. Elliptic curves over $\mathbb{Q}$.

THEOREM 3.2 ($d = 1$). *Every elliptic curve E over $\mathbb{Q}$ is modular. There is a cusp form f of weight $2 = d + 1$ on some $\Gamma_0(N)$ such that $L(E, s) = L(f, s)$.*

This is the celebrated theorem of Wiles [**48**], Taylor–Wiles [**43**] and Breuil–Conrad–Diamond–Taylor [**6**].

3.2. Singular K3 surfaces over $\mathbb{Q}$. Let S be a K3 surface defined over $\mathbb{C}$. Let $H^2(S,\mathbb{Z})$ be the K3 lattice, that is, $H^2(S,\mathbb{Z}) \simeq U_2^3 \oplus (-E_8)^2$ where U_2 denotes the hyperbolic plane and E_8 is the positive definite even unimodular root lattice of rank 8. Let $NS(S)$ denote the Néron–Severi group of S, generated by algebraic cycles on S. It is a free finitely generated abelian group, and $NS(S) = H^{1,1}(S,\mathbb{R}) \cap H^2(S,\mathbb{Z})$ so that the rank of $NS(S)$, called the *Picard number* of S and denoted by $\rho(S)$, is bounded by 20. (In characteristic zero, $\rho(S)$ can vary from 0 to 20; 0 may occur when S is not algebraic.) Let $T(S) = NS(S)^{\perp}$ be the orthogonal complement of $NS(S)$ in $H^2(S,\mathbb{Z})$. $T(S)$ has the $\mathbb{Z}$-rank $22 - \rho(S)$, and is called the group of *transcendental cycles* on S. We have the decomposition

$$H^2(S,\mathbb{Z}) \otimes_{\mathbb{Q}} \mathbb{Q}_\ell = H^2_{et}(S,\mathbb{Q}_\ell) = (NS(S) \otimes \mathbb{Q}_\ell) \oplus (T(S) \otimes \mathbb{Q}_\ell)$$

and this gives rise to the decomposition of the L-series:

$$L(S,s) = L(NS(S) \otimes \mathbb{Q}_\ell, s) L(T(S) \otimes \mathbb{Q}_\ell, s).$$

The Tate conjecture is valid for any algebraic K3 surface over $\mathbb{Q}$ (e.g., Tate [**41**]), which asserts that

$$H^2_{et}(\overline{S}, \mathbb{Q}_\ell(1))^{G_{\mathbb{Q}}} = NS(S)_{\mathbb{Q}}.$$

The L-series $L(NS(S) \otimes \mathbb{Q}_\ell, s)$ is expressed in terms of $\zeta_{\mathbb{Q}}(s-1)^{\rho(S)}$ if $NS(S)$ is generated by divisors defined over $\mathbb{Q}$, where $\zeta_{\mathbb{Q}}(s)$ denotes the Riemann zeta-function. However, this extreme situation occurs very rarely. In general, some algebraic cycles may be defined over $\mathbb{Q}$, but others are defined over finite extensions of $\mathbb{Q}$. In this case, Artin L-function should come into the picture, and the modularity of Artin L-functions is still unsolved.

For K3 surfaces, we will address the motivic modularity/automorphy, that is, we will establish the modularity/automorphy of the L-series $L(T(S) \otimes \mathbb{Q}_\ell, s)$ associated to the $\mathbb{Q}_\ell$-subspace $T(S) \otimes \mathbb{Q}_\ell \subset H^2_{et}(S, \mathbb{Q}_\ell)$.

DEFINITION 3.3. A K3 surface S over $\mathbb{Q}$ is *singular* (or *extremal*) if $\rho(S) = 20$ (so that $\mathbb{Z}$-rank of $T(S) = 2$).

For a singular K3 surface S defined over $\mathbb{Q}$, the transcendental lattice $T(S)$ is a two-dimensional $G_{\mathbb{Q}}$-module. There is a positive definite even lattice of rank 2 corresponding to $T(S)$. Shioda and Inose [**39**] have shown that there is a bijection between the isomorphism classes of singular K3 surfaces and the set of reduced quadratic forms $Q = \begin{pmatrix} 2a & b \\ b & 2c \end{pmatrix}$ where $a, b, c \in \mathbb{Z}$, $a > 0$, $-a < b \leq a \leq c$ (and $b \geq 0$ if $a = c$) and $\det(Q) = D > 0$. The $\det(Q)$ is referred as the *discriminant* of S.

Now we will consider the motivic modularity of the $L(T(S) \otimes \mathbb{Q}_\ell, s)$.

THEOREM 3.4 ($d = 2$). *Every singular $K3$ surface S of discriminant D over $\mathbb{Q}$ is motivically modular.*

There is a cusp form f of weight $3 = d + 1$ on some $\Gamma_0(N) + \chi$ such that

$$L(T(S) \otimes \mathbb{Q}_\ell, s) = L(f \otimes \chi, s).$$

This is a theorem of Livné [**25**]. Note that the space of cusp forms of weight 3 (or any odd weight) for $\Gamma_0(N)$ consists only of scalars, i.e., $\mathbb{C}$, so we ought to twist with some Dirichlet character χ to have non-trivial cusp forms for $\Gamma_0(N) + \chi$. Here χ is a quadratic character $\left(\dfrac{-D}{p}\right)$ with $D > 0$ as above. In this case, the cusp form f has complex multiplication by the imaginary quadratic field $\mathbb{Q}(\sqrt{-D})$.

3.3. Rigid Calabi–Yau threefolds.

DEFINITION 3.5. A Calabi–Yau threefold X over $\mathbb{Q}$ is said to be *rigid* if $h^{2,1}(X) = 0$ (so that $B_3(X) = 2$). Thus, the Hodge diamond of any rigid Calabi–Yau threefold is given by

$$\begin{array}{ccccccccc} & & & 1 & & & & \qquad & B_0 = 1 \\ & & 0 & & 0 & & & & B_1 = 0 \\ & 0 & & h^{1,1} & & 0 & & & B_2 = h^{1,1} \\ 1 & & 0 & & 0 & & 1 & & B_3 = 2 \\ & 0 & & h^{2,2} & & 0 & & & B_4 = h^{2,2} \\ & & 0 & & 0 & & & & B_5 = 0 \\ & & & 1 & & & & & B_6 = 1 \\ & & & & & & & & E = 2h^{1,1} \end{array}$$

Here $h^{i,j} := h^{i,j}(X)$ denotes the (i, j)-th Hodge number, $B_i := B_i(X)$ the i-th Betti number and $E := E(X)$ the Euler characteristic of X, respectively.

THEOREM 3.6 ($d = 3$). *Every rigid Calabi–Yau threefold X over $\mathbb{Q}$ is modular. There is a cusp form f of weight $4 = d + 1$ on some $\Gamma_0(N)$ such that $L(X, s) = L(f, s)$. Here N is divisible only by bad primes.*

This is a theorem of Gouvêa–Yui [**18**], and Dieulefait [**12**], obtained independently, around the same time. The proof is based on the validity of the Serre's conjecture on the modularity of the residual Galois representations of dimension 2, proved by Khare–Wintenberger [**20**] and Kisin [**22**].

REMARK 3.7. (1) The modularity is established for the above Calabi–Yau varieties over $\mathbb{Q}$. However, we do not know conceptual reasons "why" they are modular. What would be physics implications of modularity? Are there string theoretic explanations for the modularity?

(2) The above results (Theorems 3.2, 3.4 and 3.6) are obtained by studying ℓ-adic 2-dimensional Galois representations associated to Calabi–Yau varieties over $\mathbb{Q}$. Here 2 coincides with the d-th Betti number B_d of the Calabi–Yau variety of

dimension d for $d = 1$ and 3; while 2 is the $\mathbb{Z}$-rank of the transcendental lattice $T(S)$ of a singular K3 surface for $d = 2$.

Now we will briefly describe methods of establishing these results by studying ℓ-adic 2-dimensional Galois representations.

• $d = 1$. Let E be an elliptic curve defined over $\mathbb{Q}$. Then $B_1(E) = 2$. For a good prime p, the Lefschetz fixed point formula reads

$$\#E(\mathbb{F}_p) = \sum_{k=0}^{2} \mathrm{trace}(Fr_p^* \mid H^k_{et}(\overline{E}, \mathbb{Q}_\ell)) \quad \text{for } \ell \neq p.$$

Put

$$t_k(p) := \mathrm{trace}(Fr_p^* \mid H^k_{et}(\overline{E}_p, \mathbb{Q}_\ell)) \quad \text{for } k = 0, 1, 2.$$

Then

$$\#E(\mathbb{F}_p) = t_0(p) - t_1(p) + t_2(p) = 1 - t_1(p) + p,$$

so

$$t_1(p) = 1 + p - \#E(\mathbb{F}_p) \quad \text{with } |t_1(p)| \leq 2p^{1/2}.$$

The L-series of E is given by

$$L(E, s) = (*) \prod_{p:\mathrm{good}} (1 - t_1(p)p^{-s} + p^{1-2s})^{-1}.$$

Theorem 3.2 then asserts that there exists a modular (cusp) form of weight $d+1 = 1+1 = 2$ with Fourier expansion $f(q) = \sum_n b_n q^n$ for $\Gamma_0(N)$ (normalized so that $b_1 = 1$) such that $t_1(p) = b_p$ for all good primes p.

• $d = 3$. Let X be a rigid Calabi–Yau threefold defined over $\mathbb{Q}$. Rigid Calabi-Yau threefolds are natural generalizations of elliptic curves in terms of Betti numbers. Then $B_3(X) = 2$. The Lefschetz fixed point formula reads

$$\#X(\mathbb{F}_p) = \sum_{k=0}^{6} \mathrm{trace}(Fr_p^* \mid H^k_{et}(\overline{X}, \mathbb{Q}_\ell)).$$

Putting $t_k(p) = \mathrm{trace}(Fr_p^* \mid H^k_{et}(\overline{X}, \mathbb{Q}_\ell))$, we have

$$\begin{aligned}\#X(\mathbb{F}_p) &= \sum_{k=0}^{6}(-1)^k t_k(p) = t_0(p) - t_1(p) + t_2(p) - t_3(p) + t_4(p) - t_5(p) + t_6(p) \\ &= 1 + t_2(p)(1+p) - t_3(p) + p^3.\end{aligned}$$

Here $|t_2(p)| \leq pB_2(X) = ph^{1,1}(X)$, and the equality holds if a basis of the $H^{1,1}$-cycles (or of $H^{1,1} = H^2$) are defined over $\mathbb{Q}$ (or sufficiently over $\mathbb{F}_p$). We have

$$t_3(p) = 1 + p^3 + t_2(p)(1+p) - \#X(\mathbb{F}_p) \quad \text{with } |t_3(p)| \leq 2p^{3/2}.$$

The L-series of X is then given by

$$L(X, s) = (*) \prod_{p:\mathrm{good}} (1 - t_3(p)p^{-s} + p^{3-2s})^{-1}.$$

Theorem 3.6 then asserts that there exists a modular form of weight $d+1 = 4$ with Fourier expansion $f(q) = \sum_n b_n q^n$ for $\Gamma_0(N)$ (normalized by $b_1 = 1$) such that $t_3(p) = b_p$ for all good primes p.

• $d = 2$. As mentioned in Subsection 3.2, the motive of the transcendental lattice $T(S) \otimes \mathbb{Q}_\ell$ gives rise to the ℓ-adic 2-dimensional Galois representation. By Shioda–Inose [**39**], there is a bijection between the isomorphism classes of singular K3 surfaces over $\mathbb{Q}$ and the $SL_2(\mathbb{Z})$-homothety classes of reduced quadratic forms $Q = \begin{pmatrix} 2a & b \\ b & 2c \end{pmatrix}$ with determinant $\det(Q) = D = 4ac - b^2 > 0$, where $a > 0$, $-a < b \leq a \leq c$ (and $b \geq 0$ if $a = c$). A positive definite even lattice $T(S)$ of rank 2 and discriminant D determines Q.

Livné [**25**] associates to a compatible family of 2-dimensional ℓ-adic Galois representation of $T(S) \otimes_{\mathbb{Q}} \mathbb{Q}_\ell$ a modular form f of weight 3 with the Dirichlet character χ. Here χ is the quadratic character uniquely determined by $\det(Q)$, i.e., $\chi(p) = \left(\frac{-D}{p}\right)$.

The Euler p-factor of the motivic L-series $L(T(S) \otimes \mathbb{Q}_\ell, s)$ is given by

$$(1 - \beta_p p^{-s})(1 - \overline{\beta}_p p^{-s}) = 1 - B(p)p^{-s} + \chi(p)p^{2-2s}$$

which coincides with the Euler p-factor of the modular form $f \otimes \chi$. A weight 3 modular form $f \otimes \chi$ has complex multiplication by $\mathbb{Q}(\sqrt{-D})$.

See Livné [**25**], and also see Livné–Yui [**27**] for more detailed discussion.

3.4. Higher dimensional Galois representations. Higher dimensional Galois representations associated to Calabi–Yau varieties defined over $\mathbb{Q}$ will occur in the following situations:

• $d = 2$: Let S be a K3 surface and let $T(S)$ be the transcendental lattice. For instance, when the $\mathbb{Z}$-rank of $T(S) \geq 3$ (so S is not a singular K3 surface).

• $d = 3$: Let X be a non-rigid Calabi–Yau threefold. When $h^{2,1}(X) \geq 1$ (so that $B_3(X) = 2(1 + h^{2,1}(X)) \geq 4$).

The modularity/automorphy question for these higher dimensional Galois representations is currently out of reach in the general setting.

For $d = 2$, we require for K3 surfaces S to have more structures, for instance,

- large Picard numbers (though this is not strictly necessary as some of the entries of Table 1 below show),
- elliptic fibrations,
- Shioda–Inose structures ([**39**] and Morrison [**29**]),
- non-trivial automorphisms.

An elliptic K3 surface is a K3 surface S with a surjective morphism $\pi : S \to \mathbb{P}^1$ such that the generic fiber of π is a smooth curve of genus 1. In many cases, a defining equation for S may be given by a Weierstrass equation. The Picard number of S and the Mordell–Weil rank of π are related by the formula of Shioda–Tate.

$$\rho(S) = 2 + \mathrm{rank}_{\mathbb{Z}} MW(\pi) + \sum_{P \in \Sigma} (m(P) - 1)$$

where Σ denotes the set of points P such that $\pi^{-1}(P)$ is singular, and $m(P)$ denotes the number of irreducible components of $\pi^{-1}(P)$. The motivic modularity of some elliptic K3 surfaces with non-symplectic automorphisms is established by Livné–Schütt-Yui [**26**], and will be discussed below.

An involution ι on S is called a *Nikulin involution* if $\iota^*(\omega) = \omega$ for every $\omega \in H^{2,0}(S)$. The rational quotient of S by a Nikulin involution ι is a K3 surface. We say that a K3 surface S admits a *Shioda–Inose structure* if there is a Nikulin involution ι on S with rational quotient map $\pi : S \to S_0$ such that S_0 is a Kummer surface and that π_* induces a Hodge isometry $T(S)(2) \cong T(S_0)$. By a theorem of Morrison [**29**], the following statements are all equivalent:

(i) the existence of a Shioda–Inose structure on S;
(ii) the existence of an abelian surface A and a Hodge isometry $T(S) \cong T(A)$;
(iii) the existence of a primitive embeddng $T(S) \hookrightarrow U_2^3$;
(iv) the existence of an embedding $E_8(-1)^2 \hookrightarrow NS(S)$.

In this case, the Picard number $\rho(NS(S))$ takes values 17, 18, 19, or 20. Knowing the (lattice) structure of $NS(S)$ and also that of $T(S)$, the the modularity of $T(S)$ may be plausible, see below for examples.

When S has a large automorphism group, under the action of automorphisms, $H^2_{et}(\overline{S}, \mathbb{Q}_\ell)$ may decompose into a product of smaller dimensional subspaces, and the motivic modularity of $T(S)$ may be plausible. Some examples of this kind will be discussed below.

For $d = 3$, we require that X has nice geometric or algebraic structures, e.g.,

- elliptic or K3 fibrations, or both,
- $H^3_{et}(\overline{X}, \mathbb{Q}_\ell)$ decomposes into a product of motives of small dimensions.

For instance, the latter may happen when X has a large automorphism group. There are some results along this line, for instance, see Hulek–Verrill [**15**], Bini–van Geemen [**3**] and Schütt [**40**]. We will discuss below some examples of this kind.

3.5. Calabi–Yau varieties of CM type. There is a distinguished class of Calabi–Yau varieties, for which the modularity can be established. These are varieties of CM (Complex Multiplication) type. In this section we will be focused on Calabi–Yau varieties of CM type. First we define Calabi–Yau varieties of CM type, following Borcea [**4**]; see also Rohde [**35**].

DEFINITION 3.8. A Calabi–Yau variety X defined over $\mathbb{Q}$ of dimension d is said to be of *CM type*, or *has CM* if the Hodge group $\mathrm{Hdg}(X)$ associated to the rational Hodge structure on $H^d(X, \mathbb{Q})$ is commutative, so $\mathrm{Hdg}(X) \otimes \mathbb{C} \simeq$ copies of $\mathbb{G}_m$.

REMARK 3.9. (1) The Hodge group $\mathrm{Hdg}(X)$ is, in general, a very difficult object to compute.

(2) A Calabi–Yau threefold X has CM if and only if the Hodge groups of the weight 1 Hodge structures corresponding to the Griffith intermediate Jacobian $J_G(X)$ and the Weil intermediate Jacobian $J_W(X)$ are tori and commute. (Borcea [**4**].)

EXAMPLE 3.10. Calabi–Yau varieties of CM type have been discussed in Borcea [**4**] and also in Yui [**51**]. Here are some examples of surfaces of CM type.

(1) Every singular (extremal) K3 surfaces defined over $\mathbb{Q}$. (Shioda and Inose [**39**].)

(2) The Fermat surface of degree $m (\geq 4)$:

$$x_0^m + x_1^m + x_2^m + x_3^m = 0 \subset \mathbb{P}^3.$$

(Weil [**46**].)

(3) Delsarte surfaces. Here a Delsarte surface is defined by four-term irreducible equation over $\mathbb{Q}$ of the form:

$$\sum_{i=0}^{3}\prod_{j=0}^{3} x_j^{a_{ij}} = 0$$

with the matrix (a_{ij}) of exponents having non-zero determinant. (Shioda [**37**].)

(4) More generally, invertible polynomials over $\mathbb{Q}$ in weighted projective 3-spaces $\mathbb{WP}^3(w_0, w_1, w_2, w_3)$. By an invertible polynomial, we mean non-degenerate weighted quasihomogeneous polynomial in 4 variables that contains exactly 4 monomials. If we write $f(x_0, x_1, x_2, x_3) = \sum_{i=0}^{3}\prod_{j=0}^{3} x_j^{a_{ij}}$, then the matrix of exponent (a_{ij}) is invertible, and $\sum_{j=0}^{3} a_{ij}w_j = \deg(f)$.

For (3) and (4), these surfaces are of CM type as they can be realized as finite Fermat quotients. See below for some explanation. For fuller details, see, for instance, Shioda [**37**], Livné–Schütt–Yui [**26**], Goto–Kloosterman–Yui [**16**] and Goto–Livné–Yui [**17**].

CONJECTURE 3.11 (Shafarevich and Piatetski-Shapiro [**36**]; see also Yui [**50**]). *Let Z be an algebraic variety of CM type. Then Z has a model defined over a number field F, and its L-series is expressed by means of L-series of one-dimensional characters of a finite extension of F.*

We will give affirmative answers to the above conjecture for certain Calabi–Yau varieties of CM type in the next two sections.

4. Modularity/automorphy of K3 surfaces with non-symplectic automorphisms

In this section, we consider certain algebraic K3 surfaces over $\mathbb{Q}$ with non-symplectic automorphisms, and establish their motivic modularity. Since we will only be considering algebraic K3 surfaces, we simply say K3 surfaces dropping "algebraic" in what follow.

4.1. K3 surfaces with non-symplectic automorphisms. Let S be a K3 surface. Let $\mathrm{Aut}(S)$ be the automorphism group of S.

Let ω_S be a non-zero holomorphic 2-form on S, fixed once and for all. Then $H^{2,0}(S) = \mathbb{C}\omega_S$. Let $g \in \mathrm{Aut}(S)$. Then g induces a map

$$g^* : H^{2,0}(S) \to H^{2,0}(S) : g^*\omega_S = \alpha(g)\omega_S$$

for some $\alpha(g) \in \mathbb{C}^*$. We say that g is *non-symplectic* if $\alpha(g) \neq 1$. Nikulin [**31**] proved that $\alpha(\mathrm{Aut}(S))$ is a finite cyclic group.

Let $NS(S)$ and $T(S)$ denote, respectively, the lattice of algebraic cycles on S, and the lattice of transcendental cycles on S. Then $\mathrm{Aut}(S)$ acts on these lattices. Let $O(L)$ denote the group of isometries of a lattice L. As a consequence of the Torelli Theorem, the the natural representation

$$\rho : \mathrm{Aut}(S) \to O(H^2(S, \mathbb{Z}))$$

is faithful, that is, the induced map $\mathrm{Aut}(S) \to O(NS(S)) \times O(T(S))$ is injective. We are interested in automorphisms which act trivially on $NS(S)$. We define the subgroup $H_S \subseteq \mathrm{Aut}(S)$ by the following exact sequence

$$1 \to H_S \to \mathrm{Aut}(S) \to O(NS(S)) \to 1.$$

Since ρ is faithful, H_S can be identified with its image under the map α. Nikulin's theorem asserts that H_S is a finite cyclic group with order $k = |H_S|$, then $\phi(k)$ divides the $\mathbb{Z}$-rank of $T(S)$. Here ϕ is the Euler function.

There are only finitely many values of k arising in this way.

THEOREM 4.1. *Let S be a K3 surface over $\mathbb{C}$ with non-symplectic automorphisms. Assume that $T(S)$ is unimodular (i.e., $det(T(S)) = \pm 1$). Let H_S be a finite cyclic subgroup of $Aut(S)$ with $k := |H_S|$.*

Let

$$\Sigma = \{66, 44, 42, 36, 28, 12\}.$$

Then the following assertions hold.

(1) *k is a divisor of an element in Σ. In particular $k \leq 66$.*
(2) *Furthermore, assume that $\phi(k) = rank(T(S))$. Then $k \in \Sigma$.*
(3) *Conversely, for each $k \in \Sigma$, there exists a unique K3 surface S (up to isomorphism) satisfying the properties that $|H_S| = k$ and $rank(T(S)) = \phi(k)$. For these values of k, $NS(S)$ and $T(S)$ are given in Table 1 below.*

TABLE 1. Unimodular non-symplectic K3 surfaces

k	$NS(S)$	rank($NS(S)$)	$T(S)$	rank($T(S)$)
66	U_2	2	$U_2^2 \oplus (-E_8)^2$	20
44	U_2	2	$U_2^2 \oplus (-E_8)^2$	20
42	$U_2 \oplus (-E_8)$	10	$U_2^2 \oplus (-E_8)$	12
36	$U_2 \oplus (-E_8)$	10	$U_2^2 \oplus (-E_8)$	12
28	$U_2 \oplus (-E_8)$	10	$U_2^2 \oplus (-E_8)$	12
12	$U_2 \oplus (-E_8)^2$	18	U_2^2	4

The possible values of k were announced in Vorontsov [**45**], and theorem was proved, independently by Vorontsov, and Kondō [**23**] using results of Nikulin [**31**].

THEOREM 4.2. *The K3 surfaces in the above theorem can all be defined over $\mathbb{Q}$.*

This theorem is due to Kondō [**23**], who exhibited defining equations in terms of Weierstrass models over $\mathbb{Q}$. We reproduce them in the following table up to some signs. We also list the action of a generator $\alpha(g) := g_k$ of H_S on S. In terms of the local coordinates of the Weierstrass model, the action of $\alpha(g) = g_k$ is given by

$$(x, y, t) \mapsto (\zeta_k^{\alpha} x, \zeta_k^{\beta} y, \zeta_k^{\gamma} t)$$

where ζ_k is a primitive k-th root of unity, and $\alpha, \beta, \gamma \in \mathbb{Z}/k\mathbb{Z}$.

Note that all these equations are defined by four monomials. This fact will play an essential role later.

REMARK 4.3. We also have the corresponding theorems in the case that $T(S)$ is non-unimodular. There are exactly ten values for k, namely,

$$k \in \{3, 9, 27, 5, 25, 7, 11, 13, 17, 19\}.$$

TABLE 2. Elliptic fibrations of unimodular K3 surfaces

k	S	$g = g_k$
66	$y^2 = x^3 - t(t^{11} + 1)$	$(x,y,t) \mapsto (\zeta_{66}^2 x, \zeta_{66}^3 y, \zeta_{66}^6 t)$
44	$y^2 = x^3 + x + t^{11}$	$(x,y,t) \mapsto (-x, \zeta_{44}^{11} y, \zeta_{44}^2 t)$
42	$y^2 = x^3 - t^5(t^7 + 1)$	$(x,y,t) \mapsto (\zeta_{42}^2 x, \zeta_{42}^3 y, \zeta_{42}^{18} t)$
36	$y^2 = x^3 - t^5(t^6 + 1)$	$(x,y,t) \mapsto (\zeta_{36}^2 x, \zeta_{36}^3 y, \zeta_{36}^{30} t)$
28	$y^2 = x^3 + x + t^7$	$(x,y,t) \mapsto (-x, \zeta_{28}^7 y, \zeta_{28}^2 t)$
12	$y^2 = x^3 + t^5(t^2 + 1)$	$(x,y,t) \mapsto (\zeta_{12}^2 x, \zeta_{12}^3 y, -t)$

For each value of k in this set, the lattice structures for $NS(S)$ and $T(S)$ are explicitly determined. Also these K3 surfaces have defining equations in terms of Weierstrass models over $\mathbb{Q}$ except for $k = 25$. For $k = 25$, the K3 surface is defined by a double sextic over $\mathbb{Q}$. See Kondō [**23**] and/or Livné–Schuett–Yui [**26**] for the lattices and explicit defining equations.

4.2. Delsarte surfaces. The K3 surfaces listed in Remark 4.3 and Theorem 4.2 are all Delsarte surfaces. Indeed, each surface is defined by a sum of four monomials in the affine 3-space. Hence such a surface is covered by a Fermat surface of an appropriate degree. From now on, we let

$$\mathcal{F}_m \,:\, x_0^m + x_1^m + x_2^m + x_3^m = 0 \subset \mathbb{P}^3$$

denote the Fermat surface of degree $m \geq 4$ over $\mathbb{Q}$. The numerical invariants of $\mathcal{F}_m$ are given as follows: the geometric genus of $\mathcal{F}_m$ equals $p_g(\mathcal{F}_m) = \binom{m-1}{3}$, $h^{1,0}(\mathcal{F}_m) = B_1(\mathcal{F}_m) = q(\mathcal{F}_m) = 0$, and $B_2(\mathcal{F}_m) = m^3 - 4m^2 + 6m - 2$. As $K_{\mathcal{F}_m}^2 = m(m-4)^2$, $\mathcal{F}_m$ is a surface of general type for any $m > 4$, and is a K3 surface when $m = 4$.

THEOREM 4.4. *Let S be one of the K3 surfaces corresponding to the values of k in Theorem 4.1 an Remark 4.3. Then S is a Delsarte surface, that is, S is covered by the Fermat surface defined over $\mathbb{Q}$ of the following degree m:*

$$\begin{cases} m = k, & \text{if } T(S) \text{ is unimodular,} \\ m = 2k, & \text{if } T(S) \text{ is non-unimodular.} \end{cases}$$

Thus, each of these K3 surfaces is birationally equivalent to a finite quotient of a Fermat surface $\mathcal{F}_m$ over $\mathbb{Q}$ with a finite covering map $\pi : \mathcal{F}_m \to S$.

Consequently each of these K3 surfaces is of CM type.

For proofs, the reader should refer to the paper of Livné–Schütt–Yui [**26**].

REMARK 4.5. For $k = 3$, the defining equation for S is given by $y^2 = x^3 + t^5(t-1)^2$. If we put $x = (t-1)u$ and $y = (t-1)v$, $v^2 = (t-1)u^3 + t^5$. This is defined by four monomials. Hence it is a Delsarte surface. If we take $v = a^3 s^{15} b^{-3}$, $u = s^{10} b^{-2}$ and $t = s^6$, then S is a finite quotient of the Fermat curve of degree $m = 2 \cdot 3 = 6$. The case $k = 3$ was treated as an exception in [**26**]. We thank the anonymous referee for pointing out that the $k = 3$ case is also a Delsarte surface.

EXAMPLE 4.6. For $k = 66$, the defining equation for S is given by

$$y^2 = x^3 - t(t^{11} + 1) = x^3 - t^{12} - t.$$

We work in the chart where $t \neq 0$. So we choose the local parameter $s = 1/t$. Then the equation becomes

$$y^2 = x^3 - \frac{1}{s^{12}} - \frac{s^{11}}{s^{12}}.$$

Introduce new coordinates $\xi := s^4 x;$, $\eta := s^6 y$. Then the resulting equation is

$$\eta^2 = \xi^3 - 1 - s^{11}.$$

The covering from the affine Fermat surface of degree 66 is given by

$$\eta \mapsto U^{33},\ \xi \mapsto -V^{22},\ s \mapsto W^6.$$

A finite quotient group G is given by $G = \mu_{33} \times \mu_{22} \times \mu_6$.

For all the remaining values of k, we obtain covering maps from Fermat surfaces of degree k (resp. $2k$) for unimodular (resp. non-unimodular) cases. However, the determination of finite quotient groups G explicitly would require more work.

4.3. Motivic decomposition. We will apply the algorithm for computing Picard numbers of Delsarte surfaces developed by Shioda [**37**] to our K3 surfaces.

Consider the Fermat surface $\mathcal{F}_m$ over a finite field $\mathbb{F}_p$ where $p \equiv 1 \pmod{m}$. Then we can define the ℓ-adic étale cohomology groups of $\mathcal{F}_m$ for a prime ℓ such that $\ell \neq p$ and $\ell \equiv 1 \pmod{m}$. We will be interested in only the second ℓ-adic étale cohomology group $H^2_{et}(\mathcal{F}_m, \mathbb{Q}_\ell)$. To ease the notation, we simply write $H^2(\mathcal{F}_m)$ for $H^2_{et}(\mathcal{F}_m, \mathbb{Q}_\ell)$.

Let μ_m be the group of m-th roots of unity and Δ the image of the diagonal inclusion $\mu_m \hookrightarrow \mu_m^4$. Then the quotient group

$$M := \mu_m^4/\Delta = \{\, \mathbf{g} = (g_0, g_1, g_2, g_3)) \in \mu_m^4\}/\Delta$$

operates on the Fermat surface $\mathcal{F}_m$ by coordinate-wise multiplication by elements of μ_m. This group operation induces a decomposition of the second cohomology $H^2(\mathcal{F}_m)$ into one-dimensional $\mathbb{Q}_\ell$-vector spaces. This result is due to Weil [**47**], see also Yui [**51**]. The decomposition is best described in terms of a specific subset of the character group $\hat{M}$ of M. Consider a subset $\mathfrak{A}_m$ of $\hat{M}$:

$$\mathfrak{A}_m := \{\mathbf{a} = (a_0, a_1, a_2, a_3) \in (\mathbb{Z}/m\mathbb{Z})^4 \mid a_i \not\equiv 0 \pmod{m}\, \forall i, \sum_{i=0}^{3} a_i \equiv 0 \pmod{m}\}.$$

Then $H^2(\mathcal{F}_m)$ decomposes into the subspace V_0 of the hyperplane class H (corresponding to $\mathbf{a} = (0,0,0,0)$) and one-dimensional $\mathbb{Q}_\ell$-subspaces $V(\mathbf{a})$ for each $\mathbf{a} \in \mathfrak{A}_m$:

$$H^2(\mathcal{F}_m) \otimes \bar{\mathbb{Q}}_\ell = V_0 \oplus \bigoplus_{\mathbf{a} \in \mathfrak{A}_m} V(\mathbf{a}). \tag{4.1}$$

We now describe a criterion when $V(\mathbf{a})$ is algebraic: If we choose representatives $0 < a_i < m$ for each i, then this gives a well-defined map

$$|\mathbf{a}| = \frac{1}{m} \sum_{i=0}^{3} a_i.$$

Let $u \in (\mathbb{Z}/m\mathbb{Z})^*$ operate on $\mathbf{a} \in \mathfrak{A}_m$ by coordinate-wise multiplication, i.e., if $a = (a_0, a_1, a_2, a_3) \in \mathfrak{A}_m$ and $u \in (\mathbb{Z}/m\mathbb{Z})^*$, then $u \cdot \mathbf{a} = (ua_0, ua_1, ua_2, ua_3) \in \mathfrak{A}_m$. Then we define a subset $\mathfrak{B}_m$ of $\mathfrak{A}_m$ as follows:

$$\mathfrak{B}_m := \{\mathbf{a} \in \mathfrak{A}_m \mid |u \cdot \mathbf{a}| = 2 \ \ \forall u \in (\mathbb{Z}/m\mathbb{Z})^*\}.$$

Criterion for algebraicity: *The $\mathbb{Q}_\ell$-eigenspace $V(\mathbf{a})$ is algebraic if and only if $\mathbf{a} \in \mathfrak{B}_m$.*

The criterion is based on the following two facts:

(1) The action of $u \in \mathrm{Gal}\,(\mathbb{Q}(\zeta_m)/\mathbb{Q}) \simeq (\mathbb{Z}/m\mathbb{Z})^*$ sends $V(\mathbf{a})$ to $V(u \cdot \mathbf{a})$.
(2) The eigenspace $V(\mathbf{a})$ with character $\mathbf{a} \in \mathfrak{A}_m$ contributes to the Hodge cohomology group $H^{|\mathbf{a}|-1,3-|\mathbf{a}|}(\mathcal{F}_m)$.

Using the above criterion, we can easily determine the transcendental and algebraic part of $H^2(\mathcal{F}_m)$ and in particular the Picard number. We shall now see how this carries over to our K3 surfaces S.

We define the following subgroup G of the automorphism group M (viewed as μ_m^3 operating on the affine coordinates of $\mathcal{F}_m$:

$$G := \{\mathbf{g} \in M \mid \pi = \pi \circ \mathbf{g}\}.$$

(Here we identify $\mathbf{g}$ with $(g_1, g_2, g_3) := (1, g_1, g_2, g_3) \in M$.) Then it is immediate that S is birationally equivalent to $\mathcal{F}_m/G$, that is, G is the finite quotient group from $\mathcal{F}_m$ to S.

Let $\lambda(S) := B_2(S) - \rho(S)$. This is called the Lefschetz number of S, and we see that $\lambda(S)$ gives the dimension of the transcendental subspace $T(S) \otimes \mathbb{Q}_\ell$ of $H^2_{et}(S, \mathbb{Q}_\ell)$. Since $\lambda(S)$ is a birational invariant, we have

$$\lambda(S) = \lambda(\mathcal{F}_m/G).$$

This identifies the transcendental subspace $T(S) \otimes \mathbb{Q}_\ell$ with the transcendental space of $H^2(\mathcal{F}_m)$ that is invariant under G. Then using the decomposition of $H^2(\mathcal{F}_m)$, we have

Criterion for G-invariance: *Let $\mathbf{a} = (a_0, a_1, a_2, a_3) \in \mathfrak{A}_m$. Then $V(\mathbf{a})$ is G-invariant if and only if*

$$\prod_{i=1}^{3} g_i^{a_i} = 1 \quad \forall \mathbf{g} = (g_1, g_2, g_3) \in G.$$

Let $\mathfrak{S}_G = \{\mathbf{a} \in \mathfrak{A}_m \mid V(\mathbf{a}) \text{ is } G\text{-invariant}\}$. Then

$$H^2(\mathcal{F}_m)^G \otimes \bar{\mathbb{Q}}_\ell = V_0 \oplus \bigoplus_{\mathbf{a} \in \mathfrak{S}_G} V(\mathbf{a}).$$

Consequently, the transcendental part $T(S) \otimes \mathbb{Q}_\ell$ of $H^2_{et}(S, \mathbb{Q}_\ell)$ can be identified with

$$T(S) \otimes \bar{\mathbb{Q}}_\ell = \bigoplus_{\mathbf{a} \in \mathfrak{S}_G \setminus \mathfrak{B}_m} V(\mathbf{a}).$$

REMARK 4.7. (1) For each k, it is easily checked that the set $\mathfrak{S}_G \setminus \mathfrak{B}_m$ constitutes a single $(\mathbb{Z}/m\mathbb{Z})^\times$-orbit.

(2) We also see that for each value of k there is the unique character $\mathbf{a} \in \mathfrak{S}_G \setminus \mathfrak{B}_m$ of Hodge type $(0,2)$, i.e., $|\mathbf{a}| = 1$. This follows from the fact that S is a K3 surface. For complete tables of characters, see Livné–Schütt–Yui **[26]** (Tables 6 and 7).

DEFINITION 4.8. Here we will recall the definition of Fermat motives. This can be done for Fermat or diagonal hypersurfaces of any dimension (see Shioda **[38]**, also see Gouvêa–Yui [18], and Yui [51]). Here we will focus on Fermat surfaces $\mathcal{F}_m$. Recall that the character group $\hat{M}$ is identified with the set

$$\left\{\mathbf{a}=(a_0,a_1,a_2,a_3)\mid a_i\not\equiv 0 \pmod m,\ \sum_{i=0}^{3}a_i\equiv 0 \pmod m\right\}$$

under the pairing

$$\hat{M}\times M\to\mathbb{Q}(\zeta_m):(\mathbf{a},\mathbf{g})\mapsto \mathbf{a}(\mathbf{g}):=\prod_{i=0}^{3}g_i^{a_i}.$$

The group $(\mathbb{Z}/m\mathbb{Z})^*$ acts on $\hat{M}$ by

$$u\cdot(a_0,a_1,a_2,a_3)=(ua_0,ua_1,ua_2,ua_3).$$

For $\mathbf{a}\in\hat{M}$, define

$$p_{\mathbf{a}}:=\frac{1}{\#M}\sum_{\mathbf{g}\in M}\mathbf{a}(\mathbf{g})^{-1}g=\frac{1}{m^3}\sum_{\mathbf{g}\in M}\mathbf{a}(\mathbf{g})^{-1}\mathbf{g}\in\mathbb{Z}[\frac{1}{m},\zeta_m][M].$$

Further, for the $(\mathbb{Z}/m\mathbb{Z})^*$-orbit of $\mathbf{a}$ denoted by $A=[\mathbf{a}]$ define

$$p_A:=\sum_{\mathbf{a}\in A}p_{\mathbf{a}}\in\mathbb{Z}[\frac{1}{m}][M].$$

Then $p_{\mathbf{a}}$ and p_A are idempotents, that is,

$$p_{\mathbf{a}}\cdot p_{\mathbf{b}}=\begin{cases}p_{\mathbf{a}} & \text{if } \mathbf{a}=\mathbf{b}\\ 0 & \text{if } \mathbf{a}\neq\mathbf{b}\end{cases}$$

and

$$p_A\cdot p_B=\begin{cases}p_A & \text{if } A=B\\ 0 & \text{if } A\neq B\end{cases}.$$

Furthermore, we have the decomposition

$$\sum_{\mathbf{a}\in\hat{M}}p_{\mathbf{a}}=\sum_{A\in O(\hat{M})}p_A=1$$

where $O(\hat{M})$ denotes the set of $(\mathbb{Z}/m\mathbb{Z})^*$-orbits in $\hat{M}$. Then $(H^2(\mathcal{F}_m),p_A)$ defines a motive $\mathcal{M}_A$, called a *Fermat motive.* The motivic Hodge numbers and the motivic Betti numbers for $\mathcal{M}_A$ can be defined. For details, see Gouvêa and Yui [14].

PROPOSITION 4.9. *$\mathcal{F}_m$ decomposes into the sum of Fermat motives*

$$\mathcal{F}_m\otimes\mathbb{Q}_\ell=\oplus_{A\in O(\hat{M})}\mathcal{M}_A.$$

Furthermore, the motivic decomposition of $\mathcal{F}_m$ gives rise to the decomposition of the cohomology group

$$H^2(\mathcal{F}_m,\mathbb{Q}_\ell)=\oplus_{A\in O(\hat{M})}H^2(\mathcal{M}_A).$$

4.4. Galois representations and modularity. In this subsection, we will prove the modularity of all K3 surfaces from Theorem 4.2 and Remark 4.3. It goes back to Weil [**46**] that the Fermat surface $\mathcal{F}_m$ is of CM type in the following sense: Over the cyclotomic field $\mathbb{Q}(\zeta_m)$, the Galois representation of $H^2(\mathcal{F}_m)$ splits into one-dimensional sub-representations corresponding to the $\mathbb{Q}_\ell$-eigenspaces $V_0, V(\mathbf{a})$ $(\mathbf{a} \in \mathfrak{A}_m)$. These sub-representations are associated to Hecke Grössencharacters and can be described in terms of Jacobi sums. Here the Jacobi sums are determined by the characters $\mathbf{a} \in \mathfrak{A}_m$, and Jacobi sums $j(\mathbf{a})$ are algebraic integers in $\mathbb{Q}(\zeta_m)$. The Fermat motive $\mathcal{M}_A$ is defined over the prime field $\mathbb{Q}$ or $\mathbb{F}_p$.

We now study the Galois representations of dimension $\phi(k)$ associated to the transcendental parts of the K3 surfaces from Theorem 4.1 and Remark 4.3. Here k runs through the values in Theorem 4.2 and Remark 4.3, and $\phi(k)$ denotes the Euler function.

First we recall the result of Nikulin [**30**], and others. Most of the following statements hold true in full generality, but we only state the special case of minimal rank of the transcendental lattice which is relevant to our issues.

THEOREM 4.10. *Let S be an algebraic K3 surface discussed in Section 4.1. Let H_S be a finite cyclic group of order k. Assume that $\phi(k) = rank(T(S))$. Thus k takes the values listed in Theorem 4.2 and Remark 4.3.*

*(1) For those values of k, the ring $\mathbb{Z}[\zeta_k]$ is a PID, i.e. is of class number one. (See also Masley–Montgometry [**28**].)*

(2) The representation of $H_S = \langle g_k \rangle$ in $T(S) \otimes \mathbb{Q}$ is isomorphic to a direct sum of irreducible representations of H_S of dimension one.

(3) Let $\Phi_k(x)$ denote the k-th cyclotomic polynomial. Regard $T(S) \otimes \mathbb{Q}$ as a $\mathbb{Z}[\langle g_k \rangle]$-module via the natural action of g_k on $T(S)$. Then $T(S) \otimes \mathbb{Q}$ is a torsion free $\mathbb{Z}[\langle g_k \rangle]/\langle \Phi_k(g_k) \rangle$-module. Identifying $\mathbb{Z}[\langle g_k \rangle]/\langle \Phi_k(g_k) \rangle$ with the ring of integers $\mathbb{Z}[\zeta_k]$, we derive an isomorphism

$$T(S) \otimes \mathbb{Q} \simeq \mathbb{Q}[\zeta_k] \quad \textit{as } \mathbb{Z}[\zeta_k]\textit{-modules.}$$

We also recall the result of Zarhin [**52**] on the Hodge groups of complex K3 surfaces. Recall that $T(S)$ is the orthogonal complement of $NS(S)$ with respect to the cup product which we denote by $< \cdot, \cdot >$.

THEOREM 4.11. *Let S be a complex K3 surface, and let $\mathrm{Hdg} \subset \mathrm{Aut}(\mathrm{H}^2(\mathrm{S}, \mathbb{Q}))$ be the Hodge group. Then*

(1) $T(S)$ is an irreducible Hdg*-module,*

(2) $E := \mathrm{End}_{\mathrm{Hdg}}(\mathrm{T(S)})$ is a field over $\mathbb{Q}$. E has an involution induced by $< \cdot, \cdot >$ with totally real fixed field E_0, and either $E = E_0$ or E is a totally complex quadratic extension of E_0, and $< \cdot, \cdot >$ induces a symmetric, respectively, Hermitian form $\Phi : T(S) \times T(S) \to E$,

(3) $\mathrm{Hdg} = \mathrm{SO}(\mathrm{T(S)}, \Phi)$, respectively, $U(T(S), \Phi)$,

(4) Let $U(S)$ be the image of $\mathrm{Aut(S)}$ in $\mathrm{Aut(T(S))}$. Then $U(S)$ is contained in the roots of unity of E, and hence is cyclic of order n for some n, and

$$\phi(n) | [E : \mathbb{Q}] | \dim_{\mathbb{Q}}(\mathrm{T(S)}).$$

We will elaborate on this theorem for our examples of K3 surfaces.

COROLLARY 4.12. *Let S be one of our K3 surfaces (in Theorem 4.1 and Remark 4.3). Then $[E:\mathbb{Q}] = \phi(k) = \dim_{\mathbb{Q}}(\mathrm{T(S)})$, so that E is a cyclotomic field, and hence a CM field, over $\mathbb{Q}$ of degree $\phi(k)$.*

PROOF. We know that S is realized as a finite Fermat quotient. The field E that corresponds to the transcendental cycles is isomorphic to $T(S) \otimes \mathbb{Q} \simeq \mathbb{Q}(\zeta_k)$ and it is generated by Grössencharacters of $\mathbb{Q}(\zeta_k)$. This is because the Galois representation defined by $T(S)$ is a sum of one-dimensional representations induced from the Jacobi sum Grössencharacters corresponding to the unique character $\mathbf{a}$ of Hodge type $(0,2)$. (See Livné–Schuett–Yui **[26]** the first characters listed in Tables 6 and 7 for each value of k.) □

Now we will study the Galois representations associated to our K3 surfaces. The main point is the comparison of the Hodge structure and some piece of the Galois representation.

THEOREM 4.13. *Let S be one of our K3 surfaces (in Theorem 4.1 and Remark 4.3). Then for each value of k, the Galois representation associated to $T(S)$ has dimension $\phi(k)$, and is irreducible over $\mathbb{Q}$. In fact, this $\mathrm{Gal}(\overline{\mathbb{Q}}/\mathbb{Q})$ representation is induced from the Jacobi sum Grössencharacters corresponding to the unique character $\mathbf{a} \in \mathfrak{G}_G \setminus \mathfrak{B}_m$ of Hodge type $(0,2)$.*

PROOF. Since $H^2_{et}(S, \mathbb{Q}_\ell)$ is a submotive of $H^2(\mathcal{F}_m)$, the $\mathrm{Gal}(\overline{\mathbb{Q}}/\mathbb{Q}(\zeta_m))$ representation defined by $T(S)$ is a sum of one dimensional representations which are simply transitively permuted by $\mathrm{Gal}(\mathbb{Q}(\zeta_m)/\mathbb{Q})$. The claim follows. The existence of the unique character $\mathbf{a} \in \mathfrak{G}_G \setminus \mathfrak{B}_m$ of Hodge type $(0,2)$ uses the fact that S is K3 surface. □

REMARK 4.14. This motivic modularity result is obtained by invoking the automorphic induction of Arthur and Clozel **[1]** and Rajan **[32]**. The first result we proved is the $GL(1)$-modularity for a CM motive over $\mathbb{Q}(\zeta_k)$. The automorphic induction then takes this $GL(1)$ automorphic representation to a $GL(\varphi(k))$ automorphic cuspidal representation over $\mathbb{Q}$ having the same L-function (for the standard representation). Thus this justifies the Galois representation associated to $T(S)$ to be automorphic.

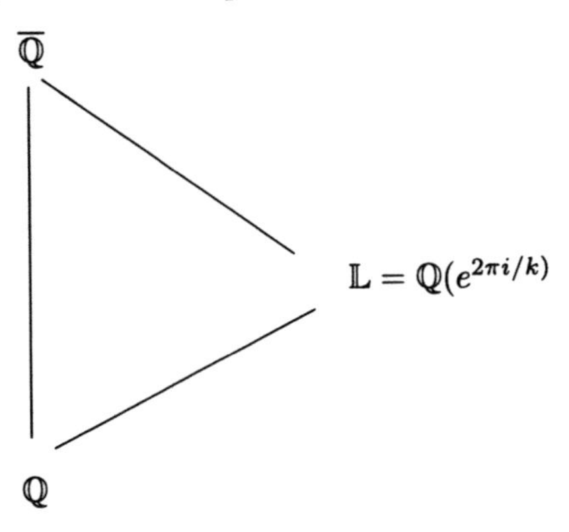

4.5. Arithmetic mirror symmetry. In this section, we will show that for all values except for $k = 3$, a mirror K3 surfaces exist. For some of them, we interpret mirror symmetry arithmetically. In the literature, there are several variants of mirror symmetry for K3 surfaces. Here we employ the notion of mirror symmetry for lattice polarized K3 surfaces introduced by Dolgachev [**13**], based on the Arnold strange duality. There is also a mirror symmetry for K3 surfaces due to Nikulin [**31**] which we will discuss below.

Let $\Lambda_{K3} = U^3 \oplus E_8(-1)^2$ be the K3 lattice. Let $M \subset \Lambda_{K3}$ be a fixed primitive sublattice. A lattice polarization on S is given by a primitive lattice embedding $M \hookrightarrow NS(S)$ whose image contains a pseudo-ample class.

DEFINITION 4.15 (Dolgachev [**13**]). Let S be an algebraic lattice polarized K3 surface. Then a K3 surface $S^\vee$ is called a *mirror* of S if

$$T(S) = U_2 \oplus NS(S^\vee).$$

Comparing the ranks of the Néron–Severi lattices, $S^\vee$ is a mirror of S if

$$\text{rank}_{\mathbb{Z}}\, T(S) = 22 - \text{rank}_{\mathbb{Z}}\, NS(S) = 2 + \text{rank}_{\mathbb{Z}}\, NS(\hat{S}),$$

that is,

$$\text{rank}_{\mathbb{Z}}\, NS(S) + \text{rank}_{\mathbb{Z}}\, NS(\hat{S}) = 20.$$

There are several variants of mirror symmetry for K3 surfaces in the literature. The above version of mirror symmetry for lattice polarized K3 surfaces is due to Dolgachev [**13**], based on the Arnold strange duality.

Usually mirror symmetry is exhibited for families of K3 surfaces. Here we are only interested in the existence of a mirror K3 surface. Then we will investigate arithmetic properties of mirror K3 surfaces.

PROPOSITION 4.16. *For each of the K3 surfaces corresponding to the values of k except for $k = 3$, a mirror K3 surface exists in the sense of Dolgachev.*

Furthermore, except for $k = 3, 7, 11, 17, 19$, there is a mirror K3 surface which is a Delsarte surface. In particular, this mirror K3 surface is again of CM type and is motivically modular, that is, $L(T(S^\vee) \otimes \mathbb{Q}_\ell, s)$ is modular/automorphic.

5. Calabi–Yau threefolds of Borcea–Voisin type

5.1. K3 surfaces with non-symplectic involution. Let (S, σ) be a K3 surface with a non-symplectic involution σ such that the induced map

$$\sigma^* : H^{2,0}(S) \to H^{2,0}(S) \quad \text{satisfies} \quad \sigma^*(\omega_S) = -\omega_S.$$

Decompose $H^2(S, \mathbb{C})$ into the $(+)$- and $(-)$-eigenspaces under the action of $\sigma^* : H^2(S, \mathbb{C}) \to H^2(S, \mathbb{C})$:

$$H^2(S, \mathbb{C}) = H^2(S, \mathbb{C})^+ \oplus H^2(S, \mathbb{C})^-.$$

Set

$$H^2(S, \mathbb{Z})^+ := H^2(S, \mathbb{C})^+ \cap H^2(S, \mathbb{Z})$$

and

$$H^2(S, \mathbb{Z})^- := H^2(S, \mathbb{C})^- \cap H^2(S, \mathbb{Z}).$$

Let

$$r := \text{rank}_{\mathbb{Z}} H^2(S, \mathbb{Z})^+.$$

Then the primitive sublattices $H^2(S, \mathbb{Z})^+$ and $H^2(S, \mathbb{Z})^-$ of $H^2(S, \mathbb{Z})$ have signatures $(1, r-1)$ and $(2, 20-r)$, respectively. We may identify $H^2(S, \mathbb{Z})^+$ with the Néron–Severi group $NS(S)$ and $H^2(S, \mathbb{Z})^-$ with the group $T(S)$ of transcendental cycles, respectively.

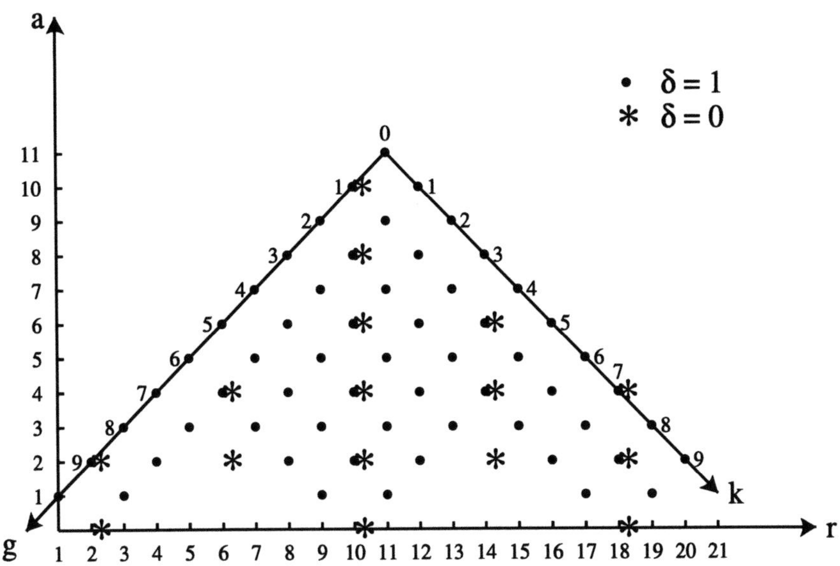

FIGURE 1. Nikulin's Pyramid

Nikulin [**31**] has classified such pairs (S, σ) of K3 surfaces S with non-symplectic involutions σ, up to deformation. For any lattice L, let L^* denote the dual lattice $\mathrm{Hom}(L, \mathbb{Z})$.

THEOREM 5.1. *There are 75 deformation classes of pairs (S, σ) of K3 surfaces S with non-symplectic involutions σ, and they are completely determined by the triplet of integers*

$$(r, a, \delta)$$

where r is as above, a is the integer determined by

$$(H^2(S, \mathbb{Z})^+)^*/H^2(S, \mathbb{Z})^+ \simeq (\mathbb{Z}/2\mathbb{Z})^a,$$

and δ is the invariant such that $\delta = 0$ if a quadratic form q on $(H^2(S, \mathbb{Z})^+)^*$ has integer values, and 1 otherwise. Here the intersection pairing on $H^2(S, \mathbb{Z})^+$ gives rise to a quadratic form q with values in $\mathbb{Q}$.

Nikulin [**31**] also determined the fixed locus of S under the action of σ.

THEOREM 5.2. *Let (S, σ) be a pair of K3 surface with non-symplectic involution σ. Let S^σ be the fixed locus of S under σ. Then*

(1) If $(r, a, \delta) \neq (10, 10, 0), (10, 8, 0)$, then

$$S^\sigma = C_g \cup L_1 \cup \ldots \cup L_k \quad \textit{(disjoint union)}$$

where C_g is a genus $g(\geq 0)$ curve, and $L_i (i = 1, \cdots, k)$ are rational curves.

(2) If $(r, a, \delta) = (10, 10, 0)$, then $S^\sigma = \emptyset$.

(3) If $(r, a, \delta) = (10, 8, 0)$, then $S^\sigma = C_1 \cup \bar{C}_1$ (disjoint union) where C_1 and $\bar{C}_1$ are elliptic curves.

REMARK 5.3. In the generic case (1) in Theorem 5.2, put

$$N := \text{the number of components of } S^\sigma = 1 + k$$

and

$$N' := \text{the sum of genera of components of } S^\sigma = g.$$

Note that

$$g = 11 - \frac{1}{2}(r+a), \quad N = 1 + k = 1 + \frac{1}{2}(r-a).$$

5.2. Calabi–Yau threefolds of Borcea–Voisin type. Now we will construct Calabi–Yau threefolds of Borcea–Voisin type following Borcea [**5**] and Voisin [**44**].

Let (E, ι) be an elliptic curve with a non-symplectic involution ι such that the induced map

$$\iota^* : H^{1,0}(E) \to H^{1,0}(E) \quad \text{satisfies} \quad \iota^*(\omega_E) = -\omega_E.$$

Here ω_E is a non-zero holomorphic 1-form on E.

Let (S, σ) be a pair of a K3 surface S with a non-symplectic involution σ such that the induced map

$$\sigma^* : H^{2,0}(S) \to H^{2,0}(S) \quad \text{satisfies} \quad \sigma^*(\omega_S) = -\omega_S.$$

DEFINITION 5.4. We will define Calabi–Yau threefolds of Borcea–Voisin type by carrying out construction in the generic case (1) in Theorem 5.2.

Take the product $E \times S$. Then the product $\iota \times \sigma$ is an involution on $E \times S$ such that the induced map

$$(\iota \times \sigma)^* : H^{3,0}(E \times S) \to H^{3,0}(E \times S)$$

is the identity map. Write

$$E^\iota = \{P_1, P_2, P_3, P_4\}$$

and

$$S^\sigma = \{C_g, L_1, \cdots, L_k\} := \{C_1, C_2, \cdots, C_N\} \quad \text{with } N = 1 + k$$

(where we use C_1 for C_g and $C_2, \cdots, C_N$ for $L_1, \cdots, L_k$ to unify the notation.) Then the fixed locus of $\iota \times \sigma$ consists of

$$\{P_i\} \times C_j \quad (i = 1, \cdots, 4; j = 1, \cdots, N).$$

The involution $\iota \times \sigma$ lifts naturally to an involution on the blow-up of $E \times S$ along these $4N$ curves. The quotient $E \times S/\iota \times \sigma$ and its crepant resolution $E \times \widetilde{S/\iota} \times \sigma$ is our Calabi–Yau threefold of Borcea–Voisin type, and will be denoted by

$$X = X(r, a, \delta).$$

Note that the exceptional divisors on X are 4 copies of ruled surfaces

$$S^\sigma \times \mathbb{P}^1 := (C_g \times \mathbb{P}^1) \cup (L_1 \times \mathbb{P}^1) \cup \cdots \cup (L_k \times \mathbb{P}^1).$$

THEOREM 5.5 (Borcea [**5**], Voisin [**44**]). *The Hodge numbers of* $X = X(r, a, \delta)$ *are given by*

$$\begin{aligned} h^{1,1}(X) &= 5 + 3r - 2a = 11 + 5N - N' \\ h^{2,1}(X) &= 65 - 3r - 2a = 11 + 5N' - N \end{aligned}$$

and

$$E(X) = 12(r - 10) = 12(N - N').$$

The topological mirror symmetry conjecture holds for Calabi–Yau threefolds of Borcea–Voisin type: $X(r,a,\delta)$ *and* $X(20-r,a,\delta)$ *form a mirror symmetric pair in the sense that the Hodge numbers are interchanged :*

$$h^{1,1}(X(r,a,\delta)) = h^{2,1}(X(20-r,a,\delta))$$

and

$$h^{2,1}(X(r,a,\delta)) = h^{1,1}(X(20-r,a,\delta)).$$

In other words, the topological mirror symmetry for Calabi–Yau threefolds of Borcea–Voisin type interchanges r *and* $20-r$*, or equivalently,* N *and* N'*, and is inherited from mirror symmetry of the K3 surface components.*

REMARK 5.6. Note that

$$B_3(X) = 2(1+(20-r)+4g) = 2(12+5N'-N) \geq 4.$$

So these Calabi–Yau threefolds of Borcea–Voisin type are all nonrigid, and mirror families do exist. In fact, mirrors of Calabi–Yau threefolds are induced from mirrors of K3 surface components.

5.3. Choices for K3 surfaces. Since any elliptic curve E defined over $\mathbb{Q}$ is modular by Theorem 3.2, the modularity/automorphy of our Calabi-Yau threefolds $X = X(r,a,\delta)$ depends on the modularity/automorphy of the K3 surface component S. We ought to choose appropriate K3 surfaces for S in order to get something provable. Our choice of K3 surfaces is from singularity theory.

We choose K3 surfaces from Yonemura's hypersurface simple K3 singularities. The notion of simple K3 singularity was introduced by Watanabe and Ishii in 1988. The reader is referred to the article of Yonemura [**49**] or the book by Ishii [**19**] for the definition and characterization of hypersurface simple K3 singularities. A three-dimensional normal isolated singularity (Z,z) is called a *simple K3 singularity* if and only if a $\mathbb{Q}$-factorial minimal model $h : \tilde{Z} \to Z$ has the irreducible exceptional divisor E which is isomorphic to a normal K3 surface (i.e., a normal surface whose minimal resolution is a K3 surface). Reid [**33**] and Yonemura [**49**] classified hypersurface simple K3 singularities using toric method.

THEOREM 5.7. *There are* 95 *admissible weights* (w_0,w_1,w_2,w_3) *of hypersurface simple K3 singularities defined by non-degenerate polynomials* $F(x_0,x_1,x_2,x_3)$ *in weighted projective* 3*-spaces* $\mathbb{WP}^3(w_0,w_1,w_2,w_3)$ *over* $\mathbb{Q}$*.*

Here an admissible weight (w_0,w_1,w_2,w_3) *means that* $w_i \in \mathbb{N}$ *for each* i *with* $gcd(w_i,w_j,w_k)=1$ *for distinct* i,j,k *and* $\sum_{i=0}^{3} w_i = deg(F)$*.*

EXAMPLE 5.8. For $F(x_0,x_1,x_2,x_3)$, the zero locus $\{F=0\} \subset \mathbb{C}^4$ has a simple K3 singularity at the origin. For each weight (w_0,w_1,w_2,w_3), there are infinitely many choices for coefficients, i.e., families i for $F(x_0,x_1,x_2,x_3)$. In the examples below, we choose F such that all non-zero coefficients are normalized to 1.

Among the 95 admissible weights, there are 14 weights given by diagonal hypersurfaces in $\mathbb{WP}^3(w_0,w_1,w_2,w_3)$, meaning that defining equations $F(x_0,x_1,x_2,x_3)$ consist of 4 monomials $x_i^{d_i}$ $(0 \leq i \leq 3)$ where $d := \deg(F)$. For the remaining cases, defining equations $F(x_0,x_1,x_2,x_3)$ have more than 4 monomial's with mixed (non-diagonal) terms. We list some examples here. #1 is the only diagonal example; all other are non-diagonal examples.

#1: weight $(1,1,1,1)$ degree 4

$$F = x_0^4 + x_1^4 + x_2^4 + x_3^4.$$

#15: weight $(5,4,3,3)$ degree 15

$$F = x_0^3 + x_1^3 x_2 + x_1^3 x_3 + x_2^5 + x_3^5.$$

#53: weight $(6,5,4,3)$ degree 18

$$F = x_0^3 + x_1^3 x_3 + x_1^2 x_2^2 + x_0 x_2^3 + x_2^3 x_3^2 + x_3^6.$$

#62: weight $(8,5,4,3)$ degree 20

$$F = x_0^2 x_3 + x_1^4 + x_1 x_3^5 + x_2^5 + x_1 x_3^4 + x_2^2 x_3^4.$$

#84: weight $(9,7,6,5)$ degree 27

$$F = x_0^3 + x_0 x_2^3 + x_1^3 x_2 + x_1 x_3^4 + x_2^2 x_3^3.$$

#95: weight $(7,5,3,2)$ degree 17

$$F = x_0^2 x_2 + x_1^3 x_3 + x_1 x_2^4 + x_1 x_3^6 + x_2^5 x_3 + x_2 x_3^7.$$

Among these 95 K3 surfaces in Yonemura's list of weights and defining equations, we ought to find those with the required involutions.

THEOREM 5.9. *(1) Among the* 95 *K3 surfaces, at least the* 92 *have the required non-symplectic involution* σ.

(2) These 92 *pairs* (S,σ) *realize at least* 40 *triplets* (r,a,δ) *of Nikulin.*

(3) Furthermore, the exactly 86 *of them can be deformed to Delsarte K3 surfaces.*

Consequently for these 86 *weights, the K3 surface* S *can be chosen of CM type (that is,* S *is realized as a finite Fermat quotient).*

REMARK 5.10. (1) we were not able to find the required involution for #53 that sends ω_S to $-\omega_S$. Altogether there are 3 such K3 surfaces in Yonemura's list, namely, #15, #53 and #54. But we have not tried other methods for finding the required involutions for these 3. Hence the assertion of at least the 92.

(2) At least the 40 triplets of Nikulin were realized based on our calculation of the invariants r and a. We have not yet calculated δ so the number 40 may be replaced by a larger number once we have done that.

(3) For #95 it has the required involution $\sigma(x_0) = -x_0$, but the defining equation cannot be deformed to a quasi-smooth hypersurface in four monomials. The referee has suggested that the condition of quasi-smoothness may not be necessary, but just suffice to have a K3 resolution. I have not yet studied this suggestion in detail.

Altogether there are 6 such K3 surfaces in Yonemua's list, namely, #85, #90, #91, #93, #94 and #95.

The referee has also suggested that for defining equations with six terms, e.g., for #93 and #95 there may be an elliptic fibration with section, which gives rise to a non-symplectic involution by -1 on the fibers. We have not yet explored this suggestion, however.

(4) The assertion (3) may be interpreted as the existence of a CM point in the moduli space.

Idea for proof of Theorem 5.9: Yonemura obtained his defining equations using toric constructions. For each weight (w_0, w_1, w_2, w_3), Yonemura wrote down a defining equation using all extremal points of the convex hull determined by the weight. By case by case analysis, we find the required involutions for the 92 families out of the 95, but not yet for the remaining 3. Thus at least the 92 families have the required involutions.

For the second assertion:

• We may remove some monomials keeping in mind Yonemura's condition:

($\star$) *for each* i, $0 \leq i \leq 3$, *the defining equation must contain a monomial of the form* x_i^n *or* $x_i^n x_j$ $(i \neq j)$ *with non-zero coefficient.*

So if there is a monomial of the form $x_i^n x_j^m$ with $n, m > 1$, we can remove it from the defining equation. We must check that even after removing monomials, a defining equation must remain quasi-smooth (i.e., $\{F = 0\}$ has a simple K3 singularity at the origin $0 \in \mathbb{C}^4$, and this is the only singularity in the interior of the polytope). The quasi-smoothness and the existence of a (unique) holomorpic 2-form (up to scalar) are satisfied, we will have $K3$ surfaces. Yonemura's condition is necessary to obtain a K3 surface.

• After removing monomials, the action of the involution on a (new) holomorphic 2-form ω_S is that it should send ω_S to $-\omega_S$.

• Also the resolution picture should be invariant under deformation, that is, the defining hypersurface preserves the configuration of singularities (types and the number of singularities) before and after removing certain monomials. This is because, the invariants r and a should remain unchanged under deformation.

• For the assertion of CM type, we use a similar argument as the proof of Theorem 4.13. Since the defining equation for S has only four monomials, it is a Delsarte surface. Hence it is a finite quotient of a surface in $\mathbb{P}^3$ with a (homogeneous) diagonal equation by some diagonal action of roots of unity. We may assume without loss of generality that a diagonal surface is a Fermat surface. Then over an appropriate cyclotomic field the Galois representation on the transcendental cycles is a sum of one-dimensional representations coming form Jacobi sum Grössencharacters, on which the absolute Galois group permutes transitively. Thus, S has CM by a cyclotomic field $\mathbb{K} = \mathbb{Q}(\zeta_t)$ for some t, where ζ_t denotes a primitive t-th root of unity.

EXAMPLE 5.11. Here we give an example describing singularities and their resolutions. Consider the K3 surface #60 in Yonemura's list. Dropping several monomials, we choose the equation

$$S_0 : x_0^2 x_1 + x_1^3 + x_1 x_2^3 + x_3^{16} = 0 \subset \mathbb{WP}^3(7, 6, 4, 1).$$

The involution is given by $\sigma(x_0) = -x_0$. We see that S_0 is quasi-smooth and the minimal resolution S is a K3 surface. S_0 has three singularities Q_1, Q_2 and Q_3 given as follows:

Singularity	Type	Exceptional divisors
$Q_1 = (1:0:0:0)$	$A_{7,6}$	$E_1 + E_2 + E_3 + E_4 + E_5 + E_6$
$Q_2 = (0:0:1:0)$	$A_{4,3}$	$E_7 + E_7 + E_8$
$Q_3 = (0:-1:1:0)$	$A_{2,1}$	E_{10}

Every singularity is fixed by σ, and E_2, E_4, E_6 and E_8 are ramified under σ (acting on S).

There are two curves on S_0 fixed by σ, namely, C' defined by $x_0 = 0$ and L' defined by $x_3 = 0$. C' has genus 3 and L' is rational. Their strict transforms

$C = (C_3)$ and L on S are ramified under σ. We have

$$S^\sigma = C_3 \cup E_2 \cup E_4 \cup E_6 \cup E_8 \cup L.$$

Hence $g = 3$ and $k = 5$, which implies that $r = 13$ and $a = 3$.

5.4. Birational models defined over $\mathbb{Q}$ for Calabi–Yau threefolds of Borcea–Voisin type. We will illustrate by examples that Calabi–Yau threefolds of Borcea–Voisin type do have birational models which are hypersurfaces defined over $\mathbb{Q}$ in weighted projective 4-spaces. Even though E and S are defined by hypersurfaces with coefficients in $\mathbb{Q}$, it is not clear if the product $E \times S$ and its finite quotients are also defined by hypersurfaces over $\mathbb{Q}$. To facilitate this, we will use the so-called *twist* map to obtain birational models. The reader is referred to Goto–Kloosterman–Yui [**16**] for details.

Here we define weighted elliptic curves:

$$E_2 : y_0^2 = y_1^4 + y_2^4 \subset \mathbb{WP}^2(2,1,1),$$
$$E_3 : y_0^2 = y_1^3 + y_2^6 \subset \mathbb{WP}^2(3,2,1).$$

We see that E_2 and E_3 are elliptic curves with complex multiplication (CM) by $\mathbb{Q}(\sqrt{-1})$ and $\mathbb{Q}(\sqrt{-3})$, respectively. The involution ι on E_2 and E_3 is given by $\iota(y_0) = -y_0$, and the fixed points consists of 4 points, $E_i^\iota = \{P_1, P_2, P_3, P_4\}$ for $i = 2, 3$.

(Note that any elliptic curve $E \subset \mathbb{P}^2$ with CM can be embedded in $\mathbb{WP}^2(2,1,1)$ with equation

$$E : x_0^2 = x_2(x_1^3 + ax_0x_2 + bx_2^3) \quad \text{with } a, b, \in \mathbb{Q}.$$

For the subsequent discussion on CM type Calabi–Yau threefolds of Borcea–Voisin type, we need elliptic curves with CM.)

EXAMPLE 5.12. Suppose that S is defined by a hypersurface

$$x_0^2 = f(x_1, x_2, x_3) \subset \mathbb{WP}^3(w_0, w_1, w_2, w_3)$$

with an admissible weight (w_0, w_1, w_2, w_3). Let σ be an involution on S defined by $\sigma(x_0) = -x_0$.

If w_0 is odd, then with $E = E_2 \in \mathbb{WP}^2(2,1,1)$. $X = \widetilde{E_2 \times S/\iota \times \sigma}$ is birational to a hypersurface defined over $\mathbb{Q}$:

$$z_0^4 + z_1^4 - f(z_2, z_3, z_4) = 0 \subset \mathbb{WP}^4(w_0, w_0, 2w_1, 2w_2, 2w_3).$$

We include a variant of the above example. Suppose that S is defined by a hypersurface

$$x_0^2 x_i + f(x_1, x_2, x_3) = 0 \quad \text{for some } i \neq 0$$

with an admissible weight (w_0, w_1, w_2, w_3). Let σ be an involution on S defined by $\sigma(x_0) = -x_0$.

If w_0 is odd, we need some w_i $(i > 0)$ to be odd. Take $E = E_2$, then $X = \widetilde{E_2 \times S/\iota \times \sigma}$ is birational to a hypersurface defined over $\mathbb{Q}$:

$$z_{i+1}(z_0^4 + z_1^4) + f(z_2, z_2, z_4) = 0 \quad \text{for some } i \neq 0.$$

If w_0 is even but not divisible by 3, take $E = E_3$, then $X = \widetilde{E_3 \times S/\iota \times \sigma}$ is birational to a hypersurface over $\mathbb{Q}$:

$$z_{i+1}(z_0^3 + z_1^6) + f(z_2, z_3, z_4) = 0 \quad \text{for some } i \neq 0.$$

EXAMPLE 5.13. Suppose that S is defined by a hypersurface

$$x_0^2 + f(x_1, x_2, x_3) = 0 \subset \mathbb{WP}^3(w_0, w_1, w_2, w_3)$$

with an admissible weight (w_0, w_1, w_2, w_3). Let σ be an involution on S defined by $\sigma(x_0) = -x_0$.

If w_0 is even but not divisible by 3, then with $E = E_3 \in \mathbb{WP}^2(3,2,1)$, $X = \widetilde{E_3 \times S/\iota \times \sigma}$ is birational to a hypersurface defined over $\mathbb{Q}$:

$$z_0^3 + z_1^6 + f(z_2, z_3, z_4) = 0 \subset \mathbb{WP}^4(2w_0, w_0, 3w_1, 3w_2, 3w_3).$$

Once we have hypersurface models in weighted projective 4-spaces for Calabi–Yau threefolds of Borcea–Voisin type, we can describe also their singular loci in terms of hypersurfaces in weighted projective 2-spaces. Here are examples.

EXAMPLE 5.14. Let $E = E_2$ and S_0 be a non-smooth K3 surface

$$S_0 : x_0^2 = x_1^5 + x_2^5 + x_3^{10} \subset \mathbb{WP}^3(5,2,2,1)$$

with singular locus

$$\{(0 : x_1 : x_2 : 0) \,|\, x_1^5 + x_2^5 = 0\} = \{Q_1, Q_2, Q_3, Q_4, Q_5\}$$

where every Q_i is a cyclic quotient singularity of type A_1.

Let C' be the curve on S_0 defined by $x_0 = 0$, that is,

$$C' : x_1^5 + x_2^5 + x_3^{10} = 0 \subset \mathbb{WP}^2(2,2,1).$$

Since $\mathbb{WP}^2(2,2,1) \simeq \mathbb{WP}^2(1,1,1)$, C' is identified with

$$C' : x_1^5 + x_2^5 + x_3^5 = 0 \subset \mathbb{P}^2$$

which is a smooth curve of genus $g = 6$.

Let L' be the curve on S_0 defined by $x_3 = 0$, that is,

$$L' : x_0^2 = x_1^5 + x_2^5 \subset \mathbb{WP}^2(5,2,2).$$

Since $\mathbb{WP}^2(5,2,2) \simeq \mathbb{WP}^2(5,1,1)$, L' is identified with

$$L' : x_0 = x_1^5 + x_2^5 \subset \mathbb{WP}^2(5,1,1)$$

and hence L' is a rational curve.

We see that

$$C' \cap L' = \{Q_1, Q_2, Q_3, Q_4, Q_5\}$$

where Q_i, $(1 \le i \le 5)$ are cyclic quotient singularities (in $\mathbb{Q}(e^{2\pi i/5})$ of type A_1.

Let S be the minimal resolution of S_0. The involution σ is lifted to S. Let C_6 and L_1 be the strict transforms of C' and L' to S, respectively. Let $\mathcal{E}_1, \cdots, \mathcal{E}_5$ be the exceptional divisors on S arising from singularities $Q_1, \cdots, Q_5$, respectively. Then C_6 and L_1 are fixed by σ but the exceptional divisors $\mathcal{E}_i$, $(i = 1, \cdots, 5)$ are not. Hence

$$S^\sigma = C_6 \cup L_1$$

and hence we get $g = 6$, $k = 1$ (or $r = 6$, $a = 4$).

The quotient $E_2 \times S/\iota \times \sigma$ has singularities along $E_2^\iota \times S^\sigma$. Let X be a crepant resolution of the quotient threefold $E_2 \times S/\iota \times \sigma$. Then X is a Calabi–Yau threefold of Borcea–Voisin type corresponding to the triplet $(6,4,0)$ and its exceptional divisors are four copies of the ruled surfaces $(C_6 \times \mathbb{P}^1) \cup (L_1 \times \mathbb{P}^1)$. Furthermore, X has a (quasi-smooth) model

$$z_0^4 + z_1^4 = z_2^5 + z_3^5 + z_4^{10} \subset \mathbb{WP}^4(5,5,4,4,2).$$

The Hodge numbers of X are given by

$$h^{1,1}(X) = 15,\ h^{2,1}(X) = 39.$$

EXAMPLE 5.15. Let

$$S_0 : x_0^2 = x_1^3 x_3 + x_2^7 + x_2^{28} \subset \mathbb{WP}^3(14, 9, 4, 1).$$

Then S_0 has the involution σ given by $\sigma(x_0) = -x_0$. The surface S_0 is a non-smooth K3 surface with two singularities

$$Q := (0:1:0:0) \quad \text{of type } A_{9,8} \quad \text{and} \quad R := (1:0:1:0) \quad \text{of type } A_{2,1}.$$

Let C' be the curve on S_0 defined by

$$C' := \{x_0 = 0\} \cap S_0 : x_1^3 x_3 + x_2^7 + x_3^{28} = 0 \subset \mathbb{WP}^2(9, 4, 1).$$

Then C' has degree 28 and is a quasi-smooth curve with singularity Q. No other curves defined by $x_i = 0$ $(i \neq 0)$ are fixed by involution σ. Let S be the minimal resolution of S_0. Then S is a smooth K3 surface and the involution σ is lifted to S. Let C_6 be the strict transform of C' to S. It has genus 6. Let $\mathcal{E}_i$, $1 \leq i \leq 8$ be exceptional divisors arising from the singularity Q. Let $\mathcal{E}_9$ be the exceptional divisor arising from R. Then $\mathcal{E}_{2i}$, $1 \leq i \leq 4$ and $\mathcal{E}_9$ are fixed by σ but others are not. Put $L_i := \mathcal{E}_{2i}$, $1 \leq i \leq 4$ and $L_5 := \mathcal{E}_9$. Then

$$S^\sigma = C_6 \cup L_1 \cdots \cup L_5.$$

So $g = 6$, $k = 5$, equivalently $r = 10$, $a = 0$.

Now let $E_3 : y_0^2 = y_1^3 + y_2^6 \subset \mathbb{WP}^2(3, 2, 1)$. Then the quotient threefold $E_3 \times S/\iota \times \sigma$ has singularities along $E_3^\iota \times S^\sigma$ where $E_3^\iota = \{P_1, P_2, P_3, P_4\}$. Let $X := \widetilde{E_3 \times S/\iota \times \sigma}$ be the crepant resolution of the quotient. Then X has a hypersurface model

$$z_0^3 + z_1^6 = z_2^3 z_4 + z_3^7 + z_4^{28} \subset \mathbb{WP}^4(28, 14, 27, 12, 3).$$

The exceptional divisors on X are four copies of

$$(C_6 \times \mathbb{P}^1) \cup (L_1 \times \times \mathbb{P}^1) \cup \cdots \cup (L_6 \times \mathbb{P}^1).$$

The Hodge numbers of X are given by

$$h^{1,1}(X) = 35,\ h^{2,1}(X) = 35.$$

PROPOSITION 5.16 (Borcea [4]). *Let $X = E \times \widetilde{S/\iota} \times \sigma$ be a Calabi–Yau threefold of Borcea–Voisin type. Then X is of CM type if and only if E and S are both of CM type.*

There exist CM type Calabi–Yau threefolds of Borcea–Voisin type.

By Theorem 5.9(3), there are subsets of equations defining quasi-smooth hypersurfaces which are of Delsarte or Fermat type, with non-symplectic involution σ. With this choice of equations, S is of CM type.

REMARK 5.17. The moduli spaces of these Calabi–Yau threefolds are Shimura varieties. The proposition shows the existence of a CM point in the moduli space.

5.5. Motivic modularity/automorphy.

DEFINITION 5.18. (1) Let (S, σ) be one of the 86 pairs of $K3$ surface with a non-symplectic involution σ. Then S is of CM type. Let $\mathcal{F}_m$ be a Fermat surface that dominates S. Then S is birational to $\mathcal{F}_m/G$ for some finite group G. By applying the argument of Subsection 4.3 (with the notation remaining in force), we can identify $T(S) \otimes \mathbb{Q}_\ell$ with a specific Fermat motive. Let $\mathcal{M}_A$ denote the Fermat motive corresponding to $A = [\mathbf{a}]$ where $\mathbf{a}$ is the unique character in $\mathfrak{S}_G \setminus \mathfrak{B}_m$ of Hodge type $(0, 2)$. We call $\mathcal{M}_A \otimes \mathbb{Q}_\ell \subset H^2_{et}(S, \mathbb{Q}_\ell)$ the *K3-motive*, and denoted it by $\mathcal{M}_S$.

(2) We will call the submotive $H^1(E, \mathbb{Q}_\ell) \otimes (\mathcal{M}_S \otimes \mathbb{Q}_\ell)$ of $H^3_{et}(X, \mathbb{Q}_\ell)$ the *Calabi-Yau motive* of X, and denoted by $\mathcal{M}_X$.

THEOREM 5.19. *Let (S, σ) be one of the 86 pairs of K3 surfaces represented by a Delsarte surface in Theorem 5.9. Let (E, ι) be an elliptic curve over $\mathbb{Q}$ with a non-symplectic involution ι. Let $X = \widetilde{E \times S/\iota \times \sigma}$ be a Calabi–Yau threefold of Borcea–Voisin type. Then X has a hypersurface model defined over $\mathbb{Q}$, and X is motivically automorphic. More precisely,*

(1) A $K3$ surface S as above with a non-symplectic involution σ, is motivically modular/automorphic.

There exists a quadruple $(\rho_S, \mathbb{K}, \varepsilon, \chi_S)$ with the following properties:

- *ρ_S is an (Artin) Galois representation of $G_{\mathbb{Q}}$, and the degree of ρ_S is the geometric Picard number of S,*
- *$\mathbb{K}$ is a CM abelian extension of $\mathbb{Q}$,*
- *$\varepsilon : \mathbb{K} \to \mathbb{C}$ is an embedding,*
- *χ_S is a Hecke character of $\mathbb{K}$ of infinite type $z \to \varepsilon(z)^2$,*
- *$dim \rho_S + [\mathbb{K} : \mathbb{Q}] = 22$ where $[\mathbb{K} : \mathbb{Q}] = dim T(\bar{S}) \otimes \mathbb{Q}_\ell$,*

such that

$$L(S, s) = L(\rho_S, s - 1)L(\chi_S, s).$$

Furthermore, $L(\chi_S, s) = L(\mathcal{M}_S, s)$ is automorphic.

(2) Let X be a Calabi–Yau threefold of Borcea–Voisin type. Then If X is of CM type, then X is motivically automorphic. That is, let ρ_E be the Galois representation corresponding to E, and let $J(C_g)$ be the Jacobian variety of $C_g \subset S^\sigma$. Then

$$L(X, s) = L(\rho_E \otimes \chi_S, s)L(\rho_E \otimes \rho_S, s - 1)L(J(C_g), s - 1)^4.$$

Furthermore, $L(\rho_E \otimes \chi_S, s) = L(\mathcal{M}_X, s)$ is automorphic, and $L(J(C_g), s)$ is also automorphic.

This theorem is established in Goto–Livné–Yui [**17**].

Idea for proof of Theorem 5.19

• The automorphy of $L(\chi_S, s)$ is proved by using the same argument of automorphic induction as in Subsection 4.4: Remark 4.5. This is justified as S is realized as a finite Fermat quotient and hence of CM type. Thus the motive $T(S)^{\sigma=-1} \otimes \mathbb{Q}_\ell$ is automorphic.

We cannot establish the automorphy of $L(\rho_S, s)$ as this may correspond to higher dimensional Artin representations, and the automorphy of Artin representations are still open even for CM fields.

• For the Calabi–Yau threefold X of Borcea–Voisin type, the L-series is computed by making use of the Künneth formula and the resolution of singularities. The tensor product representation $\rho_E \otimes \chi_S$ is automorphic, as both ρ_E and χ are automorphic.

However, we are not able to establish the automorphy of $L(\rho_E \otimes \rho_S, s)$ as ρ_S corresponds to some Artin representations (as pointed out above) whose automorphy are still open.

• The $L(\mathbb{P}^1 \times J(C_g), s)$ is automorphic as $J(C_g)$ is also determined by Jacobi sum Grössencharacters. Since C_g is a component of S^σ where S is a finite quotient of a Fermat or a diagonal surface, C_g is also expressed in terms of a diagonal or a quasi-diagonal curve in a weighted projective 2-space. Hence the Jacobian variety $J(C_g)$ of C_g is again of CM type.

If all fixed points on E under the action of ι are defined over $\mathbb{Q}$, we get 4 copies of $J(C_g)$. If not all fixed points are defined over $\mathbb{Q}$, we pass onto a finite extension $\mathbb{K}$ of $\mathbb{Q}$ on which all such points are defined. Then we obtain 4 copies of $J(C_g) \otimes \mathbb{K}$.

5.6. An example. Mirror symmetry for Calabi–Yau threefolds is exhibited for families of Calabi–Yau threefolds. However, in the example below, we are only interested in the existence of a particular mirror Calabi–Yau threefold of Borcea–Voisin type.

EXAMPLE 5.20. Let $E = E_2 : y_0^2 = y_1^4 + y_2^4 \subset \mathbb{WP}^2(2,1,1)$. Let S_0 be a (quasi-smooth) K3 surface given by

$$S_0 : x_0^2 = x_1^3 + x_2^7 + x_3^{42} \subset \mathbb{WP}^3(21,14,6,1)$$

of degree 42. S_0 has a non-symplectic involution$\sigma(x_0) = -x_0$. Let S be the minimal resolution of S_0. Then S corresponds to the triplet $(10,0,0)$ of Nikulin. So its mirror $S^\vee$ also corresponds to the triplet $(10,0,0)$. The fixed locus of S under σ is given by

$$S^\sigma = C_6 \cup L_1 \cup \cdots \cup L_5.$$

S is of CM type as it is dominated by the Fermat surface of degree 42. Now we use Theorem 5.19. Once a Fermat cover of degree 42 is determined, we also have the characters $\mathbf{a}$, and the Jacobi sums $j(\mathbf{a})$ which live in the cyclotomic field $Q(\zeta_{42})$. Fish out those characters $\mathbf{a}$ that are compatible with the covering map. Among the surviving characters, there is a unique character $\mathbf{a}$ of Hodge type $(0,2)$. The transcendental cycles $T(S) \otimes \mathbb{Q}_\ell$ is determined by this character $\mathbf{a}$ and the corresponding Fermat motive $\mathcal{M}_A$ with $A = [\mathbf{a}]$ of dimension $\varphi(42) = 12$. The character $\mathbf{a}$ gives rise to the Jacobi sum Grössencharacter $j(\mathbf{a})$ in the cyclotomic field $\mathbb{K} = \mathbb{Q}(\zeta_{42})$ with $[\mathbb{K} : \mathbb{Q}] = \varphi(42) = 12 = 22 - 10$. Here S_0 is dominated by the Fermat surface of degree $42 = \mathrm{lcm}(3,7,42)$. The motive $\mathcal{M}_S$ is automorphic.

In general, $NS(S)^\sigma$ is a sublattice of $NS(S)$ of rank r, but for For this example, note that $10 = 22 - 12 = r$, so that $NS(S)^\sigma = NS(S)$.

Now the Calabi–Yau threefold $X = \widetilde{E_2 \times S/\iota \times \sigma}$ has a birational model defined over $\mathbb{Q}$ by a hypersurface:

$$X : z_0^4 + z_1^4 = z_2^3 + z_3^7 + z_4^{42} \subset \mathbb{WP}^4(21,21,28,12,2)$$

of degree 84. Since E_2 and S are of CM type, so is X by Proposition 5.16. The Hodge numbers are given by

$$h^{1,1}(X) = 35,\ h^{2,1}(X) = 35, \text{ and } E(X) = 0.$$

So X is its own topological mirror.

We now pass from $\mathbb{Q}(\zeta_{42})$ to $\mathbb{Q}(\zeta_{84})$ to take $\mathbb{Q}(\zeta_4)$ (corresponding to E_2) into account. The Calabi–Yau motive $\mathcal{M}_X$ has dimension $24 = \varphi(84)$. The Jacobi sum Grössencharacter of $\mathbb{Q}(\zeta_{84})$ gives rise to the $GL(24)$ irreducible automorphic representation over $\mathbb{Q}$ for $\mathcal{M}_X$. Hence $\mathcal{M}_X$ is automorphic.

5.7. Mirror Calabi–Yau threefolds of Borcea–Voisin type and arithmetic mirror symmetry. We know that the topological mirror symmetry conjecture holds for Calabi–Yau threefolds of Borcea–Voisin type, and it is inherited from the mirror symmetry for K3 surfaces (in Dolgachev's sense). However, the 95 (families of) K3 surfaces of Reid and Yonemura are not closed under mirror symmetry in the sense of Definition 4.15.

LEMMA 5.21 (Belcastro [2]). *Among the* 95 *(families of) K3 surfaces of Reid and Yonemura, the* 57 *of them have mirror partners within the list in the sense of Dolgachev* [**13**].

All the 57 *K3 surfaces S have non-symplectic involutions σ, and so do their mirror partners $S^\vee$.*

REMARK 5.22. There are at least three notions of mirror symmetry for K3 surfaces. One is the Dolgachev mirror symmetry (Definition 7). The second one is the Nikulin mirror symmetry, that is, if pair (S, σ) of a K3 surface with non-symplectic involution σ corresponds to a Nikulin's triplet (r, a, δ), then a mirror pair $(S^\vee, \sigma^\vee)$ corresponds to the triplet $(20 - r, a, \delta)$. In this version, a mirror is placed at the vertical line $r = 10$ in Picture 1.

The third one is the BHK mirror symmetry, which is formulated as follows. Let W be an invertible polynomial with a group G of diagonal automorphisms. Let Y_W be the hypersurface $\{W = 0\}$ in a weighted projective 3-space. Then the orbifold Y_W/G defines a K3 surface. Now define the polynomial W^T by transposing the exponent matrix of W. Then W^T is also invertible. Let G^T be the dual group of G. Then the orbifold Y_{W^T}/G^T is again a K3 surface. The BHK mirror symmetry is that Y_W/G and Y_{W^T}/G^T form a mirror pair of K3 surfaces. (Here BHK stands for Berglund, Hübsch and Krawitz. For details, see, for instance, Kelly [**21**].)

These seemingly different versions of mirror symmetry for K3 surfaces coincide for some K3 surfaces. We will discuss some examples below.

EXAMPLE 5.23. Let $E = E_2$ and S_0 be a (quasi-smooth) K3 surface given by

$$S_0 : x_0^2 = x_1^3 x_2 + x_1^3 x_2^2 + x_2^7 + x_3^{14} \subset \mathbb{WP}^3(7,4,2,1)$$

of degree 14. S_0 has a non-symplectic involution $\sigma(x_0) = -x_0$. Its minimal resolution S corresponds to the triplet $(7, 3, 0)$ of Nikulin. Remove the monomial $x_1^3 x_2^2$ from the defining equation for S_0, we get

$$S_0 : x_0^2 = x_1^3 x_2 + x_2^7 + x_3^{14}.$$

Let S be the minimal resolution of S_0.

Now let $E = E_2$ be the elliptic curve, and let $X = \widetilde{E_2 \times S/\iota \times \sigma}$ be the Calabi–Yau threefold. Then X has a birational model defined over $\mathbb{Q}$ by a hypersurface:

$$X : z_0^4 + z_1^4 = z_2^3 z_3 + z_3^7 - z_4^{14} \subset \mathbb{P}^4(7,7,8,4,2)$$

of degree 28. Since E_2 and S are of CM type, so is X by Proposition 5.16. The Hodge numbers are given by

$$h^{1,1}(X) = 20,\ h^{2,1}(X) = 38, \text{ and } e(X) = -36.$$

We will construct a mirror (family) of Calabi–Yau threefold of Borcea–Voisin type using the K3 surface S in Remark 5.22 above. To find a mirror Calabi–Yau threefold, we look for a mirror $S^\vee$ of S. We may take for $S_0^\vee$ the K3 surface defined by

$$S_0^\vee : x_0^2 = x_1^3 + x_1x_2^7 + x_2^9x_3^2 + x_3^{14} \subset \mathbb{P}^3(21,14,4,3)$$

of degree 42. It has a non-symplectic involution $x_0 \mapsto -x_0$. Let $S^\vee$ be the minimal resolution of $S_0^\vee$. Then the pair $(S^\vee, \sigma^\vee)$ corresponds to the triplet $(13,3,0)$. Removing the monomial $x_2^9x_3^2$ we can make S to be of CM type. So $S^\vee$ is a mirror of S in the sense of Nikulin and also of Dolgachev (see Belcastro [**2**]: S_0 is #40 and $S_0^\vee$ is #81.)

A candidate for a mirror Calabi–Yau family $X^\vee$ has a birational model over $\mathbb{Q}$ by a hypersurface:

$$X^\vee : z_0^4 + z_1^4 = z_2^3 + z_2z_3^7 + z_4^{14} \subset \mathbb{P}^4(21,21,28,8,6)$$

of degree 84. The Hodge numbers and the Euler characteristic are given by

$$h^{1,1}(X^\vee) = 38,\ h^{2,1}(X^\vee) = 20, \text{ and } e(X^\vee) = 36.$$

We observe here that $S_0^\vee$ discussed above is also a BHK mirror of S_0. That is, the matrices of exponents for the K3 surfaces in Example 5.20 and the above K3 surface are given by

$$\begin{pmatrix} 2 & 0 & 0 & 0 \\ 0 & 3 & 1 & 0 \\ 0 & 0 & 7 & 0 \\ 0 & 0 & 0 & 14 \end{pmatrix} \quad \text{and} \quad \begin{pmatrix} 2 & 0 & 0 & 0 \\ 0 & 3 & 0 & 0 \\ 0 & 1 & 7 & 0 \\ 0 & 0 & 0 & 14 \end{pmatrix}.$$

The latter is the transpose of the former.

REMARK 5.24. Usually, mirror symmetry for Calabi–Yau threefolds is exhibited for families of Calabi–Yau threefolds. That is, a mirror partner $X^\vee$ of a Calabi–Yau threefold X of Borcea–Voisin type appears in a family, e.g., multi-parameter family of hypersurfaces defined over $\mathbb{Q}$. However, we do not have effective methods of counting numbers of rational points for these hypersurfaces with parameters. At the moment, we have no efficient method of determining the zeta-functions and L-series of these families.

What we can do is to compute the L-series of a mirror family at a CM point (an isolated point), where the Calabi–Yau threefold becomes of CM type. However, we are not yet able to interpret mirror symmetry at the level of L-series.

REMARK 5.25. There are several methods to compute zeta-functions and L-series of families of Calabi–Yau hypersurfaces. Goto–Kloosterman–Yui [**16**], Sections 10, 11 and 12 used p-adic cohomology and hypergeometric series trying to compute zeta-functions. Candelas, et al. [**7**] and [**8**] used Picard–Fuchs differential equation to compute the zeta-function of the 1-parameter family of Calabi–Yau quintic hypersurfaces. Harvey, Costa, Kedlaya and others have developed algorithmic approaches via p-adic cohomology for computing zeta-functions of K3 surfaces. (Several slides of Kedlaya's talks are available by web search, however, it appears that papers are yet available.)

In this article, we have not included discussions on computing zeta-functions and L-series for families of Calabi–Yau varieties; this is left as a topic for our future endeavour.

6. Acknowledgments

I would like to thank the anonymous referee for carefully reading through the several versions of the manuscript and giving very constructive criticisms. I did address the comments of the referee, and revise the paper taking in those comments within my capability.

References

[1] James Arthur and Laurent Clozel, *Simple algebras, base change, and the advanced theory of the trace formula*, Annals of Mathematics Studies, vol. 120, Princeton University Press, Princeton, NJ, 1989. MR1007299

[2] Sarah-Marie Belcastro, *Picard lattices of families of $K3$ surfaces*, Comm. Algebra **30** (2002), no. 1, 61–82, DOI 10.1081/AGB-120006479. MR1880661

[3] Gilberto Bini and Bert van Geemen, *Geometry and arithmetic of Maschke's Calabi-Yau three-fold*, Commun. Number Theory Phys. **5** (2011), no. 4, 779–820, DOI 10.4310/CNTP.2011.v5.n4.a2. MR2905137

[4] Ciprian Borcea, *Calabi-Yau threefolds and complex multiplication*, Essays on mirror manifolds, Int. Press, Hong Kong, 1992, pp. 489–502. MR1191439

[5] Ciprian Borcea, *$K3$ surfaces with involution and mirror pairs of Calabi-Yau manifolds*, Mirror symmetry, II, AMS/IP Stud. Adv. Math., vol. 1, Amer. Math. Soc., Providence, RI, 1997, pp. 717–743. MR1416355

[6] Christophe Breuil, Brian Conrad, Fred Diamond, and Richard Taylor, *On the modularity of elliptic curves over $\mathbf{Q}$: wild 3-adic exercises*, J. Amer. Math. Soc. **14** (2001), no. 4, 843–939, DOI 10.1090/S0894-0347-01-00370-8. MR1839918

[7] Philip Candelas and Xenia de la Ossa, *The zeta-function of a p-adic manifold, Dwork theory for physicists*, Commun. Number Theory Phys. **1** (2007), no. 3, 479–512, DOI 10.4310/CNTP.2007.v1.n3.a2. MR2385244

[8] Philip Candelas, Xenia de la Ossa, and Fernando Rodriguez-Villegas, *Calabi-Yau manifolds over finite fields. II*, Calabi-Yau varieties and mirror symmetry (Toronto, ON, 2001), Fields Inst. Commun., vol. 38, Amer. Math. Soc., Providence, RI, 2003, pp. 121–157. MR2019149

[9] Adrian Clingher and Charles F. Doran, *Lattice polarized K3 surfaces and Siegel modular forms*, Adv. Math. **231** (2012), no. 1, 172–212, DOI 10.1016/j.aim.2012.05.001. MR2935386

[10] Pierre Deligne, *La conjecture de Weil. I* (French), Inst. Hautes Études Sci. Publ. Math. **43** (1974), 273–307. MR0340258

[11] Pierre Deligne, *La conjecture de Weil. II* (French), Inst. Hautes Études Sci. Publ. Math. **52** (1980), 137–252. MR601520

[12] Luis Dieulefait, *On the modularity of rigid Calabi-Yau threefolds: epilogue* (English, with English and Russian summaries), Zap. Nauchn. Sem. S.-Peterburg. Otdel. Mat. Inst. Steklov. (POMI) **377** (2010), no. Issledovaniya po Teorii Chisel. 10, 44–49, 241, DOI 10.1007/s10958-010-0175-8; English transl., J. Math. Sci. (N.Y.) **171** (2010), no. 6, 725–727. MR2753647

[13] I. V. Dolgachev, *Mirror symmetry for lattice polarized $K3$ surfaces*, J. Math. Sci. **81** (1996), no. 3, 2599–2630, DOI 10.1007/BF02362332. Algebraic geometry, 4. MR1420220

[14] Fernando Q. Gouvêa and Noriko Yui, *Arithmetic of diagonal hypersurfaces over finite fields*, London Mathematical Society Lecture Note Series, vol. 209, Cambridge University Press, Cambridge, 1995. MR1340424

[15] Klaus Hulek and Helena Verrill, *On modularity of rigid and nonrigid Calabi-Yau varieties associated to the root lattice A_4*, Nagoya Math. J. **179** (2005), 103–146. MR2164402

[16] Yasuhiro Goto, Remke Kloosterman, and Noriko Yui, *Zeta-functions of certain $K3$-fibered Calabi-Yau threefolds*, Internat. J. Math. **22** (2011), no. 1, 67–129, DOI 10.1142/S0129167X11006726. MR2765443

[17] Yasuhiro Goto, Ron Livné, and Noriko Yui, *Automorphy of Calabi-Yau threefolds of Borcea-Voisin type over $\mathbb{Q}$*, Commun. Number Theory Phys. **7** (2013), no. 4, 581–670, DOI 10.4310/CNTP.2013.v7.n4.a2. MR3228297

[18] Fernando Q. Gouvêa and Noriko Yui, *Rigid Calabi-Yau threefolds over $\mathbb{Q}$ are modular*, Expo. Math. **29** (2011), no. 1, 142–149, DOI 10.1016/j.exmath.2010.09.001. MR2785550

[19] Shihoko Ishii, *Introduction to singularities*, Springer, Tokyo, 2014. MR3288750

[20] Chandrashekhar Khare and Jean-Pierre Wintenberger, *Serre's modularity conjecture. II*, Invent. Math. **178** (2009), no. 3, 505–586, DOI 10.1007/s00222-009-0206-6. MR2551764

[21] Tyler L. Kelly, *Picard ranks of K3 surfaces of BHK type*, Calabi-Yau varieties: arithmetic, geometry and physics, Fields Inst. Monogr., vol. 34, Fields Inst. Res. Math. Sci., Toronto, ON, 2015, pp. 45–63, DOI 10.1007/978-1-4939-2830-9_2. MR3409772

[22] Mark Kisin, *Modularity of 2-adic Barsotti-Tate representations*, Invent. Math. **178** (2009), no. 3, 587–634, DOI 10.1007/s00222-009-0207-5. MR2551765

[23] Shigeyuki Kondō, *Automorphisms of algebraic K3 surfaces which act trivially on Picard groups*, J. Math. Soc. Japan **44** (1992), no. 1, 75–98, DOI 10.2969/jmsj/04410075. MR1139659

[24] R. P. Langlands, *L-functions and automorphic representations*, Proceedings of the International Congress of Mathematicians (Helsinki, 1978), Acad. Sci. Fennica, Helsinki, 1980, pp. 165–175. MR562605

[25] Ron Livné, *Motivic orthogonal two-dimensional representations of* $\mathrm{Gal}(\overline{\mathbf{Q}}/\mathbf{Q})$, Israel J. Math. **92** (1995), no. 1-3, 149–156, DOI 10.1007/BF02762074. MR1357749

[26] Ron Livné, Matthias Schütt, and Noriko Yui, *The modularity of K3 surfaces with non-symplectic group actions*, Math. Ann. **348** (2010), no. 2, 333–355, DOI 10.1007/s00208-009-0475-9. MR2672305

[27] Ron Livné and Noriko Yui, *The modularity of certain non-rigid Calabi-Yau threefolds*, J. Math. Kyoto Univ. **45** (2005), no. 4, 645–665, DOI 10.1215/kjm/1250281650. MR2226623

[28] J. Myron Masley and Hugh L. Montgomery, *Cyclotomic fields with unique factorization*, J. Reine Angew. Math. **286/287** (1976), 248–256. MR0429824

[29] D. R. Morrison, *On K3 surfaces with large Picard number*, Invent. Math. **75** (1984), no. 1, 105–121, DOI 10.1007/BF01403093. MR728142

[30] V. V. Nikulin, Integral Symmetric Bilinear Forms and Some of their Applications. *Math. USSR Izvestija* **14** (1980), No. 1, 103–167.

[31] V. V. Nikulin, *Finite groups of automorphisms of Kählerian K3 surfaces* (Russian), Trudy Moskov. Mat. Obshch. **38** (1979), 75–137. MR544937

[32] C. S. Rajan, *On the image and fibres of solvable base change*, Math. Res. Lett. **9** (2002), no. 4, 499–508, DOI 10.4310/MRL.2002.v9.n4.a9. MR1928869

[33] Miles Reid, *Canonical 3-folds*, Journées de Géometrie Algébrique d'Angers, Juillet 1979/Algebraic Geometry, Angers, 1979, Sijthoff & Noordhoff, Alphen aan den Rijn—Germantown, Md., 1980, pp. 273–310. MR605348

[34] Jan Christian Rohde, *Cyclic coverings, Calabi-Yau manifolds and complex multiplication*, Lecture Notes in Mathematics, vol. 1975, Springer-Verlag, Berlin, 2009. MR2510071

[35] Jan Christian Rohde, *Maximal automorphisms of Calabi-Yau manifolds versus maximally unipotent monodromy*, Manuscripta Math. **131** (2010), no. 3-4, 459–474, DOI 10.1007/s00229-009-0329-5. MR2592090

[36] I. I. Pjateckiĭ-Šapiro and I. R. Šafarevič, *The arithmetic of surfaces of type* K3 (Russian), Proceedings of the International Conference on Number Theory (Moscow, 1971), Trudy Mat. Inst. Steklov. **132** (1973), 44–54, 264. MR0335521

[37] Tetsuji Shioda, *An explicit algorithm for computing the Picard number of certain algebraic surfaces*, Amer. J. Math. **108** (1986), no. 2, 415–432, DOI 10.2307/2374678. MR833362

[38] Tetsuji Shioda, *Some observations on Jacobi sums*, Galois representations and arithmetic algebraic geometry (Kyoto, 1985/Tokyo, 1986), Adv. Stud. Pure Math., vol. 12, North-Holland, Amsterdam, 1987, pp. 119–135. MR948239

[39] T. Shioda and H. Inose, *On singular K3 surfaces*, Complex analysis and algebraic geometry, Iwanami Shoten, Tokyo, 1977, pp. 119–136. MR0441982

[40] Matthias Schütt, *Modularity of Maschke's octic and Calabi-Yau threefold*, Commun. Number Theory Phys. **5** (2011), no. 4, 827–847, DOI 10.4310/CNTP.2011.v5.n4.a3. MR2905138

[41] J. Tate, *Number theoretic background*, Automorphic forms, representations and L-functions (Proc. Sympos. Pure Math., Oregon State Univ., Corvallis, Ore., 1977), Proc. Sympos. Pure Math., XXXIII, Amer. Math. Soc., Providence, R.I., 1979, pp. 3–26. MR546607

[42] Richard Taylor, *Galois representations* (English, with English and French summaries), Ann. Fac. Sci. Toulouse Math. (6) **13** (2004), no. 1, 73–119. MR2060030

[43] Richard Taylor and Andrew Wiles, *Ring-theoretic properties of certain Hecke algebras*, Ann. of Math. (2) **141** (1995), no. 3, 553–572, DOI 10.2307/2118560. MR1333036

[44] Claire Voisin, *Miroirs et involutions sur les surfaces* $K3$ (French), Astérisque **218** (1993), 273–323. Journées de Géométrie Algébrique d'Orsay (Orsay, 1992). MR1265318

[45] S. P. Vorontsov, *Automorphisms of even lattices arising in connection with automorphisms of algebraic* $K3$*-surfaces* (Russian, with English summary), Vestnik Moskov. Univ. Ser. I Mat. Mekh. **2** (1983), 19–21. MR697215

[46] André Weil, *Numbers of solutions of equations in finite fields*, Bull. Amer. Math. Soc. **55** (1949), 497–508, DOI 10.1090/S0002-9904-1949-09219-4. MR0029393

[47] André Weil, *Jacobi sums as "Grössencharaktere"*, Trans. Amer. Math. Soc. **73** (1952), 487–495, DOI 10.2307/1990804. MR0051263

[48] Andrew Wiles, *Modular elliptic curves and Fermat's last theorem*, Ann. of Math. (2) **141** (1995), no. 3, 443–551, DOI 10.2307/2118559. MR1333035

[49] Takashi Yonemura, *Hypersurface simple* $K3$ *singularities*, Tohoku Math. J. (2) **42** (1990), no. 3, 351–380, DOI 10.2748/tmj/1178227616. MR1066667

[50] Noriko Yui, *Update on the modularity of Calabi-Yau varieties*, Calabi-Yau varieties and mirror symmetry (Toronto, ON, 2001), Fields Inst. Commun., vol. 38, Amer. Math. Soc., Providence, RI, 2003, pp. 307–362, DOI 10.1090/fic/038/17. With an appendix by Helena Verrill. MR2019160

[51] Noriko Yui, *The L-series of Calabi-Yau orbifolds of CM type*, Mirror symmetry. V, AMS/IP Stud. Adv. Math., vol. 38, Amer. Math. Soc., Providence, RI, 2006, pp. 185–252. With an appendix by Yasuhiro Goto. MR2282961

[52] Yu. G. Zarhin, *Hodge groups of* $K3$ *surfaces*, J. Reine Angew. Math. **341** (1983), 193–220, DOI 10.1515/crll.1983.341.193. MR697317

Department of Mathematics and Statistics, Queen's University, Kingston, Ontario K3L 3N6 Canada

E-mail address: yui@mast.queensu.ca